Benchmark Papers in Electrical Engineering and Computer Science

Series Editor: John B. Thomas
Princeton University

Published Volumes

SYSTEM SENSITIVITY ANALYSIS
J. B. Cruz, Jr.
RANDOM PROCESSES: Multiplicity Theory and Canonical Decompositions
Anthony Ephremides and John B. Thomas
ALGEBRAIC CODING THEORY: History and Development
Ian F. Blake
PATTERN RECOGNITION: Introduction and Foundations
Jack Sklansky
COMPUTER-AIDED CIRCUIT DESIGN: Simulation and Optimization
S. W. Director
ENVIRONMENTAL MODELING: Analysis and Management
Douglas Daetz and Richard H. Pantell
AUTOMATIC CONTROL: Classical Linear Theory
George J. Thaler
CIRCUIT THEORY: Foundations and Classical Contributions
M. E. Van Valkenburg

Additional volumes in preparation

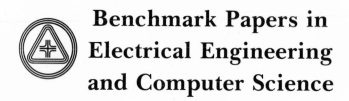

Benchmark Papers in Electrical Engineering and Computer Science

A *BENCHMARK* ® Books Series

CIRCUIT THEORY:
Foundations and Classical Contributions

Edited by
M. E. VAN VALKENBURG
University of Illinois at Urbana-Champaign

Dowden, Hutchinson & Ross, Inc.
Stroudsburg, Pennsylvania

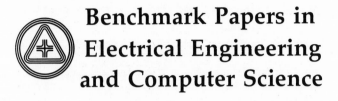

Benchmark Papers in Electrical Engineering and Computer Science, Volume 8
Library of Congress Catalog Card Number: 74–2475
ISBN: 0–87933–084–8

Manufactured in the United States of America

Exclusive distributor outside the United States and Canada: John Wiley & Sons, Inc.

74 75 76 5 4 3 2 1

Library of Congress Cataloging in Publication Data

Van Valkenburg, Mac Elwyn, 1921- comp.
Circuit theory: foundations and classical contributions.

(Benchmark papers in electrical engineering and computer science, v. 8)
1. Electric networks. 2. Electric circuits.
I. Title.
TK454.2.V35 621.319'2 74-2475
ISBN 0-87933-084-8

Acknowledgments and Permissions

ACKNOWLEDGMENT

BROWN UNIVERSITY—*Quarterly of Applied Mathematics*
The Transfer Function of General Two Terminal-Pair RC Networks

S. BUTTERWORTH—*Wireless Engineer*
On the Theory of Filter Amplifiers

PERMISSIONS

The following papers have been reprinted with permission of the authors and copyright holders.

AMERICAN INSTITUTE OF PHYSICS
Journal of Applied Physics
A Definition of Passive Linear Networks in Terms of Time and Energy
Impedance Synthesis Without Use of Transformers
Transmission Losses in $2n$-Terminal Networks
Journal of Mathematics and Physics
Synthesis of a Finite Two-Terminal Network Whose Driving-Point Impedance Is a Prescribed Function of Frequency
Synthesis of Reactance 4-Poles Which Produce Prescribed Insertion Loss Characteristics

AMERICAN TELEPHONE AND TELEGRAPH COMPANY—*The Bell System Technical Journal*
Effect of Feedback on Impedance
Physical Theory of the Electric Wave-Filter
A Reactance Theorem
Regeneration Theory
Relations Between Attenuation and Phase in Feedback Amplifier Design

INSTITUTE OF ELECTRICAL AND ELECTRONICS ENGINEERS, INC.
IEEE Transactions on Circuit Theory
A New Theory of Broad-Band Matching
IRE Transactions on Circuit Theory
The *A* Matrix, New Network Description
Bounded Real Scattering Matrices and the Foundations of Linear Passive Network Theory
A Practical Method of Designing RC Active Filters
Realizability Theorem for Mid-Series or Mid-Shunt Low-Pass Ladders Without Mutual Induction
Regenerative Modes of Active Networks
A Survey of Network Realization Techniques
Proceedings of the IRE
Feedback Theory—Further Properties of Signal Flow Graphs
A Note on the Ladder Development of RC–Networks
Some Fundamental Properties of Transmission Systems
Summary of the History of Circuit Theory

PHILIPS RESEARCH LABORATORIES, THE NETHERLANDS—*Philips Research Reports*
A General Network Theorem, with Applications
The Gyrator, a New Electric Network Element

Series Editor's Preface

This Benchmark Series in Electrical Engineering and Computer Science is aimed at sifting, organizing, and making readily accessible to the reader the vast literature that has accumulated. Although the series is not intended as a complete substitute for a study of this literature, it will serve at least three major critical purposes. In the first place, it provides a practical point of entry into a given area of research. Each volume offers an expert's selection of the critical papers on a given topic as well as his views on its structure, development, and present status. In the second place, the series provides a convenient and time-saving means for study in areas related to but not contiguous with one's principal interests. Last, but by no means least, the series allows the collection, in a particularly compact and convenient form, of the major works on which present research activities and interests are based.

Each volume in the series has been collected, organized, and edited by an authority in the area to which it pertains. In order to present a unified view of the area, the volume editor has prepared an introduction to the subject, has included his comments on each article, and has provided a subject index to facilitate access to the papers.

We believe that this series will provide a manageable working library of the most important technical articles in electrical engineering and computer science. We hope that it will be equally valuable to students, teachers, and researchers.

This volume, *Circuit Theory: Foundations and Classical Contributions*, has been edited by M. E. Van Valkenburg of the University of Illinois at Urbana-Champaign. It contains twenty-five papers that trace the emergence and development of modern circuit theory. Professor Van Valkenburg is particularly well qualified to edit this material. He has occupied a prominent position in the field of circuit theory for the past twenty years and has contributed approximately thirty technical papers, supervised over forty doctoral students, and written three well-known books. One of these, *Network Analysis*, has served as a basic text for a generation of circuit theorists, and his *Introduction to Modern Network Synthesis* remains the most widely used introduction to the "classical" techniques of passive network synthesis.

John B. Thomas

Contents

III. FEEDBACK AMPLIFIER THEORY

IV. ACTIVE NETWORKS

V. THE SCATTERING MATRIX AND APPLICATIONS

VI. SOME FOUNDATION CONCEPTS

VII. SURVEYS AND HISTORICAL SUMMARIES

Contents by Author

Introduction

This volume is intended to fill needs in research and teaching for a source book of classical papers from which present-day circuit theory has emerged. My interest in this project stemmed from an experience at the University of Illinois in the mid-1950s: There the late Sundaram Seshu and I provided copies of the original Brune paper to our students, with gratifying results in terms of student interest. From this came my own conviction that study and research in circuit theory is enhanced by reference to the original contributions, which frequently offer insight and clarity not found in textbook treatments.

Circuit theory as a field has its roots in the distant past, which is not true of the theory in many other engineering fields and specialties. It may be thought of as having started with Ohm's law (1827) and Kirchhoff's laws (1845). Belevitch has pointed out (Paper 24) that the number of publications per year in circuit theory averaged under one until 1910 and passed one hundred in 1954. At present, there are about four hundred publications per year, excluding patents, theses, and reports, and no sign that this number will diminish.

With such a vast literature to draw upon, how can the "classic" papers be determined? Clearly, the choices are strictly personal ones for which the volume editor must take full responsibility. Within the publisher's space limitations, I have tried to select those papers which in my opinion have introduced ideas that have greatly influenced the development of circuit theory. I have decided also to limit the number of papers by a given author and to offer for each, instead, a representative sample.

I have decided to exclude some papers that are not in English, the important ones by Cauer, for example. By way of apology for not including all important papers that should be ranked as classics, I can only express the hope that this volume will stimulate the reader's interest to the extent that he will consult the literature stored in his library.

The papers have been organized into seven parts. Part I is devoted to network synthesis, which may be regarded as the theoretical basis for important parts of circuit theory. This subject has been popularized since World War II, particularly by Guillemin and his students, and it is a subject now familiar to many engineers.

The next two parts relate to the motivation for circuit theory and the historically important applications. Part II is on filter design, an important reason for the development of circuit theory in the 1920s, and later for applications in problems related to the long-distance telephone. Feedback amplifier design, the subject of Part III, became important in the 1930s in the application of wide-band highly linear amplifiers for multichannel telephone systems.

The subject of Part IV, active networks, is closely coupled with feedback amplifier theory. The objective of papers in this part is to exploit active elements in changing the network characteristics of passive elements, alone or in combination. The procedures that have evolved make it possible to realize an active network that is equivalent to one realized with passive elements. These topics have now assumed significance because of the ease with which active elements may be realized by use of integrated circuit technology.

The scattering matrix formulation, the topic of Part V, came from microwave applications developed during and after World War II. Since a microwave system is most readily explained in terms of transmitted and reflected waves, the scattering matrix was developed for these systems. It has been applied to circuit theory in a unified way that gives new insight into old subjects such as filter design.

Several foundation concepts have had impact on the development of the field of circuit theory, and a few of these are the subjects of Part VI. Two concepts were introduced by Tellegen: The gyrator is useful as an artifice leading to network realizations, especially using active elements; and it may be realized physically in either microwave applications or with active elements. Tellegen's theorem has found wide application in explaining circuit phenomena and has worked its way into elementary textbooks with unusual speed. Raisbeck's time–energy approach to the qualitative characterization of n-port networks provided new understanding and quickly attained the status of a classic. Finally, the A Matrix as a network description which relates to the state-variable approach in circuit analysis and synthesis is a "first" that has had significant impact on the stream of circuit theory literature.

The historical survey prepared by Vitold Belevitch on the occasion of the 50th anniversary issue of the *Proceedings of the IRE,* reprinted in Part VII, is itself a classic in the coverage of the important ideas of circuit theory for the 50 years prior to 1962. Darlington's survey of network realization techniques served as an introduction to a special issue of the *IRE Transactions on Circuit Theory* devoted to realization techniques then available and problems then unsolved.

I have greatly benefited from discussions with many colleagues on the choice of papers included in this volume, but of course the responsibility for the final selections rests solely with me.

References

Kirchhoff, Gustav. "On the Solution of the Equations Obtained from the Investigation of the Linear Distribution of Galvanic Currents," *Poggendorf Ann. Phys.,* **72**, 497–508 (1847); English translation by J. B. O'Toole, *IRE Trans. Circuit Theory,* **CT-5**, 4–7 (1958).

Ohm, G. S. *The Galvanic Circuit Investigated Mathematically,* original German edition: Berlin, 1827; English translation by W. Francis, Van Nostrand Reinhold Company, New York, 2nd ed., 1905.

I
Network Synthesis

Editor's Comments on Papers 1 Through 5

1 Foster: *A Reactance Theorem*

2 Brune: *Synthesis of a Finite Two-Terminal Network Whose Driving-Point Impedance Is a Prescribed Function of Frequency*

3 Bott and Duffin: *Impedance Synthesis Without Use of Transformers*

4 Guillemin: *A Note on the Ladder Development of RC-Networks*

5 Fialkow and Gerst: *The Transfer Function of General Two Terminal-Pair RC Networks*

The paper popularly regarded as the beginning of network synthesis is that of Foster concerning *LC* one-port networks, published in 1924 (Paper 1). The solution for the *RC* and *RL* cases was given by Cauer (1926), thus completing the two-element one-port problem. In 1931 Brune provided a procedure for the synthesis of *RLC* one-port networks and introduced the fundamental concept of a positive real function (Paper 2). For about 20 years following the work of Brune, it was generally believed that an ideal transformer was unavoidable in the realization of the general *RLC* one-port network. This restriction was removed by Bott and Duffin in 1949 (Paper 3).

Although these are some of the cornerstones upon which network synthesis was built, many others have contributed to a better understanding of the one-port problem. The work of Bott and Duffin, for example, made use of a result due to Richards (1947) concerning positive real functions. The result cited from Bott and Duffin is a small part of their more general study (1953), which has not received as much attention as that of their short letter to the editor. Another interesting class of synthesis procedures for the one-port network which makes use of properties of the real part of the driving-point impedance is credited to Miyata (1952) and Fialkow and Gerst (1954).

Studies of one-port networks pose interesting theoretical problems, but are also critical as related to two-port synthesis methods. The principal application of two-port theory is in the design of filters, which is the subject of Part II. Some contributions do not immediately relate to filters but are fundamental in nature. One such area is that of the synthesis of *RC* two-port networks, which was first treated by Guillemin (1949) and extended in a number of papers by Guillemin (Paper 4) and others. The most general and complete results on the *RC* two-port problem were given by Fialkow and Gerst in a number of papers published in the early 1950s, especially Paper 5 and a companion paper (1954). These results are important in showing that networks that can be realized in general may be realized in *RC* form, except for a constant multiplier of the transfer function. *RC* two-port networks have grown in importance with the years, since in combination with operational amplifiers or other active elements they are used to realize inductorless filters with advantages over classical *LC* filters.

References

Bott, Raoul, and R. J. Duffin. "On the Algebra of Networks," *Trans. Amer. Math. Soc.*, **74**, 99–109 (1953).

Cauer, Wilhelm. "The Realization of Impedances with Prescribed Frequency Dependence," *Arch. Electrotech.,* **15**, 355–388 (1926).
Fialkow, A. D., and I. Gerst. "The Transfer Function of Networks Without Mutual Reactance," *Quart. Appl. Math.,* **12**, 117–131 (1954).
Guillemin, E. A. "Synthesis of *RC* Networks," *J. Math. Phys.,* **28**, 22–42 (1949).
Miyata, F. "A New System of Two-Terminal Synthesis," *J. Inst. Elec. Engrs. (Japan),* **35**, 211–218 (1952).
Richards, P. I. "A Special Class of Functions with Positive Real Part in Half-Plane," *Duke Math. J.,* **14**, 777–786 (1947).

Reprinted with permission from *Bell System Tech. J.*, **3**, 259–267 (Apr. 1924)

A Reactance Theorem

By RONALD M. FOSTER

SYNOPSIS: The theorem gives the most general form of the driving-point impedance of any network composed of a finite number of self-inductances, mutual inductances, and capacities. This impedance is a pure reactance with a number of resonant and anti-resonant frequencies which alternate with each other. Any such impedance may be physically realized (provided resistances can be made negligibly small) by a network consisting of a number of simple resonant circuits (inductance and capacity in series) in parallel or a number of simple anti-resonant circuits (inductance and capacity in parallel) in series. Formulas are given for the design of such networks. The variation of the reactance with frequency for several simple circuits is shown by curves. The proof of the theorem is based upon the solution of the analogous dynamical problem of the small oscillations of a system about a position of equilibrium with no frictional forces acting.

AN important theorem [1] gives the driving-point impedance [2] of any network composed of a finite number of self-inductances, mutual inductances, and capacities; showing that it is a pure reactance with a number of resonant and anti-resonant frequencies which alternate with each other; and also showing how any such impedance may be physically realized by either a simple parallel-series or a simple series-parallel network of inductances and capacities, provided resistances can be made negligibly small. The object of this note is to give a full statement of the theorem, a brief discussion of its physical significance and its applications, and a mathematical proof.

THE THEOREM

The most general driving-point impedance S obtainable by means of a finite resistanceless network is a pure reactance which is an odd rational function of the frequency $p/2\pi$ and which is completely determined, except for a constant factor H, by assigning the resonant and anti-resonant frequencies, subject to the condition that they alternate and include both zero and infinity. Any such impedance may be physically

[1] The theorem was first stated, in an equivalent form and without his proof, by George A. Campbell, *Bell System Technical Journal*, November, 1922, pages 23, 26, and 30. By an oversight the theorem on page 26 was made to include unrestricted dissipation. Certain limitations, which are now being investigated, are necessary in the general case of dissipation. The theorem is correct as it stands when there is no dissipation, that is, when all the R's and G's vanish; this is the only case which is considered in the present paper.

A corollary of the theorem is the mutual equivalence of simple resonant components in parallel and simple anti-resonant components in series. This corollary had been previously and independently discovered by Otto J. Zobel as early as 1919, and was subsequently published by him, together with other reactance theorems, *Bell System Technical Journal*, January, 1923, pages 5–9.

[2] The driving-point impedance of a network is the ratio of an impressed electromotive force at a point in a branch of the network to the resulting current at the same point.

constructed either by combining, in parallel, resonant circuits having impedances of the form $iLp+(iCp)^{-1}$, or by combining, in series, anti-resonant circuits having impedances of the form $[iCp+(iLp)^{-1}]^{-1}$. In more precise form,

$$S=-iH\frac{(p_1^2-p^2)\,(p_3^2-p^2)\,\ldots\,(p_{2n-1}^2-p^2)}{p(p_2^2-p^2)\,\ldots\,(p_{2n-2}^2-p^2)}, \qquad (1)$$

where $H\geq 0$ and $0=p_0\leq p_1\leq p_2\leq \ldots \leq p_{2n-1}\leq p_{2n}=\infty$.[3] The inductances and capacities for the n resonant circuits are given by the formula,

$$L_j=\frac{1}{C_jp_j^2}=\left(\frac{ipS}{p_j^2-p^2}\right)_{p=p_j} \quad (j=1,3,\ldots,2n-1), \qquad (2)$$

and the inductances and capacities of the $n+1$ anti-resonant circuits are given by the formula,

$$C_j=\frac{1}{L_jp_j^2}=\left(\frac{ip}{S(p_j^2-p^2)}\right)_{p=p} \quad (j=0,2,4,\ldots,2n-2,2n), \qquad (3)$$

which includes the limiting values,

$$C_0=\frac{p_2^2\ldots p_{2n-2}^2}{Hp_1^2p_3^2\ldots p_{2n-1}^2},\quad L_0=\infty,\ C_{2n}=0,\ L_{2n}=H.$$

Formula (1) may be stated in several mutually equivalent forms.[4] This particular form is the driving-point impedance of the most general symmetrical network in which every branch contains an inductance and a capacity in series, with mutual inductance between each pair of branches. This includes as special cases the driving-point impedances of every other finite resistanceless network.

[3] Since the impedance S is an odd function of the frequency, resonance or anti-resonance for $p=P$ implies resonance or anti-resonance for $p=-P$. In enumerating the resonant and anti-resonant frequencies it is customary, however, to exclude negative values of the frequency. Thus, in the present case, we say that there are n resonant points ($p_1, p_3, \ldots, p_{2n-1}$) and $n+1$ anti-resonant points ($p_0=0, p_2, p_4, \ldots, p_{2n-2}, p_{2n}=\infty$).

[4] The expression for S given by formula (1) may be written in the mutually equivalent forms,

$$\left[-iH\frac{(p_1^2-p^2)\,(p_3^2-p^2)\ldots(p_{2n-1}^2-p^2)}{p\,(p_2^2-p^2)\ldots(p_{2n-2}^2-p^2)}\right]^{\pm 1} \text{ and } \left[iHp\frac{(p_2^2-p^2)\ldots(p_{2n-2}^2-p^2)}{(p_1^2-p^2)\ldots(p_{2n-3}^2-p^2)}\right]^{\pm 1}$$

If the constant H and all the p_j's of these formulas are restricted to finite values greater than zero, the four cases, obtained by separating the plus and minus exponents, are mutually exclusive, but together they cover the entire field. If p_1 is allowed to be zero, either the first or the second pair covers the entire field. Finally, if in addition p_{2n-1} or p_{2n-2} is allowed to become infinite, while Hp_{2n-1}^2 or Hp_{2n-2}^2 is maintained finite, any one of the four expressions covers the entire field. Sometimes one, sometimes another way of covering the field is the more convenient. Formulas (2) and (3) apply to all of these expressions for S provided the p_j's include all the resonant points and all the anti-resonant points, respectively.

PHYSICAL DISCUSSION

The variation of the reactance $X=S/i$ with frequency is illustrated by the curves of Fig. 1 in all the typical cases of formula (1) for $n=1$ and for $n=2$. For every curve the reactance increases with the frequency,[5] except for the discontinuities which carry it back from a positive infinite value to a negative infinite value at the anti-resonant points. Thus between every two resonant frequencies there is an anti-resonant frequency, no matter how close together the two resonant frequencies may be. The effect of increasing n by one unit is to add one resonant point, and thus to introduce one additional branch to the reactance curve, this branch increasing from a negative infinite value through zero to a positive infinite value.

That formula (1) includes several familiar circuits is seen by considering the most general network with one mesh, that is, an inductance and a capacity in series, with the impedance $iLp+(iCp)^{-1}$. This expression is given immediately by (1) upon setting $n=1$, $H=L$, and $p_1=1/\sqrt{LC}$. Since L and C are both positive these constants satisfy the conditions stipulated under (1), thus verifying the theorem for circuits of one mesh. This general one-mesh circuit includes as special cases a single inductance L by setting $H=L$ and $p_1=0$, and a single capacity C by setting $H=0$ and $p_1=\infty$ such that $Hp_1^2=1/C$.

In Fig. 1 the reactances shown by the curves on the right are the negative reciprocals of those on the left. Fig. 1 also shows networks which give the several reactance curves, the networks being computed by means of formulas (2) and (3). The networks are arranged in pairs with reciprocal driving-point impedances and with the networks themselves reciprocally related, that is, the geometrical forms of the networks are conjugate,[6] and inductances correspond to capacities of the same numerical value and vice versa. This relation is a natural consequence of the reciprocal relation between an inductance and a capacity of the same numerical value, these being the elements from which the networks are constructed.

For $n=1$, formulas (2) and (3) give identical networks, as illustrated by the reactances A, B, A', and B' of Fig. 1, each of which is realized by a single network. For the reactances C and C' the two formulas give distinct networks, c_1 and c_2, c_1' and c_2', respectively, these

[5] This has been proved by Otto J. Zobel (loc. cit., pp. 5, 36), using the formula for the most general driving-point impedance given by George A. Campbell (loc. cit., p. 30).

[6] For a further treatment of conjugate or inverse networks, see **P. A. MacMahon,** *Electrician*, April 8, 1892, pages 601, 602, and Otto J. Zobel, loc. cit., pages **5, 36, and 37,**

two being the only networks with the minimum number of elements which give the specified impedance. In general, however, there are four ways of realizing a given impedance when $n=2$, as illustrated by D and D' of Fig. 1; formulas (2) and (3) give only the first two

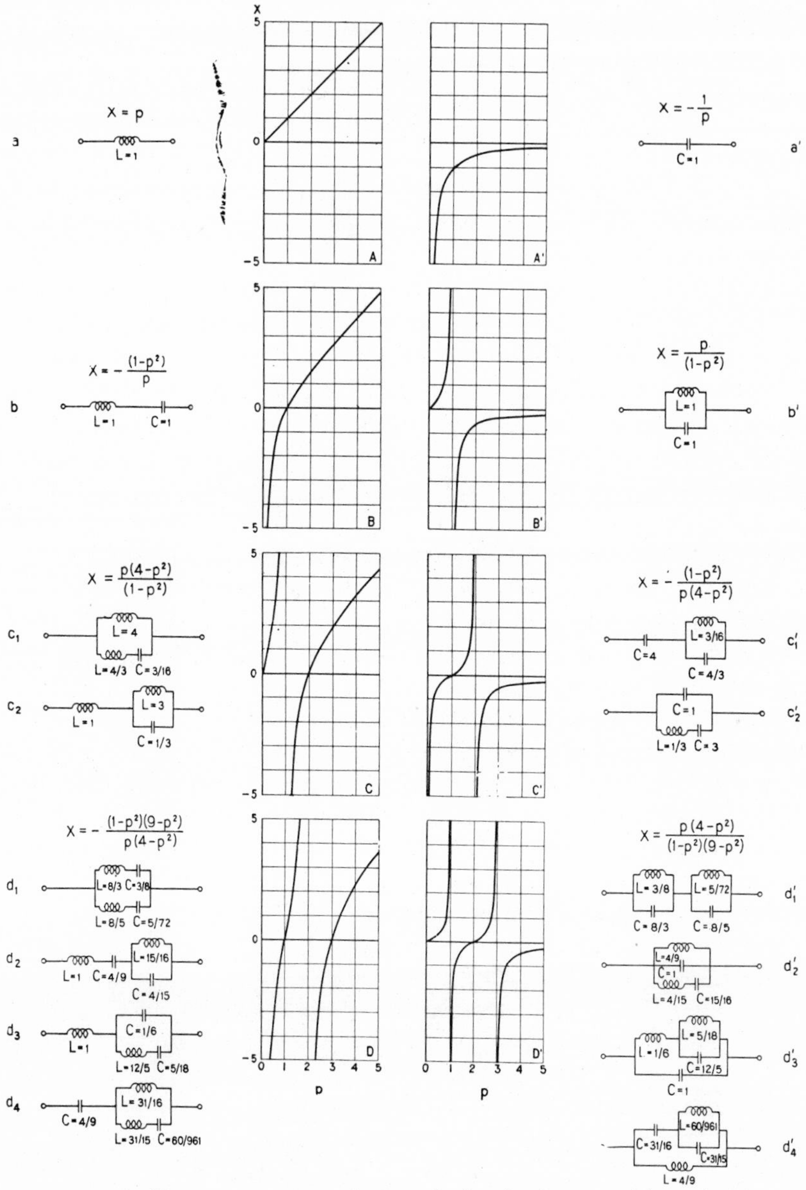

1—Reactance curves and networks for simple cases of formula (1).

networks, d_1 and d_2, d_1' and d_2', respectively. The total number of possible ways of realizing a given impedance increases very rapidly for values of n greater than 2; for $n=3$, there are, in general, 32 distinct networks giving a specified impedance.

Formulas (2) and (3) are to be used for determining the constants of the circuits which have certain specified characteristics, whereas most network formulas are for the determination of the characteristics of the circuit from the given constants of the circuit. The application of these formulas is illustrated by the following numerical problem:

To design a reactance network which shall be resonant at frequencies of 1000, 3000, 5000, and 7000 cycles; anti-resonant at 2000, 4000, and 6000 cycles, as well as at zero and infinite frequencies; and have a reactance of 2500 ohms at a frequency of 10,000 cycles.

By formula (1) the reactance of such a network must be

$$X=-H\frac{(p_1^2-p^2)\,(p_3^2-p^2)\,(p_5^2-p^2)\,(p_7^2-p^2)}{p\,(p_2^2-p^2)\,(p_4^2-p^2)\,(p_6^2-p^2)}, \tag{4}$$

where p_1, p_3, p_5, and p_7 are determined by the resonant frequencies to be $1000\times 2\pi$, $3000\times 2\pi$, $5000\times 2\pi$, and $7000\times 2\pi$, respectively; p_2, p_4, and p_6 are determined by the anti-resonant frequencies to be $2000\times 2\pi$, $4000\times 2\pi$, and $6000\times 2\pi$, respectively; and H must be made equal to 0.0596 in order that the reactance at $p=10{,}000\times 2\pi$ may be 2500. The variation of the reactance with the frequency is shown by the curve of Fig. 2.

A network having this reactance may be constructed by combining $n=4$ simple resonant circuits in parallel, or $n+1=5$ simple anti-resonant circuits in series. These two networks are shown by Fig. 2. The numerical values of the elements are determined as follows: Applying formula (2) we have

$$L_1=\frac{1}{C_1p_1^2}=H\frac{(p_3^2-p_1^2)\,(p_5^2-p_1^2)\,(p_7^2-p_1^2)}{(p_2^2-p_1^2)\,(p_4^2-p_1^2)\,(p_6^2-p_1^2)}=0.349,$$

$$L_3=\frac{1}{C_3p_3^2}=H\frac{(p_1^2-p_3^2)\,(p_5^2-p_3^2)\,(p_7^2-p_3^2)}{(p_2^2-p_3^2)\,(p_4^2-p_3^2)\,(p_6^2-p_3^2)}=0.323,$$

$$L_5=\frac{1}{C_5p_5^2}=H\frac{(p_1^2-p_5^2)\,(p_3^2-p_5^2)\,(p_7^2-p_5^2)}{(p_2^2-p_5^2)\,(p_4^2-p_5^2)\,(p_6^2-p_5^2)}=0.264,$$

$$L_7=\frac{1}{C_7p_7^2}=H\frac{(p_1^2-p_7^2)\,(p_3^2-p_7^2)\,(p_5^2-p_7^2)}{(p_2^2-p_7^2)\,(p_4^2-p_7^2)\,(p_6^2-p_7^2)}=0.142;$$

and applying formula (3) we have

$$C_0=\frac{p_2^2p_4^2p_6^2}{Hp_1^2p_3^2p_5^2p_7^2}=0.0888\times10^{-6},\ L_0=\infty,$$

$$C_2=\frac{1}{L_2p_2^2}=\frac{-p_2^2(p_4^2-p_2^2)(p_6^2-p_2^2)}{H(p_1^2-p_2^2)(p_3^2-p_2^2)(p_5^2-p_2^2)(p_7^2-p_2^2)}=0.0461\times10^{-6},$$

$$C_4=\frac{1}{L_4p_4^2}=\frac{-p_4^2(p_2^2-p_4^2)(p_6^2-p_4^2)}{H(p_1^2-p_4^2)(p_3^2-p_4^2)(p_5^2-p_4^2)(p_7^2-p_4^2)}=0.0523\times10^{-6},$$

$$C_6=\frac{1}{L_6p_6^2}=\frac{-p_6^2(p_2^2-p_6^2)(p_4^2-p_6^2)}{H(p_1^2-p_6^2)(p_3^2-p_6^2)(p_5^2-p_6^2)(p_7^2-p_6^2)}=0.0725\times10^{-6},$$

$$C_8=0,\qquad L_8=H=0.0596.$$

These formulas give the numerical values of the inductances in henries and the capacities in farads. The entire set of numerical values is shown in Fig. 2. It is to be noted that the anti-resonant circuit corresponding to $p_0=0$ consists of a simple capacity since the inductance is infinite and thus does not appear in the network, whereas for $p_8=\infty$ the anti-resonant circuit consists of a simple inductance, the capacity being zero and thus not appearing in the network.

Mathematical Proof

We shall first prove that the driving-point impedance S, as given by (1), may be physically realized by either a simple parallel-series or a simple series-parallel network of inductances and capacities, provided resistances can be made negligibly small.

The rational function $1/S$ can be expanded in partial fractions,

$$\frac{1}{S}=\frac{iH_1p}{p_1^2-p^2}+\frac{iH_3p}{p_3^2-p^2}+\ .\ .\ .\ +\frac{iH_{2n-1}p}{p_{2n-1}^2-p^2},$$

where $$H_j=\left(\frac{p_j^2-p^2}{ipS}\right)_{p=p_j}\qquad(j=1,3,\ .\ .\ .\ ,2n-1).$$

Hence S is equal to the impedance of the parallel combination of the n circuits having the impedances $(p_j^2-p^2)/(iH_jp)=iH_j^{-1}p+[i(H_jp_j^{-2})p]^{-1}$, that is, n simple resonant circuits in parallel, each circuit consisting of an inductance and a capacity in series, with the numerical values given by (2). Furthermore, these numerical values of the inductances and capacities given by (2) are all positive, an even number of negative factors being obtained upon substituting $p=p_j$, since in every case $p_j\leq p_{j+1}$. Hence the network defined by (2) has the impedance S as given by (1) and is physically realizable.

Likewise, by expanding S in partial fractions, it can be shown that the network defined by (3) has the impedance S as given by (1) and is physically realizable.

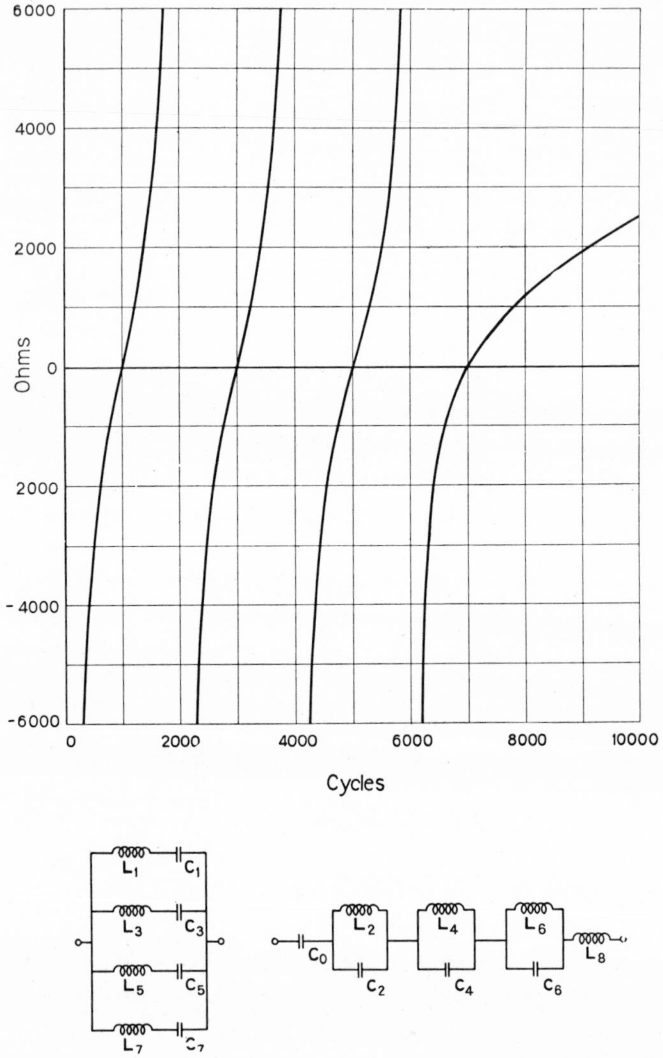

Fig. 2—Reactance curve and networks for formula (4).

The values of the inductances and capacities are (in henries and microfarads):

$L_1=0.349$	$C_1=0.0726$	$L_2=0.137$	$C_0=0.0888$
$L_3=0.323$	$C_3=0.00872$	$L_4=0.0302$	$C_2=0.0461$
$L_5=0.264$	$C_5=0.00384$	$L_6=0.00971$	$C_4=0.0523$
$L_7=0.142$	$C_7=0.00363$	$L_8=0.0596$	$C_6=0.0725$

The electrical problem of the free oscillations of a resistanceless network is formally the same as the dynamical problem of the small oscillations of a system about a position of equilibrium with no frictional forces acting. The proof of formula (1) may be derived from the treatment of this dynamical problem as given, for example, by Routh.[7]

In any network the driving-point impedance in the qth mesh, S_q, is equal to the ratio A/A_q, where A is the determinant[8] of the network and A_q the principal minor of this determinant obtained by striking out the qth row and the qth column. The determinant of a network has the element Z_{jk} in the jth row and kth column, Z_{jk} being the mutual impedance between meshes j and k (self-impedance when $j=k$), the determinant including n independent meshes of the network.

Hence the determinant A has the element $Z_{jk}=iL_{jk}p+(iC_{jk}p)^{-1}$, where L_{jk} is the total inductance and C_{jk} the total capacity common to the meshes j and k. Upon taking the factor $(ip)^{-1}$ from each row and substituting $-p^2=x$, the expression for A may be put in the form $A=(ip)^{-n}D$, where D is a determinant with $L_{jk}x+1/C_{jk}$ as the element in the jth row and the kth column. This is of exactly the same form as the determinant given by Routh [9] for the solution of the dynamical problem; it is proved there that this determinant, regarded as a polynomial, has n negative real roots which are separated by the $n-1$ negative real roots of every first principal minor of the determinant.

Hence, we may write $D=E(x_1+x)(x_3+x)\ldots(x_{2n-1}+x)$, where $x_1, x_3, \ldots, x_{2n-1}$ are all positive and arranged in increasing order of magnitude, and where E is also positive since D must be positive for $x=0$. The determinant D_q may be expressed in similar manner since it is of the same form as D but of lower order.

[7] E. J. Routh, "Advanced Rigid Dynamics," sixth edition, 1905, pages 44–55. In the notation of the dynamical problem as presented here, the coefficients A_{jk} correspond to the inductances, $1/C_{jk}$ to the capacities, $p/(i2\pi)$ to the frequency, and θ', ϕ', etc., to the branch currents in the electrical problem.

A complete proof of formula (1) has been worked out for the electrical problem, without depending in any way upon the solution of the corresponding dynamical problem. This proof has not been published here in view of the great simplification made by using the results already worked out for the dynamical problem.

[8] A complete discussion of the solution of networks by means of determinants has been given by G. A. Campbell, Transactions of the A. I. E. E., 30, 1911, pages 873–909.

[9] The determinant given by Routh (loc. cit., p. 49) has the element $A_{jk}p^2+C_{jk}$.

The driving-point impedance is given by

$$S_q = \frac{A}{A_q} = (ip)^{-1}\frac{D}{D_q} = (ip)^{-1}\frac{E(x_1+x)(x_3+x)\ldots(x_{2n-1}+x)}{E_q(x_2+x)\ldots(x_{2n-2}+x)},$$

where $0 \leq x_1 \leq x_2 \leq x_3 \leq \ldots \leq x_{2n-2} \leq x_{2n-1}$, since the roots of D are separated by the roots of D_q. Upon substituting $x = -p^2$ and introducing the notation $H = E/E_q$ and $p_1^2, p_2^2, \ldots, p_{2n-1}^2 = x_1, x_2, \ldots, x_{2n-1}$, respectively, we see that formula (1) is completely verified as the most general driving-point impedance obtainable by means of a finite resistanceless network.

2

Reprinted from *J. Math. Phys.*, **10**(3), 191–236 (1931)

SYNTHESIS OF A FINITE TWO-TERMINAL NETWORK WHOSE DRIVING-POINT IMPEDANCE IS A PRESCRIBED FUNCTION OF FREQUENCY

BY OTTO BRUNE [1]

CONTENTS

PART I. INTRODUCTION

1. *Statement of the Problem*

In the well known methods of analysing the performance of linear passive electrical networks with lumped network elements it is usual to derive from the given structure of the network a scalar function $Z(\lambda)$ known as the impedance function of the network; this function determines completely the performance

[1] Containing the principal results of a research submitted for a doctor's degree in the Department of Electrical Engineering, Massachusetts Institute of Technology. The author is indebted to Dr. W. Cauer who suggested this research.

of the network as far as current and voltage at one pair of terminals is concerned.

The converse problem is assuming increasing importance in electrical engineering practice, namely: Given the behavior of the network at its terminals, to find the structure of the network. The solution of this problem is not unique; for the present we shall concern ourselves mainly with finding at least one solution.

The behavior of a network at a pair of terminals may be prescribed in several different ways; we shall assume that it is prescribed by specifying a function $Z(\lambda)$ to be the impedance function of the network, all other methods being reducible to this one.

It is evident that certain restrictions are imposed on the possible forms of $Z(\lambda)$ by the physical conditions inherent in a passive network with lumped linear elements. A concise statement of the necessary and sufficient conditions is an essential part of the solution of the problem.

The problem may therefore be stated as follows:—

To find the necessary and sufficient conditions to be satisfied by the impedance function of a finite passive network, and to construct a network corresponding to any function satisfying these conditions.

2. *Contributions of previous Investigators*

The first important contribution to the theory of synthesis of two-terminal networks was probably made by R. M. Foster when, in his "Reactance Theorem" (1) he gave the necessary and sufficient conditions which must be satisfied by the impedance function of a *purely reactive network*. They are:

The poles and zeros of $Z(\lambda)$ are pure imaginary and mutually separate each other (this includes the necessity for a pole or zero at the origin and at infinity).

Foster gives two methods of constructing a network when the prescribed impedance function satisfies these conditions. These methods correspond to a development of $Z(\lambda)$ or its reciprocal in partial fractions; the networks so obtained are shown in Fig. 1 (*a*) and (*b*). Cauer (3) has amplified these results by pointing out a continued fraction development corresponding to the equivalent networks shown in Fig. 1 (*c*) and (*d*).

Cauer (3) has also adapted the results for purely reactive networks to all networks containing only two kinds of elements, i.e. such networks as contain only resistance and capacitance or

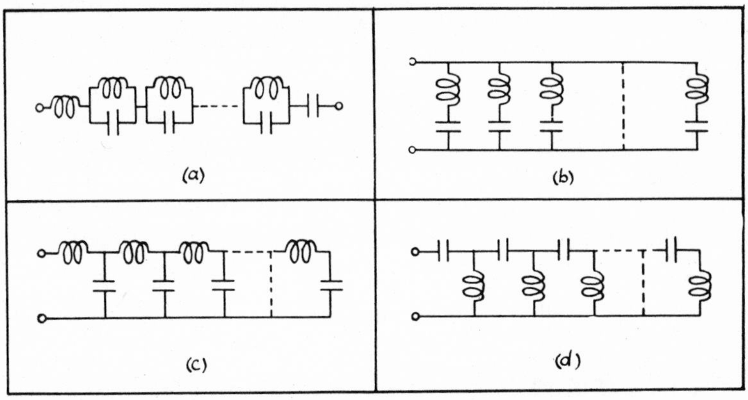

FIG. 1

only resistance and inductance. For such networks the prescribed $Z(\lambda)$ *must have all its poles and zeros on the negative real axis and the poles and zeros must mutually separate each other.* Corresponding networks may be constructed similar to those in Fig. 1 but having either coils or condensers (as the case may be) replaced by resistances.

The particular case of our problem in which only two kinds of elements occur has thus been completely solved.

Both Foster (2) and Cauer (3) have studied the case when $Z(\lambda)$ is capable of representation by a two-mesh network having three kinds of elements. Such a function must have the form

$$Z(\lambda) = \frac{a_0 + a_1\lambda + a_2\lambda^2 + a_3\lambda^3 + a_4\lambda^4}{b_1\lambda + b_2\lambda^2 + b_3\lambda^3},$$

where the values of the coefficients in the numerator and denominator are restricted by certain inequalities (some of the coefficients may of course be zero).

The network shown in Fig. 2 however has an impedance function of the same form as a two-mesh network but with coefficients which do not satisfy the inequalities necessary for the impedance function of a two-mesh network. Consequently the

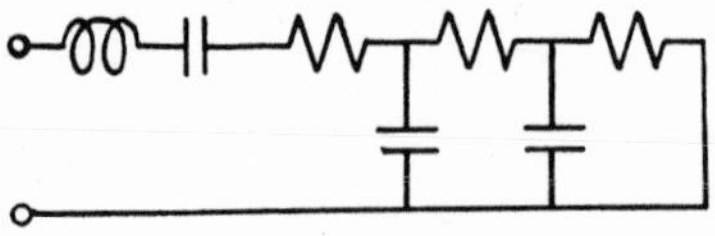

FIG. 2

conditions discussed by Foster and Cauer are sufficient but not necessary in the general problem.

T. C. Fry (4) has discussed in some detail the possibility of using the Stieltjes' continued fraction as a means of developing the impedance function in a form usable for the construction of a corresponding network. He is guided in this by the fact that a continued fraction development of the function is the counterpart of a ladder structure in the network. The conditions for the applicability of a Stieltjes continued fraction expansion of a function are sufficient for the function to be an impedance function of an electric network.[2] Fry also succeeds in extending the method to functions which do not immediately satisfy these conditions by means of several ingenious transformations. The method is not general however, and since it leads to infinite networks in many cases, does not seem to be the natural method of attack.

In all the above cases the conditions imposed on $Z(\lambda)$ are sufficient for the construction of a network but not necessary from the general point of view.

3. *Necessary Conditions to be Satisfied by the Impedance Function of a Finite Network*

Since the impedance function of every finite network is a rational function, it is clear that we must be able to write every

[2] See also Cauer, loc. cit. (3).

prescribed impedance function in the form

$$Z(\lambda) = \frac{f(\lambda)}{g(\lambda)}, \tag{1}$$

where

$$f(\lambda) = a_0 + a_1\lambda + a_2\lambda^2 + \cdots a_n\lambda^n, \tag{1a}$$

$$g(\lambda) = b_0 + b_1\lambda + b_2\lambda^2 + \cdots b_m\lambda^m. \tag{1b}$$

Now if the network be disturbed it is a well known fact that the transient current is given by

$$I_t = \sum_k C_k \epsilon^{\alpha_k t}, \tag{2}$$

where α_k are the roots of $f(\lambda)$; and since transient currents cannot increase without limit in a passive network, we must have [3]

$$\operatorname{Re} \alpha_k \leq 0. \tag{2a}$$

In the above the terminals of the network were regarded as closed, i.e. the transient currents are such as to cause no voltage drop across the terminals. Similar considerations for open terminals give the condition

$$\operatorname{Re} \beta_k \leq 0, \tag{2b}$$

where β_k are the roots of $g(\lambda) = 0$.

A further necessary condition is obtained by considering a steady alternating voltage $E = E_m \cos \omega t$ applied. We may then write

$$E = E_m \operatorname{Re} \epsilon^{i\omega t},$$

$$I_s = E_m \operatorname{Re} \frac{\epsilon^{i\omega t}}{Z(i\omega)}$$

$$= \frac{E_m}{|Z(i\omega)|} \cos(\omega t - \varphi) \quad \text{where} \quad \varphi = \arg Z(i\omega).$$

From the fact that the average power taken by the network cannot be negative, it follows that $|\varphi| \leq \pi/2$ or

$$\operatorname{Re} Z(i\omega) \geq 0. \tag{3}$$

[3] The notation Re means "real part of" throughout.

We may also consider an applied voltage

$$E = \epsilon^{\gamma t} \cos \omega t$$
$$= \text{Re } \epsilon^{(\gamma + i\omega)t}.$$

The "steady state" current may be calculated as before as

$$I_s = \text{Re } \frac{\epsilon^{(\gamma+i\omega)t}}{Z(\gamma + i\omega)}$$
$$= \frac{\epsilon^{\gamma t}}{|Z(\gamma + i\omega)|} \cos(\omega t + \varphi) \quad \text{where} \quad \varphi = \arg Z(\gamma + i\omega).$$

If γ is *positive* the transient currents can, after a sufficiently long time, be neglected (even if the network is purely reactive). If on the other hand γ is *negative*, the transient term may become more important than the steady state term, and no conclusions concerning the energy taken by the network can be drawn by considering the "steady state" current alone; in fact the application of the term "steady state" is misleading in this case. Assuming γ to be positive, however, let us determine the energy W taken by the network after the transient current has become negligible, and consider $t = 0$ at the beginning of this time, the voltage then being unity. Thus

$$E = \epsilon^{\gamma t} \cos \omega t; \qquad I = \frac{1}{|Z(\gamma + i\omega)|} \epsilon^{\gamma t} \cos(\omega t - \varphi).$$

Integrating the product EI over a time τ

$$W = \frac{1}{|Z(\gamma + i\omega)|} \int_0^{\tau} \epsilon^{2\gamma t} \cos \omega t \cos(\omega t - \varphi) dt$$
$$= \frac{1}{4|\gamma + i\omega| \cdot |Z(\gamma + i\omega)|} \left\{ \epsilon^{2\gamma\tau} \cos(2\omega\tau - \varphi - x) \right.$$
$$\left. - \cos(\varphi + \chi) + \frac{\cos \varphi}{\cos x} (\epsilon^{2\gamma\tau} - 1) \right\}$$
$$\text{where} \qquad \chi = \arg(\gamma + i\omega).$$

It is evident that as τ increases the value of W will depend entirely on the terms having $\epsilon^{2\gamma\tau}$ as a factor. Consequently if W is never to be negative we must have $\dfrac{\cos\varphi}{\cos\chi} \geqq 1$ (the equality sign holding only for the non-dissipative case). This apparently requires besides

$$\text{Re } Z(\lambda) \geqq 0 \quad \text{where Re } \lambda \geqq 0; \tag{4a}$$

also

$$|\arg Z(\lambda)| \leq |\arg \lambda| \quad \text{when } |\arg \lambda| \leq \frac{\pi}{2}. \tag{4b}$$

We have enumerated a number of necessary conditions, but it is evident that they are not all independent. It will next be shown that all other conditions are consequences of either (4*a*) or (4*b*). Several other important properties of functions satisfying these conditions will also be noted; these will be of value in finding a network corresponding to a given impedance function.

Part II—Properties of "Positive Real" Functions

1. *Definitions.* For the sake of brevity we define the following terms:

A "*positive function*" $Z(\lambda)$ is a function whose real part is positive when the real part of λ is positive, i.e. a function which satisfies condition (4*a*). If in addition $Z(\lambda)$ is real when λ is real, it will be called a "*positive real*" function. The quotation marks will be retained as a reminder of the special sense in which these terms are being used.

In general the notation $Z(\lambda)$ will imply that the function falls under the above definition. If a function does not or is not known to satisfy these conditions, an asterisk will be added or another notation employed.

If a function is the impedance function of a physical network, it will be said to have a *network representation.* The process of finding such a network will be called finding a network representation of the function.

The proof that every "positive real" function has a network

representation is reserved for treatment in Part III of this paper. In the meantime we shall frequently anticipate this fact in interpreting mathematical properties of " positive real " functions.

The discussions which follow depend for the most part on well known properties of functions of a complex variable. For more ready reference however they are stated in the form of theorems.

2. *Theorems on " Positive Real " Functions*

The idea of a " positive " function can be interpreted very conveniently in terms of the transformation of one complex plane into another. By the equation $Z = Z(\lambda)$ every point in the λ plane determines uniquely a point in the Z plane. The property of " positiveness " as defined by (4*a*) means that if λ is chosen in the right half plane, Z will fall in the right half plane, but not necessarily conversely. Consequently if $Z(\lambda)$ is a " positive " function, the transformation $Z = Z(\lambda)$ will transform the right half of the λ plane into a part of the right half of the Z plane (except in special cases where the right half λ plane

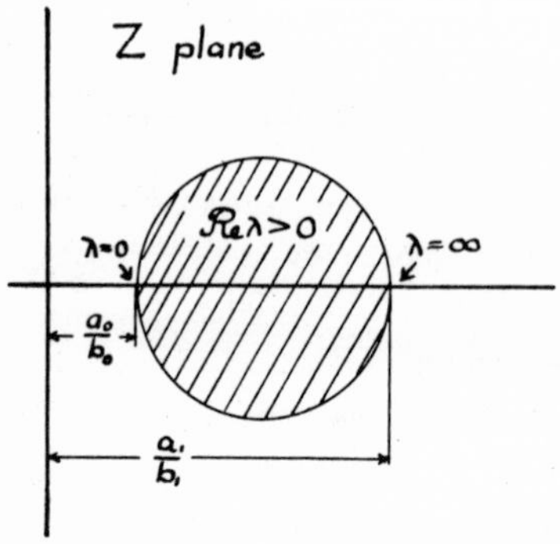

FIG. 3

goes into the whole right half Z plane). A simple example of this is shown in Fig. 3 where $Z = \dfrac{a_0 + a_1\lambda}{b_0 + b_1\lambda}$.

We thus have

THEOREM I. If

$$Z = Z(\lambda)$$

and

$$W' = W(\lambda),$$

Z and W both being "positive functions," then

$$W[Z(\lambda)]$$

will be a "positive" function of λ.

This theorem can obviously be applied in two ways illustrated by the two corollaries:

THEOR. I, COROLL. 1. If $Z(\lambda)$ is a "positive" function, $Y(\lambda) = \frac{1}{Z(\lambda)}$ will be a "positive" function.

THEOR. I, COROLL. 2. If $Z(\lambda)$ is a "positive" function, $Z_1(\lambda) = Z\left(\frac{1}{\lambda}\right)$ will be a "positive" function.

The physical interpretation of Corollary 1 is: to every network corresponds a "reciprocal" network whose admittance characteristic is the impedance characteristic of the first network.

On the other hand Coroll. 2 means that to every network corresponds another whose impedance characteristic is that of the first network but with the reciprocal frequency scale (e.g. frequency is replaced by wave-length).

Taken more generally Theorem I states that if in $Z(\lambda)$, the impedance function of a given network N, λ be replaced by the impedance function $z(\lambda)$ of another network n, the impedance function of a third network N' will result. In certain cases N' can be derived from N by replacing inductances by networks proportional to n, and capacities by networks proportional to the network $(1/n)$, reciprocal to n. If the conception of an ideal transformer be admitted a representation of mutual elements in terms of n is also possible and N' can be constructed directly and completely from N by such substitions.

We next come to a discussion of the poles and zeroes of a "positive" function. We note first of all the

Lemma 1: A zero of multiplicity n is surrounded by $2n$ sectors of equal angles (Fig. 4) in which the real part of the function is

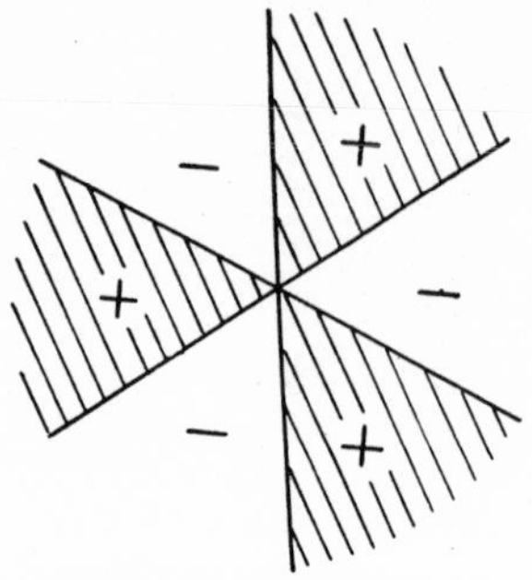

FIG. 4

alternately positive and negative. For at such a point $\lambda = c$ the Taylor's series for the function becomes

$$Z(\lambda) = \left\{ \frac{d^n}{d\lambda^n} \frac{Z(\lambda)}{n!} \right\}_{\lambda=c} (\lambda - c)^n + \text{term of higher degree}$$

In the immediate neighbourhood of the point $\lambda = c$ the first term predominates. Placing

$$(\lambda - c) = \rho\epsilon^{i\theta} \quad \text{and} \quad \left\{ \frac{d^n}{d\lambda^n} \frac{Z(\lambda)}{n!} \right\}_{\lambda=c} = k\epsilon^{i\varphi} \quad (k, \varphi \text{ real constants})$$

we have

$$Z(\lambda) = Z(\rho, \theta) \approx k\rho^n \epsilon^{i(n\theta+\varphi)}$$

$$\text{Re}\, Z(\lambda) \approx k\rho^n \cos(n\theta + \varphi), \tag{5}$$

whence the truth of the lemma is evident.

A zero within the right half plane would therefore obviously violate the condition of "positiveness." Moreover on the boundary of this region (i.e. on the imaginary axis) only two sectors are possible; the real part of the function must be positive on the one side of the boundary, negative on the other. This means that zeros on the imaginary axis must be simple, and that φ in equation (5) must be zero, i.e. $(dZ/d\lambda)_{\lambda=i\sigma}$ = a real positive constant.

Corresponding conditions for the poles are immediately evident from Theorem I, Coroll. 1. For a pole, however, the condition $\left(\frac{d}{d\lambda}\frac{1}{Z}\right)_{\lambda=i\sigma}$ = a real positive constant, can be expressed more simply by saying that the residue of the function at $\lambda = i\sigma$ must be real and positive. All these facts are summarized in

THEOREM II. If $Z(\lambda)$ is a "positive function" then

(i) its zeros and poles must lie in the left half λ plane or or the boundary;
(ii) zeros and poles on the imaginary λ axis must be simple;
(iii) at a zero on the imaginary λ axis $dZ/d\lambda$ is a real positive constant while at a pole on the boundary the residue is a real positive constant.

Of especial interest are the zeros and poles at $\lambda = 0$ and $\lambda = \infty$. These points lie on the boundary; consequently at them only simple poles and zeros can exist.

If $\lambda = 0$ is a simple zero or a pole, λ will be a factor in the numerator or denominator respectively. If $\lambda = \infty$ is a simple pole or zero, the degree of the numerator or denominator will be greater by one than the degree of the denominator or numerator respectively. This consequence of Theorem II (ii) can therefore be summarized as

THEOREM II, COROLL. 1. The degree in λ or in $1/\lambda$ of numerator and denominator of $Z(\lambda)$ cannot differ by more than 1.

We interpose here a well-known theorem in function theory, which finds special application in the theory of potentials. This theorem states that a function $R(x, y)$ which is regular within a certain region attains its maximum and minimum value on the boundary of that region. If poles occur on the boundary they may be excluded by arbitrarily small semi-circles, the theorem then holding for the modified boundary. In the case of "positive" functions the value of the real part becomes arbitrarily large on such semi-circles if the radius is made arbitrarily small, but remains positive. Hence we have

THEOREM III. If $Z(\lambda)$ is a "positive" function the real part of $Z(\lambda)$ in the right half plane attains its *minimum* value on the axis of imaginaries.

THEOREM III, COROLL. 1. If R_1 is equal to or less than the minimum value of the real part of $Z(i\omega)$ then

$$Z'(\lambda) = Z(\lambda) - R_1$$

is a "positive" function.

We shall apply this theorem immediately to the separation of poles on the axis of imaginaries from the function.

Consider first a pole at $\lambda = \infty$. The numerator is of degree one higher than the denominator, i.e.

$$Z(\lambda) = \frac{a_{n+1}\lambda^{n+1} + a_n\lambda^n + \cdots a_1\lambda + a_0}{b_n\lambda^n + \cdots b_1\lambda + b_0}.$$

By ordinary division of numerator into denominator

$$\begin{aligned} Z(\lambda) &= L\lambda + \frac{a_n'\lambda^n + \cdots a_1'\lambda + a_0'}{b_n\lambda^n + \cdots b_1\lambda + b_0}, \\ &= L\lambda + Z'(\lambda) \end{aligned}$$

where

$$L = \frac{a_{n+1}}{b_n}.$$

Note that the real part of $L\lambda$ on the axis of imaginaries is zero, consequently $\operatorname{Re} Z'(i\omega) = \operatorname{Re} Z(i\omega) \geq 0$. Since $\operatorname{Re} Z'(\lambda)$ is regular in the right half plane however (i.e. has no poles there) it is everywhere within this region greater than the minimum value on the boundary; i.e. > 0. hence $Z'(\lambda)$ is a "positive" function.

The case when $\lambda = 0$ is a pole follows from the above by making the substitution $\lambda' = 1/\lambda$ everywhere. Then

$$\begin{aligned} Z(\lambda) &= \frac{a_0 + a_1\lambda + a_2\lambda^2 + \cdots a_n\lambda^n}{b_1\lambda + b_2\lambda^2 + \cdots b_m\lambda^m} \\ &= \frac{D}{\lambda} + \frac{a_1' + a_2'\lambda + \cdots a_n'\lambda^{n-1}}{b_1 + b_2\lambda + \cdots b_m\lambda^{m-1}} \end{aligned}$$

$$= \frac{D}{\lambda} + Z'(\lambda), \quad \text{where } D = \frac{a_0}{b_1}.$$

and $Z'(\lambda)$ is a "positive real" function.

If $Z(\lambda)$ has a pole on the boundary at $\lambda = i\sigma$ (say) the condition of reality on the real axis necessitates the presence of the conjugate pole. These poles can always be removed in the following way, which is merely an application of partial fractions.

Let

$$Z(\lambda) = \frac{f(\lambda)}{g(\lambda)} = \frac{f(\lambda)}{(\lambda^2 + \sigma^2)g_1(\lambda)}.$$

Then

$$Z(\lambda) = \frac{k}{\lambda + i\sigma} + \frac{k}{\lambda - i\sigma} + \frac{f_1(\lambda)}{g_1(\lambda)}$$

(k being the real positive residue of $Z(\lambda)$ at $\lambda = \pm i\sigma$)

$$Z(\lambda) = \frac{2k\lambda}{\lambda^2 + \sigma^2} + Z'(\lambda). \tag{6}$$

The important difference between equation (6) and the ordinary partial fraction development is that the numerator of the first partial fraction would ordinarily have been $2(k\lambda + h\sigma)$; here however because the residue at $\lambda = i\sigma$ is real, $h = 0$. This fact enables us to conclude that the real part of the first partial fraction in equation (18) is zero on the axis of imaginaries whence it follows as before that $Z'(\lambda)$ is a "positive" function. Summing up we have

THEOREM IV. If $Z(\lambda)$ is a "positive real" function which has poles (residue $= k_r/2$) at $\lambda = i\sigma_0, i\sigma_2, \cdots i\sigma_n, \infty$. Then

$$Z(\lambda) = L\lambda + \sum_{r=0}^{n} \frac{k_r\lambda}{\lambda^2 + \sigma_r^2} + Z'(\lambda), \tag{7}$$

where each term is a "positive real" function and

$$\operatorname{Re} Z'(i\omega) = \operatorname{Re} Z(i\omega).$$

THEOREM IV, COROLL. 1. If $Z(\lambda)$ is a "positive real" function which has zeros at $\lambda = 0, \cdots i\sigma_r \cdots \infty$ then

$$\frac{1}{Z(\lambda)} = C\lambda + \sum_{r=0}^{n} \frac{k_r\lambda}{\lambda^2 + \sigma_r^2} + \frac{1}{Z'(\lambda)}, \tag{7a}$$

where each term is a "positive real" function and

$$\text{Re}\left[\frac{1}{Z'(i\omega)}\right] = \text{Re}\left[\frac{1}{Z(i\omega)}\right].$$

THEOREM IV, COROLL. 2. If $Z(\lambda)$ is a "positive real" function all of whose poles lie on the imaginary axis then $Z'(\lambda)$ in (7) is a real positive constant. A similar statement holds when all the zeros lie on the axis of imaginaries by applying Corollary 1.

Corresponding to Theorem III above is the well-known fact that a function $R(x, y)$ which is regular within a certain region is completely specified within that region by the values which it takes on the boundary. There is obviously a connection between this fact and the possibility of stating the necessary and sufficient conditions in terms of conditions on the boundary.

We note first of all that the condition of regularity in the right-half plane excludes the possibility of poles within this region, i.e. is synonymous with equation (2*b*). We have seen further that poles on the imaginary axis are independent of the real part of $Z(i\omega)$ (Theorem IV). Consequently the condition to be satisfied by poles on the boundary must be stated separately. If in addition to this we state that the real part of $Z(i\omega)$ is positive, all conditions for a "positive" function are fulfilled. We thus have

THEOREM V. If $Z(\lambda)$ is a function such that

(i) No poles lie in the right half plane.

(ii) Poles on the boundary are simple and have positive real residues.

(iii) $\text{Re}\, Z(i\omega) \geq 0$.

then $Z(\lambda)$ is a "positive" function. In particular the zeros $\lambda = \alpha_r$ will satisfy the condition (2*a*).

By Theorem I, Coroll. 1 we have also

THEOREM V, COROLL. 1. If $Z(\lambda)$ satisfies the conditions

(i) No zeros lie within the right half plane.

(ii) Zeros on the boundary are simple and at them $dZ/d\lambda$ = a positive real constant.

(iii) Re $Z(i\omega) \geq 0$.

then $Z(\lambda)$ is a "positive" function. In particular the poles $\lambda = \beta_r$ will satisfy the condition (2*b*).

It is always possible to reduce a given function to one in which the second restriction in Theor. V is not necessary by the procedure of Theor. IV.

So far we have dealt entirely in rectangular coordinates in the λ and Z planes. Equation (46) however involves an inequality between λ and $Z(\lambda)$ in polar coordinates. This is essentially a question of conformality, and like most questions of this nature can be made to depend on a fundamental lemma of Schwarz.[4] For our purposes the following generalization of this lemma is convenient.

THEOREM DUE TO PICK (13): If the function W of γ has no essential singularities for values of γ within the circle K_γ, and takes on values which lie only in the interior of another circle K_W, then all non-euclidean distances, elements of arcs and arcs are shortened in the conformal mapping by $W(\gamma)$. If one such measure remains unchanged, all remain unchanged and W is a linear function of γ.

In this theorem a non-euclidean distance between two points γ_1, γ_2 is defined by the logarithm of a cross-ratio in the usual manner of projective measurement as follows:—Through the points γ_1, γ_2 construct a circle orthogonal to the circle K_γ and cutting it in the points h and k; the non-euclidean distance is then

$$\mathcal{P}(\gamma_1, \gamma_2) = \log \frac{\gamma_1 - h}{\gamma_1 - k} \cdot \frac{\gamma_2 - k}{\gamma_2 - h}. \tag{8}$$

This theorem of Pick also holds for the limiting case when the interior of the circle K_γ or K_W becomes a half-plane.

[4] H. A. Schwarz 1869. See also Bibliogr (13) and (14) Vol. II, p. 114.

14

Now W will be a " positive " function of γ if K_γ and K_W· become the right half of the γ and W planes respectively. The further restriction that W be real for real values of γ makes the correspondence between W and γ on the one hand and a " positive real " function $Z(\lambda)$ and λ on the other hand complete.

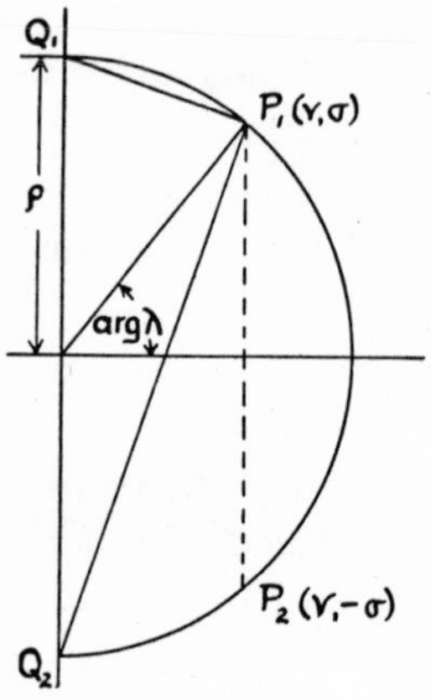

FIG. 5

All that remains is to interpret the non-euclidean distances. We shall do this only for the more interesting cases:—

(i) Two conjugate points λ, $\bar{\lambda}$ will transform into two other conjugate points Z, $\bar{Z}$ because of the condition of " reality." In Fig. 5 let $P_1(\lambda)$, $P_2(\bar{\lambda})$ be a pair of conjugate points and let

$$\lambda = \nu + i\sigma, \quad |\lambda| = \rho.$$

The circle through λ, $\bar{\lambda}$ orthogonal to the boundary of the right half plane will cut this boundary in the points $Q_1(0 + i\rho)$ and $Q_2(0 - i\rho)$

$$\begin{aligned} \mathcal{P}(\lambda, \bar{\lambda}) &= \log \frac{(\nu + i\overline{\sigma - \rho})}{(\nu + i\overline{\sigma + \rho})} \cdot \frac{(\nu - i\overline{\sigma - \rho})}{(\nu - i\overline{\sigma + \rho})} \\ &= \log \frac{\nu^2 + (\sigma - \rho)^2}{\nu^2 + (\sigma + \rho)^2} \\ &= 2 \log \frac{P_1Q_1}{P_1Q_2} \cdot \end{aligned} \tag{9}$$

As $\mathcal{P}(\lambda, \bar{\lambda})$; decreases towards zero (in absolute magnitude since sign does not enter into the idea of distance) the ratio P_1Q_1/P_1Q_2 must approach unity. A ratio of lengths will involve only angles and in Fig. 5 an approach of the ratio P_1Q_1/P_1Q_2 to unity is readily seen to mean a decrease in the angle P_1OP_2 towards zero, i.e. $|\arg \lambda|$ decreases.

(ii) Consider two points $\lambda_1\lambda_2$ on the real axis. The orthogonal circle in this case is the real axis itself. The non-euclidean distance, obtained by considering a limit, becomes

$$\mathcal{P}(\lambda_1, \lambda_2) = \log \frac{\lambda_1}{\lambda_2} \cdot \tag{10}$$

From this we see that if for one value of λ_0 on the positive real axis we have $\lambda_1 = Z(\lambda_1) = Z_1$, this will be the only point for which such a relation holds unless $\lambda \equiv Z(\lambda)$. For if there be another point where $\lambda_2 = Z(\lambda_2) = Z_2$ we shall have

$$\mathcal{P}(\lambda_1, \lambda_2) = \log \frac{\lambda_1}{\lambda_2} = \log \frac{Z_1}{Z_2} = \mathcal{P}(Z_1 Z_2),$$

which by the theorem of Pick is possible only if this is true for all values of λ. The same is obviously true if we substitute $L\lambda$ for λ (L being a real positive constant).

Hence we have

THEOREM VI. (i) if $Z(\lambda)$ is a "positive real" function $|\arg Z(\lambda)| \leq |\arg \lambda|$ for all values of λ satisfying $0 < |\arg \lambda| \leq \pi/2$. The equality signs can only hold simultaneously, unless they hold identically.

(ii) On the positive real axis (where $\arg \lambda = 0$) the equation $L\lambda = Z(\lambda)$ cannot be satisfied at more than one internal point. (L a real positive constant.).

THEOREM VI, COROLL. 1. The function $(Z(\lambda) - L\lambda)$, (which is not necessarily a positive function) cannot have more than one zero within the right-half plane.

Theorem VI may be interpreted in the light of the physical interpretation given to Theorem I. Thus if λ be replaced by a

" real positive " function of λ the resulting network will have a power factor closer to unity than the original network, for corresponding frequencies.

The converse of Theorem VI is at once evident for if $|\arg \lambda| \leq \pi/2$, λ lies in the right-half plane and $|\arg Z(\lambda)|$ being $\leq |\arg \lambda|$, $Z(\lambda)$ will likewise lie in the right-half plane. Furthermore when $\arg \lambda = 0$, $\arg Z(\lambda) = 0$, i.e. Z is a " real " function. Hence we have

THEOREM VII. If $|\arg Z(\lambda)| \leq |\arg \lambda|$ when $|\arg \lambda| \leq \pi/2$, $Z(\lambda)$ will be a " positive real " function.

It is now apparent that if $Z(\lambda)$ is a " positive real " function it will satisfy all the necessary conditions mentioned in Part I, section 3; for conditions 2(*a*) and 2(*b*) follow from Theorem II; condition (3) is merely (4*a*) when $\operatorname{Re} \lambda = 0$; and condition (4*b*) is fulfilled by Theorem VI.

Theorem V gives two sets of conditions equivalent to (4*a*) which are easier to handle numerically and more readily interpreted electrically.

From Theorem VII we see that (4*b*) is an alternative way of defining a " positive real " function.

We next proceed to show that a " positive real " function not only satisfies all the necessary conditions immediately derivable from physical considerations, but that these conditions are also *sufficient* in order that a network representation of the function may be found. For this purpose the above theorems will prove particularly useful.

PART III. SYNTHESIS OF NETWORKS WITH PRESCRIBED IMPEDANCE FUNCTION

1. *Functions with not more than Two Poles (or Zeros) in the Interior of the Left Half Plane*

With the more detailed formulation of the properties of " positive real " functions given in Part II we proceed to consider some of the simpler cases of such functions and their network representations. These have been treated in part by Foster, Cauer and Fry, but the greater power of the methods now at our disposal will be abundantly clear even for these cases.

For convenience in calculation we use the conditions for a "positive real" function as stated in Theorem V. We may restrict ourselves to functions which have no poles or zeros on the axis of imaginaries for if such poles or zeros are present they may be removed in the manner demonstrated in Theorem IV; they correspond to pure reactance elements in series or parallel with each other and with the remaining network.

(a) Functions with One Pole (or Zero)

Consider the function [5]

$$Z(\lambda) = \frac{a_0 + a_1\lambda}{b_0 + b_1\lambda}, \tag{11}$$

$$\operatorname{Re} Z(i\omega) = \frac{a_0 b_0 + a_1 b_1 \omega^2}{b_0^2 + b_1^2\omega^2}. \tag{12}$$

We may arbitrarily assume a_0 positive without imposing any restrictions on $Z(\lambda)$. For $\operatorname{Re} Z(i\omega) \geq 0$ we must have $b_0 \geq 0$ and $a_1 b_1 \geq 0$. The condition for the pole requires $b_1 \geq 0$ hence also $a_1 \geq 0$.

Let us now plot $\operatorname{Re} Z(i\omega)$ as a function of ω^2 (negative values of ω^2 are included in the figure for the sake of completeness although they do not enter into the present discussion). The curve is a hyperbola which takes one of two forms depending on whether $\frac{a_0}{b_0} <$ or $> \frac{a_1}{b_1}$ (see Fig. 6). In the case $\frac{a_0}{b_0} < \frac{a_1}{b_1}$ (Fig. 6*a*) the smallest value of $\operatorname{Re} Z(i\omega)$ in the right half plane occurs at the origin. This value may be subtracted from $Z(\lambda)$ and still leave a "positive" function (Theorem III, Coroll. 1). Then

$$\begin{aligned} Z(\lambda) &= \frac{a_0}{b_0} + \frac{(a_1 b_0 - b_1 a_0)\lambda}{b_0^2 + b_0 b_1 \lambda} \\ &= R_1 + Z'(\lambda). \end{aligned} \tag{13}$$

[5] This function was discussed by Fry, Bibliogr (4).

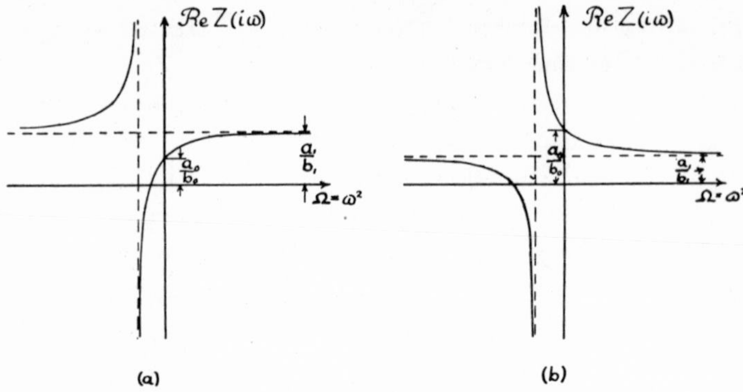

FIG. 6

$Z'(\lambda)$ may be immediately recognized as the impedance function of a network, or we may proceed by noting that it has a zero at the origin. Hence applying Theorem IV

$$\frac{1}{Z'(\lambda)} = \frac{b_0^2}{(a_1b_0 - b_1a_0)\lambda} + \frac{b_0b_1}{(a_1b_0 - b_1a_0)}. \tag{14}$$

The corresponding network is shown in Fig. 7 (*a*).

An exactly similar procedure is possible when $\frac{a_0}{b_0} > \frac{a_1}{b_1}$ (see Fig. 6*b*). Then

$$Z(\lambda) = \frac{a_1}{b_1} + \frac{(a_0b_1 - a_1b_0)}{b_0b_1 + b_1^2\lambda}$$

and so on. (The results are obtainable from those of section 40 by the substitution $\lambda' = 1/\lambda$ or by an interchange of the subscripts 0 and 1.) The corresponding network is shown in Fig 7 (*b*).

The network representation of (10) thus contains resistance and inductance or resistance and capacitance according as

$$\frac{a_0}{b_0} < \quad \text{or} \quad > \frac{a_1}{b_1} \tag{15}$$

When $\frac{a_0}{b_0} = \frac{a_1}{b_1}$, $Z(\lambda)$ obviously reduces to $R = \frac{a_0}{b_0} = \frac{a_1}{b_1}$.

A corresponding discussion which is dual to the above in the sense of Theorem 1, Coroll. 1 can be carried through with $1/Z(\lambda)$ giving rise to the networks in Fig. 7 (c) and (d). The networks

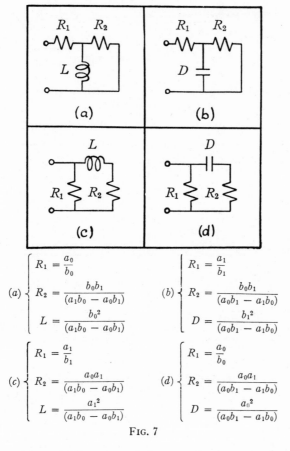

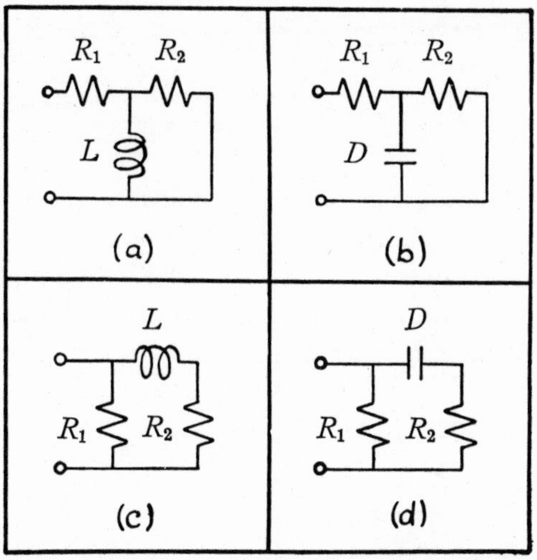

$$(a)\begin{cases} R_1 = \dfrac{a_0}{b_0} \\ R_2 = \dfrac{b_0 b_1}{(a_1 b_0 - a_0 b_1)} \\ L = \dfrac{b_0^2}{(a_1 b_0 - a_0 b_1)} \end{cases} \qquad (b)\begin{cases} R_1 = \dfrac{a_1}{b_1} \\ R_2 = \dfrac{b_0 b_1}{(a_0 b_1 - a_1 b_0)} \\ D = \dfrac{b_1^2}{(a_0 b_1 - a_1 b_0)} \end{cases}$$

$$(c)\begin{cases} R_1 = \dfrac{a_1}{b_1} \\ R_2 = \dfrac{a_0 a_1}{(a_1 b_0 - a_0 b_1)} \\ L = \dfrac{a_1^2}{(a_1 b_0 - a_0 b_1)} \end{cases} \qquad (d)\begin{cases} R_1 = \dfrac{a_0}{b_0} \\ R_2 = \dfrac{a_0 a_1}{(a_0 b_1 - a_1 b_0)} \\ D = \dfrac{a_0^2}{(a_0 b_1 - a_1 b_0)} \end{cases}$$

FIG. 7

of Figs. 7 (c) and (d) are *equivalent* to the networks of Figs. 7 (a) and (b) respectively.

It is interesting to note that the complete set of equivalent networks (having only two meshes) is obtained by combining the processes leading to Figs. 7a and 7c in different ways. For example if $\dfrac{a_0}{b_0} < \dfrac{a_1}{b_1}$ any value less than a_0/b_0 can be separated as a series resistance (Fig. 7a) and thereafter the remaining network constructed as in Fig. 7 (c).

The above discussion disposes immediately of such forms as

$$\left.\begin{array}{ll}\text{(i) } Z(\lambda)=\dfrac{a_0+a_1\lambda+a_2\lambda^2}{b_0+b_1\lambda}, & \text{(iii) } Z(\lambda)=\dfrac{a_0+a_1\lambda+a_2\lambda^2}{b_1\lambda+b_2\lambda^2}, \\ \text{(ii) } Z(\lambda)=\dfrac{a_0+a_1\lambda}{b_0+b_1\lambda+b_2\lambda^2}, & \text{(iv) } Z(\lambda)=\dfrac{a_1\lambda+a_2\lambda^2}{b_0+b_1\lambda+b_2\lambda^2}.\end{array}\right\} \quad (16)$$

The corresponding networks are shown in Fig. 8 for future reference. The coefficients in these functions may not be arbitrary positive constants as in (10), the additional conditions

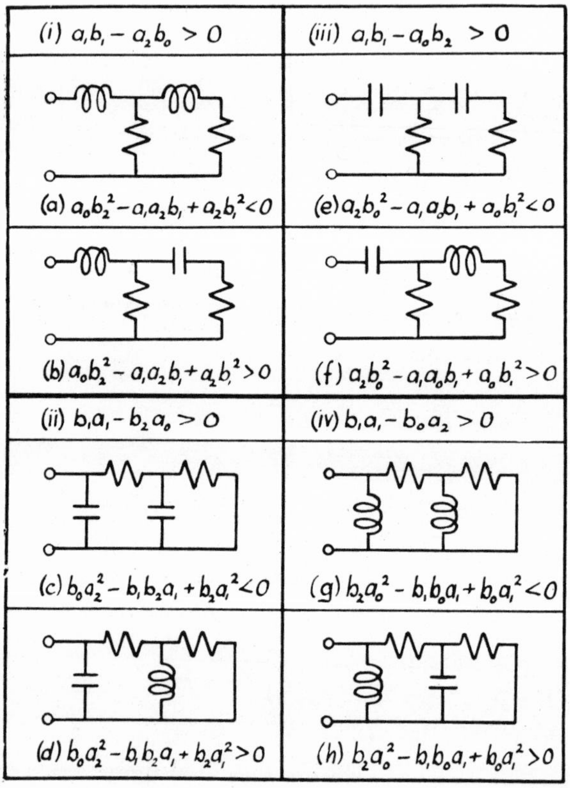

FIG. 8

which they must fulfill being shown in the Figure for each case. These conditions are readily obtained by applying the conditions for (10) to the reduced functions which result on separating the pole (or zero) on the imaginary axis from the forms (16).

It is interesting to note the dualities in Fig. 8. The forms (ii) and (iv) are derivable from (i) and (iii) respectively by the substitution $Z' = 1/Z$ while (iii) and (iv) are derivable from (i) and (ii) respectively by the substitution $\lambda' = 1/\lambda$ (compare Theorem I, Coroll. 1 and 2). A discussion of any one of them is therefore readily adaptable to all the others, by a mere change of notation.

Equivalents to the networks in Fig. 8 similar to those in Fig. 7 are also possible.

(*b*) *Two Poles and Two Zeros not on the Boundary*

Consider next the case

$$Z(\lambda) = \frac{a_0 + a_1\lambda + a_2\lambda^2}{b_0 + b_1\lambda + b_2\lambda^2}, \tag{17}$$

$$\text{Re}\, Z(i\omega) = \frac{a_0b_0 + (a_1b_1 - a_2b_0 - a_0b_2)\omega^2 + a_2b_2\omega^4}{b_0^2 + (b_1^2 - 2b_0b_2)\omega^2 + b_2^2\omega^4}. \tag{18}$$

The condition for the poles to have negative real part is given by

$$\frac{b_1}{b_0} \text{ and } \frac{b_2}{b_0} \geq 0. \tag{19}$$

In (18) the denominator cannot be negative for real values of ω (this is best seen by expressing it as the sum of squares). Hence the numerator must always be positive, i.e. it must have no real positive roots when considered as a function of ω^2. Hence we must have (when $b_0 > 0$) $a_0 > 0$, and either

$$a_2 > 0, \quad a_0b_2 - a_1b_1 + a_2b_0 \leq 0 \tag{20a}$$

or

$$(a_0b_2 - a_1b_1 + a_2b_0)^2 - 4a_0a_2b_0b_2 \leq 0. \tag{20b}$$

Assuming that the coefficients satisfy the conditions (19), and (20*a*)or (20*b*), let us consider the possible manners of variation of Re $Z(i\omega)$ according to equation (18). Let us write

$$\operatorname{Re} Z(i\omega) = R(\Omega) = \frac{A_0 + A_1\Omega + A_2\Omega^2}{B_0 + B_1\Omega + B_2\Omega^2}, \tag{18a}$$

where $\Omega = \omega^2$.

The discussion of such functions is an elementary problem in algebra, and is usually carried through in the following way. From equation (18a)

$$(RB_0 - A_0) + (RB_1 - A_1)\Omega + (RB_2 - A_2)\Omega^2 = 0.$$

For real values of Ω we must have

$$(RB_1 - A_1)^2 - 4(RB_0 - A_0)(RB_2 - A_2) \geq 0,$$

i.e.

$$R^2(B_1^2 - 4B_0B_2) - 2R(A_1B_1 - 2A_0B_2 + A_2B_0) + (A_1^2 - 4A_0A_2) \geq 0. \tag{21}$$

This means that R can or cannot (for real values of Ω) take on values lying between the roots of the expression (21) according as the sign of $(B_1^2 - 4B_0B_2)$ is negative or positive. (Note that this is identical with the sign of $b_1^2 - 4b_0b_2$, i.e. is determined by the poles of $Z(\lambda)$ being real or complex.)

Only in the case where the roots of this expression are equal or pure imaginary can R take on all values. Typical curves for R are shown in Fig. 9. In these figures the axes can be shifted parallel to themselves and reverse direction in any manner provided only R remains positive for positive values of Ω. Thus for example Figs. 9 (b) and (d) are of the same type, but with the direction of the R axis reversed.

As in the discussion of $Z(\lambda)$ defined by (11) the procedure depends on the frequency at which the least value of the 're-sistance' $R(\Omega)$ occurs. For this purpose we need consider only positive values of Ω in Fig. 9 since these alone correspond to points on the boundary of the right-half plane (or to real frequencies).

If this least value occurs for $\Omega = 0$ or ∞, as for example in Figs. 9 (*a*) and (*b*) the procedure is essentially the same as before. To illustrate, let the minimum value of $R(\Omega)$ occur for $\Omega = 0$

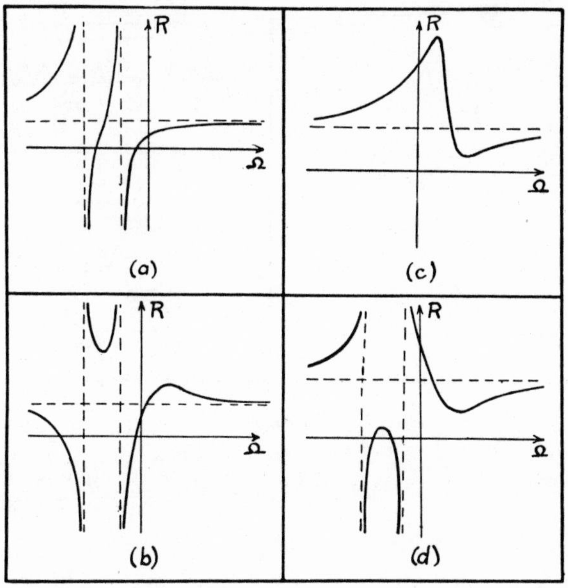

FIG. 9

$$\text{i.e.}\ \frac{a_0}{b_0} \leq R(\Omega) \quad \text{for} \quad \Omega > 0. \tag{22}$$

Then

$$\begin{aligned} Z(\lambda) &= \frac{a_0 + a_1\lambda + a_2\lambda^2}{b_0 + b_1\lambda + b_2\lambda^2} \\ &= \frac{a_0}{b_0} + \frac{(a_1b_1 - a_0b_1)\lambda + (a_2b_0 - a_0b_2)\lambda^2}{b_0(b_0 + b_1\lambda + b_2\lambda^2)} \\ &= \frac{a_0}{b_0} + Z'(\lambda), \end{aligned}$$

in which the function $Z'(\lambda)$ is a " positive " function under the assumption (22) by Theorem III, Coroll. 1. Furthermore this form of $Z'(\lambda)$ has already been discussed (Fig. 8 (iv)). Consequently $Z(\lambda)$ can be represented by such a network together with a series resistance, as shown in Fig. 10 A and B.

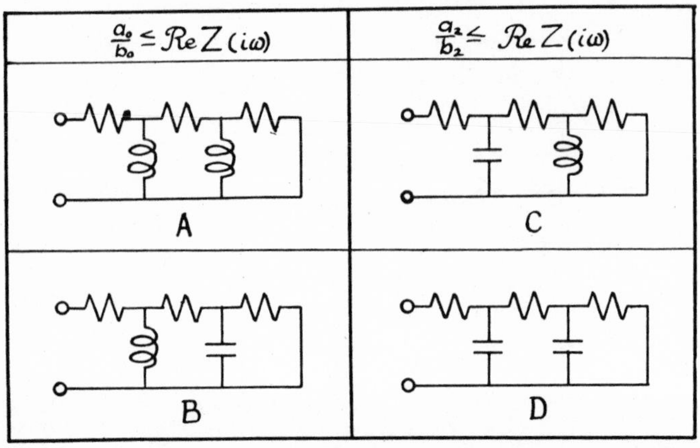

FIG. 10

The case when $R(\omega)_{\min}$ occurs for $\omega = \infty$ needs no further discussion, the corresponding networks being shown in Fig. 10 C and D.

There remains the case when $R(\omega)_{\min}$ occurs for some finite value $\omega = \sigma$. Proceeding by analogy, let this value R_σ be separated out. It will be a solution of (21) (with the equality sign). Let us write

$$Z(\lambda) = R_\sigma + \frac{a_0' + a_1'\lambda + a_2'\lambda^2}{b_0 + b_1\lambda + b_2\lambda^2} = R_\sigma + Z'(\lambda). \qquad (23)$$

The function $Z'(\lambda)$ now does not have a zero on the axes of imaginaries, as in equation 28a, but will be characterized by the fact that for some value of frequency ($\lambda = i\sigma$) $Z'(\lambda)$ will be pure imaginary. Physically this means that the corresponding network behaves like a pure reactance at this frequency. This can only mean that that part of the network which contains resistance is short circuited by (i.e. is in parallel with) a reactive branch which is in resonance for this frequency; in series with this is a pure reactance which may be either positive or negative, since the pure imaginary part may have either sign. The only type of element which can meet this requirement is inductance with mutual coupling. Accordingly let us consider the network of Fig. 11 as a possible repre-

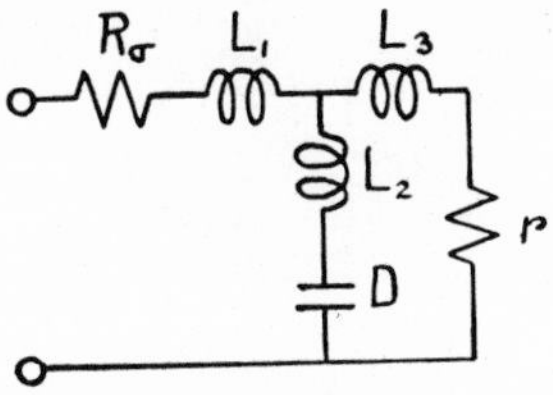

FIG. 11

sentation of the function (17) under these circumstances. In this network L_1, L_2, L_3, are an equivalent representation of two inductances with mutual coupling; one of them may be negative, with the necessary and sufficient condition for physical realizability

$$L_1 + L_2 \geq 0; \quad L_1L_2 + L_2L_3 + L_3L_1 \geq 0 \tag{24}$$

whence it follows that $L_2 + L_3 > 0$ and $L_3 + L_1 > 0$.

Now considering the network remaining after R_σ has been removed from the network of Fig. 11, its impedance function is readily found to be

$$Z''(\lambda) = \frac{(L_1L_2 + L_2L_3 + L_3L_1)\lambda^3 + (L_1 + L_2)r\lambda^2 + (L_1 + L_3)D\lambda + rD}{(L_2 + L_3)\lambda^2 + r\lambda + D} \cdot \tag{25}$$

Comparison with $Z'(\lambda)$ in equation (23) makes it apparent that

$$L_1L_2 + L_2L_3 + L_3L_1 = 0 \tag{26}$$

which is the condition that the two coils of which L_1, L_2, L_3 are the equivalent shall be perfectly coupled.

Further relations which have to be satisfied are

$$
\left.
\begin{array}{l}
r = \dfrac{a_0'}{b_0}; \quad D = \dfrac{a_0'}{b_1}\left(= b_0 \times \dfrac{a_0'}{b_1 b_0}\right) \\
\left.\begin{array}{ll}
L_1 + L_2 & = \dfrac{a_2'}{b_1} \\
L_2 + L_3 & = b_2 \cdot \dfrac{a_0'}{b_1 b_0} \\
L_1 \quad + L_3 & = \dfrac{a_1'}{b_0}
\end{array}\right\} \text{i.e.} \left\{\begin{array}{l}
L_1 = \dfrac{-a_0'b_2 + a_1'b_1 + a_2'b_0}{2b_1b_0} \\
L_2 = \dfrac{a_0'b_2 - a_1'b_1 + a_2'b_0}{2b_1b_0} \\
L_3 = \dfrac{a_0'b_2 + a_1'b_1 - a_2'b_0}{2b_1b_0}
\end{array}\right.
\end{array}
\right\} \quad (27)
$$

Consequently the relation (26) becomes

$$
\begin{aligned}
0 &= L_1L_2 + L_3(L_1 + L_2) \\
&= (-a_0'b_2 + a_1'b_1 + a_2'b_0)(a_0'b_2 - a_1'b_1 + a_2'b_0) \\
&\qquad\qquad + 2a_2'b_0(a_0'b_2 + a_1'b_1 - a_2'b_0) \\
&= 4a_0'a_2b_0b_2 - (a_0'b_2 - a_1'b_1 + a_2'b_0)^2. \qquad (28)
\end{aligned}
$$

Comparing this with equation (18) and remembering that $\text{Re}\, Z'(i\omega)$ has a double zero when $\omega^2 = \sigma^2$, it is evident that condition (28) is satisfied. Hence $Z(\lambda)$ may be represented by the network in Fig. 11 by going through the steps indicated in equations (23) and (27).

We have thus established a procedure whereby a network representation may be found for any " real positive " function of the form given in equation (17). The network will have one of the forms shown in Figs. 10 and 11.

2. *General Procedure*

The methods exemplified in the last section are, with very little modification, applicable to the general case of a " positive real " function with a finite number of poles and zeros. The general procedure for finding a network representation of such a function will be a step by step process; at each step the function is decomposed into a part which can immediately be represented by an element of the network, and another part which is again a positive real function. The new function should be simpler than the previous function if any progress is to be made towards complete representation; this will show itself in a decrease in the number of poles and zeros. (The number of poles being always equal to the number of zeros.)

Let the given function be $Z(\lambda)$. Poles and zeros on the axis of imaginaries can immediately be removed by Theorem IV and corresponding network elements constructed. When this has been repeated often enough we are left with a "positive real" function which is either a constant or has all its poles and zeros in the interior of the left-half plane. Let this function be $Z_1(\lambda)$.

Next examine the real part of $Z_1(\lambda)$ on the axis of imaginaries.

We remark in passing the $\operatorname{Re} Z(i\omega)$ is an even function of ω while $\operatorname{Im} Z(i\omega)$ is odd, in other words we may always write

$$Z(i\omega) = R(\omega^2) + i\omega N(\omega^2). \tag{29}$$

Since $\operatorname{Re} Z(i\omega)$ cannot be negative it must have a lower limit ≥ 0. Let this minimum value be R_1 and occur when $\omega = \sigma$ (σ real). Then clearly $Z_2(\lambda) = Z_1(\lambda) - R_1$ will have its real part ≥ 0 on the boundary and regular in the right-half plane, i.e. $Z_2(\lambda)$ will be a "positive real" function. Moreover $Z_2(i\sigma)$ will be pure imaginary. Let

$$Z_2(i\sigma) = iX. \tag{30}$$

Now if $\sigma = 0$ or ∞, $X = -X = 0$; hence we can immediately conclude that $Z_2(\lambda)$ has a zero at $\lambda = 0$ or ∞ respectively. We could therefore write

$$\frac{1}{Z_2(\lambda)} = \frac{D}{\lambda} + \frac{1}{Z_3(\lambda)} \tag{31a}$$

or

$$\frac{1}{Z_2(\lambda)} = L\lambda + \frac{1}{Z_3(\lambda)} \tag{31b}$$

respectively; in either case $Z_3(\lambda)$ will have one less zero and pole than $Z_2(\lambda)$ (or $Z_1(\lambda)$).

Consider next the case when σ is not zero or infinite.

We proceed to make the points $\lambda = \pm i\sigma$ zeros of a new function simply derivable from $Z_2(\lambda)$ by a step corresponding to the realization of an element in the network representation. Note that if

$$Z_2(i\sigma) = iX,$$

then

$$\frac{1}{Z_2(i\sigma)} = -i\frac{1}{X}.$$

Two conditions may arise; $X/\sigma = N(\sigma^2)$ (equation (29)) may be either positive or negative. We consider these two possibilities separately.

(i) Let $X/\sigma = N(\sigma^2) = L_1$ be negative.

Then we can write

$$Z_2(\lambda) = L_1\lambda + W(\lambda), \tag{32}$$

where $W(\lambda)$ is a "positive real" function and has zeros at $\lambda = \pm i\sigma$. For $-L_1\lambda$ is a "positive real" function and $W(\lambda) = Z_2(\lambda) - L_1\lambda$ is thus the sum of two "positive" functions; also $W(i\sigma) = Z_2(i\sigma) - iX = 0$. (Note: $W(\lambda)$ has a pole at infinity.) Hence we may write

$$\frac{1}{W(\lambda)} = \frac{K\lambda}{\lambda^2 + \sigma^2} + \frac{1}{W'(\lambda)}, \tag{33}$$

where $W'(\lambda)$ is a "positive real" function which, like $W(\lambda)$, has a pole at infinity.

Let

$$W'(\lambda) = L_3\lambda + Z_3(\lambda). \tag{34}$$

If we further write

$$L_2 = \frac{1}{K}, \quad D = \frac{\sigma^2}{K} \tag{35}$$

the relation between $Z_2(\lambda)$ and $Z_3(\lambda)$ in terms of a network representation is shown in Fig. 12. Comparing this with Fig. 11 and formula (25) and remembering that $Z_2(\lambda)$ has no pole at $\lambda = \infty$ we have as the algebraic relation

$$Z_2(\lambda) = \frac{(L_1 + L_2)\lambda^2 Z_3(\lambda) + (L_3 + L_1)D\lambda + DZ_3(\lambda)}{(L_2 + L_3)\lambda^2 + Z_3(\lambda)\cdot\lambda + D}. \tag{36}$$

Also L_2, L_3 are positive by (33), (34) and (35) (cf. equation 24) hence the network is realizable.

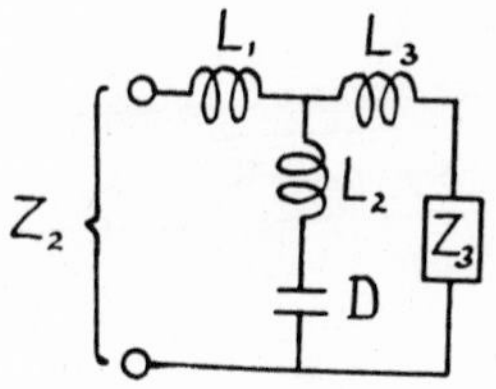

FIG. 12

(ii) Let $X/\sigma = N(\sigma^2)$ be positive. Then we may consider

$$Y_2(\lambda) = \frac{1}{Z_2(\lambda)} \quad \text{where} \quad Y_2(i\sigma) = -i\frac{1}{X} = -i\sigma\frac{1}{\sigma^2 N(\sigma^2)}$$

and

$$C_1 = -\frac{1}{\sigma^2 N(\sigma^2)}$$

is negative and proceed in an exactly analogous manner as with the preceding case. Translating the results into network elements, however, the network will be reciprocal to that in Fig. 12. This is shown in Fig. 13, L_1 being replaced by C_1, L_2 by C_2, D

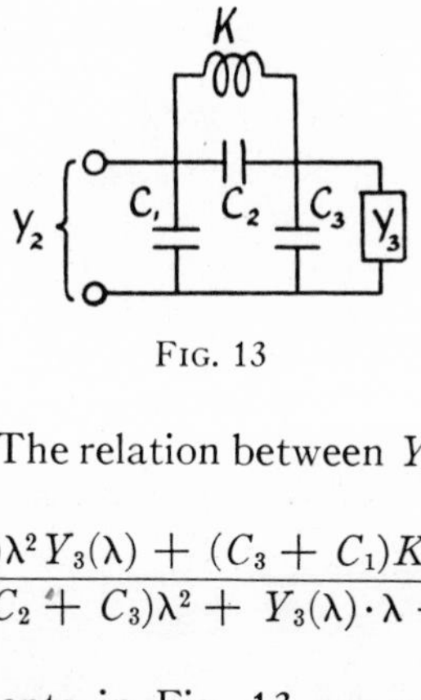

FIG. 13

by K and L_3 by C_3. The relation between $Y_2(\lambda)$ and $Y_3(\lambda)$ will be

$$Y_2(\lambda) = \frac{(C_1 + C_2)\lambda^2 Y_3(\lambda) + (C_3 + C_1)K\lambda + KY_3(\lambda)}{(C_2 + C_3)\lambda^2 + Y_3(\lambda)\cdot\lambda + K}. \tag{37}$$

The network elements in Fig. 13 are not directly realizable since they contain the negative capacity C_1. Equation (37)

15

may, however, be rewritten

$$Z_2(\lambda) = \frac{(C_3 + C_2)\lambda^2 Z_3'(\lambda) + \lambda + KZ_3'(\lambda)}{(C_1 + C_2)\lambda^2 + (C_3 + C_1)KZ_3'(\lambda)\cdot\lambda + K} \qquad (37a)$$

$$\text{where } Z_3'(\lambda) = 1/Y_3(\lambda)$$

which is of exactly the same form as (36). Let us write down the necessary relations between the coefficients of (37*a*) and (36) in order that Z_3 and Z_3' may be identical. They are

$$D = \frac{1}{C_3 + C_1}, \qquad (38a)$$

$$\left.\begin{aligned} L_1 + L_2 \quad &= \frac{1}{K}\frac{C_2 + C_3}{C_3 + C_1}, \\ L_2 + L_3 &= \frac{1}{K}, \\ L_1 \quad + L_3 &= \frac{1}{K}\frac{C_1 + C_2}{C_3 + C_1}, \end{aligned}\right\} \qquad (38b)$$

we have also (analogous to 37*a* and 31*a*)

$$\left.\begin{aligned} \frac{K}{C_2} &= \sigma^2, \\ C_1C_2 + C_2C_3 + C_3C_1 &= 0, \end{aligned}\right\} \qquad (38c)$$

and from 38*a*

$$\left.\begin{aligned} L_1 &= \frac{1}{K}\frac{C_3}{C_3 + C_1}, \\ L_2 &= \frac{1}{K}\frac{C_2}{C_3 + C_1}, \\ L_3 &= \frac{1}{K}\frac{C_1}{C_3 + C_1}, \end{aligned}\right\} \qquad (38d)$$

with the relations from (38*c*)

$$\frac{D}{L_2} = \frac{K}{C_2} = \sigma^2;$$

$$L_1L_2 + L_2L_3 + L_3L_1 = \frac{C_3C_2 + C_2C_1 + C_1C_3}{K^2(C_3 + C_1)^2} = 0. \qquad (38e)$$

It follows that the fictitious network of Fig. 13 can be made equivalent to the physically realizable network of Fig. 12 by means of the equations (38*a*) and (38*d*). Note that the signs of L_1, L_3 and C_1, C_3 are interchanged. Hence in case (ii) L_1 and L_2 are positive, and L_3 is negative.

Furthermore, since we now know the structure of the network it is evident that $Z_2(i\sigma) - iL_1\sigma$ must be zero. Consequently $L_1 = N(\sigma^2)$, and the procedure of Case (i) may be followed irrespective of the sign of $N(\sigma^2)$. (At one stage a non-" positive " function will be encountered but the final $Z_3(\lambda)$ will be a " real positive " function.)

Examining the reduction of poles and zeros by this process, we note that the first step (equation 32) increases the number by 1. The next step (equation 33) decreases the number by 2, while the last step (equation 34) effects a further reduction of 1. In all, thus, if R_1 occurs for $\lambda = i\sigma \neq 0$ or ∞, $Z_3(\lambda)$ will have two poles and two zeros less than $Z_1(\lambda)$.

This procedure can now be applied to all cases and repeated until the final function has no poles and no zeros, i.e. is a positive constant corresponding to a pure resistance. It leads uniquely to a network representation of the original " positive real " function.

The calculation of the network is made entirely in terms of the values of successive impedance functions on the axis of imaginaries. Each reduction of the function is effected by the removal of a zero (or pole) of the function on the axis of imaginaries. If a function does not have such a zero on the imaginary axis it is shown always to be possible to derive from it a function which will have such a zero; the removal of this zero together with the related steps also corresponds to the calculation of realizable elements of the corresponding network.

For more ready reference we summarize the different steps with the corresponding network interpretations in Table I, and state the conclusion as

TABLE I

GENERAL PROCEDURE FOR CONSTRUCTING A NETWORK REPRESENTATION OF ANY GIVEN "POSITIVE" REAL FUNCTION

(*a*) Remove all poles on the axis of imaginaries:— $Z(\lambda)=L_0\lambda+\Sigma\frac{K_r\lambda}{\lambda^2+\sigma_r^2}+Z'(\lambda)$ (Theor. IV)	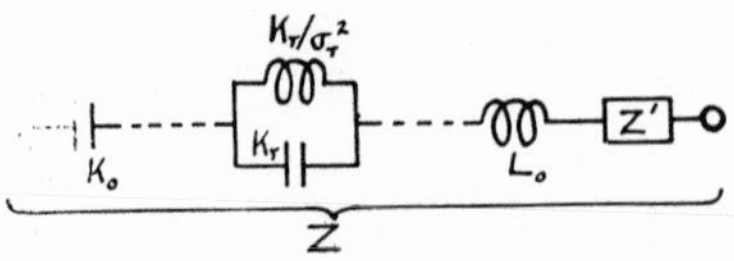
(*b*) Remove all zeros on the axis of imaginaries:— $\frac{1}{Z'(\lambda)}=C_0\lambda+\Sigma\frac{K_r'\lambda}{\lambda^2+\sigma_r'^2}+\frac{1}{Z_1(\lambda)}$ (Theor. IV, Cor. 1)	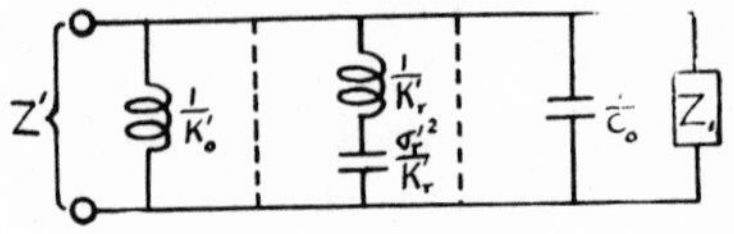
Repeat (*a*) and (*b*) until no poles or zeros of $Z_1(\lambda)$ lie on the axis of imaginaries.	
(*c*) Examine $R(\omega^2)$ in $Z_1(i\omega)=R(\omega^2)+i\omega N(\omega^2)$ Let R_1 be the least value of $R(\omega^2)$ for positive values of ω^2 and let $R_1=R(\sigma^2)$ $Z_1(\lambda)=R_1+Z_2(\lambda)$	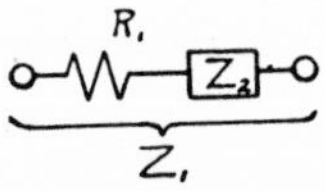
(*d*) Write $L_1=N(\sigma^2)$ (σ found in (*c*)) $Z_2(\lambda)=L_1\lambda+W(\lambda)$ $W(\lambda)$ has a zero at $\lambda=\pm i\sigma$ and hence $\frac{1}{W(\lambda)}=\frac{\lambda}{L_2\lambda^2+D}+\frac{1}{L_3\lambda+Z_3(\lambda)}$ As special cases when $\sigma=0$ or ∞, this reduces to $L_1=0$, $L_3=0$ $\frac{1}{Z_2(\lambda)}=\frac{1}{L_2\lambda}+\frac{1}{Z_3(\lambda)}$ $\frac{1}{Z_2(\lambda)}=\frac{D_2}{\lambda}+\frac{1}{Z_3(\lambda)}$	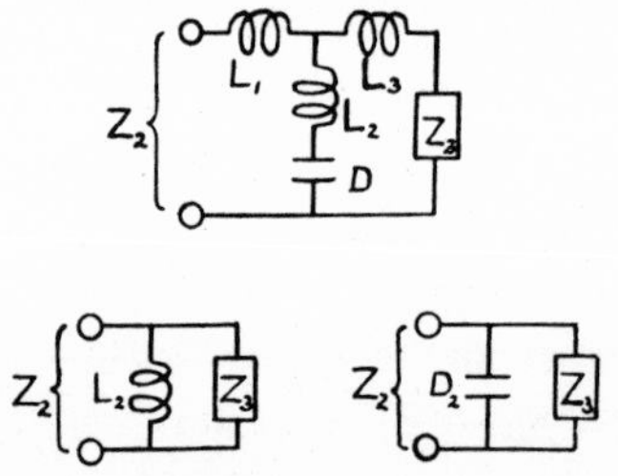

THEOREM VIII. To every " positive real " function with a finite number of poles and zeros corresponds a finite physically realizable network. Hence the terms " *positive real* " *function with a finite number of poles and zeros* and *impedance function of a finite network* with lumped linear passive elements are synonymous.

3. *Discussion of Networks Obtained*

A discussion of the general type and properties of the network obtained by the process summarized in Table I, as well as of certain special cases, is of interest.

It will be noticed first of all that the network always has a ladder structure. The ladder structure is thus capable of realizing the most general form of driving point impedance.

In general mutual inductances will be involved but this mutual inductance if present occurs always between self-inductances in adjacent meshes; in certain cases mutual inductance may be entirely absent, viz. when procedure (c), Table I, involves only minimum values R_1 at $\lambda = 0$ or ∞ throughout.

The procedure is carried out in such a way that mutual inductances, when they occur, do so with a coefficient of coupling unity; this apparently involves an ideal physical condition not generally inherent in the given impedance function. If, however, the function $Z(\lambda)$ has a pole at $\lambda = \infty$ the necessity for perfect coupling is only apparent, since a pole at $\lambda = \infty$ corresponds to a self-inductance L_0 in series with the perfectly coupled inductance L_a in Fig. 14. But this separation of L_0 and L_a is

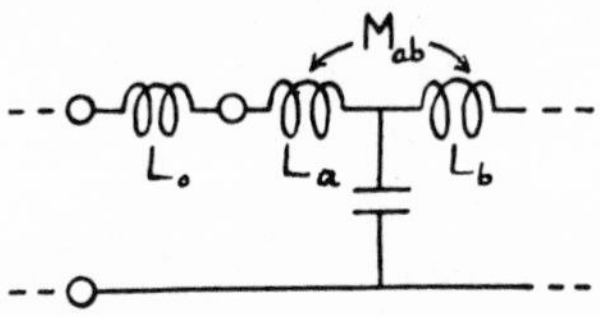

FIG. 14

quite arbitrary, the only limitation being that L_0 is the largest inductance which can be so separated and still leave the remaining network physically realizable. If L_0 (or any part of it) be combined with L_a into an inductance L_a', coupled with the inductance L_b then obviously $L_a' \cdot L_b > M_{ab}^2$ and the coupling will be " loose." This condition of " loosening " will simultaneously extend throughout the remaining network. Only in such cases where the condition of " perfect " coupling is indeed inherent in the given impedance function because of the absence of a pole at $\lambda = \infty$ will this condition be forced on the corresponding network.

In fact a network structure like that indicated in Fig. 14 would under certain conditions be eminently suitable for practical construction. The series resistances are in the proper position to be combined with the inductances, while in the parallel branches only capacities, whose series resistance is negligible, occur.

The whole question of practical usability, however, depends on the special conditions to be met in the particular application. Its full discussion would of necessity involve the consideration of equivalent networks.

In this connection it may also be pointed out that the dual circuit met with in Fig. 13 could sometimes be used, namely in such cases where the negative capacity occurs in parallel with a larger positive capacity (corresponding to a zero at $\lambda = \infty$) so that the combination will be positive.

At each stage in the general procedure described in the preceding section the determination of a given number of independent elements in the network is accompanied by a corresponding reduction in the number of independent constants in the function which remains. The resulting network therefore does not contain any syperfluous elements. Let us consider this property of the network in greater detail.

By an independent network element is meant one whose value does not depend on the values of other elements; thus in Fig. 12 the elements of the group $L_1L_2L_3$ are not independent since their values are connected by the relation

$$L_1L_2 + L_2L_3 + L_1L_3 = 0 \tag{26}$$

Similarly by the number of independent constants in the function is meant the number of such constants necessary to determine the function.

A function is completely determined by prescribing its poles, its zeros, and an arbitrary factor. Poles and zeros in the interior of the left-half plane are either real or occur in conjugate complex pairs. Hence to each such pole or zero in the interior of the left-half plane corresponds one independent constant. A pair of poles (or zeros) on the imaginary axis involves only one constant, while poles or zeros at the origin or infinity introduce no new constants. The number of independent constants in the function is thus

$$\mu = (n + m + h + 1) \tag{39}$$

where

n = number of poles in the interior of left-half plane,

m = number of zeros in the interior of left-half plane,

h = number of *pairs* of poles and zeros on the imaginary axis (excluding 0 and ∞).

We have stated this result in this way because a mere count of the number of coefficients in numerator and denominator is not always sufficient. To every conjugate pair of poles or zeros on the imaginary axis corresponds in general a relation among the coefficients. In the extreme case where all zeros (or poles) lie on the imaginary axis, these relations are simply expressed by stating that the coefficients of either the even or odd powers of λ in the numerator (or denominator) are zero. In intermediate cases the relations are not so easily seen and therefore likely to be missed unless a detailed examination of the poles and zeros is made.

With these facts in mind it is easy to trace the reduction of the independent constants in the function brought about by the removal of the various groups of elements in Table I from the network.

In special cases additional relations besides those mentioned above may exist among the coefficients, making possible a still further reduction in the number of elements in the network. The possibility of making this reduction will not always be evident from the general procedure of Table I. A complete discussion of such possibilities would involve a discussion of all equivalent networks which will not be attempted here.

Fosters " Reactance Theorem " is readily seen to be only a particular case of the general procedure in which all the poles and zeros lie on the axis of imaginaries. For the other cases of networks having only two kinds of elements the procedure of Table I leads to the types of structure shown in Fig. 1 (*b*) and (*c*) in which either inductance coils or condensers are replaced by resistances.

Part IV—Extension of Method to the Determination of Certain Equivalent Networks

1. *Change in the Order of Procedure*

Very slight consideration will show that the order of the steps (*a*), (*b*), (*c*) and (*d*) specified in Table I, is not essential. Thus, for example, in the simple case where all the poles and zeros lie on the axis of imaginaries any pole or any zero may be removed first and thereafter again any pole or any zero. The networks shown in Fig. 1 are special cases when a definite order is adopted, Fig 1 (*a*) results when all poles of $Z(\lambda)$ are removed in succession, i.e. the network given by Table I; (*b*) when all zeros are removed in succession; (*c*) results when we remove first the pole at infinity, then the zero at infinity and so on alternately; and (*d*) when poles and zeros are removed alternately at the origin. The order may, however, be entirely haphazard. In fact only " part of a pole " i.e. a term representing a pole at the same point but with residue which is a positive proper fraction of the corresponding residue of $Z(\lambda)$, may be separated, the remaining positive function can thereafter be represented in any of the ways shown in Fig. 1 (*b*), (*c*) or (*d*); this would give a network with superfluous elements.

When we are dealing with a more general type of function, the order of removing poles and zeros on the imaginary axis is still arbitrary until the procedure of Table I (c) becomes necessary. Obviously if there are any zeros on the axis of imaginaries these will also be points at which $\mathrm{Re}\, Z(i\omega)$ is zero; consequently step (d) becomes equivalent to (b) until all zeros on the axis of imaginaries have been removed. The presence of a pole in $Z_1(\lambda)$ on the imaginary axis however will not alter the value of R_1 in step (c). In the next step (d) therefore $Z_2(\lambda)$ will be a " positive " function and nothing will be altered in the argument concerning the validity of the procedure (d). In the event of a pole at infinity having been left in $Z_2(\lambda)$ we shall obtain $L_1L_2 + L_2L_3 + L_3L_1 > 0$ instead of $= 0$, which obviously still fulfills the physical requirements. In fact in this case L_1 is exactly equal to $L_0 + L_1$ (Table I (a) and (d)). Poles on the imaginary axis can therefore be left in the function or separated at will at any stage in the procedure. Another alternative is to treat $1/Z(\lambda)$ as an admittance; step (c) will correspond to the separation of a parallel conductance from $Z(\lambda)$; in this case zeros of $Z(\lambda)$ do not necessarily have to be first removed, but poles (i.e. zeros of $1/Z(\lambda)$) must.

2. *Consideration of " imaginary frequencies "*

In considering a network representation of " positive real " functions with two poles and two zeros. we were lead by certain physical considerations to the network of Fig. 12, which in turn lead to equations (27) and the relation (28). Now, while our physical reasoning depended on the fact that $R(\Omega)$ (equation 18a) had its minimum value for a positive value of Ω, equations (27) have no reference to this value of Ω. The necessary relation (28) will hold whenever $R'(\Omega)$ has a double zero, and in order that equations (27) may lead to realizable elements, it is only necessary that $Z'(\lambda)$ (equation (23)) shall be a " positive real " function. Both these conditions are satisfied if a stationary value of $R(\Omega)$ equal to R_ν (say) be deducted from $Z(\lambda)$, and if

$$R_\nu \leq R(\Omega) \text{ for positive values of } \Omega. \qquad (40)$$

This therefore gives us an alternative procedure in certain cases leading to equivalent networks. E.g. if $R(\Omega)$ varies in the manner shown in Fig. 10 (*d*) R_ν may be taken as the maximum occuring between the two vertical asymptotes in this Figure. In this case L_2 in Fig. 11 would be negative, and physical conceptions of 'resonance' would need to be extended to give a physical interpretation to this procedure. The type of variation of $R(\Omega)$ shown in Fig. 10 (*a*) on the otherhand will never permit this kind of procedure; it corresponds to networks with two kinds of elements only.

It will now be shown that a similar procedure can be justified in the general case.

Negative values of Ω correspond to imaginary values of ω which in turn means real values of λ. If therefore we write

$$Z(\lambda) = R(-\lambda^2) + \lambda N(-\lambda^2) \tag{41}$$

the interpretation in terms of real values of λ can be carried over from the discussion for pure imaginary values without any difficulty.

Let a stationary value of $R(-\lambda^2)$ satisfying (40) occur when $\lambda^2 = \nu^2$, and let $R_\nu = R(-\nu^2)$. Then if

$$Z_1(\lambda) = R_\nu + Z_2(\lambda) \tag{42}$$

it is clear that $Z_2(\lambda)$ will be a "positive real" function in virtue of (48). Now if

$$Z_2(\lambda) = (R(-\lambda^2) + R_\nu) + \lambda N(-\lambda^2)$$

we know that

$$Z_2(\nu) = \nu N(-\nu^2).$$

Let us write

$$L_1 = N(-\nu^2)$$

and

$$Z_2(\lambda) = L_1\lambda + W(\lambda). \tag{43}$$

Then it follows from equation (41) and (42) that

$$W(\lambda) = [R(-\lambda^2) - R_\nu] + \lambda[N(-\lambda)^2 - L_1], \tag{43a}$$

i.e.

$$W(\nu) = 0 \quad \text{and} \quad L_1 = Z_2(\nu)\nu. \tag{43b}$$

It is further clear that equations (43*b*) holds for $\lambda = \pm \nu$. Hence W will *not* be a "positive" function, since it has the zero $\lambda = \nu$ in the right-half plane. By Theorem VI, Coroll. 1 however we know that this is the only zero of $W(\lambda)$ within the right-half plane. Note further that

$$\operatorname{Re} W(i\omega) \geq 0 \tag{44}$$

in virtue of (40).

If therefore we separate the zero's at $\lambda = \pm \nu$ from $W(\lambda)$ by a partial fraction expansion of $1/W(\lambda)$, the remaining function may be a positive real function.

For this purpose we examine the residues of the poles of $1/W(\lambda)$ at $\lambda = \pm \nu$. We may rewrite (43*a*)

$$W(\lambda) = R'(-\lambda^2) + \lambda N'(-\lambda^2)$$

which, in virtue of (43*b*) and the fact that R_ν is a stationary value of $R(-\lambda^2)$, can again be written

$$W(\lambda) = (\lambda^2 - \nu^2)^2 \cdot r(-\lambda^2) + \lambda(\lambda^2 - \nu^2) \cdot n(-\lambda^2)$$

Hence the residue at $\lambda = \nu$ is

$$\left[\frac{\lambda - \nu}{W(\lambda)}\right]_{\lambda=\nu} = \frac{1}{2\nu^2 n(-\nu^2)} = -\frac{K}{2} \text{ (say)}. \tag{45}$$

The residue at $\lambda = -\nu$ obviously has the same value.[6]

Consequently we may write

$$\frac{1}{W(\lambda)} = \frac{-\frac{K}{2}}{\lambda + \nu} + \frac{-\frac{K}{2}}{\lambda - \nu} + \frac{1}{W_1(\lambda)}$$

$$= \frac{K\lambda}{-\lambda^2 + \nu^2} + \frac{1}{W_1(\lambda)}. \tag{46}$$

Since $\operatorname{Re} \dfrac{1}{W(i\omega)} \geq 0$ and the first term of the right-hand side

[6] This property is the exact analogy to the property that for a "positive real" function the residue at a pole on the imaginary axis is a positive real quantity.

of (46) is pure imaginary for $\lambda = i\omega$, it follows from (44) that $\text{Re}\, \frac{1}{W_1(i\omega)} \geq 0$. But $W_1(\lambda)$ has no zeros in the right-half plane and no zeros on the imaginary axis. Consequently by Theorem V, Coroll. 1, $W_1(\lambda)$ is a "positive real" function. The process is completed by writing

$$W_1(\lambda) = L_3\lambda + Z_3(\lambda). \tag{47}$$

Also let

$$L_2 = -\frac{1}{K}, \quad D = \frac{\nu^2}{K}. \tag{48}$$

It is now only necessary to show that the reduction of $Z_2(\lambda)$ to $Z_3(\lambda)$ corresponds to the calculation of elements in a physically realizable circuit.

Referring to equation (43*b*), we have

$$L_1 = \frac{Z_2(\nu)}{\nu}, \tag{43b}$$

where, if ν is positive (and real), $Z_2(\nu)$ will be positive and real, since $Z_2(\lambda)$ is a "positive real" function. Consequently L_1 is positive.

Furthermore L_3, equation (47), is positive since Z_3 is a "positive real" function with a pole at infinity; and the relation

$$L_1L_2 + L_2L_3 + L_3L_1 = 0 \tag{26}$$

can be shown to hold in exactly the same manner as for Figs. 11 and 12, consequently L_2 is negative and D is positive. The process therefore corresponds to a determination of the elements L_1, L_2, L_3, D, of a realizable network connected in the manner of Fig. 12.

3. *Relation to Equivalent Networks Obtained by Affine Transformation of Quadratic Forms*

Equivalent networks have been derived from a given network by Howitt, (6) by the application of an affine transformation to the quadratic forms[7] connected with the network. Cauer (5) has also

[7] Corresponding to the stored electromagnetic and electrostatic energies and the rate of dissipation of energy in heat.

mentioned this method. Such a transformation can be accomplished by the matrix multiplication

$$C' \cdot A \cdot C = A', \tag{49}$$

where A is the matrix $\|a_{rs}\|$ of the system of equations

$$\left.\begin{array}{l} E = a_{11}I_1 + a_{12}I_2 + a_{13}I_3 + \cdots a_{1n}I_n \\ 0 = a_{21}I_1 + a_{22}I_2 + a_{23}I_3 + \cdots a_{2n}I_n \\ \cdot \qquad\qquad \cdot \qquad \cdot \qquad \cdot \\ \cdot \qquad\qquad \cdot \qquad \cdot \qquad \cdot \\ 0 = a_{n1}I_1 + a_{n2}I_2 + \cdots \qquad a_{nn}I_n \end{array}\right\} \tag{50}$$

for the given network; in these equations the I_s are a set of independent circulating mesh currents and

$$a_{rs} = L_{rs}\lambda + R_{rs} + D_{rs}\lambda^{-1} \tag{51}$$

are generalized self- $(r = s)$ and mutual- $(r \neq s)$ impedances of the sth mesh.

A' is a similar matrix $\|a_{rs}'\|$ for an equivalent network.

C is the transformation matrix $\|c_{rs}\|$ with c_{rs} real constants and

$$c_{11} = 1, \quad c_{1s} = 0 \quad (s \neq 1). \tag{52}$$

C' is the matrix $\|c_{sr}\|$ conjugate to C.

If we denote the determinant of A by D, the determinant of A' by D', and the determinant of C by Δ we have

$$D' = \Delta^2 D \tag{53}$$

and in virtue of (52) the first minor

$$\frac{\partial D'}{\partial a_{11}'} = \Delta^2 \frac{\partial D}{\partial a_{11}} \cdot \tag{54}$$

It is on account of this property that the impedance function is invariant since

$$Z'(\lambda) = \frac{D'}{\dfrac{\partial D'}{\partial a_{11}'}} = \frac{D}{\dfrac{\partial D}{\partial a_{11}}} = Z(\lambda). \tag{55}$$

This property is however equally true of the determinants $|L_{rs}|$, $|R_{rs}|$, $|D_{rs}|$ and their first minors with respect to L_{11}, R_{11}, D_{11} respectively (equation (51)).

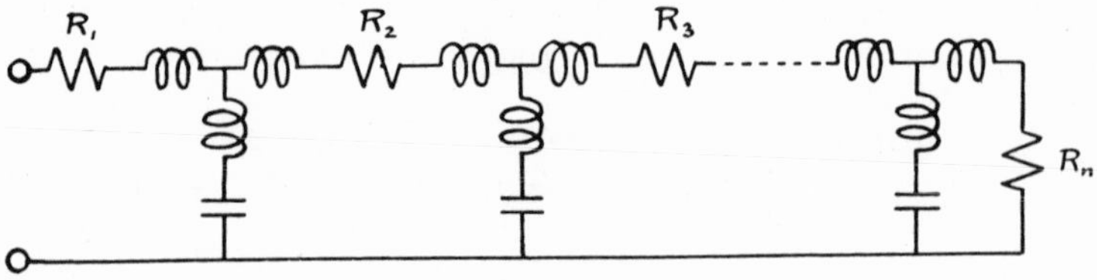

FIG. 15

Turning now to the networks obtained by the procedure outlined in the preceding sections we note that they always have the ladder structure indicated in Fig. 15 with no resistance in the parallel branches; consequently

$$|R_{rs}| = \begin{vmatrix} R_1 & 0 & 0 & \cdot & \cdot & 0 \\ 0 & R_2 & 0 & \cdot & \cdot & 0 \\ 0 & 0 & R_3 & \cdot & \cdot & 0 \\ \cdot & \cdot & \cdot & \cdot & \cdot & \cdot \\ \cdot & \cdot & \cdot & \cdot & \cdot & \cdot \\ 0 & 0 & 0 & \cdot & \cdot & R_n \end{vmatrix} \tag{56}$$

and the first minor

$$\frac{\partial}{\partial R_{11}} |R_{rs}| = \begin{vmatrix} R & 0 & \cdot & \cdot & 0 \\ 0 & R_3 & \cdot & \cdot & 0 \\ \cdot & \cdot & \cdot & \cdot & \cdot \\ \cdot & \cdot & \cdot & \cdot & \cdot \\ 0 & 0 & \cdot & \cdot & R_n \end{vmatrix} \tag{57}$$

The quotient of (56) and (57) is simply R_1 and this quotient is invariant under the transformation (49). Thus if according to the last section several values of R_ν are possible (e.g. $R_\nu R_\rho R_k$ in Fig. 16), the corresponding networks cannot be derivable from each other by a transformation like (49) although an infinite group can be derived from each one of them by such a transformation. This is true also when a usable value of $R(\Omega)$ occurs for $\Omega = 0$ or ∞. In such a case a network with a greater number of meshes results.

The question naturally presents itself whether all equivalent networks can be derived by the application of the transformation (49) to each of the networks obtained by using separate R_ν.

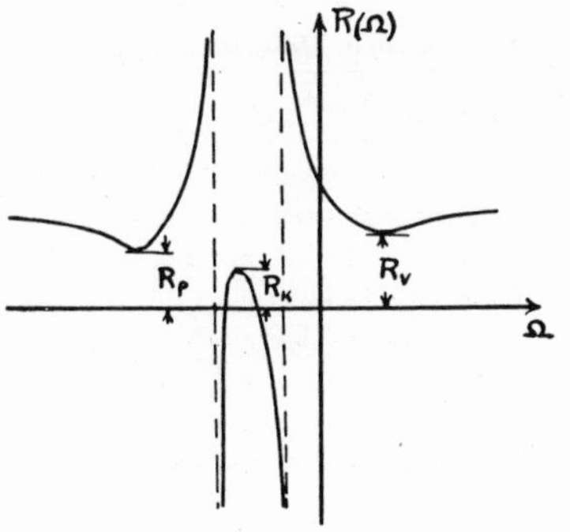

FIG. 16

This question must be answered in the negative, for examples can be constructed in which the quotient $\dfrac{|R_{rs}|}{\dfrac{\partial}{\partial R_{11}}|R_{rs}|}$ for a given network does not appear as a stationary value of the function $R(\omega^2) = \operatorname{Re} Z(i\omega)$ derived from the network. It is thus evident that a gap still remains to be bridged in the theory of equivalent networks. The results of bridging this gap should have considerable mathematical as well as physical interest.

BIBLIOGRAPHY

1. R. Foster: "A Reactance Theorem," *Bell Syst. Techn. Journ.*, 1924, vol. 3, p. 259.
2. R. Foster: "Theorems on the Driving-point Impedance of Two-mesh Circuits," *Bell Syst. Techn. Journ.*, 1924, vol. 3, p. 651.
3. W. Cauer: "Die Verwirklichung von Wechselstromwiderständen vorgeschriebener Frequenzabhängigkeit," *Archiv für Elektrot.*, 1926–27, Bd. 17, p. 355.
4. T. C. Fry: "The Use of Continued Fractions in the Design of Electric Networks," *Bull. of Am. Math. Soc.*, 35, p. 463 (1929).
5. W. Cauer: "Ueber eine Klasse von Funktionen die die Stieltjesschen Kettenbrüche als Sonderfall enthält," Jahresbericht d. Deutschen Mathematiker Ver., 38, 1929, p. 63.

6. N. Howitt: "On the Invariant Impedance Function and its Associated Group of Networks," M. I. T. doctor's thesis, 1930.—"Group Theory and the Electric Circuit" Phys. Rev. Vol. 37 Ser. 2, Num. 12, 1931, p. 1583.
7. H. G. Baerwald: "Ein enfacher Beweis des Reaktanztheorems," *Elektrische Nachrichten-Technik*, 1930, Bd. 7, p. 331.
8. W. Cauer: "Ueber die Variablen eines Passiven Vierpols," *Sitzungsber. d. Preuss. Ak. d. Wissensch.*, 1927 (Dez.), p. 268.
"Vierpole," *Elektrische Nachrichten-Technik*, 1929, Bd. 6, p. 272.
9. W. Cauer: "Untersuchungen über ein Problem das drei positiv definite quadratische Formen mit Streckenkomplexe in Beziehung setzt," *Mathematische Annalen*, 1931 (to appear shortly).
10. Yuk-Wing Lee: "Synthesis of Electric Networks by Means of the Fourier Transforms of Laguerre Functions," M. I. T. doctor's thesis, June, 1930.
11. A. Hurwitz: "Ueber die Bedingungen unter welchen eine Gleichung nur Wurzeln mit negativen reëllen Theilen besitzt," *Math. Ann.*, 1895, Bd. 46, p. 273.
12. W. Cauer: "Ein Satz über zwei zusammenhängende Hurwitz'sche Polynome," *Sitzungsberichte d. Berliner Mathem. Gesellsch.*, 1928, XXVII. Jahrg., p. 25.
13. G. Pick: "Konforme Abbildung kreisförmiger Bereiche," *Math. Ann.*, 1915–16, Bd. 77, p. 1.
14. L. Bieberbach: *Lehrbuch der Funktionentheorie*, I. u. II. B. G. Teubner, Leipzig 1927.
15. F. Klein: *Vorlesungen über Nicht-euklidische Geometrie.* (Die Grundlagen der Mathematischen Wissenschaften in Einzeldarstellungen, Band 26.) Julius Springer, Berlin, 1928.

3

Reprinted from *J. Appl. Phys.*, **20**(8), 816 (1949)

Impedance Synthesis without Use of Transformers

R. Bott and R. J. Duffin
Department of Mathematics, Carnegie Institute of Technology, Pittsburgh, Pennsylvania

December 13, 1948

Let $Z(s)$ be termed a B(rune) function if: (1) it is a rational function; (2) it is real for real s; and (3) the real part of Z is positive when the real part of s is positive. In his significant thesis, O. Brune[1] shows that the driving-point impedance of a passive network is a B function of the complex frequency variable s. Conversely, he shows that any B function can be realized by some passive network and gives rules for constructing such a network. In this synthesis he is forced to employ transformers with perfect coupling. It is recognized by Brune and others that the introduction of perfect transformers is objectionable from an engineering point of view. Prior to Brune, R. M. Foster[2] had shown how to synthesize the driving-point impedance of networks containing no resistors by simple series-parallel combinations of inductors and capacitors. This note gives a similar synthesis of an arbitrary impedance by series-parallel combinations of inductors, resistors, and capacitors.

A B function can be expressed as the ratio of two polynomials without common factor. Let the "rank" be the sum of the degrees of these polynomials. Obviously any B function of rank O can be synthesized. Suppose, then, it has been shown that all B functions of rank lower than n can be synthesized, and let $Z(s)$ be a B function of rank n. Brune gives four rules for carrying out a mathematical induction to a B function of lower rank:

(a) If Z has a pole on the imaginary axis, then Z can be synthesized by a parallel resonant element in series with an impedance Z' of lower rank; $Z=1/(cs+1/ls)+Z'$ where l^{-1}, $c\geq 0$.

(b) If Z has a zero on the imaginary axis, then Z can be synthesized by a series resonant element in parallel with an impedance Z' of lower rank; $1/Z=1/(ls+1/cs)+1/Z'$ where l, $c^{-1}\geq 0$.

(c) If the real part of Z does not vanish on the imaginary axis, $Z=r+Z_0$ where r is a positive constant (to be interpreted as resistance) and Z_0 is a B function of no greater rank than Z.

Brune's fourth rule, (d), which employs the perfect transformer, we replace by the following procedure:

(d′) If none of these reductions are possible, there exists a $w>0$ such that $Z(iw)$ is purely imaginary. First assume that $Z(iw)=iwL$ with $L>0$. We now make use of a key theorem discovered by P. I. Richards.[3] Let k be a positive number, and let

$$R(s)=\frac{kZ(s)-sZ(k)}{kZ(k)-sZ(s)}. \tag{1}$$

Then $R(s)$ is a B function whose rank does not exceed the rank of $Z(s)$. Richards states this theorem for $k=1$; the above form is an obvious modification, because $Z(ks)$ is also a B function. Let k satisfy the equation $L=Z(k)/k$. This is clearly always possible, because the function on the right varies from ∞ to 0 as k varies from 0 to ∞. With this choice of k, clearly $R(iw)=0$. Solving (1) for Z gives

$$\begin{aligned} Z(s)&=(1/Z(k)R(s)+s/kZ(k))^{-1}+(k/Z(k)s+R(s)/Z(k))^{-1}\\ &=(1/Z_1(s)+Cs)^{-1}+(1/Ls+1/Z_2)^{-1}. \end{aligned} \tag{2}$$

Here $Z_1(s)=kLR(s)$, $Z_2(s)=kL/R(s)$, $C=1/k^2L$. Since Z_1 is a B function with a zero on the imaginary axis, it can be synthesized. Likewise, Z_2 is a B function with a pole on the imaginary axis and can be synthesized. $Z(s)$ is therefore synthesized by two networks in series. The first network consists of the impedance Z_1 in parallel with a capacitor C, and the second network consists of the impedance Z_2 in parallel with an inductor L. In the case that $Z(iw)=-iwL$, similar considerations applied to the function $1/Z$ show that Z is synthesized by two networks in parallel. The synthesized network finally resulting has the configuration of a tree whose branches are ladder networks.

Richards[4] has sought necessary and sufficient conditions for the driving-point impedance of resistor-transmission-line circuits by means of an ingenious transformation of the Brune theory. The perfect transformers, which are again found to be objectionable, may be dispensed with by the above procedure.

[1] O. Brune, J. Math. and Phys. **10**, 191–236 (1931).
[2] R. M. Foster, Bell Syst. Tech. J. **3**, 259 (1924).
[3] P. I. Richards, Duke Math. J. **14**, 777–786 (1947).
[4] P. I. Richards, Proc. I.R.E. **36**, 217–220 (1948).

4

Reprinted from *Proc. IRE*, **40**, 482–485 (Apr. 1952)

A Note on the Ladder Development of RC-Networks*

E. A. GUILLEMIN†, FELLOW, IRE

***Summary*—Darlington and Cauer have described a useful way of synthesizing a lossless two-terminal-pair network from a single driving-point impedance and a knowledge of the zeros of transmission. The procedure involved has a good deal of similarity to the Brune process of synthesis, and can readily be extended to RL and RC networks provided the zeros of transmission are restricted to the negative real axis of the complex frequency plane. Although such an extension is obvious, it seems worth while to point out the details involved and to illustrate these with a numerical example.**

* Decimal classification: R143. Original manuscript received by the Institute, May 22, 1951; revised manuscript received, October 10, 1951.

† Massachusetts Institute of Technology, Cambridge, Mass.

IN A PAPER recently presented at the 1951 IRE Convention, entitled "Transfer Ratio Synthesis by RC Networks," the authors John T. Fleck and Philip F. Ordung appear to be unaware of a simple procedure for obtaining ladder realizations that parallels a well-known technique commonly applied to LC networks. Although adaptation of the method to RC networks follows a rather obvious pattern, it would seem worth while to discuss it briefly and to illustrate with a numerical example since many engineers are apparently unacquainted with this useful tool, and its porticular form in the RC case is not discussed elsewhere in the literature.

The procedure in question arises, for example, in the situation depicted in Fig. 1 for which the transfer impedance is

$$Z_{12} = \frac{E_2}{I_1} = \frac{z_{12}}{1 + z_{22}}, \tag{1}$$

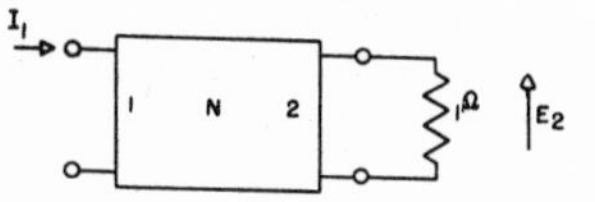

Fig. 1—A relevant physical network arrangement

in which z_{22} is the open-circuit impedance of the two-terminal-pair network N alone as measured at the end 2, and z_{12} is its open-circuit transfer impedance, that is, the ratio of open-circuit voltage at one end to input current at the other. After the functions z_{22} and z_{12} are found[1] from a specified Z_{12}, it remains to construct the network N from these two functions. In the dual situation (in which the transfer admittance $Y_{12}=I_2/E_1=y_{12}/(1+y_{22})$ is specified) the network N is to be constructed from its short-circuit driving point and transfer admittances y_{22} and y_{12}. That is to say, the successful synthesis of the coupling network N depends upon one's ability to construct it from the pair of functions z_{22} and z_{12}, or the pair y_{22} and y_{12}. (The driving-point function y_{11} or z_{11} is of no interest in this type of problem.)

In any of the two-element cases (RC, RL, LC) there are well-known methods for realizing the driving-point function z_{22} or y_{22} through developing it into an unbalanced ladder (Fig. 2). It is necessary to know how this

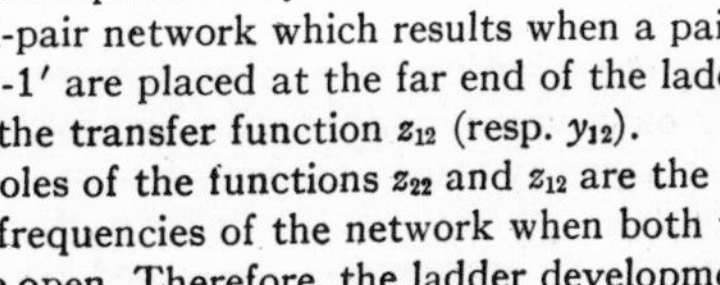

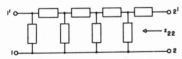

Fig. 2—Geometry of the desired development.

ladder development may be carried out so that the two-terminal-pair network which results when a pair of terminals 1-1′ are placed at the far end of the ladder, also realizes the transfer function z_{12} (resp. y_{12}).

The poles of the functions z_{22} and z_{12} are the complex natural frequencies of the network when both terminal pairs are open. Therefore, the ladder development of z_{22} automatically yields a network with a z_{12} function having the proper poles (except in degenerate cases which are easily avoided). It is only necessary to know how

[1] For a detailed discussion of how this is done see "A Summary of Modern Methods of Network Synthesis," by E. A. Guillemin, "Advances in Electronics." Academic Press, Inc., New York, N. Y., vol. III, pp. 261–303; 1951.

one can produce the proper zeros of z_{12}. In the ladder network, zeros of the transfer function require either that series branch impedances or shunt branch admittances be infinite at the appropriate frequencies. That is to say, if at a given frequency the transfer function is to be zero, then it is necessary (though not sufficient) that a series branch have an infinite impedance, or that a shunt branch have a zero impedance at that frequency. The reason that this condition does not necessarily produce a zero in the transfer function is that the transfer function for the network portion to the left of the branch in question may have a pole at the same frequency, as it will if the driving-point impedance, looking to the left at this point, contains this pole. Herein lies the key to the method of producing zeros at the desired frequencies regardless of the form of z_{22}.

Since this procedure is discussed in detail and illustrated for LC networks in the reference cited above, it will suffice here to confine our further attention to the analogous RC case. In the LC case, a branch of the ladder may be made to have infinite or zero impedance for any point on the j axis of the complex frequency plane (the plane of the variable $s=\sigma+j\omega$ as it is commonly used in the Laplace transform theory). For RC networks the same may be done for points on the negative real axis of this plane, and so the present discussion is relevant only to RC synthesis where the zeros of transmission are so restricted. More general cases are readily taken care of through a parallel combination of these restricted ladders, as is obvious from the procedure given in "Synthesis of RC Networks" by E. A. Guillemin (*Jour. Math. Phys.*, vol. 28, p. 22), or as pointed out by Fleck and Ordung in their paper referred to above.

Since it is more common to approach the design on an admittance basis, let us illustrate the details of the method by assuming the pair of functions

$$y_{22} = \frac{(s+1)(s+3)(s+5)}{(s+2)(s+4)(s+6)}$$

$$y_{12} = \frac{(s+2.5)^2(s+7)}{(s+2)(s+4)(s+6)}. \tag{2}$$

A sketch of y_{22} as a function of real values of s (i.e., for $s=\sigma+j0$) is shown in Fig. 3, from which we note that none of the zeros of y_{22} fall where y_{12} has zeros. This situation can easily be remedied, however.

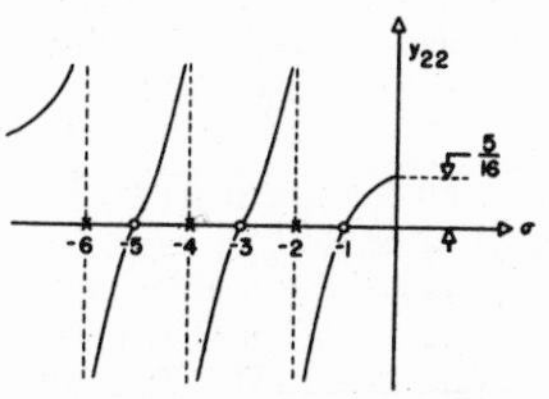

Fig. 3—The input admittance versus negative real values of the complex frequency variable.

First it is necessary to compute the value of y_{22} for $s=-2.5$, which is readily found to be

$$y_{22}(-2.5)=\frac{5}{7}\cdot \tag{3}$$

If the zero-frequency value of y_{22} (which is the minimum of its real part for $s=j\omega$) were as large or larger than 5/7, one could cause a zero of y_{22} to shift to the point $s=-2.5$ through simply subtracting an appropriate positive real constant (conductance). This scheme fails in the present example because $y_{22}(0)=5/16$, which is less than 5/7.

There is, however, another way of causing the zero at $s=-3$ to shift over to the point $s=-2.5$, and that is through the removal (subtraction) of an appropriate part of the pole at $s=-2$. This pole alone is given by the component admittance

$$y_1=\frac{k_1s}{s+2}; \tag{4}$$

and so we find the k_1 value for which

$$y_{22}(-2.5)-\left(\frac{k_1s}{s+2}\right)_{s=-2.5}=\frac{5}{7}-5k_1=0, \tag{5}$$

when $k_1=1/7$, and the appropriate admittance 4 is

$$y_1=\frac{(1/7)s}{s+2}\cdot \tag{6}$$

Thus the development of y_{22} is begun through constructing a shunt branch having the admittance y_1, and computing the remaining admittance function

$$y'=y_{22}-y_1=\frac{(6s^2+38s+42)(s+2.5)}{7(s+2)(s+4)(s+6)}=\frac{1}{z'}\cdot \tag{7}$$

It is important to observe that this shunt branch does not produce a zero of transmission because the remainder function y' still contains the pole at $s=-2$, and hence so does the transfer admittance of the as yet undeveloped network to the left of this branch. That is to say, in removing the shunt branch with admittance y_1, we have removed only part of the pole of y_{22} at $s=-2$. If we had removed all of this pole, then the shunt branch in question would produce a zero of transmission at $s=-2$, but the partial removal of a pole does not produce such a zero.

Thus a partial removal of a pole may be used as a "zero-shifting" technique, and the total removal of a pole becomes the "zero-producing" step.

By removing part of any admittance or impedance function, one cannot shift the poles of this function, but its zeros can thus be shifted. Since these zeros are the poles of the subsequently inverted function, one is able to control pole positions. In fact, through applying the shifting procedure first to the given function and then to the inverted remainder, and finally inverting again, one can shift both zeros and poles of the given function. Through appropriately repeating this technique as often as needed, one readily appreciates that a zero or a pole can be shifted to any desired position, regardless of where the critical frequencies lie originally. None of these shifting operations become zero-producing so far as the transfer function of the ultimate network is concerned so long as the pole removals used in the shifting process are partial ones.

Returning now to the remainder function in (7), and considering its reciprocal z', we next totally remove its pole at $s=-2.5$ since this is a point at which a zero of y_{12} must be produced. To this end we compute the residue of z' at $s=-2.5$ which is found to equal 1.185, and then subtract the component impedance corresponding to this pole, which is

$$z_2=\frac{1.185}{s+2.5}, \tag{8}$$

leaving the remainder

$$z''=\frac{7s^2+59.39s+114.45}{6s^2+38s+42}=\frac{7(s+2.958)(s+5.526)}{6(s+1.427)(s+4.907)}\cdot \tag{9}$$

Here z_2 is the impedance of a series branch following the shunt branch with admittance y_1, (6). This series branch produces a zero of transmission at $s=-2.5$.

At this stage in the development procedure, we have completed one cycle. That is to say, we have gone through a series of steps whereby the given y_{22} has been partially developed, and one of the zeros called for by y_{12} has been produced. It will be observed that while y_{22} in (2) is the ratio of two cubic polynomials, the admittance $y''=1/z''$, which we now have yet to develop, is the ratio of two quadratic polynomials. We have produced one zero of transmission, and have lowered the degree of our given rational function by one. The same series of steps, carried through analogous cycles, is now continued until all zeros of transmission are produced, and the remainder function is fully developed.

We begin the second cycle as we did the first, through computing the value of y'' at $s=-2.5$; thus

$$y''(-2.5)=-1.596. \tag{10}$$

Since this is a negative value, there is no possibility of carrying out the zero-shifting step simply by the removal of a positive constant (conductance). A sketch of the function y'' for $s=\sigma+j0$ similar to that for y_{22} in Fig. 3 (the factored form (9) is best used for this purpose) readily reveals that the desired zero-shifting can be accomplished through a partial removal of the pole at $s=-2.958$. That is to say, we plan to subtract from y'' the admittance

$$y_3=\frac{k_3s}{s+2.958}, \tag{11}$$

with k_3 determined by the condition

$$y''(-2.5) - \left(\frac{k_3 s}{s + 2.958}\right)_{s=-2.5}$$

$$= -1.596 + \frac{2.5k_3}{0.458} = 0, \quad (12)$$

when $k_3 = 0.2925$ and the appropriate admittance 11 is

$$y_3 = \frac{0.2925s}{s + 2.958}. \quad (13)$$

Construction of a shunt branch having this admittance is equivalent to subtraction of y_3 from y'', leaving

$$y''' = \frac{(3.952s + 16.8)(s + 2.5)}{7(s + 2.958)(s + 5.526)} = \frac{1}{z'''}. \quad (14)$$

The residue of z''' in its pole at $s = -2.5$ is found to be 1.401, so that the term representing this pole is

$$z_4 = \frac{1.401}{s + 2.5}, \quad (15)$$

from which the next series branch is constructed. The remainder $z^{iv} = z''' - z_4$ is found to be

$$z^{iv} = \frac{7s + 36.35}{3.952s + 16.8} = \frac{1.77(s + 5.193)}{(s + 4.25)} = \frac{1}{y^{iv}}. \quad (16)$$

This completes the second cycle in the development procedure.

Sketches of y^{iv} and z^{iv} are shown in parts (a) and (b), respectively, of Fig. 4, from which we observe that there

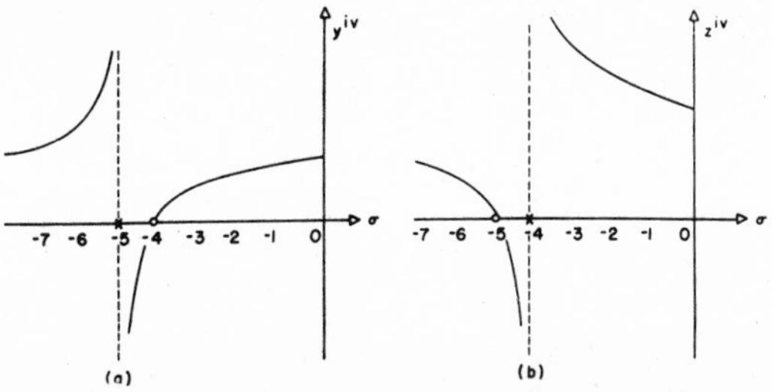

Fig. 4—General character of the remainder functions (16).

is no possibility of shifting a zero of y^{iv} to the point $s = -7$ (the remaining point at which y_{12} is to be zero), but that z^{iv} can be made to have a zero at $s = -7$ through subtracting a positive constant (resistance). That is,

$$z^{iv}(-7) = 1.163, \quad (17)$$

and this is less than $z^{iv}(\infty) = 1.77$, which is the minimum of its real part for $s = j\omega$.

So we can next remove a series resistance

$$z_5 = R_5 = 1.163 \quad (18)$$

and leave

$$z^v = z^{iv} - 1.163 = \frac{0.607(s + 7)}{(s + 4.25)} = \frac{1}{y^v}. \quad (19)$$

Since

$$y^v = \frac{0.648s}{s + 7} + 1, \quad (20)$$

we see that the next shunt branch is given by

$$y_6 = \frac{0.648s}{s + 7}, \quad (21)$$

and that the final series branch is a resistance of one ohm. The completely developed network is shown in Fig. 5. The branches producing zeros of transmission are the two parallel RC components in the series branches (these are z_2 and z_4 (8) and (15)), and the series RC

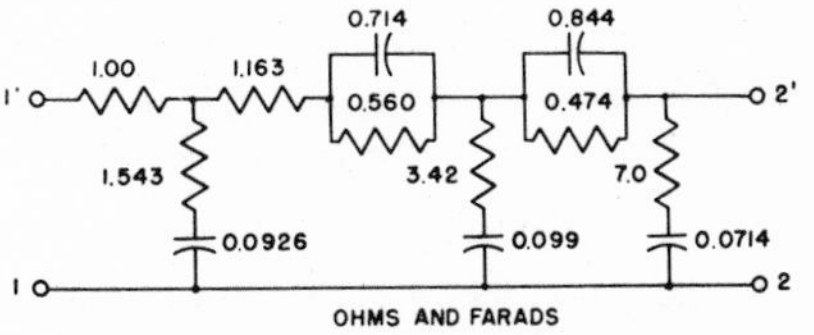

Fig. 5—A two-terminal-pair network realizing the functions given by (2).

component in the left-hand shunt branch (y_6 of (21)). Remaining branches correspond to zero-shifting steps.

Note that zero-producing branches may be either series or shunt and that one need not consistently use shunt branches for zero-shifting and series branches for zero-producing steps. Thus in the first two cycles in the above example, the shunt branches correspond to zero-shifting steps and the series branches to zero-producing ones, while in the last cycle zero-shifting is done by the series resistance 1.163 and the final shunt branch is zero-producing. Since zero-shifting may be done in several steps instead of just one (sometimes this may become necessary or desirable), it is clear the number of alternatively possible developments is infinite.

5

Reprinted from *Quart. Appl. Math.*, **10**(2), 113–127 (1952)

THE TRANSFER FUNCTION OF GENERAL TWO TERMINAL-PAIR RC NETWORKS*

BY

AARON FIALKOW

(*Polytechnic Institute of Brooklyn*)

AND

IRVING GERST

(*Control Instrument Co., Brooklyn, N. Y.*)

1. Introduction. Recent years have seen the increased use in many fields of passive electrical two terminal-pair networks lacking inductive elements. In particular, those networks having a common ground have found widespread employment in servomechanism design. In these applications, the network function of prime interest is generally the transfer function $A(p)$ defined as the ratio of steady state output voltage to input voltage in the domain of the complex frequency variable p.

Our discussion of the transfer function in the sequel will be given in terms of RC-networks. However, it is known that by means of simple transformations the results in the RC-case can be made applicable to networks containing any two kinds of elements only. We will indicate these transformations later for LC-networks, important in classical filter theory, and for RL-networks. The actual statement of our results for these two cases is left as an exercise for the reader.

Despite its importance, there have been few investigations of any generality concerned with the transfer function. Being given a class of networks, the basic questions with regard to the transfer function are: (1) by what properties may one characterize the transfer functions which arise from this class of networks; (2) given a transfer function having these properties, how does one obtain corresponding networks of the class. If, as we may, we write

$$A(p) = KN/D = K(p^n + a_1p^{n-1} + \cdots + a_n)/(p^m + b_1p^{m-1} + \cdots + b_m) \qquad (1.1)$$

where N and D have no common factors, we require a complete characterization of K, N and D for the class of networks under discussion as well as a synthesis procedure. It is important to note that since we are interested in what can be obtained from the *networks alone* without amplifiers, transformers or similar devices, the description of the multiplicative constant K is just as essential as that of N or D.

With this view of the problem in mind, we find that the bulk of the literature on RC transfer functions, besides concerning itself with quite special networks, only partially treats the questions here raised. Thus, generally, the constant K is ignored, and incomplete conditions on N and D are given. The complete problem as stated here has thus far been investigated in the case of rather specialized classes of networks. Bower and Ordung [2] have considered the transfer function of the symmetric lattice using a geometric method while we [4] have treated this network analytically as well as the L and general ladder networks.

In the present paper, we do not restrict ourselves to networks of any special internal

*Presented to the American Mathematical Society, February 24, 1951. (Cf. Bulletin of the American Mathematical Society, Vol. 57 (1951) pp. 182–3.)

structure but treat the general grounded two terminal-pair (3 external terminals) and also the general two terminal-pair (4 external terminals) RC networks (abbreviated 3 T.N., 4 T.N., respectively.) An informal description of our results follows. For more precise statements, see Theorems 1 and 2. It is well known that the transfer function of a 3 T.N. or 4 T.N. is a rational function with real coefficients, regular at infinity, and having negative real, simple poles. In addition to these properties, we find that for the 3 T.N., the zeros of N may not be positive real. For given N and D, K must be a number in the interval $0 < K < K_0$, where K_0 is a constant depending upon N and D which is explicitly determined. Conversely, a 3 T.N. may be constructed which realizes any transfer function $A(p)$ whose K, N, D satisfy the preceding conditions. In this connection, Guillemin [5] had previously shown that if $a_i \geq 0$ in N, then a 3 T.N. exists realizing the transfer function $A(p)$ for some unspecified K. In the case of the 4 T.N., we find that there is no restriction on the zeros of N. It is necessary and sufficient that K lie in an interval $-K_0 < K < K_0$, where again K_0 is given explicitly in terms of N and D.

With regard to the above results, we remark that they do not imply the corresponding complete theorems for any structural sub-classes of 3 T.N. or 4 T.N. such as were discussed in [2] and [4]. For the transfer functions associated with a sub-class have further distinctive properties which are peculiar to the particular internal network structure of that sub-class and which permit synthesis in that structure.

Finally, we note that the results of the paper may be modified to take account of any resistive load or source.

2. The grounded two terminal-pair network. *Theorem* 1. *The necessary and sufficient conditions that a real rational function $A(p)$ given by* (1.1) *be the transfer function of an RC* − 3 *T.N. are*:

(*i*) *The zeros of D are distinct negative numbers.*

(*ii*) *The zeros of N may not be positive real but are otherwise arbitrary.*

(*iii*) $m \geq n$.

(*iv*) *The number K satisfies the inequalities $0 < K < K_0$, where K_0 is the least of the three quantities* K_d, b_m/a_n, 1 if $m = n$ and of the first two quantities if $m > n$. If $K_0 \neq K_d$ then K may equal K_0. Here K_d is the least positive value of κ (if it exists) for which the equation $D - \kappa N = 0$ has a positive multiple root.*

Proof: (*a*) *Necessity*. As stated in the introduction, conditions (i) and (iii) are well known in the case of a 4 T.N. To establish these conditions it is sufficient to write the transfer function (1.1) in the equivalent form

$$A(p) = Y_{12}/Y_{22}, \tag{2.1}$$

where Y_{12} and Y_{22} are the short circuit transfer and driving point admittances, respectively, of the 4 T.N. and then apply Cauer's results on RC admittances and Brune's residue conditions. (Cf. [6], pp. 134-136, 211-212, 216-218). Since a 3 T.N. may be considered as a 4 T.N. with one input and one output terminal joined together (i) and (iii) also apply to a 3 T.N.

*If $a_n = 0$, omit b_m/a_n. It may be shown that in any case at least one of the three quantities K_d, b_m/a_n, 1 actually appears as an upper bound for K.

For the proof that the remaining conditions are necessary, the 3 T.N. shown in Fig. 1 is considered upon a nodal basis. Here the ground terminal is taken as node 0, the other input and output terminals as nodes 1 and 2 respectively. For our purpose the remaining nodes are identified so that each branch is an R and a C in parallel. Hence

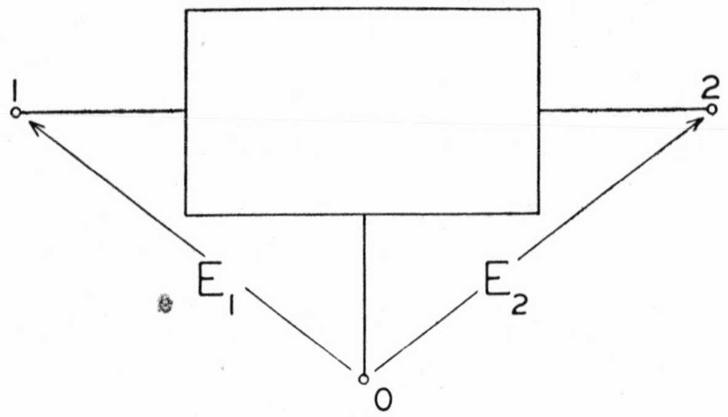

FIG. 1.

the admittance y_{ij} $(i \neq j)$ of the branch between nodes i and j is of the form $ap + b$, $a \geq 0$, $b \geq 0$. Of course $y_{ij} = y_{ji}$. Write

$$y_{ii} = \sum_{\substack{j=0 \\ i \neq j}}^{t} y_{ij} \qquad (i = 1, 2, \cdots, t),$$

where $t + 1$ is the total number of nodes. Then y_{ii} is also linear in p with non-negative coefficients. Let I be the current impressed on node 1 by the driving source and denote by E_i the voltage from node 0 to node i. As is well known [1], the equations of the nodal system are

$$\begin{aligned} I &= \quad y_{11}E_1 - y_{12}E_2 - \cdots - y_{1t}E_t \\ 0 &= -y_{21}E_1 + y_{22}E_2 - \cdots - y_{2t}E_t \\ &\cdots\cdots\cdots\cdots\cdots\cdots\cdots\cdots \\ 0 &= -y_{t1}E_1 - y_{t2}E_2 - \cdots + y_{tt}E_t . \end{aligned} \tag{2.2}$$

Since by definition $A(p) = E_2/E_1$, it follows from these equations that

$$A(p) = \Delta_1/\Delta_2 , \tag{2.3}$$

where

$$\Delta_1 = \begin{vmatrix} y_{21} & y_{23} & \cdots & y_{2t} \\ -y_{31} & y_{33} & \cdots & -y_{3t} \\ \cdots & \cdots & \cdots & \cdots \\ -y_{t1} & -y_{t3} & \cdots & y_{tt} \end{vmatrix} = c_0p^s + c_1p^{s-1} + \cdots + c_s , \tag{2.4}$$

$$\Delta_2 = \begin{vmatrix} y_{22} & -y_{23} & \cdots & -y_{2t} \\ -y_{32} & y_{33} & \cdots & -y_{3t} \\ \cdots & \cdots & \cdots & \cdots \\ -y_{t2} & -y_{t3} & \cdots & y_{tt} \end{vmatrix} = d_0 p^s + d_1 p^{s-1} + \cdots + d_s , \; d_0 \neq 0. \tag{2.5}$$

Here Δ_1 and Δ_2 may have common factors and by (iii) the degree of Δ_1 may actually be less than s.

It may be shown that*

$$0 \leq c_j \leq d_j \qquad (j = 0, 1, \cdots, s). \tag{2.6}$$

(In order not to interrupt the train of the argument at this point we defer the proof to Appendix A). It follows that none of the zeros of Δ_1 and hence of $A(p)$ can be positive real. This establishes (ii). Also since $c_j \geq 0$, $d_0 > 0$, it follows that in the reduced form of the transfer function (1.1), $K > 0$. Then by (ii), $a_n \geq 0$ in (1.1). (But note that some a_i $(i \neq n)$ may be negative).

In order to establish the necessity of (iv) let us suppose the roles of nodes 0 and 1 in Figure 1 are interchanged. This new 3 T.N. whose ground terminal is node 1 and whose other input and output terminals are nodes 0 and 2 respectively, while all other nodes are left unchanged, is termed the *complementary network* (of the original network) and its transfer function $A'(p)$ is called the *complementary transfer function.* Then for this network under the same impressed current I as before, the input voltage is $-E_1$ while the output voltage is $E_2 - E_1$. Hence

$$A'(p) = (E_2 - E_1)/-E_1 = 1 - A(p).$$

Using (1.1), we find

$$A'(p) = (D - KN)/D, \tag{2.7}$$

where it is clear that numerator and denominator are relatively prime. Hence the reduced form for $A'(p)$ corresponding to (1.1) is

$$A'(p) = K' \frac{p^q + a_1' p^{q-1} + \cdots + a_q'}{p^m + b_1 p^{m-1} + \cdots + b_m}, \tag{2.8}$$

where, as shown above, $K' > 0$, $a_q' \geq 0$. Comparison of the two expressions for $A'(p)$ given by (2.7) and (2.8) shows that these last inequalities require $K \leq 1$ if $m = n$, $K \leq b_m/a_n$ if $a_n \neq 0$ respectively.

The remaining inequality on K in (iv) is obtained by first showing that the transfer function of a 3 T.N. has what may be called the "interval property" with respect to its multiplicative constant K. That is, *if $A(p)$ as given by (1.1) is the transfer function of a 3 T.N. then*

$$A^*(p) = K^*N/D, \qquad 0 < K^* < K, \tag{2.9}$$

is also the transfer function of a 3 T.N.

To prove this, suppose $A(p)$ arises from the network of Figure 1. We may write

*That $d_j \geq 0$ is of course well known if either the conductance or capacitance matrix of the y_{ij} occurring in Δ_2 is positive definite.

$A(p)$ as in (2.1). Now modify the 3 T.N. by introducing a new branch joining nodes 0 and 2 whose admittance is $(K - K^*)Y_{22}/K^*$. Then the modified 3 T.N. will have as short circuit driving point and transfer admittances $Y_{22}^* = Y_{22} + (K - K^*)Y_{22}/K^*$ and $Y_{12}^* = Y_{12}$ respectively; so that the corresponding transfer function is given by (2.9).

Now suppose $A(p)$ in (1.1) is the transfer function of a 3 T.N. and let K_d be defined* as in Theorem 1, (iv). Then $K < K_d$. For by what has just been proved, if $K \geq K_d$, $A^*(p) = K_dN/D$ is also a 3 T.N. transfer function. But then the complementary transfer function $(D - K_dN)/D$ would have a positive real zero, in contradiction to (ii).

At first glance the bounds given for K in Theorem 1, (iv) seem quite unrelated. However, one may alternatively characterize K_0 in (iv) as the greatest lower bound of all positive κ for which the equation $D - \kappa N = 0$ has a positive real root. It may then be shown that this implies the statement as given in (iv).

(b) *Sufficiency.* Suppose now that $A(p)$ satisfies the conditions of Theorem 1. We shall construct a 3 T.N. whose transfer function is $A(p)$. For this purpose, we first write $A(p)$ in a special form suggested by the fact that, before algebraic simplification, every transfer function has the form (2.3) with $d_i \geq c_i \geq 0$. We define this special form, called an *R-function* (*realizable function*), as follows: A real rational function

$$R(p) = (g_0p^l + g_1p^{l-1} + \cdots + g_l)/(h_0p^l + h_1p^{l-1} + \cdots + h_l), \qquad h_0 \neq 0, \tag{2.10}$$

is said to be an R-function if the zeros of its denominator are negative real and distinct, and if $0 \leq g_i \leq h_i$ $(i = 0, 1, \cdots, l)$. We call l the degree of the R-function. The expression of a rational function as an R-function is not unique.

It is shown in Appendix B that a function $A(p)$ satisfying the conditions of Theorem 1 may be written as an R-function. In the sequel, we give an algorithm for realizing every R-function as the transfer function of a 3 T.N. Our method consists of an induction on the degree l of the R-function.

The case $l = 0$ is trivial. If $l = 1$, i.e., $R(p) = (g_0p + g_1)/(h_0p + h_1)$ with $h_0h_1 \neq 0$, $0 \leq g_0 \leq h_0$, $0 \leq g_1 \leq h_1$, then $R(p)$ is realized by an L-network whose series and shunt arm admittances are $Y_a = g_0p + g_1$ and $Y_b = (h_0 - g_0)p + (h_1 - g_1)$ respectively. Note that the Y_{22} of this network is $h_0p + h_1$.

Now suppose that all R-functions of degree less than $l \geq 2$ are realizable as transfer functions of 3 T.N. Also assume that for $R(p) = U/V$ of degree $f < l$, U and V as in (2.10), the Y_{22} of the corresponding network is of the form V/S where S is a polynomial of degree $f - 1$. We will show that (2.10) is realizable as the transfer function of a 3 T.N. such that its Y_{22} is of the form $\sum_{i=0}^{l} h_ip^{l-i}/T$ where T is a polynomial of degree $l - 1$.

Let the negatives of the zeros of the denominator in (2.10) be $\gamma_1 < \gamma_2 < \cdots < \gamma_l$. Choose any σ_i $(i = 1, 2, \cdots, l - 1)$ such that

$$0 < \gamma_1 < \sigma_1 < \gamma_2 < \sigma_2 < \cdots < \sigma_{l-1} < \gamma_l .$$

Then the function $Z = h_0 \prod_{i=1}^{l} (p + \gamma_i)/p \prod_{i=1}^{l-1} (p + \sigma_i)$ is an RC-impedance who canonical expansion is

$$Z = h_0 + A_0/p + \sum_{i=1}^{l-1} A_i/(p + \sigma_i), \qquad A_i > 0 \qquad (i = 0, 1, \cdots, l - 1). \tag{2.11}$$

*It follows that K_d is one of the roots of the equation in κ obtained by equating the discriminant of $D - \kappa N = 0$ to zero. For the expression of the discriminant of an algebraic equation in terms of its coefficients see [3].

In (2.11) the terms of the right member with h_0 omitted form another RC-impedance which we write as $c \prod_{i=1}^{l-1} (p + \xi_i)/p \prod_{i=1}^{l-1} (p + \sigma_i)$, $c > 0$. We have

$$0 < \xi_1 < \sigma_1 < \xi_2 < \sigma_2 < \cdots < \xi_{l-1} < \sigma_{l-1} . \tag{2.12}$$

From (2.11) it now follows that

$$h_0 \prod_{i=1}^{l} (p + \gamma_i) = h_0 p \prod_{i=1}^{l-1} (p + \sigma_i) + c \prod_{i=1}^{l-1} (p + \xi_i). \tag{2.13}$$

If we let

$$h_0 p \prod_{i=1}^{l-1} (p + \sigma_i) = \sum_{i=0}^{l-1} h_i' p^{l-i} \qquad \text{and} \qquad c \prod_{i=1}^{l-1} (p + \xi_i) = \sum_{i=1}^{l} h_i'' p^{l-i},$$

then the h_i' and h_i'' are all positive; and equating coefficients of like powers of p in (2.13) gives

$$h_0 = h_0' , \qquad h_i = h_i' + h_i'' \qquad (i = 1, 2, \cdots , l - 1), \qquad h_l = h_l' . \tag{2.14}$$

Since $0 \leq g_i \leq h_i$ in (2.10), and in view of (2.14), there exists for each i ($i = 1, 2, \cdots , l - 1$) at least one pair g_i' , g_i'' such that

$$g_i = g_i' + g_i'', \qquad 0 \leq g_i' \leq h_i' , \qquad 0 \leq g_i'' \leq h_i''. \tag{2.15}$$

Take $g_0' = g_0$ and $g_l'' = g_l$.

Now consider the functions

$$R_1(p) = \sum_{i=0}^{l-1} g_i' p^{l-i} / h_0 p \prod_{i=1}^{l-1} (p + \sigma_i) = \sum_{i=0}^{l-1} g_i' p^{l-1-i} / h_0 \prod_{i=1}^{l-1} (p + \sigma_i),$$

$$R_2(p) = \sum_{i=1}^{l} g_i'' p^{l-1} / c \prod_{i=1}^{l-1} (p + \xi_i).$$

These are R-functions of degree $l - 1$ at most. Hence by the hypothesis of induction there exist two 3 T.N. Γ_1 and Γ_2 whose transfer functions are R_1 and R_2 respectively, and whose Y_{22}'s are

$$Y_{22}^{(1)} = h_0 \prod_{i=1}^{l-1} (p + \sigma_i)/\lambda_1 S_1 \qquad \text{and} \qquad Y_{22}^{(2)} = c \prod_{i=1}^{l-1} (p + \xi_i)/\lambda_2 S_2$$

respectively. Here S_1 and S_2 are polynomials of degree $l - 2$ while λ_1 and λ_2 are arbitrary positive impedance-level constants whose values do not affect the transfer functions. We shall presently fix λ_1 and λ_2 .

Now in view of (2.12) we can choose β_i ($i = 1, 2, \cdots , l - 1$) such that

$$\xi_1 < \beta_1 < \sigma_1 < \xi_2 < \beta_2 < \sigma_2 < \cdots < \xi_{l-1} < \beta_{l-1} < \sigma_{l-1} .$$

Hence

$$\mathcal{Y}_{22}^{(1)} = h_0 p \prod_{i=1}^{l-1} (p + \sigma_i)/(p + \beta_i), \qquad \mathcal{Y}_{22}^{(2)} = c \prod_{i=1}^{l-1} (p + \xi_i)/(p + \beta_i) \tag{2.16}$$

are both RC-admittances. We are going to modify the networks Γ_1 and Γ_2 so that their transfer functions remain unchanged but their Y_{22}'s become $\mathcal{Y}_{22}^{(1)}$ and $\mathcal{Y}_{22}^{(2)}$ respectively.

In Fig. 2 consider the networks Γ_1' and Γ_2' where Z_1 and Z_2 are RC-impedances to be determined. Evidently the transfer functions of Γ_1' and Γ_2' are the same as those of Γ_1

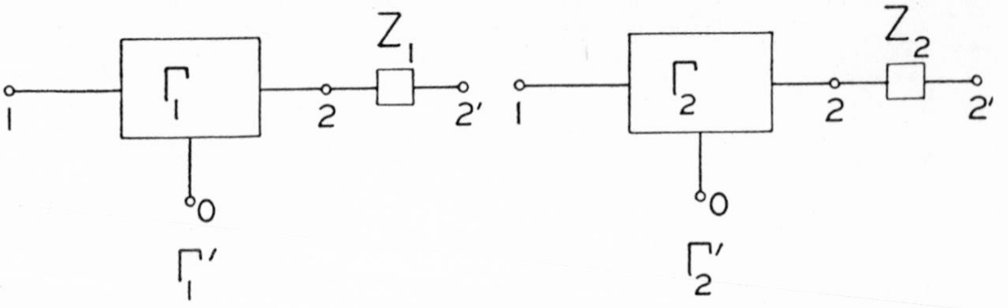

FIG. 2.

and Γ_2. However, the Y_{22}'s of the new networks are $1/(Z_1 + 1/Y_{22}^{(1)})$ and $1/(Z_2 + 1/Y_{22}^{(2)})$ respectively. If these are to equal $\mathcal{Y}_{22}^{(1)}$ and $\mathcal{Y}_{22}^{(2)}$ respectively, then it is necessary and sufficient that Z_1 and Z_2 as given by

$$Z_1 = 1/\mathcal{Y}_{22}^{(1)} - 1/Y_{22}^{(1)}, \qquad Z_2 = 1/\mathcal{Y}_{22}^{(2)} - 1/Y_{22}^{(2)} \tag{2.17}$$

be RC-impedances. Now in canonical form

$$1/\mathcal{Y}_{22}^{(1)} = B_0/p + \sum_{i=1}^{l-1} B_i/(p + \sigma_i), \qquad 1/\mathcal{Y}_{22}^{(2)} = C_0 + \sum_{i=1}^{l-1} C_i/(p + \xi_i)$$

with

$$B_i > 0, \qquad C_i > 0 \qquad (i = 0, 1, \cdots, l - 1);$$

and

$$1/Y_{22}^{(1)} = \lambda_1 \sum_{i=1}^{l-1} D_i/(p + \sigma_i), \qquad 1/Y_{22}^{(2)} = \lambda_2 \sum_{i=1}^{l-1} E_i/(p + \xi_i)$$

with

$$D_i \geq 0, \qquad E_i \geq 0 \qquad (i = 1, 2, \cdots, l - 1).$$

Hence

$$Z_1 = B_0/p + \sum_{i=1}^{l-1} (B_i - \lambda_1 D_i)/(p + \sigma_i), \qquad Z_2 = C_0 + \sum_{i=1}^{l-1} (C_i - \lambda_2 E_i)/(p + \xi_i),$$

and by taking λ_1 and λ_2 sufficiently small and positive, Z_1 and Z_2 will be RC-impedances. The Y_{12}'s of the networks Γ_1', Γ_2' in view of (2.1), are, respectively,

$$\mathcal{Y}_{12}^{(1)} = \mathcal{Y}_{22}^{(1)} R_1 = \sum_{i=0}^{l-1} g_i' p^{l-1-i} \Big/ \prod_{i=1}^{l-1} (p + \beta_i),$$

$$\mathcal{Y}_{12}^{(2)} = \mathcal{Y}_{22}^{(2)} R_2 = \sum_{i=1}^{l} g_i'' p^{l-i} \Big/ \prod_{i=1}^{l-1} (p + \beta_i).$$

Now connect the networks Γ_1' and Γ_2' in parallel to form a new 3 T.N., Γ. Then [6 p. 146] the Y_{12} and Y_{22} of Γ are given by $Y_{12} = \mathcal{Y}_{12}^{(1)} + \mathcal{Y}_{12}^{(2)}$, $Y_{22} = \mathcal{Y}_{22}^{(1)} + \mathcal{Y}_{22}^{(2)}$.

Replacing the $\mathcal{Y}_{12}$'s and $\mathcal{Y}_{22}$'s by their expressions given above and using (2.13) and (2.15) we find that the transfer function $A(p)$ of Γ is (2.10) while

$$Y_{22} = \sum_{i=0}^{l} h_i p^{l-i} / \prod_{i=1}^{l-1} (p + \beta_i).$$

This completes the induction*.

The preceding synthesis procedure may be modified to accommodate any given resistive source and load. In the general case, this may require a finite reduction in the value of the maximum realizable K. However, if the source is zero or the load is infinite, all K except possibly K_0 may be realized.

As an illustration, we outline the synthesis procedure for the case of a resistive load, which we may take as 1 ohm without loss of generality. If $K < K_0$ we may suppose the R-function in (2.10) which corresponds to the transfer function is such that $0 \leq g_i < h_i$. The procedure is now started by choosing τ_i $(i = 1, 2, \cdots, l)$ such that $0 < \tau_1 < \gamma_1 < \tau_2 < \cdots < \tau_l < \gamma_l$. Then as before we have

$$h_0 \prod_{i=1}^{l} (p + \gamma_i) = h_0 \prod_{i=1}^{l} (p + \tau_i) + c' \prod_{i=1}^{l-1} (p + \chi_i), \qquad c' > 0,$$

where $\tau_1 < \chi_1 < \tau_2 < \chi_2 < \cdots < \chi_{l-1} < \tau_l$. The original synthesis procedure is now applied to

$$\sum_{i=0}^{l} g_i p^{l-i} / h_0 \prod_{i=1}^{l} (p + \tau_i)$$

to give a 3 T.N. Γ whose Y_{22} is

$$\mu \prod_{i=1}^{l} (p + \tau_i) / \prod_{i=1}^{l-1} (p + \chi_i), \qquad \mu > 0.$$

If the τ_i are taken sufficiently close to the γ_i this will always be possible. The network consisting of Γ working into the load then realizes the original transfer function.

We conclude this section with the observation that theorems analogous to Theorem 1 exist for 3 T.N. having only resistance and (self) inductance or capacitance and (self) inductance. As shown in [4, Sec. 7] these are obtained from Theorem 1 by replacing p by $1/p$ in the RL case and by replacing p by p^2 in the LC case.

3. An illustrative example. As an illustration of the foregoing theory and synthesis procedure consider the function $A(p) = KN/D = K(p^2 - p + 9)/(p + 1)(p + 14)$. First let us determine for what range of K, $A(p)$ is the transfer function of a 3 T.N. Since $m = n = 2$, $a_2 = 9$, $b_2 = 14$, Theorem 1 tells us that $0 < K < K_0$ where $K_0 = \min [K_d, 14/9, 1]$. To calculate K_d it is best to proceed indirectly by first eliminating κ between the equations $D - \kappa N = 0$ and $d/dp\ (D - \kappa N) = 0$. This gives $16\ p^2 + 10\ p - 149 = 0$ whose roots $p_1 = -3.380$, $p_2 = 2.755$ are the multiple roots of $D - \kappa N = 0$ which occur for all variations of κ. Only the latter of these is positive and corresponding to it we find $\kappa = D(p_2)/N(p_2) = 4.546 = K_d$. Hence $K_0 = \min [4.546, 14/9, 1] = 1$.

*A variation of the above procedure may be obtained by modifying the decomposition as given in (2.13). Using either procedure an appropriate choice of (2.15) will frequently lead to simple network realizations. Still another modification of the synthesis procedure makes it possible to always choose Z_1 as a condenser and Z_2 as a resistor in Fig. 2 at each stage of the synthesis.

Let us choose (as we may by Theorem 1) $K = K_0 = 1$. To apply the synthesis procedure, we must first write $A(p)$ as an R-function. Using the ideas of Appendix B (or by inspection) we find that the introduction of the common factor $(p + 3)$ into $A(p)$ converts it into the R-function

$$R = (p^3 + 2p^2 + 6p + 27)/(p + 1)\,(p + 3)\,(p + 14).$$

Following the procedure of §2(b) choose $\sigma_1 = 2$, $\sigma_2 = 6$. Then (2.13) reads $p^3 + 18p^2 + 59p + 42 = p(p^2 + 8p + 12) + (10p^2 + 47p + 42)$, from which $\xi_1 = 6/5$, $\xi_2 = 7/2$. Now, for simplicity, in (2.15) let $2 = 2 + 0$, $6 = 0 + 6$; also choose $\beta_1 = 3/2$, $\beta_2 = 4$. Then we get

$$R_1 = (p^3 + 2p^2)/p(p + 2)(p + 6) = p/(p + 6),\ R_2 = (6p + 27)/10(p + 6/5)(p + 7/2),$$

while according to (2.16)

$$\mathcal{Y}_{22}^{(1)} = p(p + 2)(p + 6)/(p + 3/2)(p + 4)$$

and

$$\mathcal{Y}_{22}^{(2)} = 10(p + 6/5)(p + 7/2)/(p + 3/2)(p + 4)$$

respectively.

The transfer function R_1 being of first degree is realized by an L-network Γ_1 whose series and shunt arm impedances are $Z_{1a} = \lambda_1/p$ and $Z_{1b} = \lambda_1/6$ respectively, and whose $Y_{22}^{(1)} = (p + 6)/\lambda_1$. Use (2.17) to determine Z_1 taking $\lambda_1 = 3/8$ for simplicity. Then $Z_1 = 1/2\,p + 1/8(p + 2)$. Finally form Γ_1' as in Fig. 2.

As for R_2, it is of the second degree and hence the reduction procedure is repeated for it to obtain R_3 and R_4 of the first degree. We give the results, the notation being evident from what has preceded.

$$R_2 = \frac{6p + 27}{10(p + 6/5)(p + 7/2)}, \qquad \begin{matrix} \gamma_1 & \xi_1 & \beta_1 & \sigma_1 & \gamma_2 \\ 6/5, & 14/9, & 7/4, & 2, & 7/2 \end{matrix},$$

$$10p^2 + 47p + 42 = 10p(p + 2) + (27p + 42);$$

$$R_3 = 0, \qquad \mathcal{Y}_{22}^{(3)} = \frac{10p(p + 2)}{(p + 7/4)}; \qquad R_4 = \frac{6p + 27}{27(p + 14/9)}, \qquad \mathcal{Y}_{22}^{(4)} = \frac{27(p + 14/9)}{(p + 7/4)}.$$

The transfer functions R_3 and R_4 are realized by L-networks Γ_3, Γ_4 whose series and shunt arm impedances are $Z_{3a} = \infty$, $Z_{3b} = (p + 7/4)/10p(p + 2)$; $Z_{4a} = \lambda_4/(6p + 27)$, $Z_{4b} = \lambda_4/(21p + 15)$ respectively; and whose Y_{22}'s are $Y_{22}^{(3)} = \mathcal{Y}_{22}^{*(3)}$ and $Y_{22}^{(4)} =$

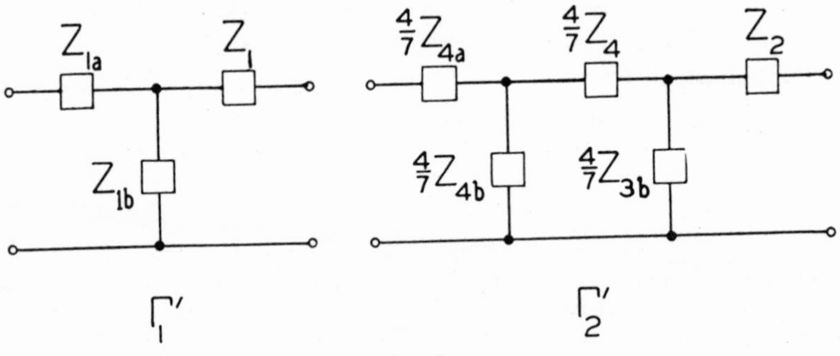

FIG. 3.

$27(p + 14/9)/\lambda_4$ respectively. Hence we need only determine a Z_4 using (2.17) and taking $\lambda_4 = 7/36$. This gives $Z_4 = 1/27$. Now form Γ_4' as in Fig. 2 and denote by Γ_2 the parallel combination of Γ_3 and Γ_4'. Introducing the impedance level factor λ_2 in Γ_2 the Y_{22} of Γ_2 is given by $Y_{22}^{(2)} = 10(p + 6/5)(p + 7/2)/\lambda_2(p + 7/4)$. Again use (2.17) to determine Z_2 taking $\lambda_2 = 4/7$. Then $Z_2 = 1/10 + 4/175(p + 6/5)$. Forming Γ_2' from Γ_2 and Z_2 as in Fig. 2, the required final network is the parallel combination of Γ_1' and Γ_2' as shown in Fig. 3.

4. The two terminal-pair network. *Theorem* 2. *The necessary and sufficient conditions that a real rational function $A(p)$ given by (1.1) be the transfer function of an $RC - 4$ T.N. are:*

(*i*) *The zeros of D are distinct negative numbers.*

(*ii*) $m \geq n$.

(*iii*) *The number K satisfies the inequalities $-K_0 < K < K_0$, where K_0 is the least of the three quantities $|K_d|$, $|b_m/a_n|$, 1 if $m = n$ and of the first two quantities if $m > n$. If $K_0 \neq |K_d|$ then K may actually equal $\pm K_0$. Here K_d is that real value of κ of smallest absolute value (if it exists) for which the equation $D - \kappa N = 0$ has a positive multiple root.*

Proof (*a*) *Necessity.* For (i) and (ii) see the remarks in Sec. 2. To prove (iii) consider the 4 T.N. of Fig. 4 on a nodal basis taking the input terminals as nodes 0 and 1, the

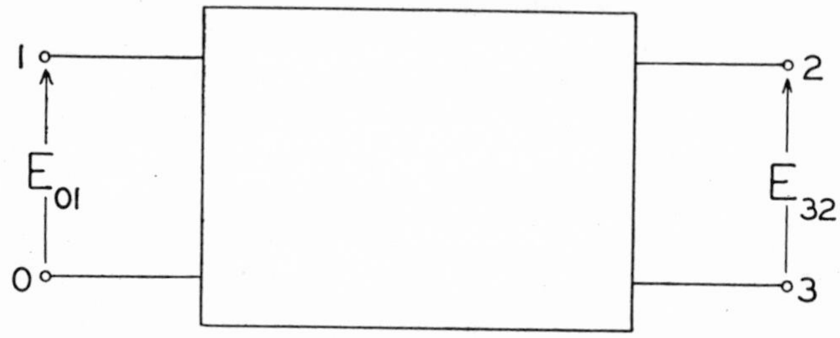

FIG. 4.

output terminals as nodes 2 and 3, and choosing the remaining nodes as in the 3 T.N. case. Let E_{01} and E_{32} be the input and output voltages, respectively. Then in the notation of Sec. 2 we have $A(p) = E_{32}/E_{01} = (E_2 - E_3)/E_1$. It follows from (2.2) that

$$A(p) = (\Delta_1 - \Delta_3)/\Delta_2 \tag{4.1}$$

with Δ_1 and Δ_2 as in (2.4), (2.5) and

$$\Delta_3 = \begin{vmatrix} y_{31} & y_{32} & y_{34} & \cdots & y_{3t} \\ -y_{21} & y_{22} & -y_{24} & \cdots & -y_{2t} \\ -y_{41} & -y_{42} & y_{44} & \cdots & -y_{4t} \\ \cdots & \cdots & \cdots & \cdots & \cdots \\ -y_{t1} & -y_{t2} & -y_{t4} & \cdots & y_{tt} \end{vmatrix} = c_0'p^s + c_1'p^{s-1} + \cdots + c_s'.$$

Since Δ_3 is of the same form as Δ_1 we have

$$0 \leq c'_j \leq d_j \qquad (j = 0, 1, \cdots, s), \tag{4.2}$$

in addition to (2.6).

If we equate the two expressions for $1 \pm A(p)$ obtained from (1.1) and (4.1), we have

$$(D \pm KN)/D = [\Delta_2 \pm (\Delta_1 - \Delta_3)]/\Delta_2 .$$

Since the fraction on the left is evidently in reduced form, and in view of (2.6) and (4.2), it follows that $|K| \leq 1$ if $m = n$ and that $D \pm KN = 0$ has no positive real roots. This in turn implies that $|K| \leq |b_m/a_n|$ if $a_n \neq 0$, and that $|K| < |K_d|$ for K_d as defined in Theorem 2. This last fact follows as in the 3 T.N. case; for the "interval property" of K holds for a 4 T.N. with $0 < |K^*| < |K|$ in (2.9).

(*b*) *Sufficiency.* Let $A(p)$ satisfy the conditions of Theorem 2. Then proceeding as in Appendix B we find that both $D + KN$ and $D - KN$ have leading coefficients positive and have no positive real zeros. Thus there exists a polynomial P having distinct negative real zeros such that $P(D + KN)$ and $P(D - KN)$ have non-negative coefficients. We may suppose P and D relatively prime. If

$$PD = \sum_{i=0}^{l} h_i p^{l-i} \qquad \text{and} \qquad KPN = \sum_{i=0}^{l} g_i p^{l-i},$$

it therefore follows that $|g_i| \leq h_i$ $(i = 0, 1, \cdots, l)$. Write $A(p) = KPN/PD = (N_1 - N_2)/PD$, where N_1 consists of those terms of KPN having positive coefficients while $-N_2$ consists of those terms of KPN having negative coefficients. Then N_1/PD and N_2/PD are R-functions. Hence by the results of Sec. 2 there exist two 3 T.N., Γ_1 and Γ_2 whose ground, input and output terminals are 0, 1, 2 and 0, 1, 2′ respectively, and whose transfer functions are $A_1 = N_1/PD$ and $A_2 = N_2/PD$ respectively. Now form the 4 T.N. shown in Fig. 5 whose input terminals are 0 and 1 and whose output

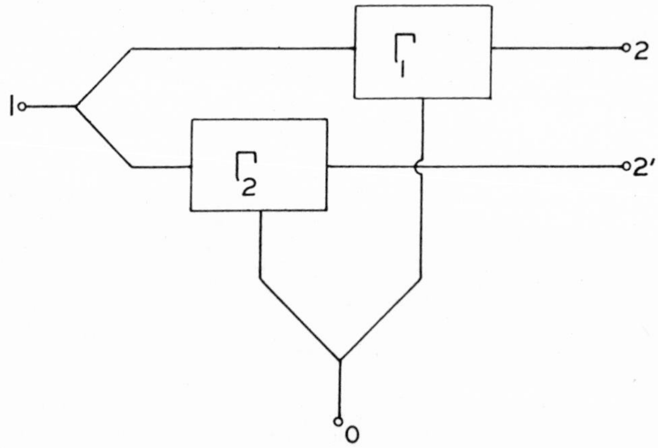

FIG. 5.

terminals are 2 and 2′. The transfer function of this network is evidently $A_1 - A_2 = A$. Remarks similar to those of Sec. 2 may also be made here concerning source and load, and the RL and LC networks.

Appendix A

Consider the determinant

$$H = \begin{vmatrix} a_{11} & a_{12} & a_{13} & \cdots & a_{1n} \\ -a_{21} & a_{22} & -a_{23} & \cdots & -a_{2n} \\ -a_{31} & -a_{32} & a_{33} & \cdots & -a_{3n} \\ \cdots & \cdots & \cdots & \cdots & \cdots \\ -a_{n1} & -a_{n2} & -a_{n3} & \cdots & a_{nn} \end{vmatrix},$$

where $a_{ii} = \sum_{i=0, i \neq j}^{n} a_{ij} (i = 2, 3, \cdots, n)$ and all the $a_{ij}(i \neq j)$ and a_{11} are independent variables. We shall prove that H *is a polynomial (multilinear form) in the* $a_{ij}(i \neq j)$ *and* a_{11} *with non-negative coefficients.*

This is evidently true for $n = 1, 2$. Suppose it is true for determinants of the form H of order less than $n (n \geq 3)$. We may write $H = H_1 + H_2$ where H_1 contains those terms which are independent of $a_{in}(i = 1, 2, \cdots, n-1)$ and H_2 those terms which are linear in each $a_{in}(i = 1, 2, \cdots, n-1)$ separately. In H, if we let $a_{in} = 0 (i = 1, 2, \cdots, n-1)$ we find

$$H_1 = a_{nn} \begin{vmatrix} a_{11} & a_{12} & \cdots & a_{1(n-1)} \\ -a_{21} & a'_{22} & \cdots & -a_{2(n-1)} \\ \cdots & \cdots & \cdots & \cdots \\ -a_{(n-1)1} & -a_{(n-1)2} & \cdots & a'_{(n-1)(n-1)} \end{vmatrix},$$

where $a'_{ii} = a_{ii} - a_{in} (i = 2, 3, \cdots, n-1)$. Hence applying the hypothesis of induction we conclude that the terms of H_1 have positive coefficients.

As for H_2 each of its terms is contained in at least one of the polynomials $a_{in}\, \partial H / \partial a_{in}$ $(i = 1, 2, \cdots, n-1)$. Now

$$\frac{\partial H}{\partial a_{1n}} = \begin{vmatrix} 0 & 0 & 0 & \cdots & 1 \\ -a_{21} & a_{22} & -a_{23} & \cdots & -a_{2n} \\ \cdots & \cdots & \cdots & \cdots & \cdots \\ -a_{n1} & -a_{n2} & -a_{n3} & \cdots & a_{nn} \end{vmatrix}$$

$$= \begin{vmatrix} a_{n1} & a_{n2} & a_{n3} & \cdots & a_{n(n-1)} \\ -a_{21} & a_{22} & -a_{23} & \cdots & -a_{2(n-1)} \\ \cdots & \cdots & \cdots & \cdots & \cdots \\ -a_{(n-1)1} & -a_{(n-1)2} & -a_{(n-1)3} & \cdots & a_{(n-1)(n-1)} \end{vmatrix};$$

and for $i = 2, 3, \cdots, n-1$,

$$\frac{\partial H}{\partial a_{in}} = \begin{vmatrix} a_{11} & a_{12} & \cdots & a_{1i} & \cdots & a_{1n} \\ -a_{21} & a_{22} & \cdots & -a_{2i} & \cdots & -a_{2n} \\ \cdots & \cdots & \cdots & \cdots & \cdots & \cdots \\ 0 & 0 & \cdots & 1 & \cdots & -1 \\ \cdots & \cdots & \cdots & \cdots & \cdots & \cdots \\ -a_{n1} & -a_{n2} & \cdots & -a_{ni} & \cdots & a_{nn} \end{vmatrix}$$

$$= \begin{vmatrix} a_{11} & a_{12} & \cdots & a_{1(i-1)} & a_{1(i+1)} & \cdots & (a_{1i} + a_{1n}) \\ -a_{21} & a_{22} & \cdots & -a_{2(i-1)} & -a_{2(i+1)} & \cdots & -(a_{2i} + a_{2n}) \\ \cdots & \cdots & \cdots & \cdots & \cdots & \cdots & \cdots \\ -a_{(i-1)1} & -a_{(i-1)2} & \cdots & a_{(i-1)(i-1)} & -a_{(i-1)(i+1)} & \cdots & -(a_{(i-1)i} + a_{(i-1)n}) \\ -a_{(i+1)1} & -a_{(i+1)2} & \cdots & -a_{(i+1)(i-1)} & a_{(i+1)(i+1)} & \cdots & -(a_{(i+1)i} + a_{(i+1)n}) \\ \cdots & \cdots & \cdots & \cdots & \cdots & \cdots & \cdots \\ -a_{n1} & -a_{n2} & \cdots & -a_{n(i-1)} & -a_{n(i+1)} & \cdots & (a_{nn} - a_{ni}) \end{vmatrix}.$$

Replacing $a_{i0} + a_{in}(i = 2, 3, \cdots, n-1)$ by new variables a'_{i0} in the first case, and $a_{ji} + a_{jn}(j = 1, 2, \cdots, n-1)$ by new variables a'_{jn} in the second case, we see that the final determinant in each case is reduced to a determinant of the same form as H but of order $n-1$. Hence the terms of H_2 also have positive coefficients and the above italicized statement about H is proved.

Now make the following specialization: $n = t-1$, $a_{i1} = y_{(i+1)1}(i = 1, 2, \cdots, t-1)$, $a_{ij} = y_{(i+1)(j+1)}(i = 1, 2, \cdots, t-1;\ j = 2, 3, \cdots, t-1;\ i \neq j)$, $a_{i0} = y_{(i+1)2} + y_{(i+1)0}(i = 1, 2, \cdots, t-1)$. Then H becomes Δ_1 and we find that Δ_1 is a polynomial in the $y_{ij}(i \neq j)$ with non-negative coefficients. This proves the $c_i \geq 0(i = 0, 1, \cdots, s)$ in (2.4).

In Δ_1 replace the elements of the first column by $y_{20}, -y_{30}, \cdots, -y_{t0}$ respectively, to form Δ'_1. Then by symmetry Δ'_1 is also a polynomial in the $y_{ij}(i \neq j)$ with non-negative coefficients. Now in Δ_2 of (2.5) add the sum of columns $2, 3, \cdots, t-1$ to column 1. We see that $\Delta_2 = \Delta_1 + \Delta'_1$. This proves that $d_j \geq c_j(j = 0, 1, \cdots, s)$.

Appendix B

Lemma: Let* $F(p) = \alpha p^a + \cdots$, $\alpha > 0$ *be a real polynomial having no positive real zeros. Then there exists a polynomial* $P = \prod(p + \delta_i)$ *with the* $\delta_i > 0$ *and distinct, such that* PF *is a polynomial with non-negative coefficients.*

---

*For a more general result which includes this one see [7]. We include the following simple constructive proof so as to keep the paper self-contained with regard to the synthesis procedure.

The real irreducible factors of F are either of the form $p + \tau$ with $\tau \geq 0$ or $p^2 - 2\rho \cos \varphi p + \rho^2$ with $\rho > 0$, $0 < \varphi < 2\pi$. It clearly suffices to prove the lemma for F equal to this latter quadratic polynomial with $0 < \cos \varphi < 1$. Replacing p by ρp we may without loss of generality suppose that $\rho = 1$.

Let $P_1 = (p + 1)^b$ where b is an integer such that

$$b > 2 \cos \varphi/(1 - \cos \varphi). \tag{B.1}$$

We have

$$P_1F = \sum_{i=0}^{b+2} \theta_i p^i,$$

where

$$\theta_i = \binom{b}{i} - 2 \cos \varphi \binom{b}{i-1} + \binom{b}{i-2} \qquad (i = 0, 1, \cdots, b + 2),$$

and $\binom{b}{i}$ is the binomal coefficient, with $\binom{b}{i}$ defined as zero if $i > b$ or $i < 0$.

Now $\theta_0 = \theta_{b+2} = 1 > 0$; and $\theta_1 = \theta_{b+1} = b - 2 \cos \varphi > 0$ by (B.1). If $2 \leq i \leq b$,

$$\theta_i = \binom{b}{i-1}\left[\frac{(b - i + 1)(b - i + 2) + i(i - 1)}{i(b - i + 2)} - 2 \cos \varphi\right]. \tag{B.2}$$

We have $i(b - i + 2) \leq (i + b - i + 2)^2/4 = (b + 2)^2/4$; and $(b - i + 1)(b - i + 2) + i(i - 1) = \{b(b + 2) + [b - 2(i - 1)]^2\}/2 \geq b(b + 2)/2$. Hence in (B.2), the bracket satisfies the inequality, $[\qquad] \geq 2b/(b + 2) - 2 \cos \varphi > 0$ by (B.1). Since the coefficients θ_i of P_1F are all positive we may by continuity considerations form $P_1' = \prod_{i=1}^{b}(p + \delta_i)$ with the δ_i sufficiently close to 1 and distinct, such that $P_1'F$ has positive coefficients. This proves the lemma. We note that the method used in the above proof may not necessarily lead to the simplest polynomial P.

Now let $A(p)$ satisfy the conditions of Theorem 1. Then evidently KN is a polynomial such as F in the lemma. But $D - KN$ is also such a polynomial! To prove this we first show that $D - KN$ has no positive real zeros. Consider the zeros $\zeta_i(\kappa)$ of $D - \kappa N = 0$ where κ is a non-negative, real parameter. When $\kappa = 0$ these zeros are the zeros of D and hence are negative real and distinct. Then for small positive κ the ζ_i will be negative real and distinct. As κ increases positively, positive real ζ_i may arise only if one of the following has occurred: (a) a ζ_i has gone through zero, (b) a ζ_i has gone through ∞, (c) two or more complex ζ_i have become equal on the positive real axis. In case (a) we must have $\kappa > b_m/a_n (a_n \neq 0)$; in case (b) $\kappa > 1$ if $m = n$; in case (c) $\kappa \geq K_d$. Hence, by (iv) in Theorem 1, if $\kappa = K$, $D - KN$ has no positive real zeros.

We must still show that the leading coefficient of $D - KN$ is positive. This is evident unless $m = n$, $K = K_0 = 1$. In this case, we have $D - N = \sum_{i=0}^{r} (b_{m-i} - a_{m-i})p^i$, $b_{m-r} - a_{m-r} \neq 0$, $r < m$. Suppose the leading coefficient $b_{m-r} - a_{m-r} < 0$. Consider the polynomial $Q_1 = \sum_{i=0}^{r} e_{m-i}p^i$, where $e_{m-i} = b_{m-i} - a_{m-i}$ if $a_{m-i} < 0$ and $e_{m-i} = b_{m-i} - (1 - \epsilon_1)a_{m-i}$, $\epsilon_1 > 0$ if $a_{m-i} \geq 0$. For ϵ_1 sufficiently small the leading coefficient of Q_1 will be negative so that there exists a $p_0 > 0$ such that $Q_1(p_0) < 0$. Now

$$D(p_0) - (1 - \epsilon)N(p_0) = \epsilon \sum_{i=r+1}^{m} a_{m-i}p_0^i + \sum_{i=0}^{r} [b_{m-i} - (1 - \epsilon)a_{m-i}]p_0^i$$

$$< \epsilon \sum_{i=r+1}^{m} a_{m-i}p_0^i + Q_1(p_0) \qquad \text{for} \qquad 0 < \epsilon < \epsilon_1 .$$

Hence for ϵ sufficiently small, say $\epsilon = \epsilon_2 < \epsilon_1$, $D(p_0) - (1 - \epsilon_2)N(p_0) < 0$ while $D(\infty) - (1 - \epsilon_2)N(\infty) > 0$. Thus $D(p) - (1 - \epsilon_2)N(p) = 0$ has a positive real root, which contradicts what has been proved above about $D - \kappa N = 0$ for $\kappa < K_0$. Thus by the lemma, there exist two polynomials P_1 and P_2 such that KP_1N and $P_2(D - KN)$ have non-negative coefficients. Then P_1P_2 will simultaneously convert KN and $D - KN$ into polynomials with non-negative coefficients. We may suppose P_1, P_2 and D are relatively prime in pairs. Now let

$$A(p) = KN/D = KP_1P_2N/P_1P_2D = \sum_{i=0}^{l} g_ip^{l-i} / \sum_{i=0}^{l} h_ip^{l-i}.$$

Then the zeros of the denominator of the last fraction are negative real and distinct, and by considering KP_1P_2N and $P_1P_2(D - KN)$ it follows that $0 \le g_i \le h_i (i = 0, 1, \cdots, l)$. Hence the last fraction is an R-function.

We remark that if $K_0 = 1$ or b_m/a_n and if R_0 is an R-function corresponding to K_0, then KR_0/K_0 corresponds to any K in the range $0 < K \le K_0$. However, if $K_0 = K_d$, no such R-function R_0 exists and the degrees of the R-functions corresponding to K in the range $0 < K < K_d$ must increase indefinitely as $K \to K_d$.

References

1. H. W. Bode, *Network analysis and feedback amplifier design*, D. Van Nostrand, Inc., New York, 1945, pp. 11–12.
2. J. L. Bower and P. F. Ordung, *The synthesis of resistor-capacitor networks*, Proc. I. R. E. **38**, 263–269 (1950).
3. L. E. Dickson, *First course in the theory of equations*, John Wiley and Sons, New York (1922), pp. 146, 152.
4. A. Fialkow and I. Gerst, *The transfer function of an RC ladder network*, J. Math. Physics, **30**, 49–72, (1951).
5. E. A. Guillemin, *Synthesis of RC-networks*, J. Math. Physics, **28**, 22–42 (1949).
6. E. A. Guillemin, *Communication networks*, Vol. 2, John Wiley and Sons, New York, 1935.
7. G. Pólya and G. Szegö, *Aufgaben und Lehrsätze aus der Analysis*, vol. 2, Dover, New York, 1945, p. 73, ex. 190.

II
Filter Design

Editor's Comments on Papers 6 Through 9

6 Campbell: *Physical Theory of the Electric Wave-Filter*

7 Darlington: *Synthesis of Reactance 4-Poles Which Produce Prescribed Insertion Loss Characteristics*

8 Fujisawa: *Realizability Theorem for Mid-Series or Mid-Shunt Low-Pass Ladders Without Mutual Induction*

9 Butterworth: *On the Theory of Filter Amplifiers*

Filter technology had its origins shortly after the turn of this century in the pioneer work of such men as Heaviside and Pupin relating to the design of transmission lines for communications, especially for telephony. Starting with the inductive loading coils, new techniques were invited to improve transmission through the use of networks for balancing, equalizing, and separating one band of frequencies from another. Wave filters were invented in 1915 independently by Campbell (1937) in the United States and Wagner in Germany. Their filters are in the form of iterative ladder structures and lattice sections, and design is accomplished by cascading sections of network to achieve overall specifications. The state of the art in 1922 is summarized by Campbell in Paper 6.

In 1923 Otto Zobel of Bell Telephone Laboratories invented m-derived filter sections in an important contribution to the theory of filters, making filter design routinely accomplished from a catalog of elementary sections. Zobel and others explored the relationship between image attenuation and insertion loss, described low-pass to band-pass frequency transformations, and otherwise completed the theory. It is interesting to note that the Zobel theory became the standard method of filter design in the 1920s, is still used on occasion, and still appears in new undergraduate textbooks.

The next major step forward came from an important discovery by Norton in 1937 on constant impedance filter pairs. Norton suggested that design start from a prescribed insertion loss and that a reactive two-port network be synthesized from this specification. The method by which this is accomplished was given by a number of workers in the field: Darlington (Paper 7), Cauer, Cocci, and Piloty, all in the period 1939–1940. The new method, generally known as the Darlington method, is the modern one in comparison with the image-parameter method, which uses m-derived sections.

Although the Darlington method is generally thought to be superior to the image-parameter method, its use was hampered by the amount of computation required. It was not until the advent of inexpensive computation facilities in the 1950s that these filters came into wide usage, aided especially through the publication of extensive tables such as those of Saal in 1963.

We have just traced the mainstream of the evolution of the theory of design of filters. There have been many other contributions that have significantly improved our understanding of the efficient design of filters. One of the problems is that of avoiding the necessity for transformers in a realization. Necessary and sufficient conditions for ladder realizability have been elusive in general, but they have been solved for important classes of networks; an important one is due to Fujisawa in 1955 (Paper 8), with later contributions from Watanabe (1958) and Meinguet and Belevitch (1958).

An important aspect of the filter design problem is that of approximation: the selection of a rational function to approximate specifications over a range of frequencies. Two functions have become standard in the low-pass filter approximation. Their use is extended to other classes of specifications through frequency transformation techniques. Both of these were introduced in network applications in 1930. Butterworth introduced the maximally flat approximation in his studies of the design of multistage amplifiers (Paper 9). At the same time, Cauer recognized the optimal property of the Chebyshev approximation and used it in the design of image-parameter filters. Details were given by Cauer in his first book, published in 1931 and translated into English in 1958.

References

Campbell, G. A. *The Collected Papers of George Ashley Campbell,* American Telephone & Telegraph Co., New York, 548 pp., 1937.

Cauer, Wilhelm. "Two-Terminal-Pair Networks with Prescribed Attenuation Behavior," *Telegr. Fernsprech-Tech.,* **29**, 185–192, 228–235 (1940).

Cauer, Wilhelm. *Synthesis of Linear Communication Networks,* McGraw-Hill Book Company, New York, 866 pp., 1958, a second volume of Cauer's work was published posthumously by Akademie-Verlag, Berlin, pp. 741–1140, 1960.

Cocci, G. "Rappresentazione di Bipoli Qualsiasi con Quadripoli di pure Reattanze Chiusi su Resistenze," *Alta Frequenza,* **9**, 685–698 (1940).

Meinguet, J., and V. Belevitch. "On the Realizability of Ladder Filters," *IRE Trans. Circuit Theory,* **CT-5**, 253–255 (1958).

Norton, E. L. "Constant Resistance Networks with Applications to Filter Groups," *Bell System Tech. J.,* **16**, 178–193 (1937).

Piloty, H. "Weichenfilter," *Telegr. Fernsprech-Tech.,* **28**, 291–298, 333–344 (1939).

Piloty, H. "Wellenfilter, instbesondere symmetrische and antimetrische, mit vorgeschriebenem betriebsverhalten," *Telegr. Fernsprech-Teh.,* **28**, 363–375 (1939).

Saal, R. "Der Entwurf von Filtern mit Hilfe des Kataloges normierter Tiefpasse," Telefunken GmbH, Backnang/Wurttemberg, Germany, 381 pp., 1963.

Wagner, K. W. "Theorie des Kettenleiters nebst Anwendungen," *Arch. Elektrotech.,* **3**, 315–332 (1915).

Watanabe, Hitoshi. "Synthesis of Band-Pass Ladder Networks," *IRE Trans. Circuit Theory,* **CT-5**, 256–264 (1958).

Zobel, O. J. "Theory and Design of Uniform and Composite Electric Wave Filters, *Bell System Tech. J.,* **2**, 1–46 (1923).

Reprinted with permission from *Bell System Tech. J.*, **1**, 1–32 (Nov. 1922)

Physical Theory of the Electric Wave-Filter

By GEORGE A. CAMPBELL

NOTE: The electric wave-filter, an invention of Dr. Campbell, is one of the most important of present day circuit developments, being indispensable in many branches of electrical communication. It makes possible the separation of a broad band of frequencies into narrow bands in any desired manner, and as will be gathered from the present article, it effects the separation much more sharply than do tuned circuits. As the communication art develops, the need will arise to transmit a growing number of telephone and telegraph messages on a given pair of line wires and a growing number of radio messages through the ether, and the filter will prove increasingly useful in coping with this situation. The filter stands beside the vacuum tube as one of the two devices making carrier telegraphy and telephony practicable, being used in standard carrier equipment to separate the various carrier frequencies. It is a part of every telephone repeater set, cutting out and preventing the amplification of extreme line frequencies for which the line is not accurately balanced by its balancing network. It is being applied to certain types of composited lines for the separation of the d.c. Morse channels from the telephone channel. It is finding many applications to radio of which multiplex radio is an illustration. The filter is also being put to numerous uses in the research laboratory.

The present paper is the first of a series on the electric wave-filter to be contributed to the Technical Journal by various authors. Being an introductory paper the author has chosen to discuss his subject from a physical rather than mathematical point of view, the fundamental characteristics of filters being deduced by purely physical reasoning and the derivation of formulas being left to a mathematical appendix.—*Editor.*

THE purpose of this paper is to present an elementary, physical explanation of the wave-filter as a device for separating sinusoidal electrical currents of different frequencies. The discussion will be general, and will not involve assumptions as to the detailed construction of the wave-filter; but in order to secure a certain numerical concreteness, curves for some simple wave-filters will be included. The formulas employed in calculating these curves are special cases of the general formulas for the wave-filters which are, in conclusion, deduced by the method employed in the physical theory.

All the physical facts which are to be presented in this paper, together with many others, are implicitly contained in the compact formulas of the appendix. Although only comparatively few words of explanation are required to derive these formulas, they will not be presented at the start, since the path of least resistance is to rely implicitly upon formulas for results, and ignore the troublesome question as to the physical explanation of the wave-filter. In order to examine directly the nature of the wave-filter in itself, as a physical structure, we proceed as though these formulas did not exist.

It is intended that the present paper shall serve as an introduction to important papers by others in which such subjects as transients on wave-filters, specialized types of wave-filters, and the practical design of the most efficient types of wave-filters will be discussed.[1]

Definition of Wave-Filter

A wave-filter is a device for separating waves characterized by a difference in frequency. Thus, the wave-filter differentiates between certain states of motion and not between certain kinds of matter, as does the ordinary filter. One form of wave-filter which is well known is the color screen which passes only certain bands of light frequencies; diffraction gratings and Lippmann color photographs also filter light. Wave-filters might be constructed and employed for separating air waves, water waves, or waves in solids. This paper will consider only the filtering of electric waves; the same principles apply in every case, however.

In its usual form the electric wave-filter transmits currents of all frequencies lying within one or more specified ranges, and excludes currents of all other frequencies, but does not absorb the energy of these excluded frequencies. Hence, a combination of two or more wave-filters may be employed where it is desired to separate a broad band of frequencies, so that each of several receiving devices is sup-

[1] I take pleasure in acknowledging my indebtedness to Mr. O. J. Zobel for specific suggestions, and for the light thrown on the whole subject of wave-filters by his introduction of substitutions which change the propagation constant without changing the iterative impedance.

plied with its assigned narrower range of frequencies. Thus, for instance, with three wave-filters the band of frequencies necessary for ordinary telephony might be transmitted to one receiving device, all lower frequencies transmitted to a second device, and all higher frequencies transmitted to a third device—separation being made without serious loss of energy in any one of the three bands.

By means of wave-filters interference between different circuits or channels of communication in telephony and telegraphy, both wire and radio, can be reduced provided they operate at different frequencies. The method is furthermore applicable, at least theoretically, to the reduction of interference between power and communication circuits. The same is true of the simultaneous use of the ether, the earth return, and of expensive pieces of apparatus employed for several power or communication purposes. In all cases the principle involved is the same as that of confining the transmission in each circuit or channel to those frequencies which serve a useful purpose therein and excluding or suppressing the transmission of all other frequencies. In the future, as the utility of electrical applications becomes more widely and completely appreciated, there will be an imperative necessity for more and more completely superposing the varied applications of electricity; it will then be necessary, to avoid interference, to make the utmost use of every method of separating frequencies including balancing, tuning, and the use of wave-filters.

Definition of Artificial Line

The wave-filter problem in this paper is discussed as a phase of the artificial line problem, and it is desirable to start with a somewhat generalized definition of the artificial line. The definition will, however, not include all wave-filters or all artificial lines, since a perfectly general definition is not called for here. Even if an artificial line is to be, under certain wave conditions, an imitation of, or a substitute for, an actual line connecting distant points, hardly any limitation is thereby imposed upon the structure of the device; an actual line need not be uniform but may vary abruptly or gradually along its length and may include two, three, four or more transmission conductors of which one may be the earth. Having indicated that wave-filters partake of somewhat this same generality of structure, the present paper is restricted to wave-filters coming under the somewhat generalized artificial line specified by the following definition:

An artificial line is a chain of networks connected together in sequence through two pairs of terminals, the networks being identical but other-

wise unrestricted. This generalized artificial line possesses the well-known sectional artificial line structure but it need not be an imitation of, or a substitute for, any known, real, transmission line connecting together distant points. The general artificial line is shown by Fig. 1 where *N*, *N*,. . . .are the identical unrestricted networks which may contain resistance, self-inductance, mutual inductance, and capacity.

In discussing this type of structure as a wave-filter, the point of view of an artificial line is adopted for the reason that it is advantageous to regard the distribution of alternating currents as being dependent upon both propagation and terminal conditions, which are to be separately considered. In this way the attenuation, or

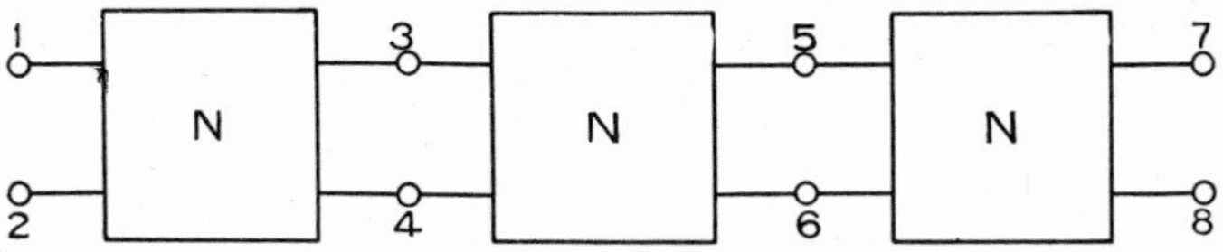

Fig. 1—Generalized Artificial Line as Considered in the Present Paper, where *N*, *N*, . . . are Identical Arbitrary Electrical Networks

falling off, of the current from section to section may be most directly studied. Terminal effects are not to be ignored, but are allowed for, after the desired attenuation effects have been secured, possibly by an increase in the number of sections to be employed.

The fundamental property of this generalized artificial line, which includes uniform lines as a special case, is the mode in which the wave motion changes from one section to the next, and may be stated as follows:

Wave Propagation Theorem

Upon an infinite artificial line a steady forced sinusoidal disturbance falls off exponentially from one section to the next, while the phase changes by a constant amount. Reversing the direction of propagation does not alter either the attenuation or phase change. When complex quantities are employed the exponential includes the phase change.[2] This theorem is proved, without mathematical equations, by observing

[2] This theorem is not new, but it is ordinarily derived by means of differential or difference equations whereas it may be derived from the most elementary general considerations, thus avoiding all necessity of using differential or difference equations, as illustrated in my paper "On Loaded Lines in Telephonic Transmission" (*Phil. Mag.*, vol. 5, pp. 313–331, 1903). In that discussion, as well as in this present one, it is tacitly assumed that the line is either an actual line with resistance, or the limit of such a line as the resistance vanishes, so that the amplitude of the wave never increases towards the far end of an infinite line.

that the percentage reduction in amplitude and the change in phase, in passing from the end of one section to the corresponding point of the next section, do not depend upon either the absolute amplitude or phase; they depend, instead, only upon the magnitudes, angles and interconnections of the impedances between the two points and of the impedances beyond the second point. These impedances are, since the line is assumed to be periodic and infinite, identically the same for corresponding points between all sections of the line, and, therefore, the relative changes in the wave will be identical at corresponding points in all sections. This proves the exponential falling off of the disturbance and the constancy of phase change; the ordinary reciprocal property shows that the wave will fall off identically whichever be the direction of propagation. By the superposition property it follows that the steady state on any finite portion of a periodic recurrent structure must be the sum of two equally attenuated disturbances, one propagated in each direction.

The fundamental wave propagation theorem may be generalized for any periodic recurrent structure irrespective of the number and kind of connections between periodic sections, provided the disturbance is such as to remain similar to itself at corresponding points of each of these connections.

Equivalent Generalized Artificial Line

Since, at a given frequency, any network employed solely to connect a pair of input terminals with a pair of output terminals may be replaced by either three star-connected impedances or three delta-connected impedances, the general artificial line of Fig. 1 may be

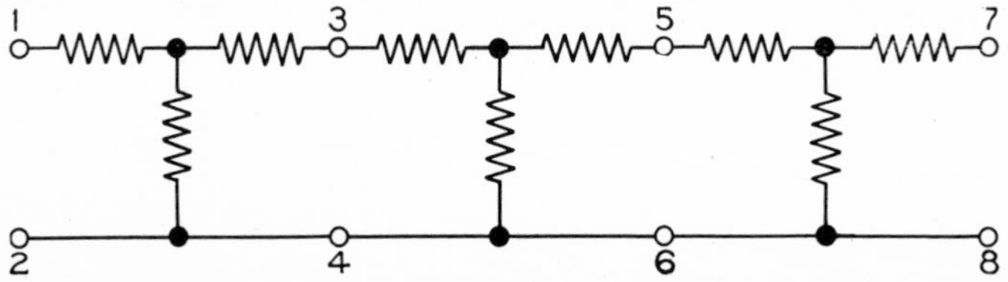

Fig. 2—Equivalent Artificial Line Obtained by Substituting Star Impedances

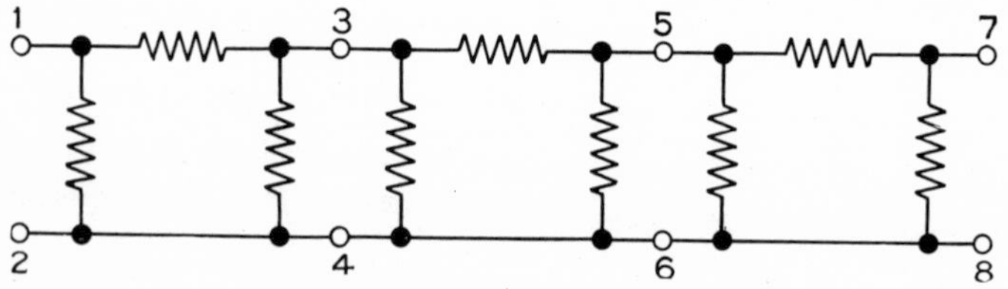

Fig. 3—Equivalent Artificial Line Obtained by Substituting Delta Impedances

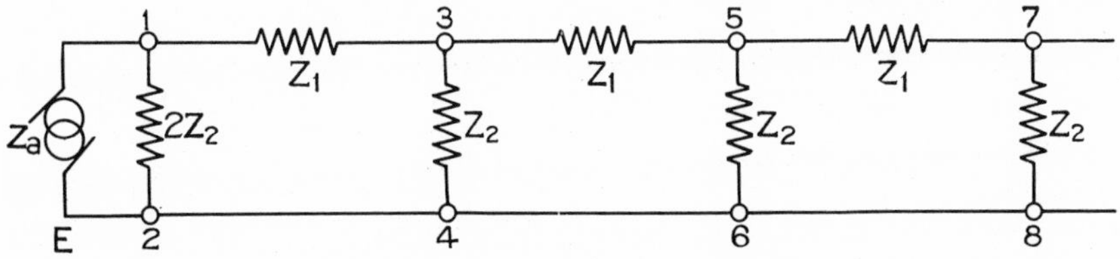

Fig. 4—Equivalent Ladder Artificial Line

replaced by the equivalent artificial line of either Fig. 2 or Fig. 3. By combining the series impedances in Fig. 2 and the parallel impedances in Fig. 3, the equivalent line in Fig. 4 is obtainable. The two ways of arriving at Fig. 4 give different values for the series and shunt impedances Z_1, Z_2, and different terminations for the line, but the propagation of the wave is the same in both cases, since the assumed substitutions are rigorously exact. While Fig. 4 may be considered as the generalized artificial line equivalent to Fig. 1, this requires including in Z_1 and Z_2 impedances which cannot always be physically realized by means of two entirely independent networks, one of which gives Z_1 and the other Z_2. This restriction is of no importance when we are discussing the behavior of the generalized artificial line at a single frequency; accordingly, the ladder artificial line is suitable for this part of the discussion. When we come to the more specific correlation of the behavior of the generalized artificial line at different frequencies, it will be found more convenient to replace the ladder artificial line by the lattice artificial line, which avoids the necessity of considering any impedances which are not individually physically realizable.

The equivalence between Figs. 1 and 4 is implicitly based upon the assumption that it is immaterial, for artificial line uses, what absolute potentials the terminals 1, 2; 3, 4; 5, 6; etc. have—this leaves us at liberty to connect 2, 4, 6, etc., together, so long as we maintain unchanged the differences in potential between 1 and 2, 3 and 4, etc. Instead of connecting 2 and 4 we might equally well connect 2 and 3, and then Z_1 would connect 1 and 4 as in Fig. 5; with these

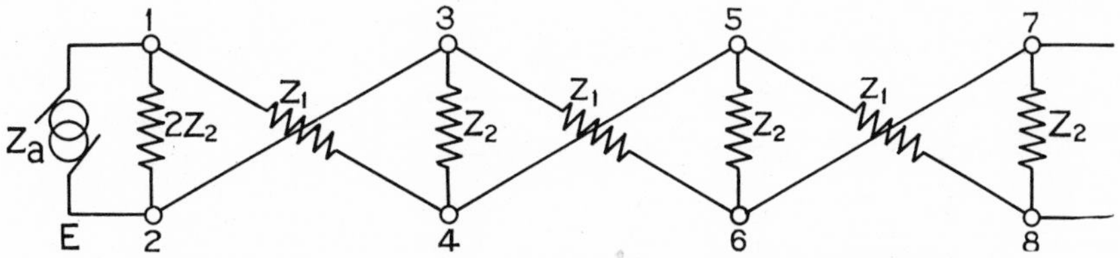

Fig. 5—Equivalent Artificial Line with Crossed Impedances

cross-connections the propagation still remains unchanged. We have again obtained Fig. 4 with no circuit difference except the interchange of terminals 3 and 7 with terminals 4 and 8; or, if this is ignored, a reversal in the sign of the current at alternate pairs of terminals. This shows that the reversal of the current in alternate sections of Fig. 4 may not be of primary significance, since networks which are essentially equivalent have reversed currents.

In order to deal, at the start, with only the simpler terminal conditions, we may consider the line to begin with only one-half of the series impedance Z_1, or only one-half of the bridged admittance $1/Z_2$. These mid-points are called the mid-series and mid-shunt points; knowing the results of termination at either of these points, the effect of termination at any other point may be readily determined. For Fig. 4 termination at mid-shunt has been chosen so that each section of the line adds a complete symmetrical mesh to the network.

An alternator, introducing an impedance Z_a, is shown as the source of the steady-state sinusoidal current in Fig. 4. Assume that the impedance Z_a is variable at pleasure, and that it is gradually adjusted to make the total impedance in the generator circuit vanish,—in this case no e.m.f. will be required to maintain the forced steady-state which becomes a free oscillation. If, in addition, it is assumed that the line has an infinite number of sections, this required value of Z_a will be the negative of the mid-shunt iterative impedance[3] of the artificial line, which will be designated as K_2. The first shunt on the line now includes $-K_2$ in parallel with $2Z_2$ so that its total impedance is, say, $Z' = -2Z_2K_2/(2Z_2-K_2)$. The infinite line with its first shunt given the special value Z' is thus capable of free oscillation.

It is possible to simplify this infinite oscillating circuit by cutting off any part of it which has the same free period as the whole circuit. The entire infinite line beyond the second shunt 3, 4 certainly has this same free period, provided its first shunt also has the impedance Z'. Conceive the shunt Z_2 at 3, 4 as replaced by the four impedances $2Z_2$, $2Z_2$, $+K_2$ and $-K_2$ all in parallel; the first and last, which together make the Z' required by the infinite line, leave $2Z_2$ and

[3] The "iterative impedance" of an artificial line is the impedance which repeats itself when one or more sections of the artificial line are inserted between this impedance and the point of measurement. It is thus the impedance of an infinite length of any actual artificial line, regardless of the termination of the remote end of the line. In general, its value is different for the two directions of propagation, but not when the line is symmetrical, as at mid-series and mid-shunt. The values at these points are denoted by K_1 and K_2. "Iterative impedance" is employed because it is a convenient term which is distinctive and describes the most essential property of this impedance; it seems to be more appropriate than "characteristic impedance," "surge impedance" and the other synonyms in use.

$+K_2$ in parallel, which have the impedance $Z''=+2Z_2K_2/(2Z_2+K_2)$. Removing Z' together with the infinite line on the right there remains on the left a closed circuit made up of the three impedances Z_1, Z' and Z'' in series.

After the division, the infinite line on the right will continue, without modification, to oscillate freely, since it is an exact duplicate of the original oscillating line, and so must maintain the free oscillation already started. Since it oscillates freely by itself, it had originally no reaction upon the simple circuit from which it was separated; this simple circuit on the left must thus also continue its own free oscillations without change in period or phase.

We might continue and subdivide the entire infinite line into identical simple circuits but it is sufficient to consider this one detached circuit, which is shown separately in two ways by Fig. 6, since from

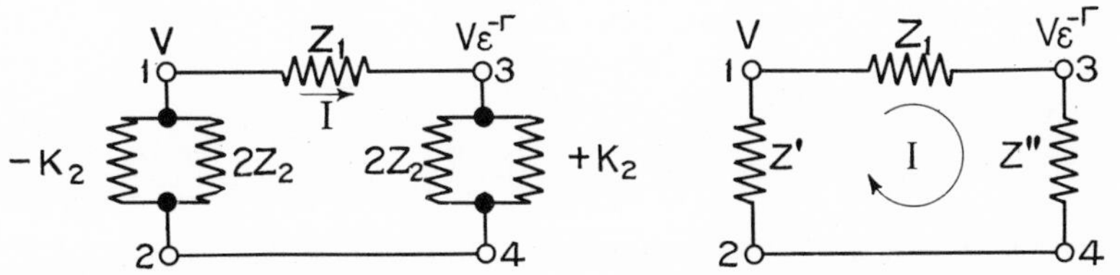

Fig. 6—Equivalent Section of Fig. 4 Terminated for Free Oscillation

its free oscillations the mathematical formulas for the steady-state propagation in the artificial line may be derived. This is deferred, however, until after the physical discussion is completed, so as to leave no room for doubt that the essentials of the physical theory are really deduced without the aid of mathematical formulas.

The generalized artificial line, if made up entirely of pure resistances, will attenuate all frequencies alike, and the entire wave will be in the same phase; this remains true, whatever be the impedance of the individual branch of the network, provided the ratio of the impedances of all branches is a constant independent of the frequency. This is precisely the condition to be avoided in a wave-filter; branches must not be similar but dissimilar as regards the variation of impedance with frequency. This calls for inductance and capacity with negligible resistance, so that there is an opportunity for the positive reactance of one branch to react upon the negative reactance of another branch, in different proportions at different frequencies. Assuming the unit network N of Fig. 1 to be made up of a finite number of pure reactances, the equivalent impedances Z_1 and Z_2 of Figs. 4 and 5 must also be pure reactances. Under this assumption let us consider the free oscillations of Fig. 6; first, with K_2 assumed to be a pure reactance; second, with K_2 assumed to be a pure resistance; and third, in order to show that this third assumption is contrary to fact, with K_2 assumed to be an impedance with both resistance and reactance.

With K_2 a reactance, the circuit contains nothing but reactances, and free oscillations are possible if, and only if, the total impedance of the circuit is zero. The end impedances Z' and Z'' being different, the potentials at the ends of the mesh will be different, and this means that the corresponding wave on the infinite line will be attenuated, since the ratio between these potentials is the rate at which the amplitudes fall off per section.

With K_2 a pure resistance, a free oscillation is possible only if the dissipation in the positive resistance at the right end of the circuit is exactly made up by the hypothetical source of energy existing in the negative resistance $-K_2$ at the left end of the circuit. An exact balance between the energy supplied at one end and that lost at the other end is possible, since the equal positive and negative resistances K_2, $-K_2$ carry equal currents. This continuous transfer of energy from the left of the oscillating circuit of Fig. 6 to the right end is the action which goes on in every section of the infinite artificial line, and serves to pass forward the energy along the infinite line.

If K_2 were complex, $-K_2$ on the left of Fig. 6 and $+K_2$ on the right would not carry the same fraction of the circulating current I, since they are each shunted by a reactance $2Z_2$ which would allow less of the current to flow through $+K_2$ than through $-K_2$, if $2Z_2$ makes the smaller angle with $+K_2$, and vice versa. No balance between absorbed and dissipated energy is possible under these conditions when the equal and opposite resistance components carry unequal currents. A complex K_2, therefore, gives no free oscillation, and cannot occur with a resistanceless artificial line.

It is perhaps more instructive to consider the transmission on the line as a whole, rather than to confine attention exclusively to the oscillations of the simple circuit of Fig. 6 and so, at this point, without following further the conclusions to be drawn directly from this oscillating circuit, the fundamental energy theorem of resistanceless artificial lines will be stated, and then proved as a property of an infinite artificial line.

Energy Flow Theorem

Upon an infinite line of periodic recurrent structure, and devoid of resistance, a sinusoidal e.m.f. produces one of two steady states, viz.:

1. *A to-and-fro surging of energy without any resultant transfer of energy; currents and potential differences each attenuated from section to section, but everywhere in the same or opposite phase and mutually in quadrature, or,*

2. *A continuous, non-attenuated flow of energy along the line to infinity with no energy surging between symmetrical sections; current and potential non-attenuated, but retarded or advanced in phase from section to section, and mutually in phase at mid-shunt and mid-series points.*

The critical frequencies separating the two states of motion are the totality of the resonant frequencies of the series impedance, the anti-resonant frequencies of the shunt impedance, and the resonant frequencies of a single mid-shunt section of the line.

To prove the several statements of this theorem let us consider first the consequences of assuming that the wave motion, in progressing along the line, is attenuated, and next the consequences of assuming that the wave motion changes its phase. If the wave is attenuated, however little, at a sufficient distance it becomes negligible, and the more remote portions of the line may be completely removed without appreciable effect upon the disturbance in the nearer portion of the line. That part of the line which then remains is a finite network of pure reactances, and in any such network all currents are always in the same, or opposite, phase; so, also, are the potential differences; moreover, the two are mutually in quadrature; there is no continuous accumulation of energy anywhere, but only an exchange of energy back and forth between the inductances, the capacities and the generator. Continuously varying the amount of the assumed attenuation will cause a continuous variation in the corresponding frequency. The motion of the assumed character may, therefore, be expected to occur throughout continuous ranges or bands of frequencies and not merely at isolated frequencies.

The question may be asked—How far does the energy surge? Is the surge localized in the individual section, or does the surge carry the energy back and forth over more than one section, or even in and out of the line as a whole? To answer this question, it would be necessary, as we will now proceed to prove, to know something about the actual construction of the individual section. If each section is actually made up as shown in Fig. 6, and this is entirely possible in the present case (since only positive and negative reactances would be called for), then the section is capable of free oscillation, as explained above, and the surging is localized within the section; twice during each cycle the amount of energy increases on the right and decreases on the left. But we do not know that the section is made up like Fig. 6; we only know that it is equivalent to Fig. 6 as regards input and output relations. As far as these external relations go, the actual network may be made exclusively of either inductances or capacities with the connections shown in Fig. 4 or with the cross-connections of Fig. 5, according as the current is to have the same or opposite signs in consecutive sections. In any network made up exclusively of inductances or of capacities, the total energy falls to zero when the current or the potential falls to zero, respectively. Twice, therefore, in every cycle the total energy surges into this line and then it all returns to the generator. With other networks, surgings intermediate between these two extremes will occur. The theorem, therefore, does not limit the extent of the surging.

Under the second assumption, the phase difference between the currents at two given points, separated by a periodic interval, is to be an angle which is neither zero nor a multiple of $\pm\pi$. The assumed difference in phase can only be due to the infinite extension of the artificial line since, as previously noted, no finite sequence of inductances and capacities can produce any difference in phase. That infinite lines do produce phase differences is well-known; in particular, an infinite, uniform, perfectly conducting, metallic pair shows a continuous retardation in phase. If the infinitely remote sections of the artificial line are to have this controlling effect on the wave motion, the wave motion must actually extend to infinity, that is, there can be no attenuation. The wave progressing indefinitely to infinity without attenuation must be supplied continuously with energy; this energy must flow along the entire line with neither loss nor gain in the reactances it encounters on the way. This continuous flow of energy can take place only provided the currents and potentials are not in quadrature; they may be in phase. In considering the free oscillations of Fig. 6 it was shown that K_2 is real if it is not pure reactance. That is, for the mid-shunt section the current and potential are in phase. It is easy to show that they are also in phase at the mid-series point which is also a point of symmetry.

This flow-of-energy state of motion thus necessarily characterizes a phase-retarded wave on a resistanceless artificial line, regardless of the amount of the assumed positive or negative retardation, which may be taken to have any value between zero and exact opposition of phase. Continuously varying the retardation throughout the 180 degrees will, in general, call for a continuous change in the frequency of the wave motion. The second state of motion occurs, therefore, throughout continuous ranges or bands of frequencies.

No other state of motion is possible. With given initial amplitude and phase any possible wave motion is completely defined by its attenuation and phase change. All possible combinations of these two elements have been included in the two states, since the excluded conditions on each assumption have been included as a consequence of the other assumption. Thus, the exclusion of no attenuation in the first assumption was found necessarily to accompany the phase change of the second assumption; currents in phase or opposed, which were excluded from the second assumption, were found to be necessary features accompanying the first assumption. There remains only to consider the critical frequencies separating the two states of motion. At these frequencies there can be no attenuation and lag angles of multiples of $\pm\pi$, including zero, only. At symmetrical points the iterative impedance of the line must be a pure reactance to satisfy the first state of motion, and a pure resistance to satisfy the second state of motion. The only iterative impedances which satisfy these conditions are zero and infinity.

Some details relating to the pass and stop bands and the critical frequencies are brought together in the following table, where "stop ($\pm$)" refers to stop bands, the current being in phase or opposed in successive sections, and where γ and k refer to the line obtained by uniformly distributing $1/Z_2$ with respect to Z_1.

TABLE I.

For Ladder Artificial Line, Fig. 4

Band	Critical Frequency	Ratio $\frac{Z_1}{4Z_2}$	Uniform Line		Artificial Line			
			γ	k	Γ	$e^{-\Gamma}$	K_1	K_2
Stop (+)		>0	+real	imag.	+real	$0<<1$	imag.	imag.
	$Z_1 = 0$ $Z_2 = \infty$	0 0	0 0	0 ∞	0 0	1 1	0 ∞	0 ∞
Pass		$)>>-1$	imag.	+real	imag.	$e^{i\theta}$	+real	+real
	$Z_1 + 4Z_2 = ($	-1	$i2$	$i2Z_2$	$i\pi$	-1	0	∞
Stop (−)		<-1	imag.	+real	$i\pi$ + rea	$-1<<($	imag.	imag.
	$Z_1 = \infty$ $Z_2 = 0$	$-\infty$ $-\infty$	∞ ∞	∞ 0	∞ ∞	0 0	∞ $\frac{1}{2}Z_1$	$2Z_2$ 0

It is not necessary to check the table item by item, many of which have already been proven, but it will be instructive to check some of the items by assuming that $Z_1/4Z_2$, called the ratio for brevity, is positive to begin with, and that a continuous increase in frequency reduces the ratio to zero and back through $\mp\infty$ to its original positive value. This cycle starts with a stop (+) band since the artificial line is in effect a network of reactances, all of which have the same sign; there is attenuation and the iterative impedances are imaginary. When the ratio decreases to zero, there must be either resonance which makes $Z_1 = 0$, or anti-resonance which makes $Z_2 = \infty$; in either case the artificial line has degenerated into a much simpler circuit; it is a shunt made up of all Z_2's combined in parallel, or a simple series circuit made up of all Z_1's, respectively; the iterative impedances are 0 and ∞, respectively; there is no attenuation in either case.

With a somewhat further increase of the frequency the ratio will assume a small negative value with the result that the artificial line will have both kinetic and potential energy. An analogy now exists between the artificial line and an ordinary uniform transmission line, which possesses both kinetic and potential energy, and is ordinarily visualized as being equivalent to many small positive reactances, in series, bridged, to the return conductor, by large negative reactances. The fact that uniform lines do freely transmit waves is a well-known physical principle, and it is not necessary to repeat here the physical theory of such transmission merely to show that the same phenomenon occurs with the identical structure when it is called an artificial line or wave-filter.

In order to determine just how far the ratio may depart from zero, on the negative side, without losing the property of free transmission, we look for any change in the action of the individual section of the artificial line which is fundamental; nothing less than a fundamental change in the behavior of the individual section can produce such a radical change in the line as an abrupt transition from the free transmission of a pass band to the to-and-fro surging of energy in a stop band. Now as the ratio is made more and more negative by the assumed increase of frequency, the value -1 is reached, at which frequency the symmetrical section (Fig. 6) of the artificial line is capable of free oscillation by itself. This is well recognized as a most fundamental change in the properties of any network, and it affords grounds for expecting a complete change in the character of the propagation over the artificial line. The change must be to a stop band with currents in opposite phase, since at resonance the potentials at the two ends of a section are in opposite phase.

Further increase in the frequency cannot make any change in the absolute difference in phase between the two ends of the other section, since opposition is the greatest possible difference in phase; the wave now adapts itself to increasing frequency by altering its attenuation.

Upon continuing the increase of frequency, so as to reduce the ratio to $-\infty$, we arrive at either anti-resonance corresponding to $Z_1 = \infty$ or resonance corresponding to $Z_2 = 0$; the artificial line has now degenerated into a row of isolated impedances Z_2, or into a series of impedances Z_1 short-circuited to the return wire; in either case the attenuation is infinite since no wave is transmitted. Passing beyond this critical frequency the ratio becomes positive, according to our assumption, and we are again in a stop (+) band.

While in this rapid survey of what happens during this frequency cycle little has been actually proven, it should have been made physically clear why abrupt changes in the character of the transmission occur at the frequencies making the ratio equal to 0, -1 or ∞, since the line degenerates into a simpler structure, or the phase change reaches its absolute maximum, on account of resonance, at these particular frequencies.

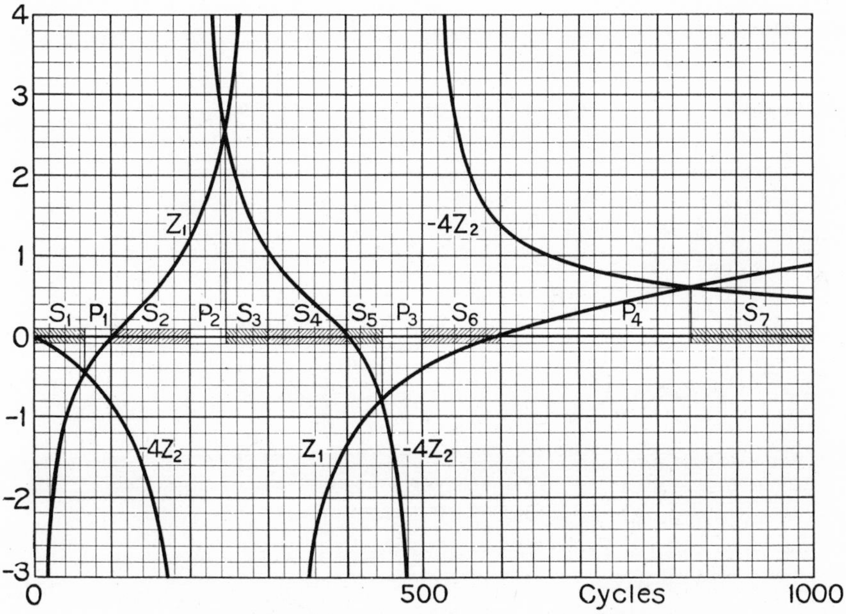

Fig. 7—Graph for Locating the Pass Bands and Stop Bands of Fig. 4
$Z_1 = -i(1^2 - x^2)(6^2 - x^2)/8x(3^2 - x^2)$,
$-4Z_2 = -i4x(4^2 - x^2)/(2^2 - x^2)(5^2 - x^2)$ and x = cycles/100

Information as to the location of the bands is often obtained most readily by plotting both Z_1 and $-4Z_2$, as illustrated in Fig. 7, and determining the critical frequencies by noting where the curves cross each other and the abscissa axis, as well as where they become infinite. Any particular band is then a pass band, a stop $(+)$ band or a stop $(-)$ band, according as Z_1, the abscissa axis, or $-4Z_2$ lies between the other two of the three lines. In Fig. 7 the pass bands are P_1, P_2, P_3, P_4; the stop $(+)$ bands are S_2, S_4, S_6; and the stop $(-)$ bands are S_1, S_3, S_5, S_7, and they illustrate quite a variety of sequences. By altering the curves the bands may be shifted, may be made to coalesce, or may be made to vanish.

Wave-Filter Curves

The pass band and stop band characteristics of wave-filters are concretely illustrated for a few typical cases by the curves of Figs. 8–13, which show the attenuation constant A, the phase constant B, and both the resistance R and reactance X components of the iterative impedance for a range of frequencies which include all of the critical frequencies, except infinity. The heavy curves apply to the ideal resistanceless case, while the dotted curves assume a power factor equal to $1/(20\pi)$ for each inductance which is a value readily obtained in practice. This value is, however, not sufficiently large to make these small scale curves entirely clear, since considerable portions of the dotted curves appear to be coincident with the heavy line curves; but this, as far as it goes, proves the value of the present discussion which rests upon a close approximation of actual wave-filters to the ideal resistanceless case.

The low pass resistanceless wave-filter, as shown by Fig. 8, presents no attenuation below 1,000 cycles; above this frequency the attenuation constant increases rapidly, in fact, the full line attenuation curve increases at the start with maximum rapidity, since it is there at right angles to the axis. The dotted attenuation curve, which includes the effective resistance in the inductance coils, follows the ideal attenuation curve closely, except in the neighborhood of 1,000 cycles, where resistance rounds off the abrupt corner which is present in the ideal A curve. The phase constant B is, at the start, proportional to the frequency, as for an ordinary uniform transmission line; its slope becomes steeper as the critical frequency 1,000 is approached where the curve reaches the ordinate π, at which value it remains constant for all higher frequencies. As shown by the dotted B curve, resistance rounds off the corner at the critical frequency, but

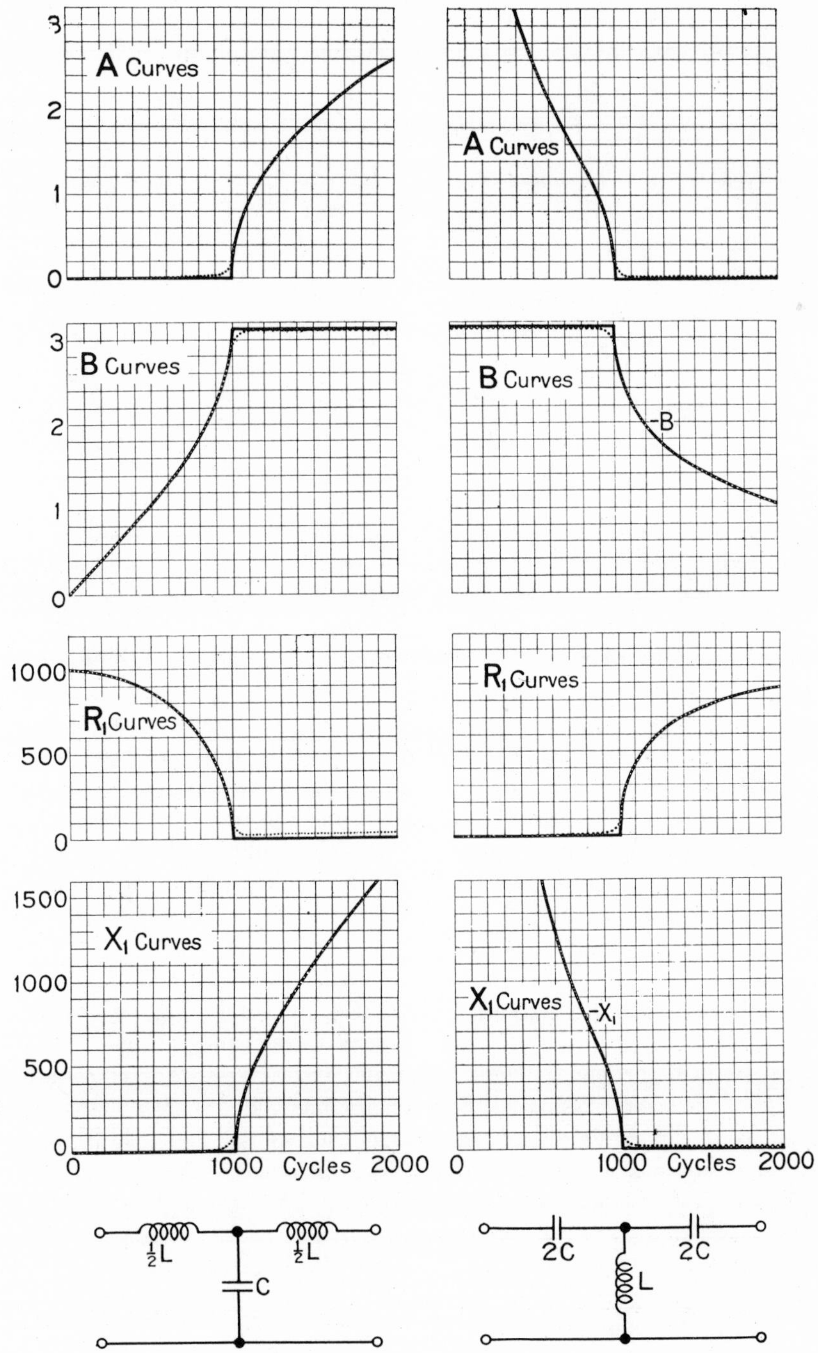

Fig. 8—Low Pass Wave-Filter: $L = 1/\pi$ Henry, $C = 1/\pi$ Microfarad
Fig. 9—Complementary High Pass Wave-Filter: $L = 1/4\pi$, $C = 1/4\pi$

otherwise leaves the curve approximately unchanged. The full line curves for R_1 and X_1 show that in the ideal case the iterative impedance is pure resistance and pure reactance in the pass and stop bands respectively, and that resistance smooths the abrupt transition at the critical frequency.

The high pass wave-filter shown by Fig. 9 passes the band which is stopped by the low pass wave-filter of Fig. 8, and vice versa. For this reason the two wave-filters are said to be complementary.

Another set of two complementary wave-filters is shown by Figs. 10 and 11, one of which passes only a single band of frequencies, not extending to either zero or infinity, while the other passes the remaining frequencies only. The single pass band of Fig. 10, embracing a total phase change 2π on the B curve, is actually a case of confluent pass bands, each of which embraces the normal angle π. The tendency of the two simple pass bands to separate, and leave a stop band between them, is shown by the hump in the dotted attenuation constant curve at 1,000 cycles. If, instead of the two simple bands having been brought together, one of them had been relegated to zero or infinity, the single remaining pass band would have exhibited the normal angular range π in the B curve, and there would have been no hump in the dotted A curve. The stop band of Fig. 11 also illustrates peculiarities which are not necessary features of a wave-filter with a single stop band in this position. This wave-filter is obtained from Fig. 7 by making all bands vanish except P_2, S_3, S_5 and P_3,—by extending P_2 to zero, P_3 to infinity, and making S_3 and S_5 coalesce, so that the attenuation becomes infinite in the stop band without passing from a stop $(-)$ to a stop $(+)$ band. The coalescing stop bands are responsible for the rapid changes in the B, R_1, and X_1 curves of Fig. 11 which would not have appeared if, in Fig. 7, the same pass band had been obtained by retaining P_1, S_2 and P_2 and making all other bands vanish.

An extreme case of complementary wave-filters is shown by Figs. 12 and 13, where no frequencies and all frequencies are passed respectively. The first result is obtained by combining inductances alone, which, as has been pointed out above, can give only an attenuated disturbance devoid of wave characteristics. The wave-filter shown for passing all frequencies has inductance coils in the line, and capacities diagonally bridged across the line. This wave-filter combines a constant iterative impedance with a progressive change in phase which is sometimes useful.[4] An outstanding char-

[4] A theoretical use of the phase shifting afforded by the lattice artificial line was made at page 253 of "Maximum Output Networks for Telephone Substation and Repeater Circuits," Trans. A. I. E. E., vol. 39, pp. 231–280, 1920.

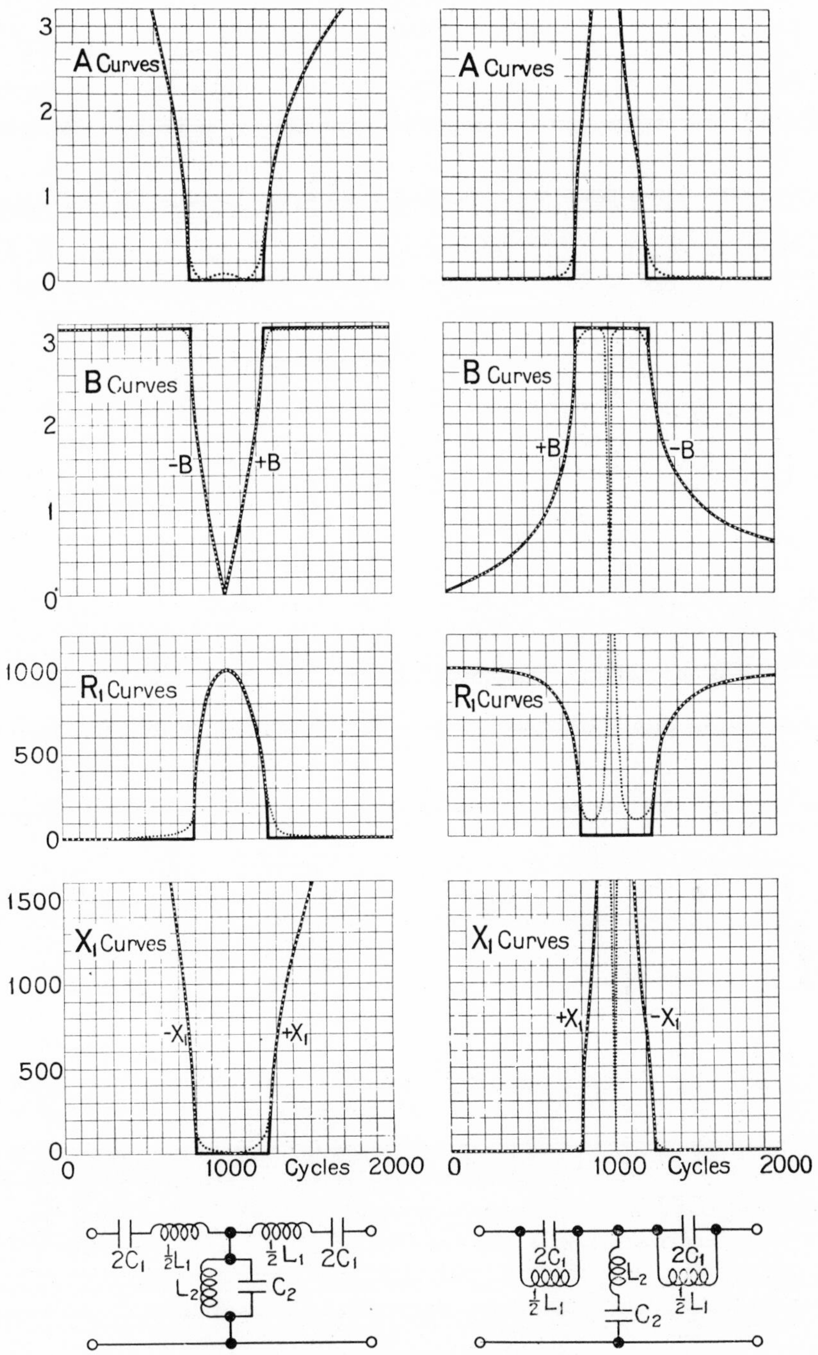

Fig. 10—Single Band Pass Wave-Filter: $L_1 = 20/9\pi$, $L_2 = 9/80\pi$, $C_1 = 9/80\pi$, $C_2 = 20/9\pi$

Fig. 11—Complementary High and Low Pass Wave-Filter: $L_1 = 9/20\pi$, $L_2 = 5/9\pi$, $C_1 = 5/9\pi$, $C_2 = 9/20\pi$

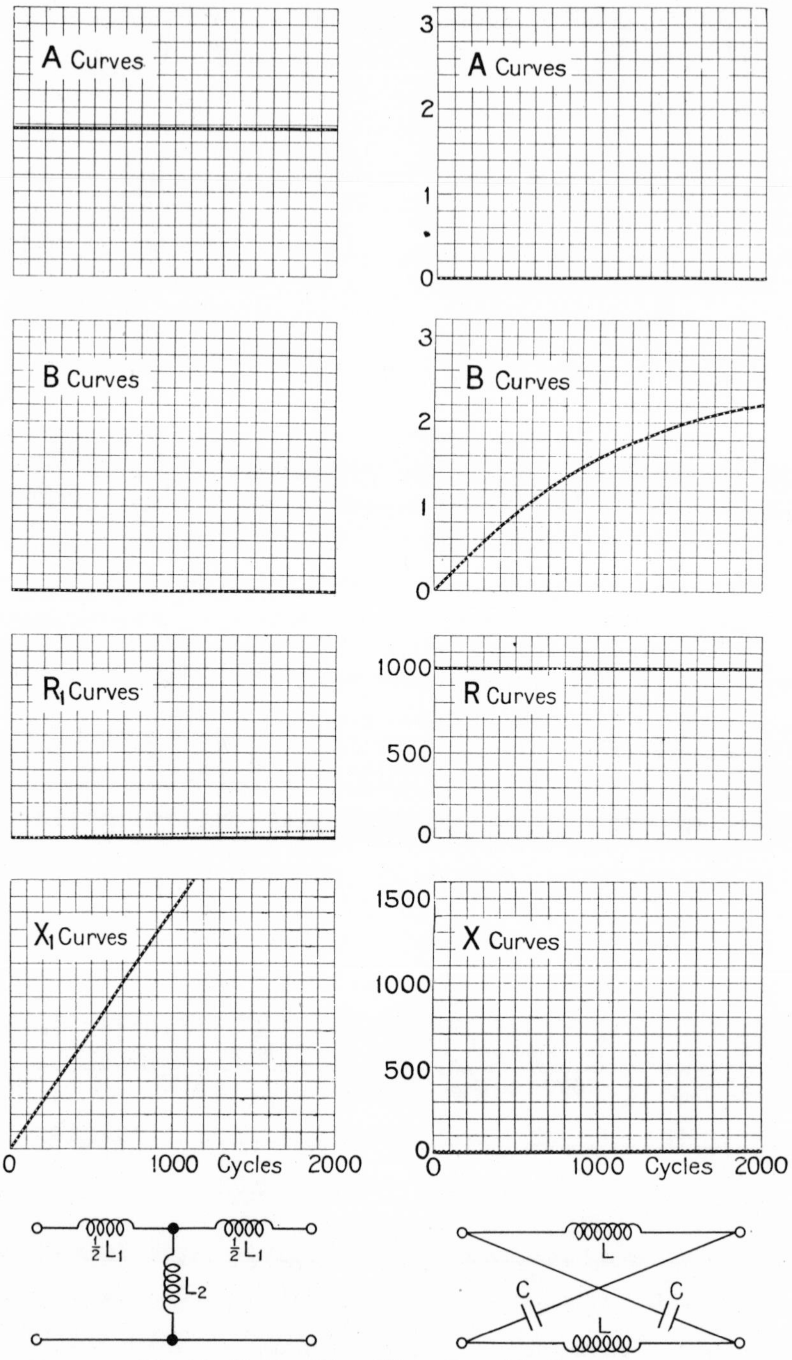

Fig. 12—No Pass Wave-Filter: $L_1 = 1/\pi$, $L_2 = 1/4\pi$
Fig. 13—Complementary All Pass Wave-Filter: $L = 1/2\pi$, $C = 1/2\pi$

acteristic of this type of artificial line is that it has, for all frequencies, the same iterative impedance as a uniform line with the same total series and shunt impedances. This artificial line will be considered in more detail in the next section of this paper.

Lattice Artificial Lines

Up to this point we have considered the properties of artificial line networks which were supposed to be given. In practice the problem is ordinarily reversed, and we ask the questions: May the locations of the bands be arbitrarily assigned? May additional conditions be imposed? How may the corresponding network be determined, and what is its attenuation in terms of the assigned critical frequencies? These questions might be answered by a study of Fig. 7, in

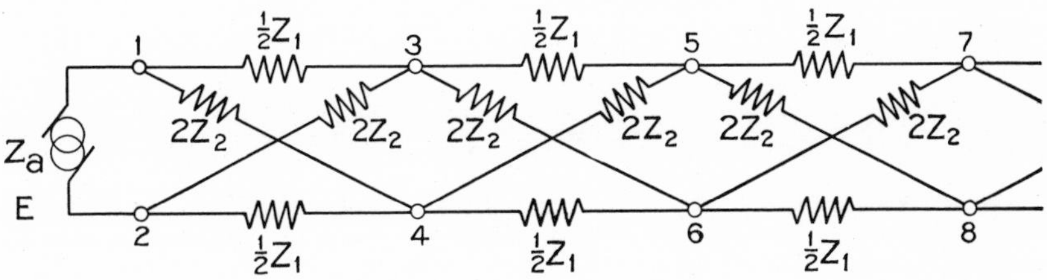

Fig. 14—Lattice Artificial Line

all its generality, but it seems simpler to base the discussion upon the artificial line shown in Fig. 14, which is to be a generalization of Fig. 13 to the extent of making the two impedances Z_1 and Z_2 any possible actual driving-point impedances. It is sometimes illuminating to regard this artificial line as a nest of bridges, one within another, as shown by Fig. 15.

On interchanging terminals 3 with 4 and 7 with 8 in Fig. 14 the network of lines remains unchanged; thus, Z_1 and $4Z_2$ may be interchanged in the formulas for the artificial line with no change in the result, except, possibly, one corresponding to a reversal of the current at alternate junction points. Another elementary feature of this artificial line is that it degenerates into a simple shunt or a simple series circuit at the resonant or anti-resonant frequencies, respectively, of either Z_1 or Z_2, and these are the critical frequencies, terminating the pass bands. At other frequencies, a positive ratio $Z_1/4Z_2$ must give a stop band, since the reactances are all of one sign. If a small negative value of this ratio gives free transmission, as we naturally expect, there will be identical transmission, except for a reversal of

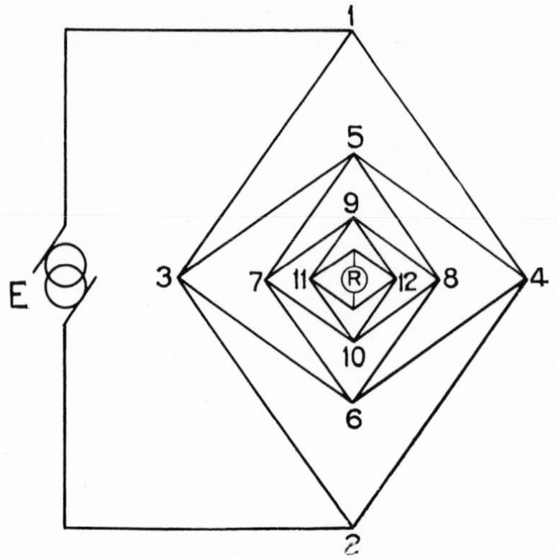

Fig. 15—Lattice Artificial Line Drawn to Show the Chain of Bridge Circuits

sign, when the ratio has the reciprocal value, which will be a large negative quantity, since we may always interchange Z_1 and $4Z_2$. The consequences of this and of other elementary properties of this artificial line are brought together in the following table:

TABLE II

For Lattice Artificial Line, Fig. 14

Band	Critical Frequency	Ratio $\frac{Z_1}{4Z_2}$	Uniform Line		Artificial Line		
			γ	k	Γ	$e^{-\Gamma}$	K
Stop (+)		$1>>0$	$2>>0$	imag.	+ real	$0<<1$	imag.
	$Z_1 = 0$	0	0	0	0	1	0
	$Z_2 = \infty$	0	0	∞	0	1	∞
Pass		<0	imag.	+ real	imag.	$e^{i\theta}$	+ real
	$Z_1 = \infty$	∞	∞	∞	$i\pi$	-1	∞
	$Z_2 = 0$	∞	∞	0	$i\pi$	-1	0
Stop (−)		<1	<2	imag.	$i\pi$ + real	$-1<<0$	imag.
	$Z_1 = 4Z_2$	1	2	$2Z_2$	∞	0	$2Z_2$

The cycle of bands: stop (+), pass, stop (−), adopted for the table, carries the attenuation factor $e^{-\Gamma}$ around the periphery of a unit semi-circle; in the stop (+) band it traverses the radius from 0 to 1, in the pass band it travels along the unit circle through 180 degrees to the value -1, completing the cycle from -1 to 0 in the stop ($-$) band. In this cycle there are four points of special interest, corresponding to ratio values 1, 0, -1 and ∞, for which the wave is infinitely attenuated, unattenuated with an angular change of 0, of 90, and of 180 degrees, respectively. It is at the 90 degree angle that resonance of the individual section occurs; the iterative impedance is then equal to $2|Z_2|$.

Graph of the Ratio $Z_1/4Z_2$ for Fig. 14

If we plot Z_1 and $4Z_2$ the pass bands are shown by the points where the curves become zero or infinite, and the intersections of the two

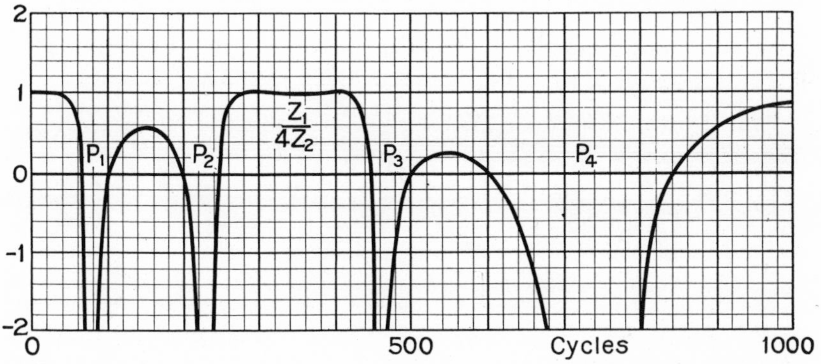

Fig. 16—Graph for Locating the Pass and Stop Bands of the Lattice Artificial Line, where $Z_1/4Z_2 = \left[(x_1^2 - x^2)\,(x_2^2 - x^2)^{-2}\,(x_3^2 - x^2)\right] \ldots$, x = cycles/100, and the resonant roots x_1, x_3, . . . are 0.650, 1, 2, 2.452, 4.442, 5, 6, 8.476 and the double anti-resonant roots x_2, x_5, . . . are 0.766, 2.301, 4.585, 7.423

curves show the frequencies at which the attenuation becomes infinite. These intersections must be at an acute angle since each branch of the two curves has a positive slope throughout its entire length; for this reason it may be desirable to plot the ratio rather than the individual curves; this is especially desirable in cases where the two curves do not intersect, but are tangent. Fig. 16 is for a lattice network equivalent to two sections of the ladder type illustrated by Fig. 7, and so cannot include a stop ($-$) band. Accordingly, the ratio does not go above unity, although it reaches unity at the two frequencies 300 and 400, corresponding to the infinite attenuation where stop ($-$) and stop (+) bands meet in Fig. 7. It is also unity at the extreme frequencies zero and infinity. The four pass bands have, of course, the same locations as in Fig. 7.

Multiplying the ratio by a constant greater than unity introduces stop (−) bands along with the stop (+) bands; multiplying it by a constant less than unity removes all infinite attenuations; these changes within the stop bands are made without altering the locations of the four pass bands.

Wave-Filter Having Assigned Pass Bands

In connection with practical applications we especially desire to know what latitude is permitted in the preassignment of properties for a wave-filter. If we consider first the ideal lattice wave-filter, its limitations are those inherent in the form which its two independent resistanceless one-point impedances[5] Z_1 and Z_2 may assume. The mathematical form of this impedance is shown by formula (7) of the appendix, which may be expressed in words as follows:

Within a constant factor the most general one-point reactance obtainable by means of a finite, pure reactance network is an odd rational function of the frequency which is completely determined by assigning the resonant and anti-resonant frequencies, subject to the condition that they alternate and include both zero and infinity.

The corresponding general expressions for the quotient and product of the impedances Z_1 and Z_2 are shown by formulas (8) and (9). Definite, realizable values for all of the $2n+2$ parameters and $2n+1$ optional signs occurring in these formulas may be determined in the following manner:

(a) Assign the location of all n pass bands, which must be treated as distinct bands even though two or more are confluent; this fixes the values of the $2n$ roots $p_1 \ldots p_{2n}$ which correspond to the successive frequencies at the two ends of the bands.

(b) Assign to the lower or upper end of each pass band propagation without phase change from section to section; this fixes the corresponding optional sign in formula (8) as + or −, respectively.

(c) Assign a value to the propagation constant at any one non-critical frequency (that is, assign the attenuation constant in a

[5] A one-point impedance of a network is the ratio of an impressed electromotive force at a point to the resulting current at the same point—in contradistinction to two-point impedances, where the ratio applies to an electromotive force and the resulting current at two different points.

stop band or the phase constant in a pass band); this fixes the value of the constant G and thus completely determines formula (8) on which the propagation constant depends.

(d) Assign to the lower or upper end of each stop band the iterative impedance zero; this fixes the corresponding optional sign in formula (9) as $+$ or $-$, respectively.

(e) Assign the iterative impedance at any one non-critical frequency (subject to the condition that it must be a positive resistance in a pass band and a reactance in a stop band); this fixes the constant H and thereby the entire expression (9) upon which the iterative impedance depends.

The quotient and product of the impedances Z_1 and Z_2 are now fully determined; the values of Z_1 and Z_2 are easily deduced and also the propagation constant and iterative impedance by formulas (11) and (12); Z_1 and Z_2 are physically realizable except for the necessary resistance in all networks.

These important results may be summarized as follows:

A lattice wave-filter having any assigned pass bands is physically realizable; the location of the pass bands fully determines the propagation constant and iterative impedance at all frequencies when their values are assigned at one non-critical frequency, and zero phase constant and zero iterative impedance are assigned to the lower or upper end of each pass band and stop band, respectively.

Lattice Artificial Line Equivalent to the Generalized Artificial Line of Fig. 1

Since any number of arbitrarily preassigned pass bands may be realized by means of the lattice network, it is natural to inquire whether this network does not present a generality which is essentially as comprehensive as that obtainable by means of any network N in Fig. 1, provided the generalized line is so terminated as to equalize its iterative impedances in the two directions. This proves to be the case.

If network N has identical iterative impedances in both directions, the lattice network equivalent to two sections of N is shown by Fig. 17; each lattice impedance is secured by using an N network; the N's placed in the two series branches of the lattice have their far terminals short-circuited so that they each give the impedance denoted by Z_0; the N's in the two diagonal branches have their far ends open and they each give the impedance denoted by Z_∞.

The lattice network of Fig. 18 has in each branch a one-point impedance obtained by means of a duplicate of the given network N and an ideal transformer. The two lattice branch impedances are $Z_q+Z_r \pm 2Z_{qr}$ where the three impedances Z_q, Z_r, Z_{qr} are the effective self and mutual impedances of the network N regarded as a transformer. This lattice network has identically the same propaga-

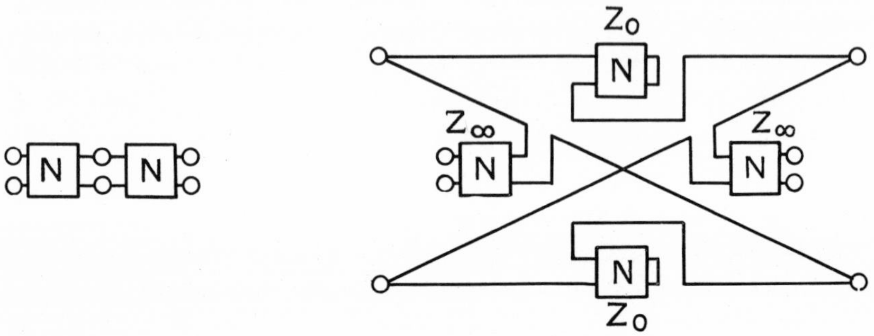

Fig. 17—Lattice Unit Equivalent to Two Sections of Fig. 1 Assumed to be Symmetrical

tion constant as the single network N shown on the left. Since the lattice cannot have different iterative impedances in the two directions, it actually compromises by assuming the sum of the two iterative impedances presented by N. A physical theory of the equival-

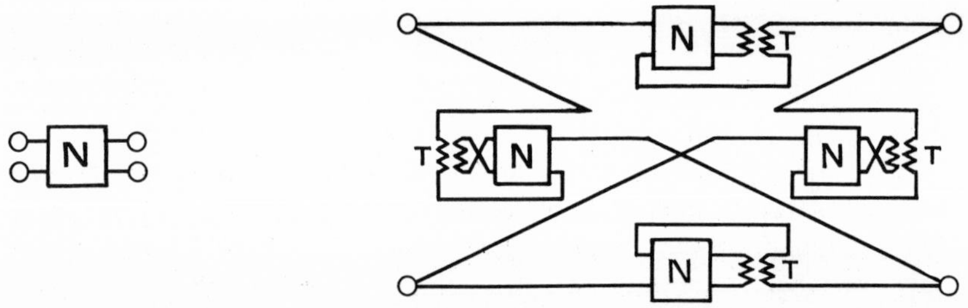

Fig. 18—Lattice Network Having the Same Propagation Constant as N and an Iterative Impedance Equal to the Sum of the Two Iterative Impedances of N

ences shown in Figs. 17 and 18 has not been worked up; the analytical proofs were made by applying the formulas given in the appendix under lattice networks.

Without going to more complex networks it is, of course, not possible to get a symmetrical iterative impedance, but that is not necessary for our present purposes where we are concerned primarily with the propagation constant. It has now been shown with complete generality that:

The lattice artificial line, with physically realizable branch impedances, is identically equivalent in propagation constant and mean iterative impedance to the chain of identical physically realizable networks connected together in sequence through two pairs of terminals.

To complete this simplification of the generalized artificial line it is necessary to know the simplest possible form of the one-point impedances employed in the branches of the lattice network. The discussion of the most general one-point impedance obtainable by means of any network of resistances, self and mutual inductances, leakages and capacities will find its natural place, together with allied theorems, in a paper on the subject of impedances. For the present purpose it is sufficient to state:

The most general branch impedance of the lattice network may be constructed by combining, in parallel, resonant circuits having impedances of the form $R+iLp+(G+iCp)^{-1}$; *or they may equally well be constructed by combining, in series, anti-resonant circuits having impedances of the form* $\left[G+iCp+(R+iLp)^{-1}\right]^{-1}$.

Summary of Physical Theory

The wave-filter under discussion approximates to a resistanceless artificial line, and such an ideal artificial line is capable of two, and only two, fundamentally distinct states of motion. In one state the disturbance is attenuated along the line, and there is no flow of energy other than a back and forth surging of energy, the intensity of which rapidly dies out along the line. In the other state there is a free flow of energy, without loss, from section to section along the line, with no surge of energy between symmetrical sections. Each state holds for one or more continuous bands of frequencies; these bands have been distinguished as stop bands and pass bands.

A high degree of discrimination, between different frequencies, may be obtained, even if each section, taken alone, gives only a moderate difference in attenuation, by the use of a sufficient number of sections in the wave-filter, since the attenuation factors vary in geometrical progression with the number of sections.

Any number of arbitrarily located pass bands may be realized by means of the lattice artificial line; furthermore, the propagation constant at one frequency, and the iterative impedance at one frequency may both be assigned, while the location of zero phase constant and zero iterative impedance at the lower or upper end of each pass band and stop band, respectively, is also optional. This completely determines the lattice artificial line. No additional condition, other than iterative impedance asymmetry, can be realized by replacing the lattice network by any four terminal network.

APPENDIX

Formulas for the Artificial Line

Formulas for the propagation constant and iterative impedance of the generalized artificial line, expressed in a number of equivalent forms, have already been given in my paper on Cisoidal Oscillations,[6] but it seems worth while to deduce the formulas anew here from the free oscillations of the detached unit circuit of Fig. 6, so as to complete the physical theory by deducing the comprehensive mathematical formulas by the same method of procedure.

Ladder Network Formulas

Notation:

Z_1, Z_2 = series impedance and shunt impedance of the section of Fig. 4, which is equivalent to the general network N of Fig. 1.

$\Gamma = A + iB$ = propagation constant per section.

K_1, K_2 = iterative impedances at mid-series and mid-shunt.

$\gamma = \alpha + i\beta = \sqrt{Z_1/Z_2}$ = propagation constant for uniform distribution of Z_1 and $1/Z_2$, per unit length.

$k = \sqrt{Z_1 Z_2}$ = iterative impedance of this same uniform line.

In Fig. 6, the current is indicated as I and the potentials at the ends of the section as $V, Ve^{-\Gamma}$. In order that the free oscillation may be possible the total impedance of the circuit $(Z_1 + Z' + Z'')$ must vanish; this determines the iterative impedance K_2. In addition to this condition it is sufficient to make use of two other simple relations: the proportionality of the potential drops in the direction of the current across Z' and Z'' to Z' and Z'', since they carry the same current (this determines the propagation constant Γ); and the equality of

[6] "Cisodial Oscillations," Trans. A. I. E. E., vol. 30, pp. 873–909, 1911. In the lowest row of squares of Table I, the iterative impedances and propagation constant of any network are given in five different ways, involving one-point and two-point impedances, equivalent star impedances, equivalent delta impedances, equivalent transformer impedances, or the determinant of the network. The only typographical errors in Table I appear to be the four which occur in the first, third and fifth squares of this row: in the values for K_q replace $(S_q - S_{qr})$ by $(S_q - S_r)$ and place a parenthesis before $U_q - U_r)$; in the first value of K_r replace S_{qr} by S_{qr}^2; in the last value for Γ_{qr} add a minus sign so that it reads $\cosh^{-1}$.

K_1, the mid-series iterative impedance of the artificial line, to the total impedance on the right of the mid-point of the series impedance Z_1. These three relations, which can be written down at once, are:

$$Z_1 + Z' + Z'' = Z_1 - \frac{4Z_2K_2^2}{4Z_2^2 - K_2^2} = 0,$$

$$\frac{Ve^{-\Gamma}}{V} = -\frac{Z''}{Z'} = \frac{2Z_2 - K_2}{2Z_2 + K_2},$$

$$K_1 = \frac{1}{2}Z_1 + Z'' = \frac{1}{2}Z_1 + \frac{2Z_2K_2}{2Z_2 + K_2},$$

from which the formulas for Γ, K_1, and K_2, in terms of Z_1, Z_2, are found to be:

$$\Gamma = 2 \sinh^{-1} \frac{1}{2}\sqrt{\frac{Z_1}{Z_2}} = 2 \sinh^{-1} \frac{1}{2}\gamma, \tag{1}$$

$$\left.\begin{matrix} K_1 \\ K_2 \end{matrix}\right\} = \sqrt{Z_1Z_2}\left(1 + \frac{Z_1}{4Z_2}\right)^{\pm\frac{1}{2}} = k\left(1 + \frac{1}{4}\gamma^2\right)^{\pm\frac{1}{2}} \text{at mid} \left\{\begin{matrix} \text{series} \\ \text{shunt,} \end{matrix}\right. \tag{2}$$

and the formulas for Z_1 and Z_2 in terms of Γ and K_1 or K_2 are likewise found to be:

$$Z_1 = 2K_1 \tanh \frac{1}{2}\Gamma = K_2 \sinh \Gamma, \tag{3}$$

$$Z_2 = K_1/\sinh \Gamma = \frac{1}{2}K_2 \coth \frac{1}{2}\Gamma. \tag{4}$$

Formulas (3) and (4) are in the nature of design formulas in that they determine the impedance Z_1 and Z_2, at assigned frequencies, which will ensure the assigned values of Γ and K at these frequencies. In general, however, it would not be evident how best to secure these required values of Z_1 and Z_2; complicated or even impossible networks might be called for, even to approximate values of Z_1 and Z_2 assigned in an arbitrary manner. Fortunately, practical requirements are ordinarily satisfied by meeting maximum and minimum values for the attenuation constant throughout assigned frequency bands. Formulas (8) and (9) may be employed for this purpose as explained below.

It is convenient to have formulas (1) and (2) expressed in a variety of ways, since no one form is well suited for calculation throughout the entire range of the variables. Accordingly, the following analytically equivalent expressions are here collected together for reference:

$$\Gamma = i\,2\sin^{-1}\frac{\gamma}{2i} = i\cos^{-1}\left(1+\frac{\gamma^2}{2}\right), \tag{5}$$

$$= 2\sinh^{-1}\frac{\gamma}{2} = 2\tanh^{-1}\frac{\frac{\gamma}{2}}{\sqrt{1+\frac{\gamma^2}{4}}} = \cosh^{-1}\left(1+\frac{\gamma^2}{2}\right)$$

$$= 2\log\left[\frac{\gamma}{2}+\sqrt{1+\frac{\gamma^2}{4}}\right], \tag{5a}$$

$$= i\pi + 2\cosh^{-1}\frac{\gamma}{2i} = i\pi + \cosh^{-1}\left(-1-\frac{\gamma^2}{2}\right)$$

$$= i\pi + 2\log\left[\frac{\gamma}{2i}+\sqrt{-1-\frac{\gamma^2}{4}}\right], \tag{5b}$$

$$= i\frac{\pi}{2} - \sinh^{-1}\left(1+\frac{\gamma^2}{2}\right)i, \tag{5c}$$

$$= \gamma - \frac{1}{24}\gamma^3 + \frac{3}{640}\gamma^5 - \frac{5}{7168}\gamma^7 + \ldots,\ \text{if } |\gamma| < 2, \tag{5d}$$

$$= 2\cosh^{-1} g + i\,2\sin^{-1}\frac{\beta}{2g}, \tag{5e}$$

$$\text{where } \gamma = \alpha + i\beta,\ 2g = \sqrt{\left(\frac{\alpha}{2}\right)^2+\left(1+\frac{\beta}{2}\right)^2} + \sqrt{\left(\frac{\alpha}{2}\right)^2+\left(1-\frac{\beta}{2}\right)^2},$$

$$= \cosh^{-1} h + i\cos^{-1}\frac{x}{h}, \tag{5f}$$

$$\text{where } 1+\frac{1}{2}\gamma^2 = x + iy,\ 2h = \sqrt{(x+1)^2+y^2}+\sqrt{(x-1)^2+y^2},$$

$$\left.\begin{matrix}K_1\\K_2\end{matrix}\right\} = k\left(1+\frac{1}{4}\gamma^2\right)^{\pm\frac{1}{2}} = k\left(\cosh\frac{\Gamma}{2}\right)^{\pm 1} = k\left(\frac{\sinh\Gamma}{\gamma}\right)^{\pm 1}$$

$$= k\left(\frac{1}{2}\gamma\coth\frac{1}{2}\Gamma\right)^{\pm 1} \text{ at mid } \begin{cases}\text{series}\\\text{shunt.}\end{cases} \tag{6}$$

The formulas leave indeterminate the signs of γ, k, Γ, and K, and also a term $\pm i2\pi n$ in γ and Γ. The signs are to be so chosen that the real parts are positive, or become positive when positive resistance is added to the system. The indeterminate $\pm\, i2\pi n$ can be made determinate only after knowing something of the internal structure of the unit network of which the artificial line is composed; the conditions to be met are—absence of phase differences when all branches of the unit network N of Fig. 1 are assumed to be pure resistances and continuity of phase as reactances are gradually introduced to give the actual network.

Formula (5) is adapted for use in the pass bands, since the expressions are real when γ^2 is real, negative and not less than -4; similarly, formulas (5a) and (5b) are adapted for use in the stop $(\pm)$ bands, that is, when γ^2 is positive and less than -4 respectively.

From the theory of impedances we know that any resistanceless one-point impedance is expressible in the form

$$Z = iD \frac{p \quad (p_2^2 - p^2) \ldots (p_{2n-2}^2 - p^2)}{(p_1^2 - p^2)\ (p_3^2 - p^2) \ldots (p_{2n-1}^2 - p^2)} \tag{7}$$

where the factor D and the roots $p_1, p_2, \ldots p_{2n}$ are arbitrary positive, reals subject only to the condition that each root is at least as large as the preceding one. This enables us to write down the forms which the quotient and product of two resistanceless one-point impedances may assume, which are as follows:

$$\frac{Z'}{Z''} = G \left(\frac{p_1^2 - p^2}{p_2^2 - p^2}\right)^{\pm 1} \left(\frac{p_3^2 - p^2}{p_4^2 - p^2}\right)^{\pm 1} \ldots \left(\frac{p_{2n-1}^2 - p^2}{p_{2n}^2 - p^2}\right)^{\pm 1} \tag{8}$$

$$Z'Z' = -H\left(\frac{p^2}{p_1^2 - p^2}\right)^{\pm 1} \left(\frac{p_2^2 - p^2}{p_3^2 - p^2}\right)^{\pm 1} \ldots \left(\frac{p_{2n-2}^2 - p^2}{p_{2n-1}^2 - p^2}\right)^{\pm 1} (p_{2n}^2 - p^2)^{\pm 1} \tag{9}$$

where G, H and the roots $p_1, p_2, \ldots p_{2n}$ are arbitrary positive reals, subject only to the condition that each root is at least as large as the preceding one, and the $2n+1$ and optional $\pm$ signs are mutually independent. Conversely, if the relations (8) and (9) are prescribed, then the required individual impedances Z' and Z'' are each of the form (7) and thus physically realizable.

If in formulas 1, 2, 5 and 6 we substitute for $Z_1/Z_2 = \gamma^2$ and $Z_1 Z_2 = k^2$ the right-hand side of formulas (8) and (9), respectively, we obtain formulas for the propagation constant and iterative impedance of an artificial resistanceless line in terms of frequencies at which the propagation constant becomes zero or infinite. Ordinarily, however, we are more interested in having expressions in terms of the frequencies which terminate the pass bands. To secure these the substitutions should be $4[8]/(4 - [8])$ and $[9]\ (1 - [8]/4)^{\pm 1}$, where [8] and [9] stand for the entire right-hand sides of formulas (8) and (9). This substitution amounts to obtaining the lattice network giving the required pass bands, and then transforming to the ladder network having the same propagation constant and the same iterative impedance at mid-series or mid-shunt.

Lattice Network Formulas Fig. 14

The impedances of a single section between terminals 1 and 2, with the far end of the section 3 and 4 either short-circuited or open, are readily seen to be

$$Z_0 = \frac{2Z_1Z_2}{\frac{1}{2}Z_1 + 2Z_2},\ Z_\infty = \frac{1}{2}\left(\frac{1}{2}Z_1 + 2Z_2\right). \tag{10}$$

Since $\sqrt{Z_0 Z_\infty}$ and $\sqrt{Z_0/Z_\infty}$ are the iterative impedance and the hyperbolic tangent of the propagation constant for any symmetrical artificial line, we have the following analytically equivalent formulas for the lattice network where $\gamma = \sqrt{Z_1/Z_2}$, and $k = \sqrt{Z_1 Z_2}$ as for the ladder type.

Lattice Formulas

$$\begin{cases} \Gamma = 2\tanh^{-1}\frac{1}{2}\sqrt{\frac{Z_1}{Z_2}} = 2\tanh^{-1}\frac{1}{2}\gamma, & (11) \\ K = \sqrt{Z_1 Z_2} = k. & (12) \end{cases}$$

$$\begin{cases} Z_1 = 2K\tanh\frac{1}{2}\Gamma, & (13) \\ Z_2 = \frac{1}{2}K\coth\frac{1}{2}\Gamma. & (14) \end{cases}$$

$$\Gamma = i2\tan^{-1}\frac{\gamma}{2i} = i\cos^{-1}\frac{1+\frac{1}{4}\gamma^2}{1-\frac{1}{4}\gamma^2}, \tag{15}$$

$$= 2\tanh^{-1}\frac{\gamma}{2} = 2\sinh^{-1}\frac{\frac{1}{2}\gamma}{\sqrt{1-\frac{1}{4}\gamma^2}} = \cosh^{-1}\frac{1+\frac{1}{4}\gamma^2}{1-\frac{1}{4}\gamma^2}$$

$$= \log\frac{1+\frac{1}{2}\gamma}{1-\frac{1}{2}\gamma}, \tag{15a}$$

$$= i\pi + 2\coth^{-1}\frac{\gamma}{2} = i\pi + \cosh^{-1}\frac{1+\frac{1}{4}\gamma^2}{-1+\frac{1}{4}\gamma^2}$$

$$= i\pi + \log\frac{1+\frac{1}{2}\gamma}{-1+\frac{1}{2}\gamma}, \qquad (15b)$$

$$= i\frac{\pi}{2} - \sinh^{-1}\frac{1+\frac{1}{4}\gamma^2}{1-\frac{1}{4}\gamma^2}\,i, \qquad (15c)$$

$$= \gamma + \frac{1}{12}\gamma^3 + \frac{1}{80}\gamma^5 + \frac{1}{448}\gamma^7 + \ldots, |\gamma| < 2, \qquad (15d)$$

$$= \frac{1}{2}\log\frac{\left(1+\frac{\alpha}{2}\right)^2 + \left(\frac{\beta}{2}\right)^2}{\left(1-\frac{\alpha}{2}\right)^2 + \left(\frac{\beta}{2}\right)^2} + i\tan^{-1}\frac{\beta}{1-\left(\frac{\alpha}{2}\right)^2 - \left(\frac{\beta}{2}\right)^2}, \qquad (15e)$$

where $\gamma = \alpha + i\beta$.

In these formulas $Z_1/Z_2 = \gamma^2$ and $Z_1 Z_2 = k^2$ might be expressed in terms of the resonant and anti-resonant complex frequencies of Z_1 and Z_2, the frequencies being made complex quantities so as to include the damping. Where there is no damping, that is, where all network impedances are devoid of resistance, the simplified forms of these expressions are given by formulas (8) and (9). The use of these formulas for designing wave filters having assigned pass bands is explained at page 23.

7

Reprinted from *J. Math. Phys.*, **18**(4), 257–353 (1939)

SYNTHESIS OF REACTANCE 4-POLES WHICH PRODUCE PRESCRIBED INSERTION LOSS CHARACTERISTICS

Including Special Applications to Filter Design

By S. Darlington*

Outline

Introduction

Of the various types of electrical networks which are frequently found useful, one of the commonest is the 4-terminal transducer of reactances, more briefly referred to as the reactance 4-pole.[1] In particular, the selective networks or filters which are commonly used for transmitting certain frequencies while blocking others are almost always reactance 4-poles, and these filters form essential parts of most communication systems.

Detailed methods of designing filters and related reactance 4-poles are well known and have been in general use for a considerable period. For the most part these fit into one general filter design scheme which may be referred to as the image parameter theory, since it is based upon

* Bell Telephone Laboratories, Inc. This paper has been accepted as a Doctor's thesis by the Faculty of Pure Science of Columbia University. The manuscript was received by the Editors May 18, 1938.

[1] Throughout this paper, the term 4-pole will be used to indicate a 4-terminal transducer—i.e. a network with two pairs of accessible terminals subject to the restriction that no external connections can be made between terminals of different pairs. The term has been widely used in this sense and also to indicate a network with four terminals to which external apparatus can be connected in any desired manner.

the concepts of the image impedances and image transfer constants of 4-terminal networks. It has been found that methods based upon this theory can be used to design practical filters with electrical characteristics meeting any ordinary engineering requirements. In recent years it has become apparent, however, that these filters are sometimes unnecessarily costly. At the same time the question of network cost has become increasingly important.

This paper describes a theory of reactance 4-pole design which differs from the image parameter theory in such a way that it sometimes leads to more advantageous choices of element values in filters of conventional image parameter configurations. While more complicated than the image parameter theory, it sometimes permits the realization of substantially greater network economy or of superior electrical characteristics with no increase in cost. Instead of starting with the concepts of image impedances and image transfer constants, this theory is based upon the general problem in network synthesis of finding reactance 4-poles yielding prescribed insertion loss versus frequency functions when inserted between prescribed resistance terminations. Because of this it has come to be called the insertion loss theory.

While the insertion loss theory applies particularly to filter design, it is capable of more general applications. As a matter of fact, it is theoretically possible to use the theory to design physical reactance 4-poles which, when terminated in prescribed resistances, produce insertion loss characteristics identical with those of any general finite passive 4-poles with the same terminations. Although the possibility of designing at least simple reactance networks on an insertion loss basis is common knowledge, when other than very simple circuits are considered extensive special theory such as that developed in this paper is necessary if hopelessly complex computations are to be avoided.

Although very general applications of the insertion loss method of design are theoretically possible, even with the theory developed here the numerical complications are such as to limit its practical usefulness. In many design problems, for instance, economies in the cost of actually constructing filters might be obtained by using insertion loss designs rather than image parameter designs but these economies frequently are off-set by the added cost involved in obtaining the insertion loss designs. As a result, the insertion loss theory applies principally to the design of filters which are to be made in such large numbers that construction economies will justify high design costs or which must meet requirements not easily satisfied with image parameter designs.

The basic insertion loss theory applies only to networks of pure reactances. Modifications are included, however, permitting the design of certain types of dissipative reactance networks with prescribed loss characteristics. These have greatly increased the practical advantages obtainable by the use of the theory.

Limitations of the Image Parameter Theory

The great simplicity of the image parameter filter theory is obtained by adopting certain definite restrictions which limit to a considerable extent the choice of the element values. It is the possibility of avoiding these restrictions by using the more complicated insertion loss theory that sometimes renders this theory advantageous as a basis of filter design. While the use of the insertion loss theory in filter design normally involves the introduction of other restrictions, these are such as to lead to the optimum choice of element values for meeting certain types of filter specifications. In order to clarify the status of the insertion loss theory, it will be best to introduce at this point a brief description of the fundamentals of the image parameter theory and of the restrictions which it involves.

The image parameter theory is, of course, an outgrowth of the artificial line theories of Pupin (1) and Campbell (2). The image impedances and image transfer constants in terms of which it is developed correspond to the characteristic impedances and propagation constants of these artificial lines. In its most familiar form the image parameter theory deals directly with the so-called composite filters introduced by Zobel (3), which are made up of chains of sections with matched image impedances but different transfer constants. In the more general form developed by Bode (4), however, it deals with the equivalent restrictions upon the image impedances and transfer constants of complete networks, without actually requiring chains of tandem sections.

The image attenuation of a non-dissipative filter is identically zero over finite ranges of frequencies. If the filter is terminated in its image impedances, the corresponding transducer loss will also be zero.[2] Although the image impedances must vary with frequency, they can

[2] By the transducer loss of a filter is meant the difference in level between the received power and the maximum power obtainable from the generator with any passive network. The image attenuation of a filter is equal to its transducer loss when terminated in its image impedances at all frequencies where these impedances are real. The transducer loss of any network terminated in resistances can be obtained from the insertion loss by adding the reflection loss corresponding to the ratio of the terminations.

approximate constant resistances over the greater parts of the ranges of zero attenuation, or theoretical pass bands. Consequently, if these resistances are used as the actual terminations the transducer loss will be small over the ranges of good approximation.

The situation described above leads to the introduction of the following requirement, which is responsible for the fundamental limitations of the image parameter theory: The image attenuation of a non-dissipative filter is required to be identically zero in continuous frequency ranges including those to be freely transmitted between the actual terminations and is required to be other than zero at all other frequencies. For the types of filters commonly encountered these restrictions reduce by almost half the constants which could otherwise be chosen arbitrarily.

Without further investigation, it might appear that good filter characteristics could be obtained only by satisfying these restrictions at least so closely that permissible departures from them would be of no practical interest. Actually, these requirements are not necessary but instead are artificial or arbitrary restrictions leading to a simple design procedure.

As an illustration, suppose that the elements of a filter designed on the image basis are changed from their design values by various amounts of the order of perhaps 10 or 20 percent. In general this will split the theoretical pass band into a number of theoretical pass bands separated by narrow theoretical attenuation bands.[3] The image attenuation corresponding to these theoretical attenuation bands, however, will ordinarily be very moderate. In addition, the actual transducer loss may be only a fraction of the image attenuation. A transducer loss of less than 0.4 decibels may be obtained, for instance, even though the corresponding image attenuation is as high as 2.5 decibels. As a result, modifying the elements may leave unchanged the frequencies included in the *effective* pass band and may result in an actual decrease in the corresponding transducer loss.

When the dissipation required in actual filters has a marked effect the arbitrariness of the restrictions of the image parameter theory becomes much more striking. Under these conditions the image attenuation is generally far from uniform over the frequency ranges in which it is zero on a non-dissipative basis. The special restrictions of the image parameter theory then amount to the requirement of transmission range

[3] In general the theoretical attenuation band will also be split up by narrow theoretical pass bands, but it will be sufficient for purposes of illustration to consider only the splitting of the pass band.

insertion losses approximating characteristics which are anything but ideal.

The above discussion indicates the arbitrariness of the restrictions forming the basis of the image parameter method of filter ***design*** or ***synthesis.*** It is well known that the ***circuit analysis*** problem of determining the operation of a known network can be solved in terms of the image impedances and image transfer constant of the network even though it does not satisfy the special restrictions of the design theory. When these restrictions are abandoned, however, the image parameters are no longer convenient in *design* problems involving the determination of resistance-terminated reactance networks with prescribed properties.

Other Previous Theories

Cauer (5), Gewertz (6) and others have studied the synthesis of perfectly general reactance 4-poles having prescribed open- and short-circuit impedances. These investigations, however, have not yielded useful methods of designing reactance 4-poles with resistance terminations except under the restrictions of the image parameter theory. On the other hand, they have produced such useful information as the necessary and sufficient conditions satisfied by sets of open- and short-circuit impedances corresponding to physical reactance 4-poles. In addition, they have shown that physical networks of certain so-called canonical configurations can readily be designed to have any specified set of impedances satisfying these conditions. The canonical configurations, however, were chosen purely for their general realizability and ease of design and are rarely of practical interest.

The development of the present insertion loss theory started with the previous theory developed by Norton (7) which permits the design of two or more filters producing a constant resistance at their common terminals when connected in series or in parallel at one end. Norton's theory involves the problem of designing a reactance 4-pole terminated in an open- or short-circuit at one end. This amounts to the special case of the insertion loss theory in which one of the specified terminations is zero or infinite. Norton's detailed design procedure, however, assumes a rather restricted type of prescribed insertion loss characteristic which is of little interest except in the design of filters to operate in constant resistance pairs.

Principal Operations Involved in the Design of Filters on an Insertion Loss Basis

Returning now to the subject of this paper, filters can readily be designed on an insertion loss basis provided certain fundamental opera-

5

tions can be carried out. There is no problem in fitting these operations together into straightforward design procedures but the operations themselves cannot readily be carried out without special mathematical machinery. The principal object of the insertion loss theory is to supply this mathematical machinery.

The logical order to follow in developing the mathematical machinery of the insertion loss theory turns out to be rather different from the order in which the various parts are used in carrying out actual designs. It will therefore be best to motivate the more detailed analysis by introducing at once a brief description of the various fundamental design operations, rather than introducing each operation at the time the corresponding mathematical machinery is taken up.

The distinguishing feature of the insertion loss theory is the use of an exactly prescribed insertion loss versus frequency function to fix the element values of the final network. Since exactly prescribed loss functions are rarely included in design specifications, the first operation in the design of a network on the insertion loss basis is usually the choice of the specific insertion loss function to be obtained. This choice is guided by the following considerations. First, the loss function must be consistent with the design specifications, e.g., it must represent sufficient suppression of the unwanted frequencies. Second, the function must be of a form leading to an economical network as regards the number of elements involved. Third, it must usually correspond to a particular type of network configuration, such as a ladder or a lattice.

After a specific insertion loss function has been chosen, the determination of a corresponding network involves two principal operations. The first of these is the determination of some or all of the open- and short-circuit impedances of the network from the loss function, the second being then the determination of the actual element values from these impedances.[4] It turns out that there are only a finite number of sets of open- and short-circuit impedances corresponding to reactance networks of minimum complexity producing a specific insertion loss function when terminated in a specific pair of resistances. After one of these sets of impedances has been selected, the method of determining the element values of a corresponding network depends upon the type of

[4] This rule, however, is not without its exceptions. In particular, symmetrical lattices with equal terminations are most easily determined from their insertion loss functions by procedures making no direct references to their open- and short-circuit impedances.

configuration chosen, equivalent networks of different configurations frequently requiring quite different methods.[5]

The parasitic dissipation which must be present in actual reactance networks sometimes influences the loss characteristics to a small enough extent to justify neglecting it in the design of the networks. On the other hand, the influence of the necessary dissipation is sometimes so important that it is highly desirable to compensate for it in some manner. The logical way to obtain this compensation is to design dissipative reactance networks which themselves produce more or less exactly prescribed loss functions. Under certain conditions this can be accomplished by modifying the first part of the non-dissipative design procedure in such a way as to obtain the open- and short-circuit impedances corresponding to the removal of the dissipation from the final network.[6] This permits the element values to be computed from the impedances on a strictly non-dissipative basis.

In order to clarify further the general design procedure outlined above, the procedure used in a simple illustrative case will now be described in more detail. It should be borne in mind, however, that the detailed design procedures used in other cases may differ considerably from this illustrative case even though they involve the same general types of operations.

Illustrative Special Case—Choice of Insertion Loss Function

Assuming for the time being that dissipation can be neglected entirely, consider the design of a low-pass filter of ideal reactance elements consistent with the following specifications. First, the insertion loss shall not be greater than a prescribed maximum α_p^0 in an effective pass band extending from zero frequency to a prescribed cut-off frequency f_1^0. Second, the insertion loss shall not be less than a prescribed minimum α_a^0 in an effective attenuation band extending from a second prescribed frequency f_2^0 to infinity. Third, these loss specifications are to be met

[5] As would be expected, when a particular configuration is desired care must usually be exercised to make sure that an appropriate set of impedances is chosen. The various sets corresponding to a single loss function, for instance, usually occur in inverse pairs leading to inverse configurations.

[6] The possibility of doing this depends upon the following situation. When the dissipation required in a network satisfies certain restrictions as regards its variation from element to element the effect of the dissipation on complex impedances and complex voltage ratios corresponds to simple transformations of the functions of frequency representing the impedances and voltage ratios, these transformations being independent of the specific network configuration and element values.

with a network which is electrically symmetrical and is terminated in equal resistances and which has the same configuration as a mid-series low-pass ladder filter of the image parameter theory even though it may have more general element values.[7]

The above specifications on the insertion loss are indicated graphically in Fig. 1A, it being required that the curve of actual insertion loss shall fall within the shaded areas at frequencies less than f_1^0 or greater than f_2^0. The required configuration is indicated in Fig. 2.

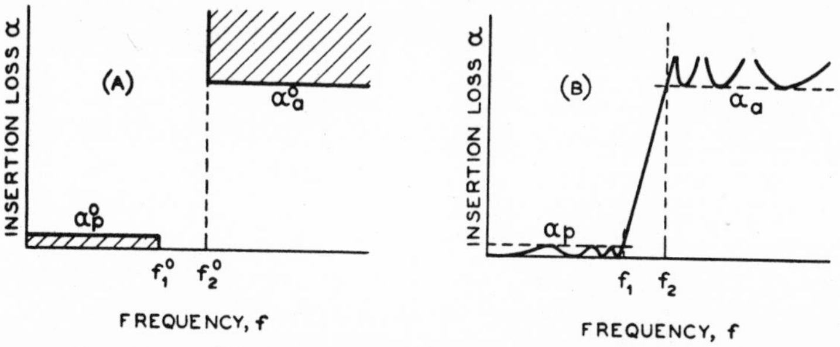

Fig. 1. A) Design specifications—If $f < f_1^0$ or $> f_2^0$, α must lie in a shaded area. B) Form of α characteristic meeting the design specifications.

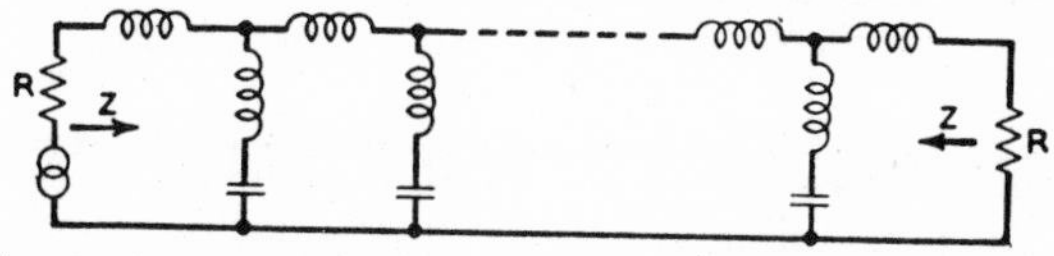

Fig. 2. Configuration to meet specifications of Fig. 1A

The theory to be developed shows that the above specifications on the form of the final network require the insertion loss α to be described by an equation of the form

$$e^{2\alpha} = 1 + \left[S_0 \frac{\omega(\omega_1^2 - \omega^2) \cdots (\omega_\eta^2 - \omega^2)}{(1 - \nu_1^2\omega^2) \cdots (1 - \nu_\eta^2\omega^2)} \right]^2 \tag{1}$$

In this equation, η represents the number of "sections," while S_0 and the ω_σ's and ν_σ's are arbitrary constants. The specific problem is to

[7] It turns out that the requirement that the network shall be either symmetrical or of a type producing inverse image impedances at the two ends tends to lead to efficient filter characteristics. The inverse impedance case is excluded here in order to simplify the illustration. Similarly, the special loss specifications described above are chosen for their simplicity rather than for their generality. Somewhat similar methods can still be used, for instance, when the minimum permissible attenuation band loss varies with frequency.

8

choose the arbitrary constants in such a way as to satisfy the specifications indicated in Fig. 1A with the smallest possible value of η, i.e., the smallest number of sections.[8]

The best choice of the ω_σ's and ν_σ's turns out to be that leading to the form of loss characteristic indicated in Fig. 1B for the special case of $\eta = 3$. A lengthy analysis will later be outlined which shows that the particular rational "power ratio" function $e^{2\alpha}$ which has the form of (1) and which corresponds to the "equal ripple" type of loss characteristic indicated in Fig. 1B can be obtained by requiring $e^{2\alpha}$ to be determined by the following set of equations relating both $e^{2\alpha}$ and ω to a new variable u.

$$e^{2\alpha} = 1 + (e^{2\alpha_p} - 1)\, sn^2\left[(2\eta + 1)u\frac{K_1}{K}, k_1\right] \qquad (2)^9$$

$$\frac{\omega}{2\pi\sqrt{f_1 f_2}} = \sqrt{k}\, sn(u, k)$$

$$k = \frac{f_1}{f_2}$$

$$q_1 = q^{2\eta+1}$$

In these equations, α_p, f_1, f_2 represent the maximum pass band loss and the limits of the effective pass and attenuation bands, as in Fig. 1B. The symbols K, K_1 and q, q_1 represent constants appearing in the general theory of elliptic functions, K, K_1 being the complete elliptic integrals of the first kind of moduli k, k_1, respectively, and q, q_1 the corresponding elliptic modular constants. These constants are uniquely related to k, k_1 in a manner represented in most elliptic function tables by tabulations of K and $\log_{10} q$ vs $\sin^{-1} k$.

[8] It turns out that choosing the constants in this way will normally lead to a physical network of the form of Fig. 2, although it is possible for difficulties to be encountered if α_p^0 is exceptionally small.

[9] The appearance of elliptic functions in a problem involving an algebraic function producing equal maxima and equal minima is not surprising to those familiar with the elliptic functions appearing in the "Tchebycheff parameter" version of the image parameter theory introduced by Cauer (16). The rational character of $e^{2\alpha}$ considered as a function of ω depends upon the equivalence of $\mathrm{sn}\left[(2\eta + 1)u\frac{K_1}{K}, k_1\right]$ to an odd rational function of sn (u, k), which can be compared with the equivalence of sin $[(2\eta + 1)u]$ to an odd polynomial in sin (u). The correspondence of (2) to the particular special case of (1) illustrated by Fig. 1B depends upon the periodic properties of α and ω considered as functions of u.

The insertion loss corresponding to (2) is completely determined by the choice of α_p, f_1, f_2 and the number of sections η, for f_1, f_2 determine k, which determines k_1 through the relation between q and q_1, while k, k_1 determine K, K_1. The corresponding value of the minimum attenuation band loss α_a indicated in Fig. 1B is thus also determined by the choice of α_p, f_1, f_2. If minor approximations are made, this relationship can be expressed in the following form:

$$\alpha_a = [10 \log_{10}(e^{2\alpha_p} - 1) - 10(2\eta + 1) \log_{10}(q) - 12.04] \text{ db} \quad (3)$$

With the help of (3) it is a simple matter to determine the minimum value of η for which the specifications of Fig. 1A can be satisfied, i.e., for which α_p, α_a, f_1 and f_2 can be chosen in such a way that

$$\begin{matrix} \alpha_p \leqq \alpha_p^0 & f_1 \geqq f_1^0 \\ \alpha_a \geqq \alpha_a^0 & f_2 \leqq f_2^0 \end{matrix} \quad (4)$$

In this way, a definite set of constants can be chosen which determine, with the help of (2), a loss characteristic leading to a final network of the required type and of the minimum number of sections permitted by the loss specifications.[10]

Illustrative Special Case—Determination of a Network Producing the Insertion Loss Chosen

The open- and short-circuit impedances of reactance networks producing the particular insertion loss function chosen cannot be expressed directly in terms of the loss function itself. They can readily be evaluated, however, in terms of the roots of the corresponding power ratio (1).[11] In the special case under consideration these roots can be computed by means of straightforward formulae derived from the elliptic function relations (2).

If reactance networks of minimum complexity are assumed, and also a definite pair of (equal) terminating resistances, it turns out that there are only four sets of open- and short-circuit impedances consistent with

[10] Any three of the four relations of (4) can be made exact equalities but the fourth will then be an equality only in very special cases. Keeping three of the parameters α_p, α_a, f_1, f_2 fixed and changing the number of sections η makes discrete changes in the fourth parameter. Since some margin is thus usually available even though η has the smallest permissible value, this margin is usually best distributed among the various relations of (4).

[11] If the requirement of a loss function appropriate for a symmetrical network is abandoned it is usually necessary to include also the roots of a linear function of the power ratio $e^{2\alpha}$.

(1). Of these only one set is appropriate for the mid-series low pass ladder configuration of Fig. 2. Formulae for computing the element values of this configuration from the impedances will be included in the more detailed development of the insertion loss theory.[12]

Illustrative Special Case—Compensation for Effects of Dissipation

Turning now to the question of compensating for effects of dissipation, consider first the problem of attempting to design a network of reactance elements of prescribed dissipativeness producing exactly the same insertion loss and phase as the non-dissipative filter considered previously. Assuming such a network to be possible, consider the insertion loss and phase that would be produced by the pure reactance network formed by simply removing the dissipation from all the elements. Provided the elements are all to be equally dissipative, it turns out that this new loss and phase can be evaluated directly from the original loss and phase and the extent of the dissipation, with no further information as to the actual element values or specific configuration.[13] Thus any pure reactance network which could be designed to produce this "predistorted" loss and phase would produce the original loss and phase upon the addition of the required dissipation.

As described above, the "predistorted" insertion loss and phase cannot actually be produced by a pure reactance network. On the other hand, if the required dissipation varies properly with frequency and is not too great, the predistortion can be modified in such a way as to lead to a physical network of the configuration of Fig. 2 which will produce a close approximation to the sum of the original insertion loss and an added constant loss when the required dissipation is added. This modified predistortion amounts to nothing more than the addition of a

[12] When the insertion loss function is not necessarily appropriate for a symmetrical network there are usually more than four sets of open- and short-circuit impedances corresponding to networks of minimum complexity. There may be a considerable variety, for instance, realizable with various networks having a single configuration, such as the configuration of Fig. 2.

In the special case of symmetry under consideration the element values of the final ladder may be computed directly from the equivalent lattice rather than from open or short-circuit impedances. The formulae for doing this, however, are derived by expressing the open-circuit impedances in terms of the impedances of the lattice arms.

[13] Actually the corresponding "complex insertion voltage ratio" is more convenient to use than the insertion loss and phase. The original voltage ratio can be readily evaluated in terms of the roots of (1) used in computing the open- and short-circuit impedances on a non-dissipative basis.

constant to each root of the complex voltage ratio function corresponding to the original loss and phase plus the addition of a constant factor representing an added constant loss.[14] Although the general configuration of Fig. 2 is obtained, however, the network yielded by the predistortion method will not satisfy the previous requirement of electrical symmetry nor will the terminations be equal.[15]

Major Divisions of the Theoretical Development

As was indicated previously, the logical order to follow in developing the insertion loss theory is rather different from the order in which the various parts are used in carrying out actual designs. Specifically, the logical procedure is to divide the development of the theory into the sequence of four parts described briefly below, which represent rather different theoretical problems even though they may all be employed in designing a single filter.

Part I deals with the determination of the open- and short-circuit impedances of reactance 4-poles corresponding to prescribed insertion loss characteristics and pairs of terminations. The necessary and sufficient conditions upon insertion loss characteristics yielding physical sets of impedances are established as well as the procedure for determining the impedances.

Part II deals with the determination of reactance 4-poles of special configurations from prescribed open- and short-circuit impedances. These include a new canonical reactance 4-pole made up of tandem sections and also networks which have the same configurations as the familiar lattices and ladders of the image parameter theory but which are not required to satisfy the restrictions on actual element values imposed by the image parameter theory.

Part III deals with the problem of choosing specific insertion loss functions for filtering purposes. In particular, it deals with certain special types of loss functions which satisfy the conditions for physical realizability established in Part I, which represent the optimum choice

[14] The roots of the voltage ratio are determined in evaluating it from the roots of (1).

[15] When the dissipation varies from element to element or varies improperly with frequency, partial compensation can be obtained by assuming dissipation of the proper type representing an average of the actual dissipation. It is also possible to reduce the restrictions on the dissipation assumed if the non-dissipative design procedure is further modified. It is possible for instance, to assume all coils to be equally dissipative and all condensers to be equally dissipative to a different extent.

of arbitrary constants for meeting certain types of filter requirements, and which lead to simplifications in the more general design procedure.

Part IV deals with modifications of the previous theory permitting the design of certain types of dissipative networks with prescribed insertion loss characteristics. It is shown that within certain severe limits it is possible to design dissipative filters with loss characteristics differing appreciably from those of non-dissipative filters of the same configuration only by added constant losses.

Because of the extent of the material to be covered, detailed derivations and proofs will be only outlined in developing the four parts of the theory described above. In addition, a variety of useful but unessential modifications of the theory will be omitted entirely. These modifications have to do with the application of the theory to various special design problems rather than forming a part of the central theory itself.

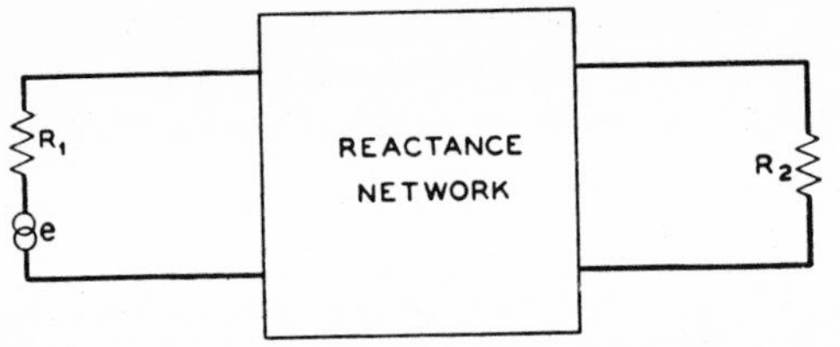

Fig. 3. Reactance 4-pole with resistance terminations

Part I. Open- and Short-Circuit Impedances of Networks with Prescribed Insertion Loss Characteristics

Statement of Problem

As is indicated in the above outline, the first two major divisions of the insertion loss theory are not restricted to the design of networks specifically for filtering purposes. Finite 4-terminal networks are assumed and these are required to be made up of pure reactances and to be terminated in constant resistances as in Fig. 3, but they are not necessarily required to have filter-like characteristics. The terminating resistances R_1, R_2 are assumed to be prescribed in each case and also the exact function of frequency representing the insertion loss α due to the presence of the reactance network. The specific problem considered is the design of the reactance network in accordance with these specifications and subject to the requirement of physical realizability.

The first division of the theory solves the problem of determining sets of open- and short-circuit impedances describing physical reactance

13

networks which would produce the prescribed insertion loss when inserted between the prescribed terminations. When combined with Cauer's method (5) of realizing prescribed impedances with reactance 4-poles of his so-called canonical configurations, it amounts to an academic solution of the general design problem described above. The design of more important equivalent configurations will be described in the second division of the theory.

To insure the possibility of physical corresponding networks the prescribed insertion loss function α must be assumed to satisfy certain necessary and sufficient conditions. By definition, α satisfies the equation

$$e^{2\alpha} = \left| \frac{V_{20}}{V_2} \right|^2 \tag{5}$$

in which V_{20} and V_2 represent the complex voltages received before and after the insertion of the network between the prescribed terminations. Since it represents the square of the magnitude of the "complex insertion voltage ratio" $\frac{V_{20}}{V_2}$, the "insertion power ratio" $e^{2\alpha}$ must be an even rational function of frequency with real coefficients and must be positive at all real frequencies. Because the power delivered by a generator of voltage e and internal resistance R_1 cannot exceed $\frac{e^2}{4R_1}$, while the received power corresponding to V_{20} is $\frac{R_2 e^2}{(R_1 + R_2)^2}$, the power ratio $e^{2\alpha}$ must not only be positive at real frequencies but must not be less than $\frac{4R_1 R_2}{(R_1 + R_2)^2}$. In other words, the relation

$$e^{2\alpha} \geqq \frac{4R_1 R_2}{(R_1 + R_2)^2} \tag{6}$$

must be satisfied at all real frequencies.

The above conditions upon $e^{2\alpha}$ are obviously necessary whether or not the network inserted between the terminations must be made up entirely of reactances. It will be shown, however, that these conditions are sufficient to insure the existence of physical corresponding networks of the pure reactance type.

Form of Solution

It turns out that the open- and short-circuit impedances of a reactance network depend upon the insertion loss in a manner involving a rather

14

complicated combination of the roots of the insertion power ratio and a related function, which are normally of fairly high degree. As a result, explicit equations expressing the impedances directly in terms of the insertion loss cannot well be obtained. On the other hand, a simple sequence of formulae has been derived which indicates the relationship and which permits the impedances to be calculated in numerical cases.

Instead of actually deriving the formulae, it will be sufficient to begin by merely stating them without proof. A derivation is unnecessary in that elementary circuit analysis can be used to show that any networks producing the impedances determined by these formulae would yield the prescribed insertion loss when inserted between the prescribed terminations. After the statement of the formulae, however, it will be necessary to demonstrate that the impedances can actually be realized physically.

Determination of Polynomials N and P from $e^{2\alpha}$

To determine the open- and short-circut impedances corresponding to the prescribed insertion loss α, the power ratio $e^{2\alpha}$ is first expressed in the form

$$e^{2\alpha} = \frac{N}{P^2} \tag{7}$$

in which N and P denote even polynomials in the familiar variable p representing $2\pi if$ and are required to have real coefficients. An expression of this type can be readily obtained, since $e^{2\alpha}$ must be an even rational function of the frequency f, while f^2 is proportional to $-p^2$. If the simplest rational fraction expression for $e^{2\alpha}$ does not have the form $\frac{N}{P^2}$ it is only necessary to multiply numerator and denominator by identical factors. As a matter of fact, N and P can always be chosen in a variety of ways, because of the possibility of associating the squares of arbitrary identical factcrs with N and P^2.

Determination of Polynomials A and B from N

After N and P have been determined, a second pair of polynomials A and B must be evaluated. These are even polynomials with real coefficients defined by the pair of equations

$$\begin{aligned} N &= J^2(p_1^2 - p^2)(p_2^2 - p^2) \cdots (p_n^2 - p^2) \\ A + pB &= J\,(p_1 - p\,)(p_2 - p\,) \cdots (p_n - p) \end{aligned} \tag{8}$$

15

in which J must be a real constant, all the p_σ's must have zero or negative real parts, and any complex p_σ's must occur in conjugate pairs. The fact that N must be positive at real frequencies in order for the power ratio $\frac{N}{P^2}$ to be positive is sufficient to insure the possibility of computing J and the p_σ's from their squares, as determined by N, in such a way as to meet the special restrictions imposed upon them.[16] As a matter of fact, there are always two solutions, although only two, because of the arbitrariness of the sign of J. The individual polynomials A and B can obviously be determined by expanding $(A + pB)$ and associating even powers of p with A and odd powers with pB.

The following expression for the complex insertion voltage ratio offers an alternative definition of A and B and indicates their physical significance:[17]

$$\frac{V_{20}}{V_2} = \frac{A + pB}{P} \tag{9}$$

That this voltage ratio corresponds to the original power ratio $e^{2\alpha}$ can be demonstrated by noting that the even polynomials A, B, P are real at real frequencies, while p is imaginary. This requires the power ratio to be

$$e^{2\alpha} = \frac{A^2 - p^2 B^2}{P^2} \tag{10}$$

[16] Since the p_σ's are determined by extracting the square roots of the p_σ^2's determined by N, the signs can be so chosen as to meet the real part requirement. Complex p_σ's with finite real parts will then automatically occur in conjugate pairs corresponding to conjugate p_σ^2's. Pure imaginary p_σ's, which have zero real parts, can also be chosen in conjugate pairs provided the corresponding negative real p_σ^2's occur in identical pairs. Single negative real p_σ^2's cannot be encountered since they represent real frequencies at which N and therefore $\frac{N}{P^2}$ change sign.

[17] In special cases, it may be possible to cancel identical factors out of the numerator and denominator of this expression. Unless the factors are constants or have roots only at real frequencies, however, their cancellation destroys the evenness of the denominator. The even and odd parts of the numerator then lose their utility in regard to the computation of the open- and short-circuit impedances.

The determination of a voltage ratio from a power ratio by (8) and (9) was introduced by Norton (7) in connection with his theory of filters terminated at one end in open- or short-circuits, which forms a part of his theory of constant-resistance groups of filters.

Since reversing the sign of p does not change the even polynomials A and B, the second equation of (8) requires

$$A - pB = J(p_1 + p)(p_2 + p) \cdots (p_n + p) \tag{11}$$

Hence forming the product of $(A + pB)$ and $(A - pB)$ indicates

$$A^2 - p^2B^2 = N \tag{12}$$

and reduces (10) to (7).

The voltage ratio (9) would correspond to the original power ratio even though roots of $(A + pB)$ had positive real parts, provided complex roots occurred in conjugate pairs. The exclusion of roots with positive real parts is merely a condition necessary for physical realizability. This condition is necessary in that the roots represent modes of free oscillation of the complete network consisting of the reactance 4-pole and its terminations.

Determination of Polynomials A' and B' from N and P

In addition to A and B, a very similar pair of polynomials A' and B' are needed. They are even polynomials with real coefficients defined by the following pair of equations, comparable to (8):

$$N - \frac{4R_1R_2}{(R_1 + R_2)^2}P^2 = J'^2(p_1'^2 - p^2)(p_2'^2 - p^2) \cdots (p_n'^2 - p^2)$$

$$A' + pB' = J'(p_1' - p)(p_2' - p) \cdots (p_n' - p) \tag{13}$$

In this case, J' must be real and any complex p_σ''s must occur in conjugate pairs but any of the p_σ''s are permitted to have positive real parts. These conditions can always be satisfied because of the fact that (6) and (7) require $\left(N - \dfrac{4R_1R_2}{(R_1 + R_2)^2}P^2\right)$ to be non-negative at real frequencies. In general, there are actually a number of solutions for A' and B' because of the arbitrariness of the signs of the p_σ''s as well as that of J'.

In the case of A' and B' there is again an alternative definition indicating physical significance, corresponding to equation (9) defining A and B. Suppose Z_1 and Z_2 represent the input impedances of the terminated network measured at the ends terminated in R_1 and R_2, respectively, as indicated in Fig. 4. Then A' and B' can be defined by the equations[18]

$$\frac{R_1 - Z_1}{R_1 + Z_1} = \frac{A' + pB'}{A + pB} \qquad \frac{R_2 - Z_2}{R_2 + Z_2} = \frac{-A' + pB'}{A + pB} \tag{14}$$

[18] In special cases, it may be possible to cancel out common factors from

17

The square of the magnitude of $(\pm A' + pB')$ is $(A'^2 - p^2B'^2)$, which in turn satisfies the equation

$$A'^2 - p^2B'^2 = N - \frac{4R_1R_2}{(R_1 + R_2)^2} P^2 \tag{15}$$

corresponding to (12). Hence (14) requires

$$\left|\frac{R_1 - Z_1}{R_1 + Z_1}\right|^2 = \left|\frac{R_2 - Z_2}{R_2 + Z_2}\right|^2 = 1 - \frac{4R_1R_2}{(R_1 + R_2)^2} e^{-2\alpha} \tag{16}$$

This expression can be checked by examining the relation between the input power and input impedance and recalling that input and output powers are identical in the case of a reactance network.

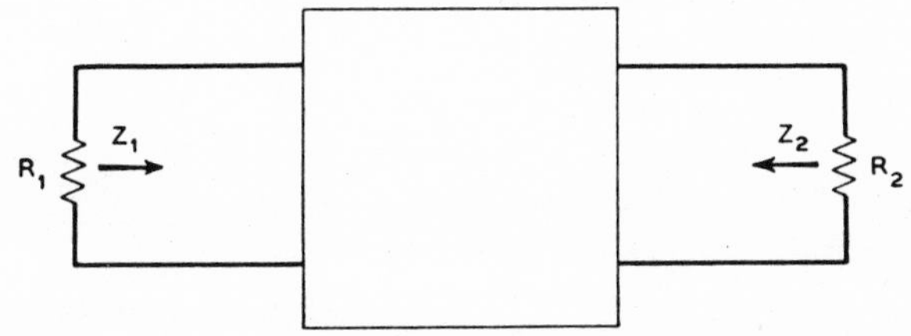

Fig. 4. Input impedances Z_1 and Z_2

The possibility of roots of $(A' + pB')$ with positive real parts is due to the fact that they are the roots of the non-physical impedance $(R_1 - Z_1)$. The simple relation between the two reflection coefficients indicated in (14) can be checked by simple circuit analysis. The following important relation between the A's and B's represents a combination of (12) and (15):

$$A'^2 - p^2B'^2 = A^2 - p^2B^2 - \frac{4R_1R_2}{(R_1 + R_2)^2} P^2 \tag{17}$$

Computation of the Open- and Short-Circuit Impedances from the A's, B's and P

In terms of the A's, B's and P the desired open- and short-circuit impedances are determined by the following formulae:

---

one or both of the equations for the reflection coefficients as in the case of the voltage ratio (9). If the denominators are not $(A + pB)$, however, the numerators lose their direct application to the computation of the open- and short-circuit impedances.

18

$$
\begin{aligned}
Z_{01} &= R_1 \frac{A - A'}{p(B + B')} & Z_{S1} &= R_1 \frac{p(B - B')}{A + A'} \\
Z_{02} &= R_2 \frac{A + A'}{p(B + B')} & Z_{S2} &= R_2 \frac{p(B - B')}{A - A'} \\
Z_{012} &= \frac{-2R_1R_2}{(R_1 + R_2)} \frac{P}{p(B + B')} & Z_{S12} &= \frac{(R_1 + R_2)}{2} \frac{p(B - B')}{P}
\end{aligned}
\tag{18}
$$

In these formulae Z_{01}, Z_{S1}, Z_{02}, Z_{S2} are the familiar open- and short-circuit driving-point impedances of the reactance network as measured at the terminals next to terminations R_1 and R_2, respectively. Z_{012} and Z_{S12} are the less familiar open- and short-circuit transfer impedances defined in Fig. 5.[19] That the impedances determined by (18) correspond to the original power ratio and to the related voltage ratio and reflection coefficients of (9) and (14) can be checked by means of simple circuit analysis.[20]

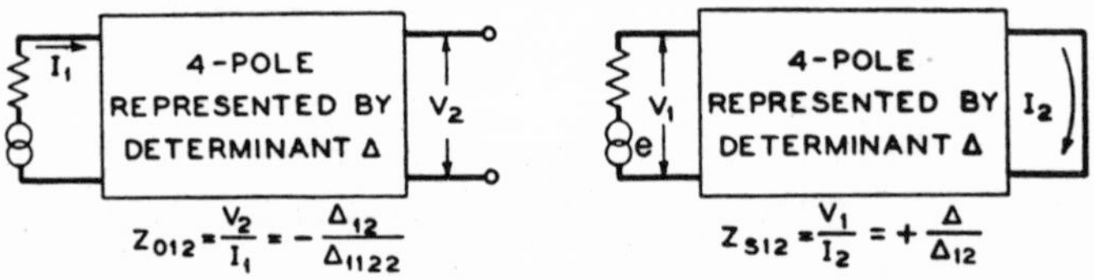

Fig. 5. Open- and short-circuit transfer impedances Z_{012} and Z_{S12}

Physical Realizability of the Impedances Determined by (18)

The necessary and sufficient conditions for the realizability of a set of open- and short-circuit impedances with a physical reactance 4-pole can be stated as follows: First, the four driving-point impedances Z_{01}, Z_{02}, Z_{S1}, Z_{S2} must be separately realizable as the impedances of reactance 2-poles. Second, the two transfer impedances Z_{012}, Z_{S12} must be odd rational functions of p with real coefficients even though they need not be realizable as 2-terminal impedances. Finally, the various

[19] In terms of more conventional notation, these impedances would be referred to as the transfer impedance and the reciprocal of the transfer admittance. The more specific description is called for in the design of tandem section networks such as those to be described in Part II, which sometimes involves the use of short-circuit impedances or open-circuit admittances.

[20] The demonstration may also require the use of equation (17) relating the A's and B's if certain possible expressions for the voltage ratio and reflection coefficients in terms of the impedances are chosen as a starting point.

impedances must be inter-related by the following identities:

$$Z_{01}Z_{02} - Z_{012}^2 = Z_{01}Z_{S2} = Z_{02}Z_{S1} = -Z_{012}Z_{S12} \tag{19}$$

The above conditions differ from the more familiar necessary and sufficient conditions introduced by Cauer (5) in that the requirement of 2-pole realizability is imposed explicitly on all four driving-point impedances, instead of on only two, while Cauer's requirement on the residues of a transfer impedance is abandoned. It turns out that the requirement of four physical 2-pole impedances, together with (19), is sufficient to insure the satisfaction of Cauer's residue requirement. This can be demonstrated by analyzing the behavior of (19) in the neighborhood of a pole of Z_{012}.

An inspection of (18) shows that the expressions for the driving-point and transfer impedances are all odd rational functions of p with real coefficients. The required identities (19) are readily shown to be satisfied by merely replacing the impedances by the corresponding expressions of (18) and then making use of (17). To prove the complete realizability of the impedances it therefore remains only to show that the four driving-point impedances are not only odd rational functions of p with real coefficients but are actually realizable as the impedances of separate reactance 2-poles.

The independent realizability of the four driving-point impedances can be demonstrated with the help of the following theorem. If A_x and B_x are even polynomials in p with real coefficients such that $(A_x + pB_x)$ has no roots with positive real parts, then $\frac{pB_x}{A_x}$ is realizable as the impedance of a physical reactance 2-pole. The truth of this theorem is indicated by the argument outlined below in terms of the theory of positive real functions, or impedances of general 2-poles, as developed by Brune (8).

In the first place, the combination $\frac{A_x}{A_x + pB_x}$ must be a positive real function because of the fact that its poles lie in the left half of the p plane while its resistance part, $\frac{A_x^2}{A_x^2 - p^2 B_x^2}$, is non-negative on the real frequency axis.[21] Hence the reciprocal, $\left(1 + \frac{pB_x}{A_x}\right)$, is a positive real

[21] The polynomial $A_x + pB_x$ is permitted to have roots on the real frequency axis as well as in the left half of the p plane but these are always roots of A_x and pB_x and produce no pole of $\frac{A_x}{A_x + pB_x}$.

function. This requires $\frac{pB_x}{A_x}$ to be a positive real function, since subtracting a real constant from a positive real function leaves a positive real function provided the real part is still non-negative on the real frequency axis. Finally, the fact that the positive real function $\frac{pB_x}{A_x}$ is odd requires it to be realizable as the impedance of a 2-pole of the reactance type.

Dividing the four driving-point impedances of (18) by R_1 or R_2 and then replacing the open-circuit functions by their reciprocals yields the four functions corresponding to different choices of the signs in $\frac{p(B \pm B')}{(A \pm A')}$. Thus the four driving-point impedances will be realizable with separate reactance 2-poles provided $[(A \pm A') + p(B \pm B')]$ has no roots with positive real parts. That this condition is satisfied can be demonstrated by showing $\frac{[(A \pm A') + p(B \pm B')]}{A + pB}$ to be a positive real function. The positive realness of this new function is proved by the following conditions: First, the poles all lie in the left half plane, since $(A + pB)$ has no roots with positive real parts; second, the real part is non-negative at real frequencies, as can be shown by recalling that the A's and B's are real at real frequencies, by evaluating the real part of the function on this basis, and by then using equation (17) relating the A's and B's.[22]

Multiplicity of Solutions

As was indicated previously, a multiplicity of solutions for the impedances can be obtained in a variety of ways. In the first place, squares of identical factors can always be associated with the polynomials N and P^2, in the power ratio $\frac{N}{P^2}$. These factors cancel out in the final impedance formulae only if they are constants or if all their roots occur at real frequencies. In addition, the dependence of A, B, A', and B' upon N and P leaves the sign of $(A + pB)$ arbitrary and normally permits a number of solutions for A' and B'.

The choice of N and P determines the complex insertion voltage ratio $\frac{A + pB}{P}$ except as to sign, while the voltage ratio in turn determines the complexity of corresponding networks, which is normally the same

[22] Roots of $(A + pB)$ on the real frequency axis turn out to be roots of A, pB, A', and pB' and produce no poles of $\frac{[(A \pm A') + p(B \pm B')]}{A + pB}$.

for all choices of A' and B' permitted by N and P.[23] Making the degree of N as low as is permitted by the prescribed insertion loss naturally makes the degree of the voltage ratio as low as possible and leads to the simplest networks. When the squares of identical factors with roots at complex frequencies are combined with N and P^2, more complicated networks producing a modified voltage ratio are obtained. Some of these networks correspond to the addition of familiar all-pass phase-shifting sections to the networks of minimum complexity. An additional multiplicity of networks is normally possible, however, corresponding to impedances which cannot be realized in this way.

The analysis outlined above indicates the existence of a number of physical solutions for the impedances, all assuming a particular form. On the other hand, it does not show that there may not be other solutions of quite different forms. A more complicated analysis, however, shows that there can be no such alternative solutions. In other words, it can be demonstrated that the proper choice of N and P and of the corresponding A's and B's permits the determination of any set of impedances realizable with a reactance network corresponding to the prescribed insertion loss function. This will not be demonstrated here, for the object is merely to show that at least some physical sets of impedances can be obtained.

Practical Design Procedure

For practical purposes it is usually desirable to restrict the insertion power ratio $e^{2\alpha}$ somewhat further than is necessary merely to insure physical realizability. Suppose that identical factors other than constants or the simple factor p^2 must be combined with numerator and denominator of the simplest expression for $e^{2\alpha}$ in order to obtain the form $\frac{N}{P^2}$. It turns out that the corresponding network complexity will then be the same as though these added factors were not identical.[24] On the other hand, requiring the factors to be identical usually leads to less desirable loss characteristics than could be obtained by making

[23] In very special cases, special choices of A' and B' may lead to networks of complexities which are less than normal for the degree of the voltage ratio.

[24] The reason for this is as follows: If Δ is the determinant of the complete network formed by the reactance 4-pole together with its terminations, the insertion power ratio is proportional to $\frac{|\Delta|^2}{|\Delta_{12}|^2}$, in which the cofactor Δ_{12} is independent of the terminations and therefore is an odd or even polynomial in p. Thus the denominator of the simplest rational fraction expression for $e^{2\alpha}$ is the square of an even or odd polynomial in the frequency except when the element values are such that Δ and Δ_{12} have roots in common.

them different. Thus it is generally desirable to require $e^{2\alpha}$ to be so chosen that no identical factors other than constants or the simple factor p^2 need be combined with the simplest corresponding rational fraction in forming $\frac{N}{P^2}$. This merely requires all the poles of $e^{2\alpha}$ to occur in identical pairs.

After the power ratio has been chosen the corresponding open- and short-circuit impedances can be computed in a straightforward manner by means of the formulae described above. Corresponding networks of the canonical type devised by Cauer (5) can then readily be designed from the partial fraction expansions of various of these impedances or of the corresponding admittances. More frequently, however, it is preferable to use equivalent tandem section networks, which can be designed by methods to be described in the next section.

In the determination of the impedances considerable labor may be encountered in the extraction of the roots of the polynomials N and $\left(N - \frac{4R_1R_2}{(R_1 + R_2)^2} P^2\right)$, which are needed in the formation of the polynomials $(A + pB)$ and $(A' + pB')$ in accordance with (8) and (13). The required root extraction can now be expedited, however, by the use of machines which have recently been constructed for the determination of the roots of any polynomials of reasonable degree.[25] In the design of networks specifically for filtering purposes, moreover, the special polynomials encountered are usually such that the greater part of the root extraction labor can be avoided by the use of special methods to be described in Part III.

The theory described above assumes non-dissipative reactance networks. The distortion due to the parasitic dissipation which must be present in actual networks can be readily estimated as soon as the complex voltage ratio is determined provided the dissipation is uniformly distributed. This can be accomplished by making use of the theory of uniformly dissipative networks as developed by Mayer (10) and Bode, which shows how the effects of uniform dissipation can be estimated from the derivative of the insertion phase.[26] Ways of avoiding distortion due to dissipation will be described in Part IV, which will be devoted to methods of designing certain types of dissipative networks producing prescribed insertion loss characteristics.

[25] See, for instance, the description of Fry's isograph recently published by Dietzold (9).

[26] Bode developed a theory similar to Mayer's independently but published only a brief reference to it. This reference is in the paper "Ideal Wave Filters" (11) which he published in collaboration with Dietzold.

One Open- or Short-Circuit Termination

Reactance networks terminated at one end in open- or short-circuits can be considered as limiting cases of the previous networks in which R_1 or R_2 approaches zero or infinity. They are of interest for two principal reasons. For one thing, it is sometimes advantageous to approximate actual open-circuit terminations by means of vacuum tubes. In addition, when groups of filters are to be combined in series or in parallel at one end, it is sometimes best to design the separate filters as though they were to be terminated in open- or short-circuits at their common terminals.[27]

Although the general methods described above can still be used, difficulties with the specific impedance formulae are experienced when one termination approaches zero or infinity. It turns out that all but one of the possible solutions for the polynomials A' and B' permitted by (13) lead to element values which approach zero or infinity with the termination. Even with the one permissible choice most of the open- and short-circuit impedance formulae become indeterminate.[28]

More careful analysis yields the following special relations.[29] If termination R_2 is infinite

$$Z_{01} = R_1 \frac{A}{pB} \qquad Z_{012} = -R_1 \frac{P}{pB} \tag{20}$$

$$R_{Z2} = R_1 e^{-2\alpha}$$

in which R_{Z2} is the resistance measured at the open-circuited end of the terminated network. If termination R_2 is zero

$$Z_{S1} = R_1 \frac{pB}{A} \qquad Z_{S12} = R_1 \frac{pB}{P} \tag{21}$$

$$G_{Y2} = \frac{1}{R_1} e^{-2\alpha}$$

[27] This is the method introduced by Norton (7) for the design of constant-resistance groups of filter. It can also be used to design more efficient filter groups producing approximately constant resistances or conductances at their common terminals at pass band frequencies, the susceptances or reactances being later corrected by the addition of reactance 2-poles.

[28] Some such indeterminacy is necessary, since any arbitrary impedance can be connected in parallel with a short-circuit termination or in series with an open-circuit termination without affecting the operation of the circuit.

[29] The formulae for Z_{O1}, Z_{S1}, R_{Z2}, G_{Y2} are exhibited by Norton (7) as a part of his theory of constant resistance groups of filters.

in which G_{Y2} is the conductance measured at the short-circuited end of the terminated network. Z_{012}, Z_{S12} are the open- and short-circuit transfer impedances while Z_{01}, Z_{S1} are the open- and short-circuit driving-point impedances corresponding to the end of the network terminated in the finite resistance R_1. The polynomials A, B and P are determined from the power ratio exactly as though termination R_2 were finite.

If minimum complexities are assumed, networks can be designed with no additional impedance data. This can be shown by ananalysis of the relations between the open- and short-circuit impedances of physical reactance 4-poles. It will also be clarified by the ladder network theory which will be considered in the next section.

Part II. Realization of Open- and Short-Circuit Impedances with Physical Configurations

Methods have been developed for designing reactance 4-poles of a variety of different configurations producing prescribed open- and short-circuit impedances, such as impedances determined from insertion loss functions by the method described in Part I. Some of these configurations are said to be canonical in that they can be designed to have any set of impedances realizable with physical reactance networks. Others are less general but can frequently be used more advantageously or else amount to special cases of the general configurations for which simpler design methods have been developed.

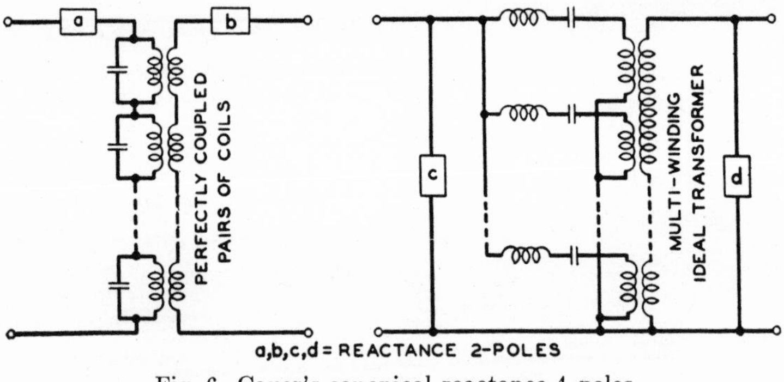

Fig. 6. Cauer's canonical reactance 4-poles

Cauer's Canonical Configurations

The simplest general design theory applies to the design of the two canonical reactance 4-poles introduced by Cauer (5). These have the configurations indicated in Fig. 6. They are designed in terms of partial fraction expansions of the three open-circuit impedances or short-circuit admittances by noting certain relations between partial fractions of the different expansions and the correspondence of these partial fractions to the various network branches. Since Cauer has described these networks very completely, it will not be necessary to describe them in more detail here.

A Canonical Tandem Section Configuration

Cauer's canonical configurations are of particular interest only in theoretical studies of the properties of reactance 4-poles. When it comes to actual construction it is almost always preferable to use equivalent circuits consisting of simple networks or sections connected in the tandem manner indicated in Fig. 7.[30] In the case of selective networks or filters the use of tandem sections is usually a practical necessity. In the first place, the use of tandem sections permits reasonable approximations to theoretical transmission characteristics to be obtained with much less precise adjustment of the elements to their theoretical values.

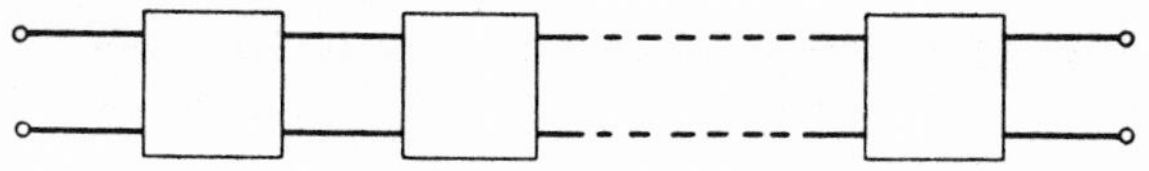

Fig. 7. Tandem 4-poles

In addition, the filters most commonly encountered can be built in tandem section form without the use of the mutual inductances required in equivalent networks of Cauer's canonical configurations.[31]

For many purposes the ladder network is the most useful form of tandem section combination. Even under its most general definition as any sequence of alternate series and shunt 2-terminal impedances, however, the ladder configuration cannot be used to realize all open- and short-circuit impedances realizable with more general reactance 4-poles. On the other hand, it can be shown that a slight modification

[30] By section is not necessarily meant a filter section with the properties of those of the image parameter theory but merely the constituent parts of a combination of 4-poles when these are to be connected in tandem.

[31] The requirement of mutual inductances adds substantially to the difficulty of building a network. If perfect coupling between coils is called for, it can be only approximated. If less than perfect coupling is required, it is difficult to obtain simultaneous adjustments of the self and mutual inductances.

or generalization of the ladder configuration can always be realized and hence can be referred to as a canonical reactance 4-pole. This configuration is defined as any tandem combination of sections included in the four types *A*, *B*, *C*, *D* indicated in Fig. 8, plus possibly an ideal transformer in tandem at one end of the network.[32]

In many cases sections of all four types are not required in the canonical network described above and frequently the ideal transformer does not appear. It also turns out that series inductances can often be included in the series branches, or networks of type *A*, in such a way that the inductive coupling in sections of types *C* and *D* can be less than perfect or can even be completely replaced by separate self-inductances in the shunt branches of these sections. In most filters, for instance,

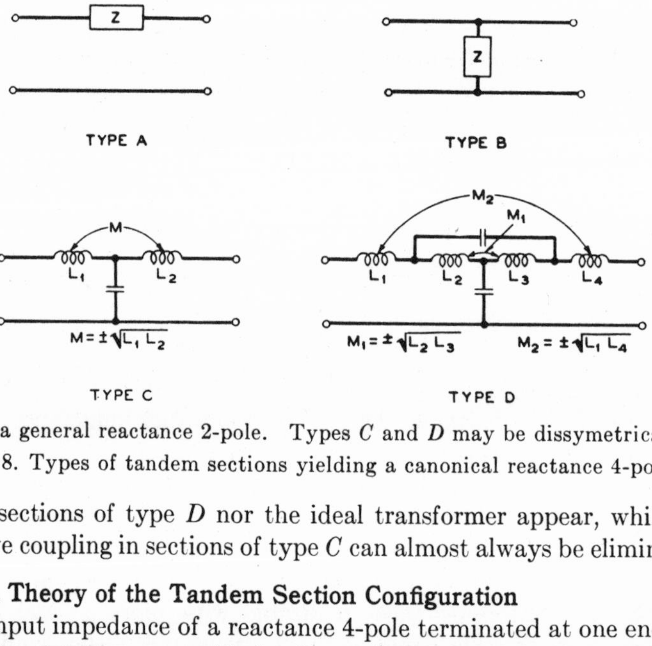

Z = a general reactance 2-pole. Types *C* and *D* may be dissymetrical.
Fig. 8. Types of tandem sections yielding a canonical reactance 4-pole

neither sections of type *D* nor the ideal transformer appear, while the inductive coupling in sections of type *C* can almost always be eliminated.

General Theory of the Tandem Section Configuration

The input impedance of a reactance 4-pole terminated at one end in a constant resistance can readily be determined from the open- and short-circuit impedances. Conversely, it can be shown that the open- and short-circuit impedances can normally be determined from the impedance function corresponding to a prescribed terminating resistance,

[32] Other canonical networks of similar tandem sections can be obtained by modifying somewhat sections of types C and D in accordance with the principles of inverse networks and frequency transformations. It will be sufficient to consider here, however, only the particular type of tandem combination described above.

except for an obvious ambiguity as to the signs of the transfer impedances corresponding to the possibility of interchanging the input or output terminals. The only exceptions to the rule correspond to values of the termination leading to impedances of reduced degree, and there can be only a finite number of such values for any one network.[33]

In accordance with the above principle, one way to obtain prescribed open- and short-circuit impedances of the reactance type is to design a resistance-terminated reactance 4-pole as a 2-pole producing an input impedance computed from the open- and short-circuit impedances. The addition of the proper ideal transformer to the terminated end of the 4-pole will yield the prescribed open- and short-circuited impedances provided the design is so carried out that the degree of the input impedance is normal for the prescribed impedances and for the final configuration. The added ideal transformer is generally necessary in order to obtain the terminating resistance assumed in computing the input impedance from the required open- and short-circuit impedances.

Brune (8) has shown how any prescribed positive real function can be realized as the input impedance of a tandem combination of sections of types A, B and C provided it is permissible to include series resistances between sections. When the roots of the resistance part of the positive real function occur only in identical pairs and are all real or imaginary, Brune's design procedure can be carried out in such a way as to eliminate all the resistances except that forming the termination.[34] The requirement of real or imaginary roots turns out to be unnecessary if Brune's procedure is replaced by a recent modification permitting sections of type D to be used. The configuration obtained can be shown to be such that impedance functions of the same degree as the prescribed function would be obtained with general values of the terminating resistance, i.e., the prescribed function does not represent a special case

[33] In terms of the determinant Δ of the reactance 4-pole and various of its cofactors, the input impedance is $\dfrac{\Delta + R_2\Delta_{22}}{\Delta_{11} + R_2\Delta_{1122}}$. Except in special cases in which the cancellation of identical factors leads to reductions in the degree of the impedance function, this requires $(\Delta + R_2\Delta_{22})$ and $(\Delta_{11} + R_2\Delta_{1122})$ to be proportional to $Kp^n(A_1 + pB_1)$ and $Kp^n(A_2 + pB_2)$, respectively, in which $(A_1 + pB_1)$ and $(A_2 + pB_2)$ are the numerator and denominator of the simplest rational fraction expression for the impedance function. The necessary evenness and oddness of the determinants permits of only two corresponding solutions for the open- and short-circuit impedances. Since it turns out that passing from one solution to the other reverses the sign of Δ_{12}^2, only one solution can be physical.

[34] The resistance part of the function is defined as the even part and represents the real part only at real frequencies.

of reduced degree.[35] Thus a tandem section equivalent can readily be found corresponding to any reactance 4-pole which can produce an input impedance such that the degree is normal for the configuration and such that all the roots of the resistance part occur in identical pairs. This turns out to be a property of all physical reactance 4-poles.[36]

The analysis outlined above proves the canonical nature of the reactance 4-pole formed by adding an ideal transformer to one end of a tandem combination of sections of types A, B, C, D. It can also be shown that the two-terminal impedance formed by closing one pair of terminals through a resistance termination constitutes a canonical general 2-pole. It was indicated above that this type of 2-pole can correspond to any positive real function provided the roots of the resistance part occur only in identical pairs. The restriction on the positive real function turns out to be unnecessary, however, when it is permissible to use special values of the terminating resistance leading to impedance functions of reduced degree.[37] Because of the canonical nature of the reactance 4-pole, this can be demonstrated by merely showing how it is possible to find a set of physical open- and short-circuit impedances of the reactance type leading to an input impedance represented by any prescribed positive real function. How this can be accomplished is indicated below.

Suppose that a reactance network is to be so designed as to produce a prescribed input impedance at one end when terminated at the far end in a prescribed resistance. Because of the identity of the input and received powers, the insertion loss that would be obtained upon terminating the input end in a second resistance can readily be computed from the prescribed impedance.[38] The general theory of Part I can then be used to determine the corresponding sets of physically realizable open- and short-circuit impedances. It is easily shown that one of these

[35] This will be clarified by the subsequent description of the more detailed design procedure.

[36] In terms of the determinant Δ of a reactance network and various of its cofactors, the resistance part of the input impedance is represented by $\dfrac{R_2\Delta_{12}^2}{\Delta_{11}^2 - R_2^2\Delta_{1122}^2}$. It is obvious that single roots of the denominator can coincide with any of the double roots of the numerator only when R_2 has special values.

[37] Because the impedance function must generally be of less than normal degree for the configuration, there are usually equivalent 2-poles representing substantially more efficient use of the elements. In spite of this practical disadvantage, a knowledge of the canonical nature of this particular 2-pole is useful in general network theory problems.

[38] Specifically, the loss can be computed with the help of (16) of Part I.

will correspond to the prescribed impedance provided it is a positive real function.

Design Procedure

Consider now the actual operations involved in the design of a reactance 4-pole of the tandem section type producing prescribed open- and short-circuit impedances. As indicated above, the first step is to compute the input impedance of either end of the network corresponding to an arbitrary far-end termination subject only to the requirement that the degree of the impedance function must be normal for the open- and short-circuit impedances. A section of type A, B, C, or D is then designed in such a way that it will produce the required input impedance when terminated in a new physical impedance of lower degree. A second section is next designed to produce the required terminating impedance when itself terminated in a new impedance of further reduced degree. This procedure is continued until the required terminating impedance is reduced to a constant resistance.[39] Finally, the required terminating resistance is replaced by the equivalent combination of an ideal transformer terminated in a resistance identical with the termination assumed in computing the original input impedance.

Brune shows how any positive real function with a pole or root on the axis of real frequencies can be realized as the input impedance of a section of type A or B, respectively, terminated in an impedance of reduced degree. Brune also shows how any positive real function with no roots or poles on the real frequency axis but such that the resistance part has a pair of identical real or imaginary roots can be realized as the input impedance of a section of type C terminated in a physical impedance of reduced degree. Finally, if the roots of the resistance part of the input impedance all occur in identical pairs, the terminating impedance required in each case turns out to have this same property.[40] Hence, to complete the explanation of the design procedure it remains only to show how the appearance of identical complex roots of the resistance part permits a section of type D to be made use of in the same

[39] The exact manner in which the degree of the terminating impedance is reduced at each stage in the design is what leads to a final configuration for which the original input impedance is of normal degree.

[40] The computation of the terminating impedance from the input impedance may eliminate a pair of identical roots of the resistance part in reducing the degree of the impedance function and may convert one or more pairs of identical real roots into roots or poles of the impedance function or vice versa. Otherwise the roots of the resistance part remain unchanged.

way that a section of type C can be used when the roots are real or imaginary.

Design of Sections of Type D

Suppose that a prescribed section of type D produces a prescribed input impedance Z when terminated in impedance Z_L. If Δ is used to represent the determinant of the section of type D and if the input and output meshes are numbered 1 and 2, respectively, Z_L is related to Z by

$$Z_L = \frac{\Delta - Z\Delta_{11}}{Z\Delta_{1122} - \Delta_{22}} = \frac{\Delta_{12}^2 Z}{\Delta_{22}(\Delta_{22} - Z\Delta_{1122})} - Z_{s2} \tag{22}$$

in which Z_{s2} represents the short-circuit impedance of the section of type D at the terminated end. If it is assumed that Z is a positive real function such that its resistance part has a pair of identical complex roots, it can be shown that the following condition always determines a physical section of type D which leads to a physical Z_L of lower degree than Z: The functions Δ_{12}^2 and $(\Delta_{22} - Z\Delta_{1122})$ are required to have coincident pairs of identical complex roots which are also coincident with a pair of identical complex roots of the resistance part of Z.

It is easily shown that the above coincidence of roots of Δ_{12}^2 and $(\Delta_{22} - Z\Delta_{1122})$ leads to the elimination of all roots of the impedance $(Z_L + Z_{s2})$ which are not also roots of Z, which excludes the possibility of roots in the right half of the p plane. If it is also assumed that the section of type D is physical, simple additional analysis shows that $(Z_L + Z_{s2})$ must actually be a positive real function. This proves the positive realness of Z_L itself except that it does not exclude real-frequency poles with negative residues covered up in $(Z_L + Z_{s2})$ by the positive residues of coincident poles of Z_{s2}. Finally, it can be shown that any coincident poles of the two impedances can be separated, without changing the corresponding residues of Z_L, by adding Lp to the original input impedance Z. The reduced degree of Z_L can be demonstrated by showing that the above conditions lead to eight coincidences of roots and poles of the expression $\dfrac{\Delta - Z\Delta_{11}}{Z\Delta_{1122} - \Delta_{22}}$ of (22).

To show that the condition on the roots of Δ_{12}^2 and $(\Delta_{22} - Z\Delta_{1122})$ actually leads to a physical section of type D, definite formulae for the element values are first derived from this condition without considering the question of physical realizability. The input impedance Z appears in these formulae only through the appearance of the values assumed by Z and its derivative at double roots of its resistance part. This permits Z to be replaced in the formulae by the open-circuit impedance of a physical reactance 4-pole which produces the input impedance Z when terminated in the proper constant resistance.[41] It can then be shown that the corresponding capacities and self inductances of the section of type D will be finite real and positive.

Conditions Necessary for Ladder Networks

When all the frequencies of infinite loss are real or imaginary, sections of type D do not have to be included in the canonical tandem section configuration.[42] This is because the frequencies of infinite loss of the complete network are the corresponding critical frequencies of the separate sections, while sections of type D are required only to realize complex frequencies of infinite loss. When sections of type D are absent the network can be considered as a general ladder of reactances, which may be defined as any combination of alternate series and shunt branches consisting of reactance 2-poles. While the coupling between the series inductances in sections of type C renders them somewhat more complicated than alternate series and shunt 2-poles the coupling can be thought of as merely a device for realizing negative inductances appearing in the equivalent T networks.[43]

The Mid-Series Low-Pass Ladder Configuration

Of the large variety of actual configurations possible in ladders of the type described above, only a few are commonly made use of. An extensive special design theory has been developed for these particular configurations in order to permit the element values to be determined

[41] As was indicated previously, the theory of Part I can be used to demonstrate that any positive real function can be realized as the input impedance of a resistance-terminated reactance 4-pole.

[42] By the frequencies of infinite loss of a network is meant the frequencies of infinite loss obtained with general finite resistance terminations excluding any specific terminations which bring roots and poles of the general expression for the insertion voltage ratio into coincidence. Each frequency of infinite loss is included in one or more of the following groups of critical frequencies: roots of the open-circuit transfer impedance, poles of the short-circuit transfer impedance, coincident roots or poles of open- and short-circuit driving-point impedances of the same end of the network, and zeros of the resistance part of an input impedance obtained with a resistance termination.

[43] In many cases the negative inductance can be eliminated without the introduction of the coupling but the conditions under which this can be done are not subject to simple statement.

without the labor involved in using the general theory of tandem sections described previously. The special design theory is best developed first for the specific type of ladder indicated in Fig. 9, which may be referred to as the mid-series low-pass configuration even though it is not necessarily a mid-series type low-pass filter of the image parameter theory.[44] The other configurations commonly encountered can be designed by means of simple modifications of the theory of this special case.

The earliest special formulae for computing the element values of ladders of the mid-series low-pass configuration indicated in Fig. 9 were developed by Norton (7) as a part of his theory of constant resistance pairs of filters. Although Norton's formulae represented an important step in the development of the theory of mid-series low-pass ladders, it has been found that the computations which they call for in numerical problems are undesirably complicated and must usually be carried to an abnormally high precision.

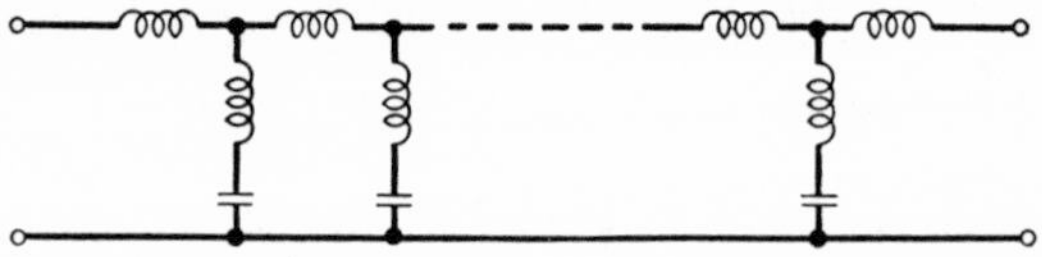

Fig. 9. The mid-series low-pass ladder configuration

As a result an extensive further analysis of the design problem has been carried out leading to the derivation of a new set of formulae. These new formulae are relatively satisfactory for numerical computations and are also useful in a variety of theoretical investigations, such as the determination of what impedances can be realized with mid-series low-pass ladders. In addition, the derivation of the formulae involves the development of an alternative set which are expressed very compactly in terms of determinants and which are useful in certain theoretical investigations even though they have the same disadvantages as Norton's when applied to ordinary numerical problems.

Assumptions and Conditions Leading to Design Formulae

The development of the design formulae for ladders of the mid-series low-pass configuration is simplified if certain simple assumptions are adopted temporarily. The procedure to be followed when the assumptions are not satisfied can best be investigated after the formulae have

[44] This type of ladder is obviously equivalent to sections of type C generally combined with one or more sections of type A consisting of simple series inductances.

been derived. In the first place, it is simplest to start by assuming a set of open- and short-circuit impedances to have been specified which are known in advance to be appropriate for the configuration, leaving until later the question of what impedances have this property. It is also best to assume further that the impedances are such that the multiplicity of solutions for the elements which are normally encountered are all possible, even though special sets of impedances can be found which require certain of the solutions to be excluded. Certain difficulties are also avoided by the temporary assumption that all the frequencies of infinite loss are different and that all the open- and short-circuit impedances are of normal degree for the configuration even though some may be of reduced degree in special cases.

The following relations, which can be shown to apply to any ladder of the mid-series low-pass configuration consistent with the above assumptions, form the basis of all known formulae for the element values.[45] First, the resonances of the shunt branches are identical with the frequencies of infinite loss except for a single infinite loss point at infinity.[46] Second, the value assumed by any of the open- and short-circuit driving-point impedances at a shunt branch resonance is independent of the elements separated from its terminals by the shunt branch, which acts as a short-circuit across the ladder. Finally, under the assumption of impedances of normal degree for the configuration the derivatives of the driving-point impedance functions with respect to the frequency have this same property.

It turns out that the relations stated above are sufficient to determine all element values from one open-circuit or short-circuit driving-point impedance together with the frequencies of infinite loss, except for the far end inductance in the case of an open-circuit impedance.[47] The

[45] These relations were introduced by Norton as the basis of his design equations.

[46] Under the present assumptions, the finite frequencies of infinite loss are the roots of the open-circuit transfer impedance and also the finite poles of the corresponding short-circuit impedance.

[47] This at first seems contrary to the well known fact that it takes three impedances to fix a 4-pole. The additional data are here supplied by the assumption of impedances appropriate for a specific configuration subject to special restrictions. Recall that Cauer's canonical reactance 4-pole of the shunt or admittance type can be designed from one short-circuit driving-point impedance and the short-circuit transfer impedance except for a two terminal shunt branch across the far end of the network. For the particular circuit under consideration the terminal shunt branch would be absent in the equivalent shunt type canonical network while the short-circuit transfer impedance could be found from a short-circuit driving-point impedance and the frequencies of infinite loss.

multiplicity of solutions which are normally obtainable are due only to the fact that the finite frequencies of infinite loss can be distributed arbitrarily among the shunt branches as their individual resonances. In order to obtain a unique solution it is expedient to assume at the outset that a particular distribution has been chosen. The problem then becomes that of realizing a known two-terminal impedance of the reactance type as an open-circuit or short-circuit impedance of a ladder of the midseries low-pass configuration with prescribed shunt branch resonances.

Continued Fraction Expansion Forming Basis of Design Problem

The development of the formulae for the element values calls for the introduction of extensive special notation so chosen as to reduce the design problem to the determination of a particularly simple continued fraction expansion of a known function. In the first place, instead of dealing directly with the element values, it is simpler to consider the

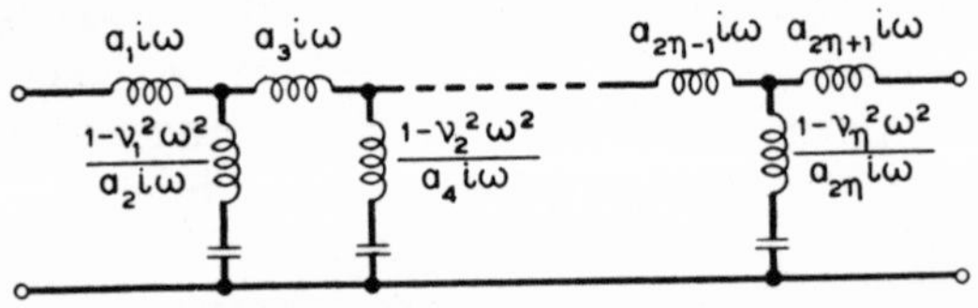

Fig. 10. Designation of the impedance branches of a mid-series low-pass ladder

constants in the designations indicated in Fig. 10 for the impedances of the various network branches, expressed in terms of ω rather than $p = i\omega$.[48] The constants $\nu_1, \cdots, \nu_\eta$ represent the reciprocals of the shunt branch resonances or finite points of infinite loss in terms of ω and are thus assumed to be known.[49] The problem under consideration therefore amounts to the determination of the so-called ladder network coefficients $a_1, \cdots, a_{2\eta+1}$, since there can be no difficulty in determining the element values from these coefficients together with the ν_σ's. In order to avoid ambiguity in the design formulae referred to previously, which are merely specific formulae for the individual a_σ's, it is convenient to include the additional requirement that the numbering of the coefficients a_1, a_2, etc., shall begin at the terminals of the particular

[48] Although the variable p is more convenient in formulating the general theory of Part I, ω turns out to be more convenient in the ladder network theory under consideration here.

[49] The reciprocals of the values of ω at infinite loss points are more convenient than the values themselves when the infinite loss points are later permitted to become infinite.

open-circuit or short-circuit driving-point impedance from which they are to be computed.

It is well known that the problem of realizing a 2-terminal impedance as the input impedance of a ladder network of prescribed configuration amounts to the problem of obtaining a prescribed form of continued fraction expansion of the impedance or of some related function.[50] For the particular network under consideration, the required continued fraction is simplest if the function F to be expanded is derived by dividing the impedance function by p, or by its equivalent $i\omega$. In other words, F is best defined by

$$F = \frac{Z}{i\omega} \tag{23}$$

where Z is the open- or short-circuit impedance from which the coefficients a_1, a_2, etc. are to be computed.

Since Z is an odd rational function of $i\omega$, the quantity F must be a function of ω^2. This suggests replacing ω^2 by a new variable in order to decrease the degree of F. It turns out, however, that a simpler continued fraction is obtained if the reciprocal of this variable is used. Hence the following additional notation is introduced:

$$z = \frac{1}{\omega^2} \tag{24}$$

$$z_\sigma = \nu_\sigma^2 \qquad \sigma = 1, \cdots, \eta$$

in which z represents the new variable, while the z_σ's indicate the values of z corresponding to the frequencies of infinite loss. In terms of this notation, the required continued fraction expansion of F takes the following form:

$$F = a_1 + \cfrac{1}{\cfrac{a_2}{z_1 - z} + \cfrac{1}{a_3 + \cfrac{1}{\cfrac{a_4}{z_2 - z} + \cdots}}} \tag{25}$$

The problem is to solve this identity for the a_σ's assuming the constants z_σ to be known and also the function F of the variable z.

In deriving the more useful formulae for the a_σ's which have been developed by solving the above problem the first part of the analysis

[50] A variety of ladders described by Fry (12), for instance, correspond to Stieltjes' Fractions.

is devoted to the derivation of the alternative formulae referred to previously as being expressed compactly in terms of determinants. The final formulae are then derived by expanding the determinants in terms of the partial fraction representation of the function F. Since the two parts of the derivation both involve long and complicated algebraic manipulation, they will be no more than briefly outlined here. Greater clarity will be obtained if the statement of the preliminary formulae in terms of determinants precedes the outline of their derivation.

Formulae for the a_σ's in Terms of Determinants

The determinants appearing in the preliminary solution for the a_σ's are formed from the quantities H_{qr} defined as follows in terms of the notation introduced above:

$$H_{qr} = \frac{F_q - F_r}{z_q - z_r} \qquad H_{qq} = F'_q \tag{26}$$

where F_q, F'_q, etc., are used to represent the values assumed by F and $\frac{dF}{dz}$ at $z = z_q$, etc. The use of the notation H_{qr} and H_{qq} is consistent since H_{qq} is the limit approached by H_{qr} as z_r approaches z_q.

The determinants themselves are of three different types. The determinant U_k is defined as $|H_{qr}|$ in which q and r take the values 1 to k. In other words,

$$U_k = \begin{vmatrix} H_{11} & H_{12} & \cdots & H_{1(k-1)} & H_{1k} \\ H_{21} & H_{22} & \cdots & H_{2(k-1)} & H_{2k} \\ \cdots & \cdots & \cdots & \cdots & \cdots \\ \cdots & \cdots & \cdots & \cdots & \cdots \\ H_{k1} & H_{k2} & \cdots & H_{k(k-1)} & H_{kk} \end{vmatrix} \tag{27}$$

The determinant V_k is obtained from U_k by changing the elements of the last column to unity. In other words,

$$V_k = \begin{vmatrix} H_{11} & H_{12} & \cdots & H_{1(k-1)} & 1 \\ H_{21} & H_{22} & \cdots & H_{2(k-1)} & 1 \\ \cdots & \cdots & \cdots & \cdots & \cdots \\ \cdots & \cdots & \cdots & \cdots & \cdots \\ H_{k1} & H_{k2} & \cdots & H_{k(k-1)} & 1 \end{vmatrix} \tag{28}$$

Finally, the determinant W_k is obtained from U_k by changing the elements of the last column from $H_{\sigma k}$ to $H_{\sigma(k+1)}$. Thus,

$$W_k = \begin{vmatrix} H_{11} & H_{12} & \cdots & H_{1(k-1)} & H_{1(k+1)} \\ H_{21} & H_{22} & \cdots & H_{2(k-1)} & H_{2(k+1)} \\ \cdots & \cdots & \cdots & \cdots & \cdots \\ \cdots & \cdots & \cdots & \cdots & \cdots \\ H_{k1} & H_{k2} & \cdots & H_{k(k-1)} & H_{k(k+1)} \end{vmatrix} \tag{29}$$

In terms of these determinants all the ladder network coefficients except for $a_{2\eta+1}$ are given by the following design equations:[51]

$$\begin{aligned} a_1 &= F_1 & a_{2k-1} &= \frac{(z_k - z_{k-1})U_{(k-1)}W_{(k-1)}}{V_{(k-1)}V_k} \\ a_2 &= \frac{-1}{U_1} & a_{2k} &= \frac{-V_k^2}{U_{(k-1)}U_k} \end{aligned} \tag{30}$$

$$k = 2 \cdots \eta$$

The coefficient $a_{2\eta+1}$ corresponds to the inductance forming the last series branch. If the function F is obtained from an open-circuit impedance, the coefficient must be determined from some other impedance. If F corresponds to a short-circuit impedance, $a_{2\eta+1}$ can be found from the value assumed by F at zero frequency, or by using the above formulae with an arbitrary additional constant $z_{\eta+1}$.[52]

The above formulae can be derived by a method of induction which is reasonably straightforward though long and tedious. The formulae for the first four coefficients are first derived from Norton's equations or directly from the behavior of the continued fraction (25) in the neighborhood of $z = z_1$ and $z = z_2$. This yields the special formulae for a_1 and a_2 indicated above and shows the formulae for a_3 and a_4 to be consistent with the general formulae indicated for a_{2k-1} and a_{2k}. The derivation is then completed by showing that if the formulae for a_{2k-1} and a_{2k} are correct, then those for a_{2k+1} and a_{2k+2} must also be correct. This is demonstrated by first using the formulae for a_{2k-1} and a_{2k} to express a_{2k+1} and a_{2k+2} in terms of the impedance that would be obtained by removing the first two branches of the network, corresponding to

[51] The formulae for a_2, a_3, a_4, involve U_1, V_1, W_1. As defined by (27), (28) and (29) these quantities appear strange in that they are determinants of only the first order. These first order determinants, however, merely represent H_{11}, 1, H_{12}, respectively.

[52] If F corresponds to a short-circuit impedance, an arbitrary shunt branch can be assumed to be connected across the short-circuit since it will not affect F. This permits $a_{2\eta+1}$ to be determined exactly as though there were a complete additional "section."

a_1 and a_2. This "reduced" impedance is then replaced by an equivalent expression in terms of the original impedance and the formulae for a_1 and a_2. Considerable manipulation of the determinants in the resulting equations finally yields the general formulae for a_{2k+1} and a_{2k+2}.

The formulae (30) are indeterminate unless all the constants $z_1, \cdots, z_\eta$ are different in accordance with the original assumption of no two identical frequencies of infinite loss. Coincident frequencies of infinite loss can be handled, however, by assuming infinitesimal differences and making use of a Taylor's series expansion of the function F representing $\frac{Z}{i\omega}$ considered as a function of $z = \frac{1}{\omega^2}$. When all the frequencies of infinite loss are identical except for the single one at infinity, the continued fraction (25) becomes a Stieltjes' fraction of a type considered by Fry (12). The known formulae for the constants in the Stieltjes' fraction expansion are undoubtedly derivable from (30) by the Taylor's series method.[53]

The effect of abandoning the original assumptions other than that of no two identical frequencies of infinite loss can best be considered after the final formulae have been developed by showing how the determinants of the first set can be expanded.

Derivation of the Final Design Formulae by Expanding the Determinants

It is readily shown that if the function F is expanded into a sum of partial fractions, this expansion will always take the following form provided the open- or short-circuit driving-point impedance from which it is derived is physically realizable:

$$F = -B_0 z + B_\infty + \frac{B_1}{z - \beta_1} + \cdots + \frac{B_\mu}{z - \beta_\mu} \tag{31}$$

In this expression the B's and β's are all positive and are also all finite except that B_0 may sometimes be zero.

In terms of the partial fraction expansion of F, the determinant element H_{qr} as defined in (26) becomes

$$H_{qr} = -B_0 - \frac{B_1}{(\beta_1 - z_q)(\beta_1 - z_r)} - \cdots - \frac{B_\mu}{(\beta_\mu - z_q)(\beta_\mu - z_r)} \tag{32}$$

[53] This derivation has not been proved rigorously but has been carried far enough to indicate the way in which the transformation of the formulae takes place.

while H_{qq} is obtained by merely equating z_r to z_q in this formula. The determinants U_k, V_k, W_k defined in (27) through (29) and appearing in the formulae (30) for the a_σ's can be expanded in terms of these partial fraction representations of the H_{qr}'s. The derivation of these expansions, however, is too long and complicated to be more than briefly outlined here.

In deriving the expansions of the determinants, the particular W_k determinant of degree identical with the number of partial fractions in the expansion of H_{qr} is first examined. It is found that this particular W_k determinant can be expressed in terms of a product of two determinants of the general form $\left|\frac{1}{x_i - y_j}\right|$, which are evaluated in well-known treatises on determinant theory.[54] It is then shown that W_k determinants of higher degree must vanish while those of lower degree are equivalent to sums of similar factorable determinants.[55] Each term in these sums is actually the determinant that would be obtained by using only k of the partial fractions of H_{qr}, i.e., by setting all but k of the B's in (32) equal to zero. There must be one such term for every possible choice of k partial fractions. After the expansions of the W_k determinants have been determined the U_k and V_k determinants can be expanded by treating them as certain limiting cases of W_k determinants.

When the expansions of the determinants are inserted in the equations (30) for the a_σ's, a variety of factors in differences between the infinite loss points z_1, z_2, etc., can be cancelled out. The formulae then take the form

$$\begin{aligned} a_1 &= F_1 & a_{2k-1} &= \frac{\Psi_{U(k-1)}\Psi_{W(k-1)}}{\Psi_{V(k-1)}\Psi_{Vk}} \\ a_2 &= \frac{1}{\Psi_{U1}} & a_{2k} &= \frac{\Psi_{Vk}^2}{\Psi_{U(k-1)}\Psi_{Uk}} \end{aligned} \tag{33}$$

in which the Ψ's represent the uncancelled parts of the expansions of the original determinants.

The Ψ's themselves are best expressed in terms of the quantities u_{qk}, v_{qk}, w_{qk} defined by the recursion formulae

54 See, for instance, the chapter on functional determinants in the treatise by Scott and Mathews (13).

55 The similar vanishing of higher degree determinants of the U_k and V_k types shows the finite nature of the continued fraction (25).

$$u_{q0} = B_q \qquad v_{qk} = \frac{u_{q(k-1)}}{(\beta_q - z_k)} \qquad u_{qk} = \frac{u_{q(k-1)}}{(\beta_q - z_k)^2} \qquad w_{qk} = \frac{u_{q(k-1)}}{(\beta_q - z_k)(\beta_q - z_{k+1})} \tag{34}$$

The statement of the Ψ's in terms of these quantities requires a very complicated summation and product notation in the general case. Hence, it will be better to avoid the necessity of a statement of the general case by listing enough specific cases to establish what such a statement would have to show.

The formulae for the simpler Ψ's are listed in Table I together with the additional previous relations necessary in the actual computation of ladder network coefficients. These are sufficient to indicate the general case and are also sufficient by themselves for ordinary design purposes, particularly when impedances corresponding to both ends of a network are known so that part of the elements can be determined from each end. Some additional simplifications can be obtained, however, by developing more specialized forms of the equations for specific numbers of sections. The behavior of the impedances at zero and infinity can also be used advantageously in determining one or two coefficients for which the standard formulae are most complicated or for checking purposes. Special simplified formulae have also been derived for the ladder equivalents of symmetrical lattice networks, which are normally computed first in the design of symmetrical circuits. These formulae can best be introduced at a later point, however, when the theory of lattice networks is considered.

In ordinary numerical problems, the expanded formulae do not require the extremely high precision of computation which is commonly necessary when Norton's equations are used or the formulae in terms of determinants. Similarly, the expanded formulae do not become indeterminate when coincident frequencies of infinite loss are encountered, as do the other formulae. They also lead to somewhat more straightforward numerical computations in ordinary design problems, although the extent to which this is true may vary widely from problem to problem. While the complexity of the formulae increases so rapidly with network complexity as to render them unsatisfactory for the design of ladders of more than four or possibly five sections, such complicated networks are rarely encountered. Finally, the formulae for ladders of general numbers of sections assume a form which renders them useful in such general studies as the investigation of the requirements on impedances of low-pass ladders or of the possibility of avoiding coupled coils, etc.

TABLE I

General Mid-Series Low-Pass Ladders

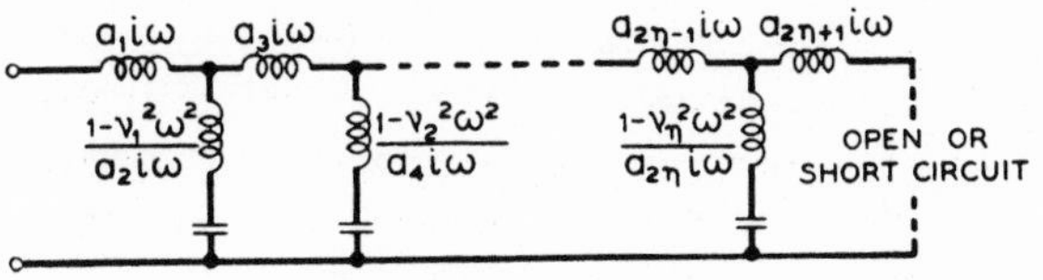

$$F = \frac{Z}{i\omega} = -B_0 z + B_\infty + \sum_{q=1}^{\mu} \frac{B_q}{z - \beta_q} \qquad z = \frac{1}{\omega^2} \qquad \alpha = \infty \text{ at } z = z_k = \nu_k^2$$

$$u_{q0} = B_q \qquad v_{qk} = \frac{u_{q(k-1)}}{(\beta_q - z_k)} \qquad u_{qk} = \frac{u_{q(k-1)}}{(\beta_q - z_k)^2} \qquad w_{qk} = \frac{u_{q(k-1)}}{(\beta_q - z_k)(\beta_q - z_{k+1})}$$

$$\Psi_{V1} = 1 \qquad \Psi_{V2} = \sum v_{q2} \qquad \Psi_{V3} = \sum v_{q3} v_{r3} (\beta_q - \beta_r)^2$$
$$\Psi_{V4} = \sum v_{q4} v_{r4} v_{s4} (\beta_q - \beta_r)^2 (\beta_q - \beta_s)^2 (\beta_r - \beta_s)^2$$

$$\Psi_{U1} = \sum u_{q1} + B_0 \qquad \Psi_{U2} = \sum u_{q2} u_{r2} (\beta_q - \beta_r)^2 + B_0 \sum u_{q2}$$
$$\Psi_{U3} = \sum u_{q3} u_{r3} u_{s3} (\beta_q - \beta_r)^2 (\beta_q - \beta_s)^2 (\beta_r - \beta_s)^2 + B_0 \sum u_{q3} u_{r3} (\beta_q - \beta_r)^2$$

$$\Psi_{W1} = \sum w_{q1} + B_0 \qquad \Psi_{W2} = \sum w_{q2} w_{r2} (\beta_q - \beta_r)^2 + B_0 \sum w_{q2}$$
$$\Psi_{W3} = \sum w_{q3} w_{r3} w_{s3} (\beta_q - \beta_r)^2 (\beta_q - \beta_s)^2 (\beta_r - \beta_s)^2 + B_0 \sum w_{q3} w_{r3} (\beta_q - \beta_r)^2$$

$$\sum u_q u_r = \sum_{q=1}^{\mu-1} \sum_{r=q+1}^{r=\mu} u_q u_r = u_1 u_2 + u_1 u_3 + u_2 u_3 + \cdots + u_\mu u_{\mu-1}$$

and similarly for the other sums of products of terms.

$$a_1 = B_\infty - B_0 z_1 - \sum v_{q1} \qquad a_{2k-1} = \frac{\Psi_{U(k-1)} \Psi_{W(k-1)}}{\Psi_{V(k-1)} \Psi_{Vk}}$$

$$a_2 = \frac{1}{\Psi_{U1}} \qquad a_{2k} = \frac{\Psi_{Vk}^2}{\Psi_{U(k-1)} \Psi_{Uk}}$$

$$k = 2, \cdots, \eta$$

Insertion Loss Functions and Impedances Realizable with Ladders of the Mid-Series Low-Pass Configuration

The mid-series low-pass ladder configuration turns out to be appropriate for the realization of insertion power ratios of the form

$$e^{2\alpha} = \frac{1 + \Gamma_1 \omega^2 + \Gamma_2 \omega^4 + \cdots + \Gamma_{2\eta+1} \omega^{4\eta+2}}{(1 - \nu_1^2 \omega^2)^2 (1 - \nu_2^2 \omega^2)^2 \cdots (1 - \nu_\eta^2 \omega^2)^2} \qquad (35)$$

in which η represents the number of shunt branches in the ladder while the Γ_σ's and ν_σ's are arbitrary constants. Corresponding open- and short-circuit impedances realizable with mid-series low-pass ladders and subject to the previous assumptions of normal degrees for the configuration and of normal multiplicity of solutions for the element values can always be found by a straightforward method except in certain special cases corresponding to discrete choices of the Γ_σ's and ν_σ's.[56]

The first step in determining the impedances is to find a solution for the polynomials A, B, A', and B' of the general impedance theory of Part I, using the numerator and denominator of (35) as the polynomials N and P^2. It is readily shown that the multiplicity of solutions for the polynomials, which was indicated in part I, is such that the signs of A' and B' can be chosen arbitrarily as far as general realizability is concerned and also the sign of $(A + pB)$. Except in the very special cases referred to above, the impedance formulae exhibited in equations (18) in Part I yield corresponding impedances realizable with ladders of the desired type provided the signs of A', B', and $(A + pB)$ are chosen in accordance with the following conditions: $\frac{A}{P}$ must be positive at zero frequency, $\frac{B'}{B}$ must be negative at infinite frequency, and $\frac{A'}{A}$ must be positive or negative at zero frequency depending upon whether termination R_1 is greater than or less than R_2.[57]

Difficulties Encountered in Special Cases

The difficulties which can be encountered in special cases are of two types. The first type can be encountered even though the impedance functions are of the form normal for the configuration. The second

[56] By discrete choices of the constants is meant choices which can always be avoided by small changes in a single constant. It is assumed, of course, that $e^{2\alpha}$ meets the physical requirement that it must be positive and no less than $\frac{4R_1R_2}{(R_1+R_2)^2}$ at all real frequencies.

[57] Actually all solutions are normally realizable if modified mid-series low-pass ladders are permitted. The requirement upon the sign of $\frac{A}{P}$ or $\frac{A'}{A}$ can be violated, for instance, if an ideal transformer is included in one end of the network. Similarly, the requirement upon the sign of $\frac{B'}{B}$ can normally be violated if a shunt condenser is added to one end of the ladder and if negative series inductances realizable with perfectly coupled coils are permitted.

type corresponds to the appearance of various open- or short-circuit impedances which are of reduced degree because of the coincidence of roots of the numerators and denominators of the corresponding general formulae.

It can be shown that if all the impedances are of normal degree, no finite frequency of infinite loss can coincide with a root or pole of an open- or short-circuit driving-point impedance. It follows that the corresponding values of the quantities v_{qk}, u_{qk}, w_{qk} appearing in the design formulae of Table I will all be finite. Of these quantities, u_{qk} will always be positive but v_{qk} and w_{qk} can be negative. As a result, the quantities Ψ_{Uk} formed from the u_{qk}'s in the manner indicated in Table I will all be finite but the quantities Ψ_{Vk} and Ψ_{Wk} may be zero. The vanishing of Ψ_{Wk} merely replaces a series inductance by a simple conductor but the vanishing of Ψ_{Vk} leads to the requirement of three network branches with infinite impedances at all frequencies, one shunt branch and the two adjacent series branches.[58]

If it is no longer assumed that all the multiplicity of solutions for the a_σ's are to be physical, difficulties of this type can normally be overcome by modifying the choice of the particular frequencies of infinite loss which are to be the resonances of the individual shunt branches or by choosing a different set of impedances corresponding to the same insertion loss. It is within the bounds of possibility, however, to encounter cases in which all solutions lead to the same difficulties. The ladder must then be modified to the extent of using at least one anti-resonant circuit as a series branch. This can be done by modifying properly the normal design procedure.

When there are finite frequencies of infinite loss which are coincident with roots or poles of open- or short-circuit driving-point impedances, some of the impedance functions will be of reduced degree. It is then normally possible to realize the impedances by adding terminal series or shunt branches or both to a mid-series loss-pass ladder with impedances of normal degree for its configuration. It is also usually possible, however, to obtain complete networks of the normal mid-series low-pass form by merely using impedances that are still of normal degree in computing the element values. When all impedances are of reduced degree, modifications of the normal design procedure can be used or else the general method applying to the canonical tandem section configuration.

[58] Ψ_{vk}, Ψ_{wk} may also be negative, of course, leading to negative elements, but these turn out to be realizable with coupled coils.

Elimination of Various Elements

There is no difficulty in permitting any or all of the frequencies of infinite loss to be placed at infinity. This merely requires the corresponding shunt branches to be simple condensers rather than resonant circuits. The design is carried out by setting the proper ν_σ's equal to zero in (35) and also the corresponding z_σ's representing their squares in the design formulae of Table I.

Certain power ratios of the general form of (35) also lead to the vanishing of one or more of the series inductances. One special case of this type is of particular importance. In this case, one terminal series branch vanishes while the next shunt branch is a simple capacity, which leaves a network of the type of Fig. 11. The proper form of power ratio is obtained from (35) by reducing the terms in the numerator by one and also the number of ν_σ's.[59] In other words,

$$e^{2\alpha} = \frac{1 + \Gamma_1\omega^2 + \Gamma_2\omega^4 + \cdots + \Gamma_{2\eta}\omega^{4\eta}}{(1 - \nu_1^2\omega^2)^2(1 - \nu_2^2\omega^2)^2 \cdots (1 - \nu_{\eta-1}^2\omega^2)^2} \tag{36}$$

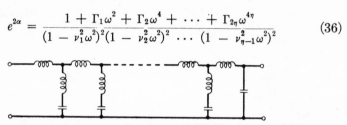

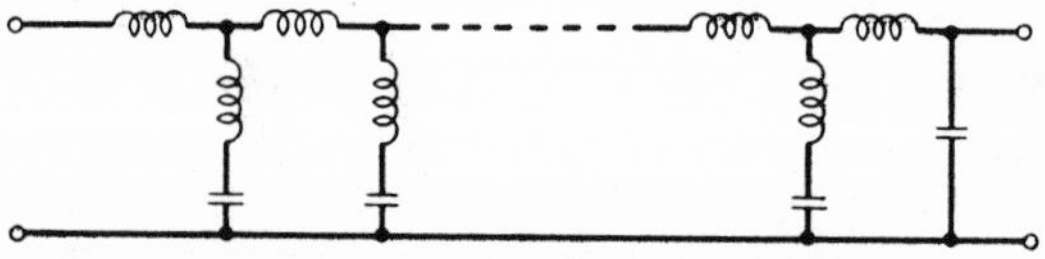

Fig. 11. A special form of mid-series low-pass ladder configuration corresponding to a special form of insertion loss function

Other Types of Ladders Commonly Encountered

Ladders of the configuration of Fig. 12, which may be described as the mid-shunt low-pass configuration, are most easily designed by redefining the coefficients $a_1, \cdots, a_{2\eta+1}$ in accordance with the designations of the branch impedances indicated in the figure. These coefficients are related to the short- and open-circuit admittances in exactly the same way that the coefficients of ladders of the mid-series type are related to the open- and short-circuit impedances. As a result, mid-series and mid-shunt ladders will produce the same insertion losses when terminated in R_1 and R_2 provided their coefficients are related by the equations

$$a_{2\sigma-1} = R_1R_2\bar{a}_{2\sigma-1} \qquad \bar{a}_{2\sigma} = R_1R_2a_{2\sigma} \tag{37}$$

[59] The vanishing of a terminal series inductance next to a resonant shunt branch, however, does not normally change the degree of the numerator in the power ratio expression in this way.

in which the a_σ's and $\bar{a}_\sigma$'s are the coefficients of the mid-series and mid-shunt ladders, respectively, and must be numbered from opposite ends in the two networks.

Negative inductances realizable with coupling in series type ladders become negative capacities in the corresponding shunt type configurations. These can be realized, however, by introducing ideal transformers in the proper way. To understand how this can be accomplished it is only necessary to note that an ideal transformer shunted by any 2-pole, such as a condenser, is equivalent to a pair of "perfectly coupled impedances" exactly similar to perfectly coupled inductances.

Other types of ladders can be designed by making use of the well-known method of frequency transformations. The previous design formulae apply to the determination of a low-pass ladder of reactance elements producing an insertion loss represented by an appropriate function of ω. Suppose some other insertion loss function is transformed

Fig. 12. The mid-shunt low-pass ladder configuration

into this same function by replacing ω by a related variable Ω. Since the reactance elements of the low-pass filter theory are merely devices for producing impedances proportional to $i\omega$ or its reciprocal, the same theory can now be used to design a corresponding ladder made up of impedances proportional to $i\Omega$ or its reciprocal. If all the elements of the original low-pass ladder are positive, the impedance branches of the transformed ladder can be realized with physical 2-poles provided $i\Omega$ represents a physical impedance function. If the original low-pass ladder includes negative inductances realizable with coupled coils, the negative elements in the transformed ladder can be realized by using ideal transformers just as in mid-shunt low-pass ladders including negative capacities.

The mid-series and mid-shunt high-pass configurations obtained by replacing inductances by capacities and vice versa in low-pass ladders can be designed by defining Ω as $-\frac{1}{\omega}$ and using power ratios represented

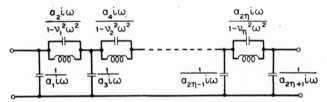

by functions of Ω identical with the functions of ω appropriate for the low-pass configurations. Band-pass configurations can be designed by defining Ω as $\frac{\omega^2 - \omega_m^2}{\omega}$ provided their insertion loss characteristics when plotted against log (ω) are to be symmetrical about log (ω_m). They can then be realized as combinations of series and parallel resonant circuits all resonating at ω_m. The situation is exactly similar in regard to band-elimination configurations, for which Ω is $\frac{-\omega}{\omega^2 - \omega_m^2}$.[60]

The only other ladder configurations commonly encountered are the more general band-pass type indicated in Fig. 13, and its inverse. The series type illustrated can normally be designed as an equivalent simpler

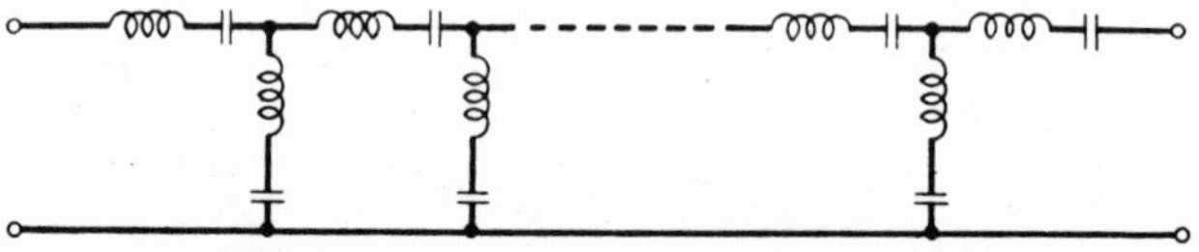

Fig. 13. The general mid-series band-pass ladder configuration

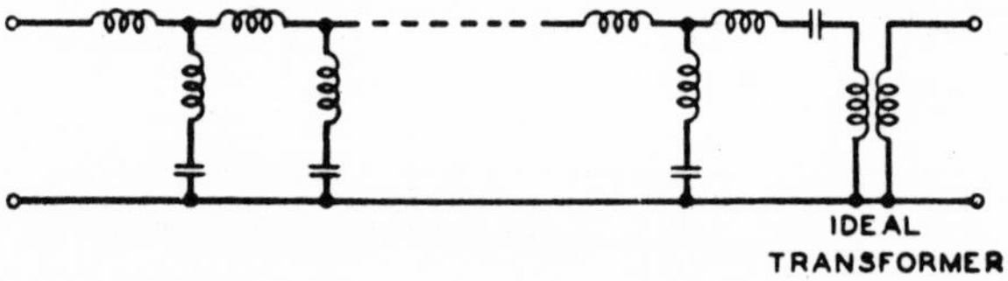

Fig. 14. An equivalent of the network of Fig. 13

network of the configuration indicated in Fig. 14.[61] It can be determined from the equivalent network by means of the so-called impedance

[60] It is sometimes more convenient, of course, to include arbitrary constant factors in the frequency transformations. In the design of a band-pass filter, for instance, it is frequently convenient to start by designing a low-pass filter with a cut-off at $\omega = 1$. Then the desired band-pass filter is obtained by replacing ω by an Ω representing

$$\frac{\omega^2 - \omega_{c1}\,\omega_{c2}}{\omega(\omega_{c2} - \omega_{c1})},$$

where ω_{c1}, ω_{c2} correspond to the two cut-offs of the band-pass filter.

[61] An equivalent of this type always exists except in limiting cases in which one or more shunt branches are simple inductances. A similar equivalent configuration obtained by replacing inductances by capacities and vice versa always exists except when one or more shunt branches are simple condensers. When simple inductance and capacity shunt branches are both encountered, special design methods must be used.

transformation indicated in Fig. 15A.[62] The equivalent network itself is of such a form that the formulae of Table I can be applied directly to its design. The short-circuit driving-point impedance measured at the terminals farther from the transformer, for instance, is determined by a configuration exactly the same as that determining an open-circuit impedance of a mid-series low-pass ladder with a far-end terminal shunt branch consisting of a simple condenser. The only operation not covered by the formulae of Table I is the determination of the impedance ratio of the transformer, which turns out to be determined by the behavior of the open- and short-circuit driving point impedances at zero frequency.

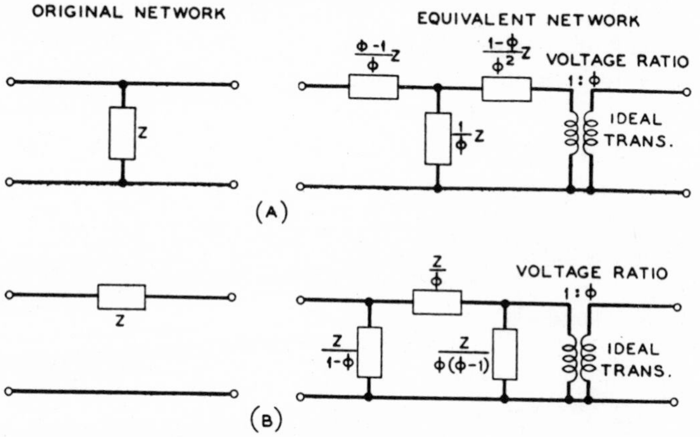

Fig. 15. Network equivalences indicating the impedance transformation principle

The configuration inverse to that of Fig. 13, which is made up of parallel resonant circuits in place of series resonant circuits, is designed by an exactly similar method with the help of the equivalence indicated in Fig. 15B.

Sufficient Conditions for Positive Elements

From the standpoint of actual construction, the perfectly coupled inductances and ideal transformers which are sometimes required for the realization of negative elements in physical ladders are highly undesirable. The two conditions stated below are sufficient to insure that

[62] The impedance transformation principle represented by Figs. 15A and 15B was discovered by Norton (14). In the determination of a network of the configuration indicated in Fig. 13 from one of the type indicated in Fig. 14 there is a considerable arbitrariness as to how the series capacity is distributed among the different series branches.

48

all the elements in a mid-series low-pass ladder will be positive, making unnecessary the use of coupled coils or ideal transformers in the ladder itself or in others related to it by frequency transformations or inverse relationships such as (37). While these conditions are not necessary, they turn out to be useful in demonstrating that most ladders encountered in filter design can be expected to be realizable without the use of coupling.

Assuming a mid-series low-pass configuration and impedances of the appropriate general form the first condition calls for frequencies of infinite loss which are all real (or infinite) and are greater than all the finite poles of at least one of the open-circuit driving-point impedances. The second condition requires the particular frequencies of infinite loss corresponding to the resonances of the shunt branches nearest the terminals to be also equal to or greater than all the roots of the corresponding open-circuit impedances. The sufficiency of these conditions is easily established by examining the formulae of Table I and recalling that the B's and β's are all positive.

One interesting special case is that in which all frequencies of infinite loss occur at infinity. It is seen that the above conditions are always satisfied in this special case. As a result, all power ratios of the form (35) with all ν_σ's reduced to zero are realizable with networks of the same *configuration* as constant-k low-pass image parameter filters provided $e^{2\alpha}$ meets the general physical requirement that it must be no less than $\frac{4R_1R_2}{(R_1 + R_2)^2}$ at all real frequencies. The same situation also holds in regard to the power ratio (36) with all ν_σ's reduced to zero, the corresponding constant-k configuration merely including an odd number of "half sections." These two special power ratios include all even polynomials in ω which have unit constant terms and which satisfy the physical limit on $e^{2\alpha}$ at real frequencies.

Symmetrical and Inverse Impedance Ladders

The ladders used as filters usually have impedances and terminations meeting one of two special conditions.[63] One condition calls for an electrically symmetrical network and equal terminations. The other requires each open-circuit impedance to be the inverse of the short-circuit impedance of the other end of the network with respect to the mean of the terminations.[64]

[63] Except when they are modified to compensate for effects of dissipation.

[64] If the image impedances and transfer constants are used in the description of the networks, these conditions require equal and inverse image impedances, respectively.

If the requirement of symmetry or of inverse impedances is to be satisfied, the polynomial A' or B', respectively, must be identically zero in the formulae (18) of Part I relating the impedances to the insertion loss. An examination of the relation of A' and B' to the insertion power ratio shows that their vanishing requires expressions of the form

$$e^{2\alpha} = 1 + \left(\frac{\omega B'}{P}\right)^2 \tag{38}$$

$$A' = 0$$

in the case of symmetrical networks and equal terminations, and of the form

$$e^{2\alpha} = \frac{4R_1R_2}{(R_1 + R_2)^2} + \left(\frac{A'}{P}\right)^2 \tag{39}$$

$$B' = 0$$

in the case of inverse impedance networks.

The specification of power ratios in the forms (38) and (39) simplifies the design procedure in that no roots need be extracted in determining A' and B', the only root extraction being that involved in finding the polynomials A and B. It is readily shown that the conditions necessary for the physical realizability of these power ratios permit A', B', and P to be any even polynomials in ω with real coefficients. In other words, there will be at least one corresponding symmetrical or inverse impedance network for every power ratio of the general form

$$e^{2\alpha} = \Lambda(1 + \Phi^2) \tag{40}$$

in which Φ is any odd or even rational function of ω with real coefficients while Λ represents $\dfrac{4R_1R_2}{(R_1 + R_2)^2}$ and must be unity when Φ is an odd function.[65]

[65] When Φ is odd and Λ is different from unity, the ratio of the open-circuit driving-point impedances, which is identical with the ratio of the corresponding short-circuit impedances, will be equal to the ratio of the terminations. This permits the use of a symmetrical network combined with an ideal transformer. It is convenient to permit unequal terminations in the case of inverse impedance networks and not in the case of symmetry, or of proportional impedances, in that unequal terminations are usually necessary in the inverse impedance case if ideal transformers are to be avoided.

When all the poles of $\frac{B'}{P}$ are real or imaginary and are also finite, the power ratio (38) leading to symmetrical networks can be expressed in the form (35) appropriate for mid-series low-pass ladders of the normal configuration indicated in Fig. 9. When $\frac{A'}{P}$ obeys the same condition except for a single pole in terms of ω^2 at infinity and when also R_1 and R_2 are so chosen that α is zero at zero frequency, the power ratio (39) leading to inverse impedance networks can be expressed in the form (36) appropriate for mid-series low-pass ladders terminated at one end in shunt condensers as in Fig. 11. Exactly similar relations exist in regard to ladders of other than the mid-series low-pass configuration.

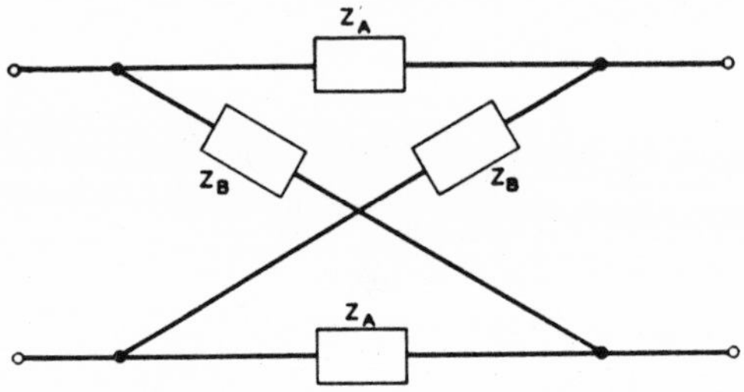

Fig. 16. The balanced lattice configuration

Lattice Networks and the Design of Symmetrical Ladders

It is well known that there is an equivalent lattice network of the type indicated in Fig. 16 corresponding to every physical network which is electrically symmetrical.[66] When the open- and short-circuit impedances are known, the impedance arms Z_A, Z_B can be very easily computed by well-known formulae. When a lattice of reactances is to be designed to produce a power ratio prescribed in the form (38), however, it is simpler to determine the impedance branches by the formulae described below rather than to first compute the open- and short-circuit impedances and then the impedance arms.

When a power ratio is prescribed in the form (38), the quantity $(P + pB')$ can obviously be expressed as the following product of two polynomials in p:

$$P + pB' = (P_1 + pB_1)(P_2 - pB_2) \tag{41}$$

[66] This was pointed out by Campbell (15).

TABLE II

Ladder Equivalents of Mid-Series Low-Pass Lattices

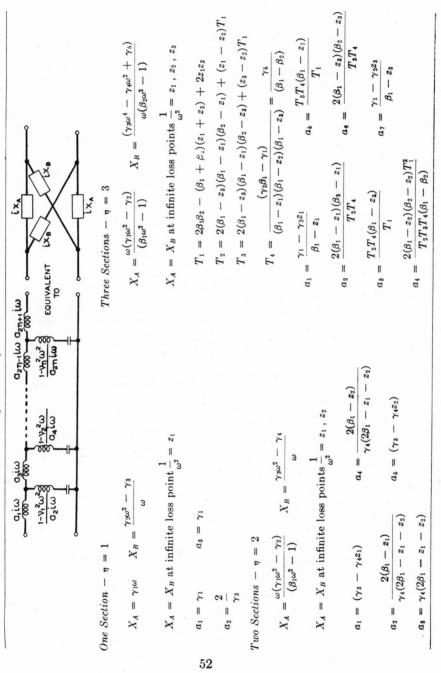

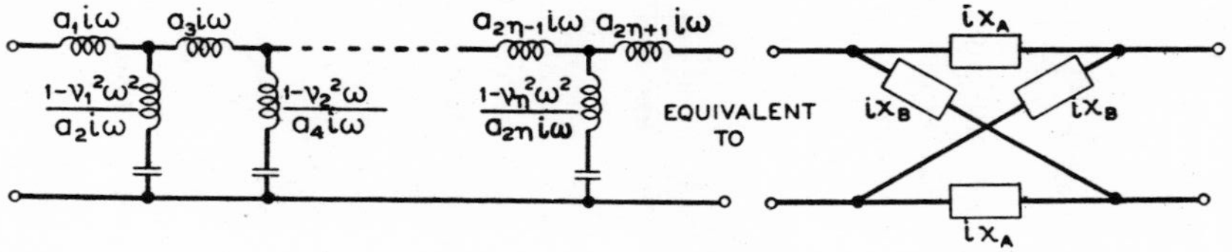

One Section $- \eta = 1$

$$X_A = \gamma_1\omega \qquad X_B = \frac{\gamma_2\omega^2 - \gamma_3}{\omega}$$

$$X_A = X_B \text{ at infinite loss point } \frac{1}{\omega^2} = z_1$$

$$a_1 = \gamma_1 \qquad a_3 = \gamma_1$$

$$a_2 = \frac{2}{\gamma_3}$$

Two Sections $- \eta = 2$

$$X_A = \frac{\omega(\gamma_1\omega^2 - \gamma_2)}{(\beta_1\omega^2 - 1)} \qquad X_B = \frac{\gamma_3\omega^2 - \gamma_4}{\omega}$$

$$X_A = X_B \text{ at infinite loss points } \frac{1}{\omega^2} = z_1 , z_2$$

$$a_1 = (\gamma_3 - \gamma_4 z_1) \qquad a_4 = \frac{2(\beta_1 - z_2)}{\gamma_4(2\beta_1 - z_1 - z_2)}$$

$$a_2 = \frac{2(\beta_1 - z_1)}{\gamma_4(2\beta_1 - z_1 - z_2)} \qquad a_5 = (\gamma_3 - \gamma_4 z_2)$$

$$a_3 = \gamma_4(2\beta_1 - z_1 - z_2)$$

Three Sections $- \eta = 3$

$$X_A = \frac{\omega(\gamma_1\omega^2 - \gamma_2)}{(\beta_1\omega^2 - 1)} \qquad X_B = \frac{(\gamma_3\omega^4 - \gamma_4\omega^2 + \gamma_5)}{\omega(\beta_2\omega^2 - 1)}$$

$$X_A = X_B \text{ at infinite loss points } \frac{1}{\omega^2} = z_1 , z_2 , z_3$$

$$T_1 = 2\beta_1\beta_2 - (\beta_1 + \beta_2)(z_1 + z_3) + 2z_1z_3$$

$$T_2 = 2(\beta_1 - z_3)(\beta_1 - z_1)(\beta_2 - z_1) + (z_1 - z_2)T_1$$

$$T_3 = 2(\beta_1 - z_3)(\beta_1 - z_1)(\beta_2 - z_3) + (z_3 - z_2)T_1$$

$$T_4 = \frac{(\gamma_2\beta_1 - \gamma_1)}{(\beta_1 - z_1)(\beta_1 - z_2)(\beta_1 - z_3)} = \frac{\gamma_5}{(\beta_1 - \beta_2)}$$

$$a_1 = \frac{\gamma_1 - \gamma_2 z_1}{\beta_1 - z_1} \qquad a_5 = \frac{T_3T_4(\beta_1 - z_1)}{T_1}$$

$$a_2 = \frac{2(\beta_1 - z_1)(\beta_2 - z_1)}{T_2T_4} \qquad a_6 = \frac{2(\beta_1 - z_3)(\beta_2 - z_3)}{T_3T_4}$$

$$a_3 = \frac{T_2T_4(\beta_1 - z_3)}{T_1} \qquad a_7 = \frac{\gamma_1 - \gamma_2 z_3}{\beta_1 - z_3}$$

$$a_4 = \frac{2(\beta_1 - z_2)(\beta_2 - z_2)T_1^2}{T_2T_3T_4(\beta_1 - \beta_2)}$$

where P_1, B_1, P_2, B_2 are even polynomials such that the roots of $(P_1 + pB_1)$ are all those roots of $(P + pB')$ which have negative real parts. It turns out that the impedance arms Z_A and Z_B of one lattice producing the prescribed power ratio are related to P_1, B_1, etc., by the formulae

$$Z_A = R\frac{pB_1}{P_1} \qquad Z_B = R\frac{P_2}{pB_2} \tag{42}$$

in which R represents the equal terminating resistances. The only three other lattices of reactances corresponding to the prescribed power ratio are obtained by interchanging these impedances and replacing them by their inverses with respect to R.[67]

In the design of symmetrical ladders with prescribed insertion losses, it is usually easier to determine the element values from the equivalent lattices rather than from the open- and short-circuit impedances. In the design of mid-series low-pass ladders of one, two or three shunt branches or of related networks, for instance, the special design formulae listed in Table II can be used. These formulae can be derived in much the same way as the general formulae of Table I by using an open-circuit impedance in formulating the partial fraction expansions of the determinants in (30) and noting relations between the constants required for symmetry. The relations between the constants are due to the fact that the open-circuit impedance is proportional to the sum of the two impedance arms, which requires the sums of the corresponding partial fractions to be equal at frequencies of infinite loss. Even when no other special relations are used, the initial determination of the open-circuit impedance as a sum of two functions simplifies the computation of the partial fraction expansion necessary for the use of Table I.

Part III. Special Insertion Loss Functions for Filtering Purposes

From the standpoint of filter design the general theory of Parts I and II is incomplete in two principal respects. In the first place, it gives no indication as to how to choose general insertion loss functions in such a way as to obtain efficient filter characteristics. In addition, the general design procedure is extremely complicated in numerical

[67] The polynomials B' and P are somewhat arbitrary as far as the power ratio of the form (38) is concerned. It turns out, however, that there are only four corresponding lattices in spite of this arbitrariness.

problems, involving the determination of the roots of two high degree polynomials. Part III rectifies this situation by introducing special types of realizable insertion loss functions which represent efficient filter characteristics and which also lead to relatively simple special design procedures.

Assumption of Insertion Loss Characteristics Appropriate for Symmetrical or Inverse Impedance Networks

The special insertion loss functions used for filtering purposes are appropriate for the symmetrical and inverse impedance networks described in the closing paragraphs of Part II. It turns out that any loss function representing an efficient filter characteristic at least approximates one of the types required for these particular networks. In addition, as was indicated previously, these networks can be designed with only half the root extraction which is more generally required.

Recall that insertion power ratios which are to be appropriate for symmetrical networks with equal terminations or for networks with impedances which are inverse with respect to the mean of the terminations must be of the form

$$e^{2\alpha} = \Lambda(1 + \Phi^2) \tag{43}$$

where Φ is any odd or even rational function of frequency in the two cases, respectively, and Λ represents $\frac{4R_1R_2}{(R_1 + R_2)^2}$ (being thus unity in the case of symmetrical networks with equal terminations). It is obvious that the poles of Φ are frequencies of infinite loss while the roots are frequencies of maximum possible insertion gain. Hence rough filter-like characteristics can be obtained by merely requiring all poles of Φ to occur in the desired attenuation bands and all roots in the desired pass bands. In order to avoid the necessity of ideal transformers or perfectly coupled inductances, however, it is best to add the restriction that Φ must be such as to make the insertion loss zero or infinite at each of the limiting frequencies zero and infinity.[68]

[68] This additional requirement is automatically satisfied when the network is to be symmetrical but not when it is to have inverse impedances. The requirement is called for in that failure to obey it always leads to the necessity of an ideal transformer or perfectly coupled inductances. On the other hand, it does not appear to be sufficient to insure that inverse impedance networks can be realized without these devices, although they are not required in the filters commonly encountered.

By way of illustration consider the special case of the low-pass filter. For a symmetrical low-pass filter

$$e^{2\alpha} = 1 + \left[S_0 \frac{\omega(\omega_1^2 - \omega^2) \cdots (\omega_\eta^2 - \omega^2)}{(1 - \nu_1^2 \omega^2) \cdots (1 - \nu_\eta^2 \omega^2)} \right]^2 \tag{44}$$

in which the ω_σ's, ν_σ's and S_0 are arbitrary constants. Similarly, for a low-pass filter of the inverse impedance type producing infinite loss at infinite frequency,

$$e^{2\alpha} = \frac{4R_1R_2}{(R_1 + R_2)^2} \left\{ 1 + \left[S_0 \frac{(\omega_1^2 - \omega^2) \cdots (\omega_\eta^2 - \omega^2)}{(1 - \nu_1^2 \omega^2) \cdots (1 - \nu_{\eta-1}^2 \omega^2)} \right]^2 \right\}. \tag{45}$$

Making each ω_σ in these expressions less than ω_c and each ν_σ less than $\frac{1}{\omega_c}$, where ω_c is the desired effective cutoff, leads to the type of low-pass filter characteristics illustrated in Figs. 17A and 17B for the case of $\eta = 3$. The insertion loss characteristics illustrated can normally be realized with the ladder networks indicated in Figs. 18A and 18B, respectively.

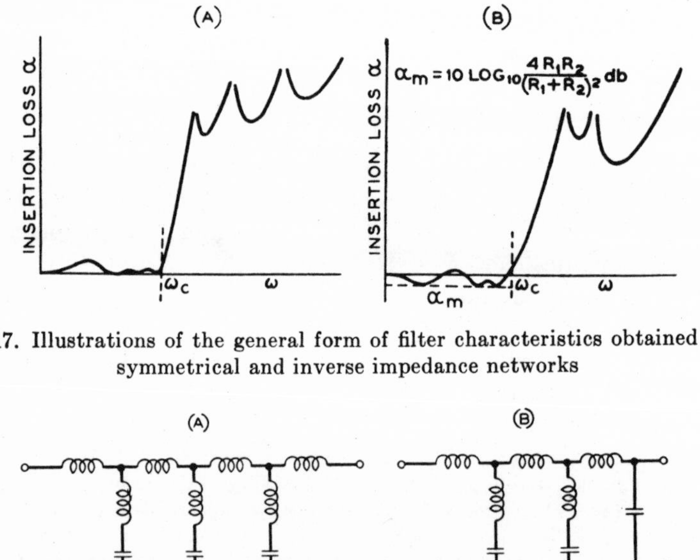

Fig. 17. Illustrations of the general form of filter characteristics obtained with symmetrical and inverse impedance networks

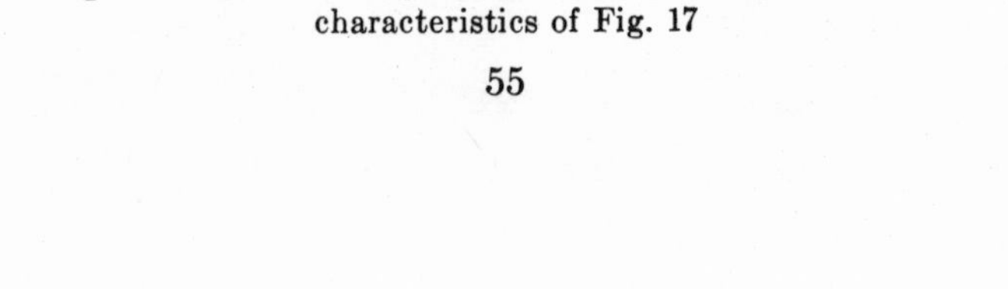

Fig. 18. Network configurations corresponding to the filter characteristics of Fig. 17

Decreasing the constant S_0 in the above expressions always decreases the pass band distortion but also decreases the suppression of attenuation band frequencies other than frequencies of infinite loss. This constant is normally made as large as is consistent with the permissible pass band distortion in order that the required suppression of unwanted frequencies may be obtained with a network of minimum complexity.[69] Zero loss at zero frequency, which is not necessarily produced by filters of the inverse impedance type, can always be obtained by properly choosing the terminations R_1 and R_2.

Tchebycheff Pass Band Parameters—Definition

Insertion loss functions of the rough filter type described above are characterized by the appearance of equal minima at arbitrary pass band frequencies and of infinite loss points at arbitrary attenuation band frequencies. Particularly useful filter characteristics and also particularly simple design procedures can be obtained by requiring the frequencies of minimum loss to be so chosen that the maxima occurring between the equal minima are themselves all equal. This leads to the type of loss characteristics illustrated in Figs. 19A and 19B for the special low-pass filters considered previously.

It is usually expedient to require also that the loss at each effective cutoff shall have the same value as at the equal maxima, as indicated in Fig. 19. In the case of inverse impedance filters transmitting zero or infinite frequencies, it is also normally advantageous to require the insertion loss to have the value zero at the equal maxima as well as at zero or infinite frequency, as in Fig. 19B. In the design of all types of filters commonly encountered, the addition of these restrictions to the requirement of equal maxima leaves the effective cutoffs and frequencies of infinite loss all arbitrary but fixes the pass band frequencies of minimum loss uniquely in terms of these parameters.[70] The characteristics obtained are the best for meeting efficiently the common type of filter requirements setting limits on permissible distortion which are constant at all pass band frequencies.

[69] The choice of S_0 is also sometimes influenced by the desire to realize the insertion loss with a ladder with no coupled inductances. Normally, however, coupled inductances are unnecessary even though S_0 is given the most advantageous value from the standpoint of loss characteristic. This will be explained later.

[70] The situation becomes more complicated only in such rare cases as multiband filters not derivable from simpler filters by means of frequency transformations.

56

The "equal ripple" pass band characteristics described above can be said to approximate constant losses in the Tchebycheff sense, while the corresponding pass band frequencies of minimum loss can be referred to as Tchebycheff pass band parameters. This means merely that the frequencies of minimum loss are chosen in such a way as to give the least maximum deviation from constant losses at pass band frequencies, other parameters being considered fixed.

The use of Tchebycheff parameters in network theory was first introduced by Cauer (16), who applied them to the design of image parameter filters. Filters satisfying the Tchebycheff pass band requirements of the insertion loss theory, however, are more nearly analogous to general image parameter filters than to Cauer's special Tchebycheff parameter type. A closer parallel to Cauer's use of Tchebycheff parameters is

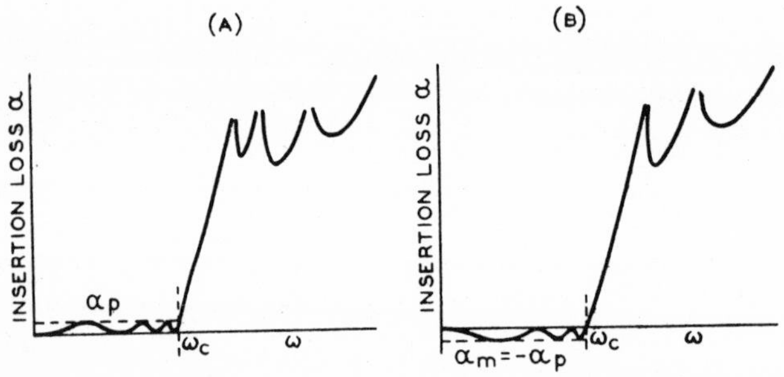

Fig. 19. Illustrations of the form of filter characteristics obtained by using Tchebycheff pass band parameters

offered by the simultaneous requirement of Tchebycheff attenuation band and pass band parameters, which will be described later.

Tchebycheff Pass Band Parameters—Theory

Tchebycheff pass band parameters can be obtained by using the following type of insertion power ratio.

$$e^{2\alpha} = \Lambda[1 + (e^{2\alpha_p} - 1)\cosh^2(\Theta_I)]. \tag{46}$$

In this expression, Λ is $\dfrac{4R_1R_2}{(R_1 + R_2)^2}$ as before; α_p is the difference between the equal maxima and equal minima of the pass band insertion loss characteristic, as indicated in Fig. 19; and Θ_I is a function of frequency meeting the following requirements: First, Θ_I must be such that $\cosh(\Theta_I)$ is an odd or even rational function of frequency, depending upon

57

whether the power ratio is to be appropriate for symmetrical or inverse impedance filters; second, Θ_I must be pure imaginary at all pass band frequencies and must have the form $(\alpha_I + n\pi i)$ at all attenuation band frequencies, where α_I is not only real but also is infinite at various arbitrary frequencies.

It is almost obvious that the above requirements on Θ_I lead to the desired type of insertion loss characteristics. At pass band frequencies, Θ_I takes the form $i\beta_I$ leading to the power ratio

$$e^{2\alpha} = \Lambda[1 + (e^{2\alpha_p} - 1) \cos^2 (\beta_I)]. \tag{47}$$

Since the value of $\cos^2 (\beta_I)$ must lie between zero and unity, the corresponding value of α must lie in a range of amplitude α_p. On the other hand, at attenuation band frequencies

$$e^{2\alpha} = \Lambda[1 + (e^{2\alpha_p} - 1) \cosh^2 (\alpha_I)]. \tag{48}$$

This requires α to approach infinity when α_I approaches infinity. The only real problem is how to choose Θ_I in such a way that $\cosh (\Theta_I)$ is an odd or even rational function of frequency.

The forms $i\beta_I$ and $(\alpha_I + n\pi i)$ assumed by Θ_I at pass band and attenuation band frequencies, respectively, suggest that Θ_I may be similar to the image transfer constant of an image parameter filter. An analysis of the properties of general image transfer constants shows the following statements to be correct. First, if Θ_I is obtained by adding $\frac{\pi}{2}i$ to the image transfer constant of any filter (of pure reactances) with inverse image impedances, $\cosh (\Theta_I)$ will be an odd rational function of frequency. Second, if Θ_I by itself represents an image transfer constant of any symmetrical filter, $\cosh (\Theta_I)$ will be an even rational function of frequency.[71] If the theoretical pass bands of these

[71] These relations follow directly from the following equations, which can readily be derived by means of elementary network theory:

$$\cosh\left(\Theta + i\frac{\pi}{2}\right) = i\,\frac{\sqrt{Z_{I1}Z_{I2}}}{-Z_{O12}}$$

$$\cosh\,(\Theta) = \sqrt{\frac{Z_{I1}}{Z_{I2}}}\left[\frac{-Z_{O2}}{Z_{O12}}\right]$$

in which Z_{I1} and Z_{I2} are the image impedances of a filter with transfer constant Θ while Z_{O12} and Z_{O2} are corresponding open-circuit transfer and driving-point impedances. Bode exhibits equations very similar to these. See, for instance, equation (30) on page 43 of his "General Theory of Electric Wave Filters (4)."

58

image transfer constants coincide with the desired effective pass bands of the insertion loss filters, Θ_I will also assume the required forms $i\beta_I$ and $(\alpha_I + n\pi i)$ at pass band and attenuation band frequencies. As a result, the well established theory of the image transfer constants of image parameter filters of the inverse impedance and symmetrical types can be applied directly to the design of insertion loss filters of the symmetrical and inverse impedance types, respectively.

Tchebycheff Pass Band Parameters—Fundamental Design Procedure

In accordance with the above principles, the design of a filter with Tchebycheff pass band parameters on the insertion loss basis is carried out in terms of the image transfer constant of a hypothetical image parameter filter. It should be borne in mind, however, that there is no direct connection between the element values of the reference filter and those of the actual filter. The reference filter involves sections with matched image impedances, or the equivalent, while the actual filter does not. In addition the reference filter is normally of somewhat greater complexity than the actual filter.[72] The reference filter is introduced at all only because its transfer constant is well understood and happens to have a functional form which facilitates the determination of the insertion loss of the actual filter.

The configuration of the reference filter corresponding to an actual filter of prescribed configuration is chosen in accordance with the requirement of equal or inverse image impedances and of a one to one correspondence between the attenuation peaks of the reference filter and all the frequencies of infinite loss of the actual filter. Since the image impedances of the reference filter are only required to be properly related, a number of configurations are always possible.

By way of illustration, the two low-pass configurations used previously as illustrative insertion loss filters are shown in Fig. 20 in comparison with the simplest corresponding reference filters. It is seen that each reference filter is obtained by adding a constant-k half-section to an image parameter filter of the same configuration as the corresponding actual filter. The added half-sections supply the infinite loss points at infinity due to the corresponding poles of the series branches of the actual filters. More generally, in the case of all types of filters commonly encountered the reference filters can be obtained by adding

[72] This is because the total insertion loss of the actual filter is determined by the transfer constant of the image parameter filter, including infinite loss points corresponding to the reflection peaks of the image parameter theory.

similar constant-k half-sections of the proper types to reference filters of the same configurations as the actual filters.[73]

When it comes to the choice of the arbitrary constants of a reference filter in such a way that the corresponding actual filter will meet design specifications of the type ordinarily encountered, the procedure is actually simpler than if the reference filter itself constituted the final network. In the first place, the theoretical pass band of the reference filter is coincident with the effective pass band of the actual filter, it being unnecessary to make any allowances for ranges of high reflection losses near theoretical cut-offs. In addition, the pass band distortion α_p can be chosen directly, rather than reflection losses due to variations in the image impedances which must be corrected for interaction effects. Finally, at attenuation band frequencies at which even moderate in-

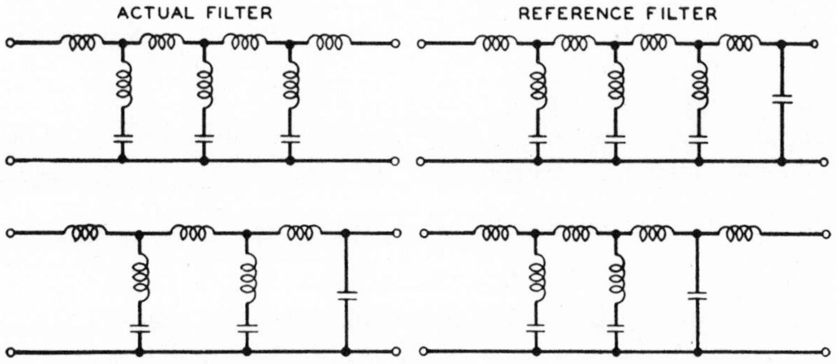

Fig. 20. Illustrations of reference image parameter filters

sertion losses are obtained, a good approximation to the loss can be computed more easily than in the case of an image parameter filter, which requires reflection losses to be added to the image attenuation. This computation is carried out by means of the following approximate formula which can be derived from equation (48) by first assuming that $\frac{1}{\Lambda}e^{2\alpha}$ and $e^{2\alpha_I}$ are both large compared with unity and then expressing α_I in terms of decibels rather than napiers:

$$\alpha - \alpha_m = [\alpha_I + 10 \log_{10} (e^{2\alpha_p} - 1) - 6.02]\, db \tag{49}$$

where α_m is the minimum pass band insertion loss, which is negative or zero since it amounts to $10 \log_{10} (\Lambda)\, db$, while α_I is identical with the image attenuation of the reference filter.

[73] Exceptions to this general rule are possible but are not apt to be encountered in ordinary filter design problems.

60

After the arbitrary constants have been chosen, the function cosh (Θ_I) appearing in (46) can readily be replaced by the equivalent odd or even rational function of frequency. In the case of low-pass filters, for instance, if χ is used to represent $\sqrt{1 - \frac{\omega_c^2}{\omega^2}}$ the function cosh (Θ_I) turns out to be

$$\cosh(\Theta_I) = \frac{\prod_\sigma (m_\sigma + \chi) + \prod_\sigma (m_\sigma - \chi)}{2 \prod_\sigma \sqrt{m_\sigma^2 - \chi^2}} \tag{50}$$

In this equation, ω_c represents the value of ω at the cutoff while the m_σ's represent the "m's" determining the attenuation peaks corresponding to the various half-sections of the reference image parameter filter, there being one factor in each product corresponding to each half-section. When cosh (Θ_I) is a rational function of ω, m_σ's different from unity occur in identical pairs, corresponding to halves of m-derived full sections in the reference filter. Each such pair is related to one of the ν_σ's in the expression of the form (44) or (45) for the power ratio of the actual filter by the expression

$$\nu_\sigma = (1/\omega_c)\sqrt{1 - m_\sigma^2} \tag{51}$$

Special Formulae Aiding Root Extraction

After cosh (Θ_I) has been replaced in (46) by the equivalent rational function, corresponding networks can always be designed by means of the general theory of symmetrical and inverse impedance networks described previously. The use of Tchebycheff pass band parameters, however, permits the more general design procedure to be considerably simplified by the introduction of additional special formulae. In particular, the required root extraction can be expedited by the use of special formulae giving good approximations to the required roots. How this is accomplished is described briefly below.

The required roots are those of equations of the form

$$1 \pm i\sqrt{e^{2\alpha_p} - 1}\cosh(\Theta_I) = 0 \tag{52}$$

expressed as a function of $p = i\omega$.[74] In other words, the problem is to calculate the values of p corresponding to the values of Θ_I determined by

[74] These are obviously factors of the function $\frac{1}{\Lambda} e^{2\alpha}$ determined by (46). In the case of symmetrical filters only one choice of sign need be considered, the

$$\Theta_I = \cosh^{-1}\left[\frac{\pm i}{\sqrt{e^{2\alpha_p} - 1}}\right] \tag{53}$$

If the reference image parameter filter is made up of identical m-derived sections or half-sections, Θ_I will take the form $(n\Theta_0)$ or $\left(n\Theta_0 + \frac{\pi}{2}i\right)$, in which Θ_0 represents the transfer constant of a single section or half-section. When the reference filter is not actually made up of identical sections, Θ_I can normally be approximated over a portion of the p plane including the roots of (52) by means of the Θ_I function representing a reference filter of this special type. In ordinary design problems it is only necessary to replace the "m's" describing the various sections of the reference filter with a single new "m" representing their mean.[75] Since p can normally be expressed as a simple function of Θ_0, it follows that approximations to the roots of (52) in terms of p can ordinarily be computed by replacing Θ_I by $(n\Theta_0)$ or $\left(n\Theta_0 + \frac{\pi}{2}i\right)$ in (53). When Θ_0 is the transfer constant of an m-derived low-pass half-section of "m" equal to m_0, for instance, the approximations to the roots of (52) can be calculated by means of the following formula in terms of the values of Θ_0 determined by (53):

$$p = \frac{\omega_c \sinh(\Theta_0)}{\sqrt{1 - (1 - m_0^2)\cosh^2(\Theta_0)}} \tag{54}$$

in which the sign of the square root must be such that its real part is positive.[76]

It can readily be shown that the approximations to the roots described above can be expected to be reasonably good in ordinary filter design

---

corresponding roots being those of $(P + pB')$, which are required in forming (41). In the case of inverse impedance filters the roots corresponding to one choice of signs are the conjugates of those corresponding to the other.

[75] The approximating Θ_I function does not have to represent a symmetrical or inverse impedance filter as does the actual Θ_I function. In the case of low-pass or high-pass filters, the harmonic mean of the "m's" appears to have particular advantages. The same thing is true of band pass or band elimination filters if they have symmetrical loss characteristics permitting the m's to represent the derivation of confluent sections from constant-k sections.

[76] Pure imaginary values of the square root, for which the sign is not defined, are not encountered in problems in which (54) gives reasonable approximations to the required roots.

problems. Experience has shown that they are normally good enough for well-known root improvement methods to be applied directly to the determination of the actual roots to any desired precision. It also turns out that the approximate roots are of such a form that special methods can be used to obtain second approximations before the application of general root improvement methods.[77]

Other Special Formulae

Various other special formulae can be derived to facilitate the design of filters on the Tchebycheff pass band parameter basis besides those expediting the root extraction. The reflection coefficients, for instance, which measure the departure of the driving-point impedances from the corresponding terminating impedances satisfy the following limit at pass band frequencies:

$$\left|\frac{R - Z}{R + Z}\right| \leqq \sqrt{1 - e^{-2\alpha_p}} \tag{55}$$

in which Z is either driving-point impedance and R is the corresponding termination.

A rough estimate of the variation α_d of the insertion loss over the pass band which will be produced by parasitic dissipation in a low-pass or high-pass filter is usually furnished by the highly approximate relation,[78]

$$e^{\alpha_d} = 1 + \frac{2n}{Q\pi m_0^2} \coth\left[\frac{1}{n}\coth^{-1}(e^{\alpha_p})\right] \tag{56}$$

In this expression Q is the harmonic mean of the magnitudes of the reactance-resistance ratios of all the inductances and capacities, evaluated at the cut-off frequency; m_0 represents the harmonic mean of the "m's" of all the half-sections in the reference image parameter filter, and n is the total number of half sections in the reference filter.

[77] This is accomplished by using a simple transformation of variable which transforms the approximate roots into the n-th roots of a constant. One method of obtaining second approximations amounts to the use of Newton's method in conjunction with Fry's isograph. Although this machine is primarily intended for general root extraction purposes it is also convenient for applying Newton's method to the improvement of roots approximated by quantities which are all of the same magnitude.

[78] This equation furnishes even rough estimates only in the case of filters meeting ordinary specifications, particularly as regards sharpness of cutoffs and non-dissipative pass band loss.

The actual loss α_σ produced by a low-pass or high-pass ladder filter at a frequency of infinite theoretical loss not too far removed from the cutoff frequency can be estimated by means of the following formula:

$$e^{\alpha_\sigma} = \frac{2m_\sigma^2 Q}{(1 - m_\sigma^2)} e^{\alpha'} \tag{57}$$

In this equation, m_σ is the "m" corresponding to the particular peak frequency considered and α' is calculated by subtracting the attenuation of the corresponding section of the reference filter from the non-dissipative insertion loss of the actual filter and evaluating the difference at the peak frequency. Q is now best chosen as the mean "Q" of the particular elements of the ladder network whose resonance produces the peak, the ratios being evaluated at the peak frequency.

Ordinarily, equation (56) can be applied to band-pass and band-elimination filters with narrow pass or attenuation bands by merely multiplying Q by $\dfrac{f_{c2} - f_{c1}}{2\sqrt{f_{c2}f_{c1}}}$, where f_{c2} and f_{c1} are the cutoff frequencies and Q is now evaluated at the mean frequency of the pass or attenuation band. This assumes, however, that the loss characteristics are at last approximately symmetrical on a logarithmic frequency scale. Equation (57) can also be applied to these filters provided the peak frequencies considered are not too far removed from cutoff frequencies. This application is accomplished by multiplying Q by $\dfrac{|f_\infty^2 - f_{c2}f_{c1}|}{2f_\infty\sqrt{f_{c2}f_{c1}}}$, in which f_∞ is the peak frequency.

Comparison of Actual Performance of Tchebycheff Pass Band Parameter Filters with That of Image Parameter Filters

From the standpoint of actual performance, a Tchebycheff pass band parameter filter can best be compared with an image parameter filter producing the same order of magnitude of distortion at frequencies in the effective pass band. When the constant α_p measuring the pass band distortion of the insertion loss filter is of the order of tenths of a decibel, for instance, the best comparison is usually obtained by assuming the image parameter filter to have image impedances of the constant-k type. When α_p is of the order of hundredths of a decibel or smaller, on the other hand, it is usually best to assume that the reflection effects produced by the image parameter filter at pass band frequencies are reduced by the use of image impedances including impedance controlling factors.

The comparison is most clear cut when α_p is of the order of the distortion due to an image parameter filter with constant-k image impedances. Recall that the reference filter used in designing an insertion loss filter is usually obtained by adding a constant-k half-section to an image parameter filter of the same configuration as the insertion loss filter. In other words, equation (49) approximating the insertion loss of an actual filter at frequencies of high loss can usually be written in the form

$$\alpha - \alpha_m = \alpha_I' + [\alpha_0 + 10 \log_{10} (e^{2\alpha_p} - 1) - 6.02] \, db \tag{58}$$

in which α_I' is the attenuation of an image parameter filter of the same configuration as the actual filter while α_0 is the attenuation of a constant-k half-section. The bracketed term in this equation is analogous to the reflection losses that would be added to the attenuation of the image parameter filter in computing its insertion loss from its attenuation α_I'. These two corrections are of the same order of magnitude when α_p is of the order of the distortion that would be produced by the image parameter filter at frequencies in its effective pass band. This is illustrated by actual curves of the two corrections in Fig. 21, assuming low-pass filters.

Although the insertion loss filter described above and the corresponding image parameter filter of the same configuration produce insertion losses of the same order of magnitude at frequencies of high loss, their pass band loss characteristics differ in one important respect. The effective cutoff of the insertion loss filter coincides with the theoretical cutoff of the image parameter filter, while the effective cutoff of the image parameter filter normally occurs at a frequency at least 10 per cent lower.[79] This means that if the image parameter filter were to be redesigned so as to have the same effective cutoff as the insertion loss filter the theoretical cutoff would have to be moved much nearer to the attenuation peaks. If the theoretical cutoff were originally within 20 per cent of an attenuation peak, for instance, it would normally have to be moved to within 10 per cent of the peak. This change would produce a marked reduction in the attenuation band loss, as is indicated by a glance at a set of attenuation curves for filters of different peak frequencies, such as that exhibited by Shea (17).

As is indicated by the above discussion, insertion loss filters of the

[79] The effective cutoff of the image parameter filter can of course be moved closer to the theoretical cutoff if impedance controlling factors are used but this requires an increase in network complexity.

Tchebycheff pass band parameter type normally represent a substantially more efficient use of the elements than image parameter filters when the pass band distortion is to be of the order of tenths of a decibel. The same general situation usually exists when the pass band distortion is to be substantially lower, calling for an image parameter filter in-

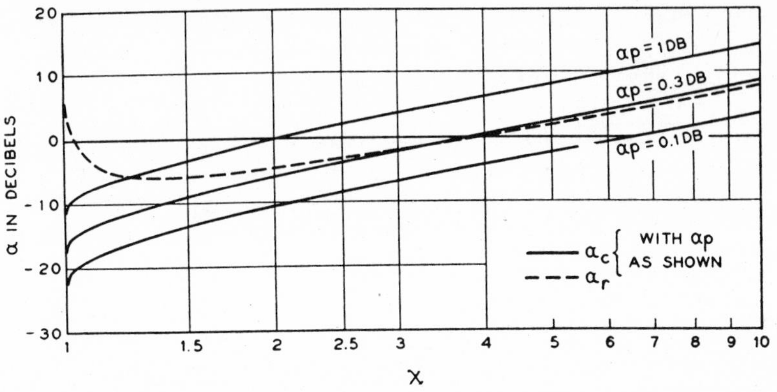

$$\alpha_c = \alpha_0 + 10 \log_{10} (e^{2\alpha_p} - 1) - 6.02 \text{ db}$$

$$\alpha_0 = \tanh^{-1} \sqrt{1 - \frac{1}{x^2}}$$

= attenuation of a constant-k low-pass half-section.

$$\alpha_r = 20 \log_{10} \left(\frac{x^2}{4\sqrt{x^2 - 1}} \right)$$ = reflection loss between two constant-k low-pass image impedances and their nominal impedances.

$$x = \frac{\omega}{\omega_c}$$

Fig. 21. A comparison of the corrections involved in the determination of the insertion losses of low-pass filters of the Tchebycheff pass band parameter and image parameter types

volving impedance controlling factors, but not necessarily to the same extent.

Dependence Upon α_p of Positiveness of Ladder Elements

In Part II certain conditions were pointed out which are sufficient to insure that a mid-series low-pass ladder will not include negative inductances requiring the use of coupled coils. The first of these requires the frequencies of infinite loss to be greater than all the finite poles of at least one of the open-circuit driving-point impedances. The

66

second requires the particular frequencies of infinite loss corresponding to the shunt branches nearest the terminals to be also equal to or greater than all the roots of the corresponding open-circuit impedances. As would be expected, the open-circuit impedances of a Tchebycheff pass band parameter filter vary in much the same way as those of a corresponding image parameter filter producing a somewhat similar loss characteristic. Specifically, a good comparison is usually obtained when the two filters produce the same order of magnitude of pass band distortion and are described by impedances which vary in the same manners at zero and infinite frequencies, rather than in inverse manners. Thus a study of the impedances of image parameter filters gives an indication of when coupled coils may be required in insertion loss filters.

The open-circuit impedances of an image parameter filter are given by

$$Z_0 = Z_I \coth(\Theta) \tag{59}$$

where Z_0 is either open-circuit impedance, Z_I is the corresponding image impedance, and Θ is the image transfer constant. Since coth (Θ) is finite in attenuation bands, the only roots and poles of the open-circuit impedances occurring in attenuation bands are roots and poles of the image impedances.

A mid-series low-pass image impedance of the constant-k type has no root above the cutoff frequency and has no pole except at infinity. It follows that low-pass filters of the insertion loss type and filters related to them by frequency transformations can be expected to be realizable with ladders without coupled coils whenever the pass band distortion is of the order of that produced by image parameter filters with constant-k impedances.

When a single impedance controlling factor is considered, a mid-series type of low-pass insertion loss filter must be compared with an image parameter filter of the so-called mid-shunt m-terminated type, which produces an image impedance with a pole at infinity.[80] In this case,

[80] The well known ladder form of a symmetrical mid-shunt m-terminated image parameter filter has the general mid-series low-pass configuration except that the terminal series inductances do not appear. The insertion power ratio of a general mid-series low-pass ladder, when expressed in the form (35), has as many constants as there are elements in the network. It turns out that removing the terminal series branches leaves the form and degree of the power ratio unchanged, the only result being changes in the numerical values of the constants, which are greater in number than the elements of the "reduced" network. Thus the mid-shunt m-terminated image parameter low-pass filter can be described as a special case of the general mid-series low-pass configuration in which the terminal series inductances have the value zero.

the open-circuit impedances of the image parameter filter will have roots in the attenuation band but no poles except at infinity. Thus when an insertion loss filter produces a pass band distortion comparable with that obtained with image impedances each including a single impedance controlling factor, it is to be expected that no coupled coils will be required provided the frequencies of infinite loss produced by the terminal shunt branches are sufficiently high.

When two or more impedance controlling factors are included in each image impedance, there will be poles in attenuation bands and the sufficient conditions for elimination of coupling will be violated unless all infinite loss frequencies are sufficiently high. This does not mean, however, that coupling will necessarily be required whenever there are infinite loss frequencies near the cutoff, since the above conditions are sufficient but not necessary for its elimination. It merely means that the possibility of required coupling must be considered.

The general rule is obviously that reducing the pass band distortion α_p always makes the necessity of coupled coils more probable. For the range of values of α_p normally of interest it can usually be assumed that no coupling will be necessary, but unusually small values of α_p may lead to its necessity.[81]

Simultaneous Attenuation Band and Pass Band Tchebycheff Parameters

In many filter design problems, the minimum permissible suppression of unwanted frequencies is constant over prescribed effective attenuation bands at the same time that the maximum permissible distortion is constant over prescribed effective pass bands. In many such problems, the most efficient use of the arbitrary constants is obtained by requiring Tchebycheff attenuation band parameters as well as Tchebycheff pass band parameters. In other words, the frequencies of infinite loss of the previous theory are required to be so chosen that the insertion loss characteristics exhibit equal minima between the frequencies of infinite loss as well as equal maxima and equal minima in the pass bands. The simplest illustration is the design of a low-pass filter with an effective pass band extending from zero frequency to some prescribed frequency f_1 and an effective attenuation band extending

[81] The values of α_p leading to the requirement of coupling are normally extremely small, but very small values are sometimes required in order to meet restrictions on permissible impedance variations. As is indicated by (55), the impedance variations occurring at pass band frequencies approach zero only as $\sqrt{\alpha_p}$.

from a second prescribed frequency f_2 to infinity. In this special case, the simultaneous use of attenuation band and pass band Tchebycheff parameters described above leads to a loss characteristic of the type illustrated in Fig. 22 for a three-section symmetrical filter.

Mathematically, the theory of simultaneous attenuation band and pass band Tchebycheff parameters is very similar to the Tchebycheff version of the image parameter theory introduced by Cauer (16). Design formulae for the essential constants such as the frequencies of infinite loss can actually be derived by combining Cauer's theory with the theory of general Tchebycheff pass band parameters described above. Cauer's theory leads to image parameter filters characterized by equal minima of attenuation between attenuation peaks. If one of these is used as the reference image parameter filter in the design of an insertion

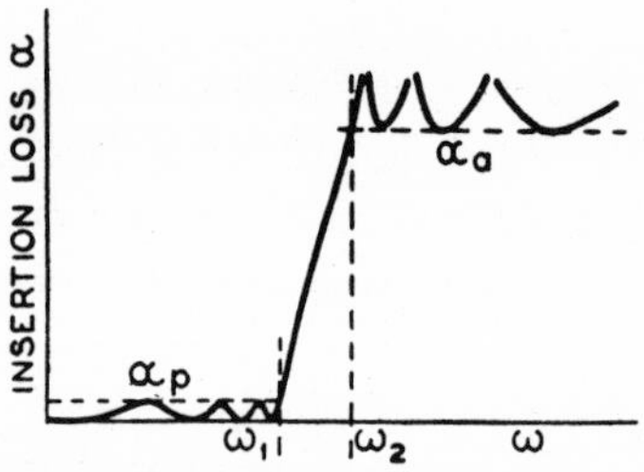

Fig. 22. An illustration of the form of filter characteristics obtained by the use of both attenuation band and pass-band Tchebycheff parameters

loss filter with Tchebycheff pass band parameters, equal minima of insertion loss between the infinite loss points are obtained. The requirement that cosh (Θ_I) must be an even or odd rational function of frequency and must be infinite at a zero or infinite attenuation band frequency, however, requires Cauer's basic theory to be modified somewhat in designing the image parameter filter.[82]

The similarity of the insertion loss filters to Cauer's image parameter filters does not extend beyond the theory involved in the choice of the design parameters. The insertion loss characteristics produced by the two types of filters are quite different. Cauer's theory would lead to insertion loss characteristics with "equal ripples" in pass bands and with equal minima between infinite loss points in attenuation bands only if

[82] The required modification can be determined by examining Cauer's various methods of modifying his theory to apply it to various image parameter problems.

interaction effects could be neglected at pass band frequencies and reflection effects at attenuation band frequencies. In actual filters, the effects of interaction and reflections are sufficient to produce marked changes in the loss characteristics.[83] The insertion loss filters, on the other hand, produce the ideal characteristics exactly, there being no corrections for such effects as reflections and interaction.[84]

Direct Derivation of the Theory

In spite of the fact that Cauer's Tchebycheff version of the image parameter theory can be used to derive the essential formulae, it is more convenient to develop the corresponding insertion loss theory by means of the more direct analysis described below. This analysis yields the same formulae as Cauer's theory and also additional useful relationships. It was used in the original derivation of the theory, which was carried out before the development of the simple theory of general Tchebycheff pass band parameters made Cauer's image parameters available for the purpose. Since other filters call for a very similar method of analysis although somewhat different formulae, only symmetrical low-pass filters will be described in detail.

Statement of Algebraic Form of the Power Ratio

It will be simplest to begin with the algebraic statement of the insertion power ratio required for simultaneous attenuation band and pass band Tchebycheff parameters, leaving its justification until later. An expression for the power ratio exhibiting a simple type of symmetry is obtained by replacing ω by a new variable Ω even though it is a low-pass filter which is under consideration. In the design of a low-pass filter, Ω is defined by

$$\Omega = \frac{\omega}{\sqrt{\omega_1 \omega_2}} \tag{60}$$

[83] At a pass band frequency at which the phase is an integral multiple of π radians, for instance, interaction effects in a symmetrical image parameter filter are such as to produce zero insertion loss whatever the reflection losses. At attenuation band frequencies the reflection effects vary from 6db reflection gain to infinite reflection loss.

[84] Even when dissipation in the elements is considered, it is frequently possible to modify the insertion loss filters in such a way that the insertion loss characteristics differ from those of non-dissipative filters only by constant losses. How this can be done will be explained in Part IV.

where ω_1 and ω_2 are the values of ω at the cutoff frequency f_1 and the limit of the effective attenuation band f_2. The values of Ω at f_1 and f_2 are then $\sqrt{k}$ and $\frac{1}{\sqrt{k}}$ where

$$k = \frac{f_1}{f_2} \tag{61}$$

The corresponding power ratio required for a symmetrical filter turns out to be of the form

$$e^{2\alpha} = 1 + \left[S_0 \frac{\Omega(\Omega^2 - \Omega_1^2) \cdots (\Omega^2 - \Omega_\eta^2)}{(1 - \Omega_1^2\Omega^2) \cdots (1 - \Omega_\eta^2\Omega^2)} \right]^2 \tag{62}$$

Consider the bracketed expression in (62). If the constant S_0 is omitted, replacing Ω by its reciprocal will replace this expression by its reciprocal. It follows that the requirement of Tchebycheff attenuation band parameters will be satisfied provided the Ω_σ's are chosen in such a way that Tchebycheff pass band parameters are obtained. As would be expected from Cauer's image parameter theory, the required values of the Ω_σ's are most simply expressed in terms of Jacobean elliptic functions. Specifically

$$\Omega_\sigma = \sqrt{k}\, \mathrm{sn}\left[\frac{2\sigma K}{2\eta + 1}, k\right] \qquad \sigma = 1, \cdots, \eta \tag{63}$$

in which the constant k representing the modulus still has the value f_1/f_2, while K is the corresponding complete integral.

Proof of Correctness of the Algebraic Expression for the Power Ratio

Suppose the power ratio (62) is transformed by replacing Ω by $\sqrt{k}\, \mathrm{sn}\, (u, k)$. The function of u obtained in this way can be simplified by replacing the Ω_σ's by their equivalents of (63) and then making use of the summation law for elliptic sines. This procedure permits the bracketed expression in (62) to be replaced by a product of elliptic sines showing the power ratio to be determined by the following pair of equations

$$e^{2\alpha} = 1 + S_0^2 \prod_{\sigma=-\eta}^{\sigma=+\eta} \left\{ k\, \mathrm{sn}^2 \left[\left(u + \frac{2\sigma K}{2\eta + 1} \right), k \right] \right\} \tag{64}$$

$$\Omega = \sqrt{k}\, \mathrm{sn}\, (u, k).$$

If u is replaced by $\left(u + \frac{2K}{2\eta + 1}\right)$ in the product in the above power ratio the net effect is to replace the factor in $\sigma = -\eta$ by one in $\sigma = \eta + 1$, the other factors being merely interchanged. The arguments of the factors in $\sigma = -\eta$ and $\sigma = \eta + 1$ differ by exactly $2K$ and must therefore be identical since this is the period of $\mathrm{sn}^2(u, k)$. In other words, the power ratio is unchanged by replacing u by $\left(u + \frac{2K}{2\eta + 1}\right)$, which amounts to saying it is periodic in u with the period $\frac{2K}{2\eta + 1}$. The frequency variable Ω is also periodic in u but has the longer period $4K$.

For real values of u the function $\mathrm{sn}(u, k)$ covers the range -1 to $+1$. In other words, the corresponding values of Ω coincide with the pass band and its image at negative frequencies. Thus as u increases Ω will vary cyclically over the pass band and its negative image, while in each cycle of Ω the power ratio will pass through several complete cycles, i.e., through a series of equal maxima and equal minima. A more detailed analysis of this situation shows that it requires the power ratio under consideration to represent at least a special case of Tchebycheff pass band parameters.[85] As indicated above, it then follows from the reciprocal nature of the bracketed expression in (62) that Tchebycheff attenuation band parameters are also obtained.[86]

Determination of the Roots of the Power Ratio and of Related Functions

The periodicity of the power ratio and of Ω considered as functions of u simplifies the calculation of the roots of the power ratio involved in the determination of corresponding networks. Since the period of the power ratio is a fraction of that of Ω, as soon as any one root of the

[85] The general principle of using periodic transformations in Tchebycheff parameter problems was introduced by Schelkunoff. Schelkunoff's application of the principle to Cauer's Tchebycheff theory is described by Guillemin (18).

[86] The above analysis merely proves the correctness of a particular solution for the choice of the Ω_σ's yielding attenuation band and pass band Tchebycheff parameters. The uniqueness of this solution can be demonstrated, however, by making use of the following relation, which can be shown to be necessarily satisfied:

$$\frac{\pm dy}{\sqrt{(k_1 - y^2)(1 - k_1 y^2)}} = (2\eta + 1)\frac{K_1}{K}\frac{d\Omega}{\sqrt{(k - \Omega^2)(1 - k\Omega^2)}}$$

in which y represents the bracketed expression in (62) with the constant S_0 omitted.

power ratio in terms of $i\Omega$ has been determined a number of others can be computed by merely adding multiples of the period of the power ratio to the corresponding value of u. Actually, it turns out that all the roots can be found in this way as soon as any one has been determined. If $\pm a_0$ is used to represent the only real roots, for instance, the complex roots take the form

$$i\Omega_\sigma = \frac{\pm a_0 cd_\sigma \pm iW\Omega_\sigma}{1 + a_0^2\Omega_\sigma^2} \qquad \sigma = 1, \cdots, \eta \qquad (65)$$

where Ω_σ is again $\sqrt{k}\,\mathrm{sn}\left[\frac{2\sigma K}{2\eta+1}, k\right]$ while W and cd_σ are defined by

$$\begin{aligned} W &= \sqrt{(1 + ka_0^2)\left(1 + \frac{1}{k}a_0^2\right)} \\ cd_\sigma &= \mathrm{cn}\left[\frac{2\sigma K}{2\eta+1}, k\right]\cdot \mathrm{dn}\left[\frac{2\sigma K}{2\eta+1}, k\right] \end{aligned} \qquad (66)$$

Assuming a_0 to be the *positive* real root of the power ratio, the roots of the related polynomial $(A + pB)$ appearing in the general theory of Part I are found by using $-a_0$ as the real root and replacing $\pm a_0$ by $-a_0$ in the above formulae for the complex roots.[87] It also turns out that the roots of $(P + pB')$, which are required in the formation of equation (41) in the determination of corresponding lattices, are obtained by using $+a_0$ as the real root and replacing $\pm a_0$ by $(-)^\sigma a_0$ in the above formulae.

Introduction of a Modular Transformation on the Elliptic Functions

Formulae for computing the real zeros $\pm a_0$ and other extremely useful design relations can be obtained by the introduction of a modular transformation on the elliptic functions.[88] By means of this trans-

[87] These formulae of course give the roots in terms of $i\Omega$, defined as $\frac{i\omega}{\sqrt{\omega_1\omega_2}}$, rather than in terms of $i\omega$. The transformation from one variable to the other is carried out at any convenient point in the design procedure, commonly after the impedances of the final network branches have been determined as functions of $i\Omega$, as in the case of more complicated frequency transformations.

[88] The general theory of modular transformations (other than Landen's transformation and Jacobi's imaginary transformation) is described in very few treatises on elliptic functions. It is described in detail by Cayley (19), however, and also by Jacobi (20), who was the first to introduce it. The use of modular transformations in connection with the insertion loss theory was introduced by Norton, who also showed that their use simplifies Cauer's Tchebycheff version of the image parameter theory.

formation, the product of elliptic functions appearing in the power ratio expression (64) can be replaced by a single elliptic function of a different modulus. This transformation yields an expression of the form

$$e^{2\alpha} = 1 + (e^{2\alpha_p} - 1)\,\mathrm{sn}^2\left[(2\eta + 1)u\frac{K_1}{K}, k_1\right] \tag{67}$$
$$\Omega = \sqrt{k}\,\mathrm{sn}\,(u, k)$$

where K_1 represents the complete integral of the new modulus k_1 while α_p is again the magnitude of the equal pass band ripples of the insertion loss characteristic. The moduli k_1 and k are uniquely related in the manner described below.

Suppose K' and K_1' represent the complete elliptic integrals of moduli $\sqrt{1 - k^2}$ and $\sqrt{1 - k_1^2}$, respectively, just as K and K_1 designate the complete integrals of moduli k and k_1. Then the so-called modular constants q and q_1 are defined as $\exp\left(-\pi\frac{K'}{K}\right)$ and $\exp\left(-\pi\frac{K_1'}{K_1}\right)$, respectively. They are uniquely related to k and k_1 and are required to satisfy the equation

$$q_1 = q^{2\eta+1}. \tag{68}$$

The above relation depends upon the fact that the equivalent power ratios (64) and (67) must both have the periods $\frac{2K}{2\eta + 1}$ and $2iK'$ in terms of u. The equivalence of the power ratios can be established by showing that when the periods are the same the roots, singularities, and behavior at infinity of the functions $(e^{2\alpha} - 1)$ are identical, which requires them to be proportional in accordance with Liouville's theorem. The constant or proportionality, which relates S_0 of (64) to α_p of (67), can be found by ordinary elliptic function analysis.

Relation between Arbitrary Design Parameters

One useful formula that can be derived from (67) is a very simple relation between the arbitrary design parameters. In the first place, well-known elliptic function relations combined with the identity of q_1 and $q^{2\eta+1}$ show (67) to be equivalent to

$$e^{2\alpha} = 1 + \frac{e^{2\alpha_p} - 1}{k_1^2\,\mathrm{sn}^2\left[(2\eta + 1)u_h\frac{K_1}{K}, k_1\right]} \tag{69}$$
$$\Omega = \frac{1}{\sqrt{k}\,\mathrm{sn}\,(u_h, k)}$$
$$u_h = u + iK'.$$

74

For real values of u_h, the elliptic sines in these equations will vary between $+1$ and -1. This requires the corresponding values of Ω to coincide with the effective attenuation band and its image at negative

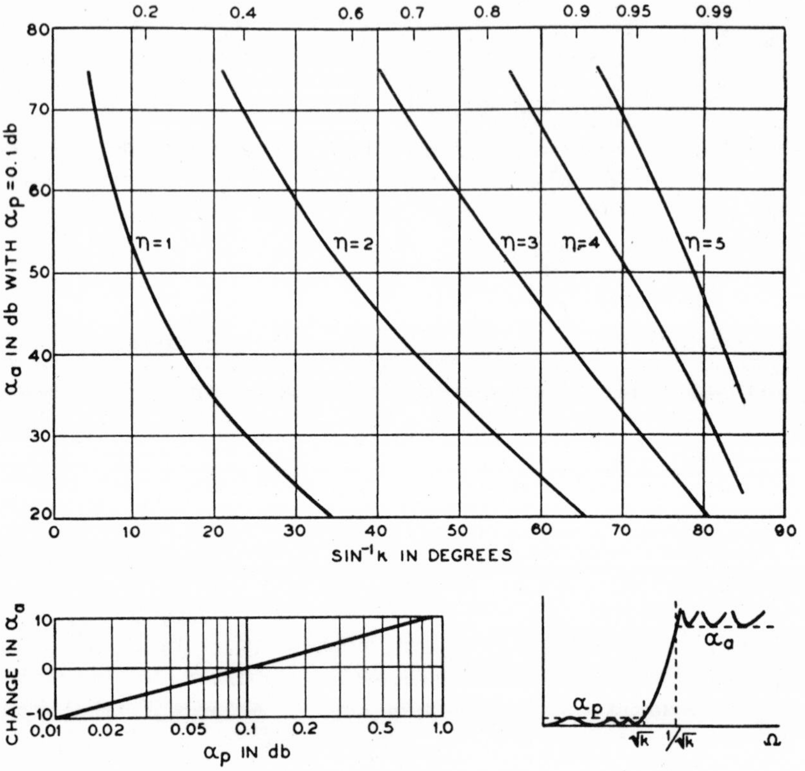

$\alpha_a = [10 \log_{10} (e^{2\alpha_p} - 1) - 10(2\eta + 1) \log_{10} q - 12.04]$ db
q = a function of k
η = number of ladder sections

Fig. 23. Tchebycheff attenuation band and pass band parameters—approximate relation between the design parameters

frequencies. It then follows that the loss α_a at the equal minima of the effective attenuation band will be such that

$$e^{2\alpha_a} = 1 + \frac{e^{2\alpha_p} - 1}{k_1^2} \tag{70}$$

It is obvious from this equation that k_1^2 will be quite small in the case of ordinary filters, for which α_p is small and α_a large. When k_1^2

is small, moreover, the corresponding modular constant q_1 approximates $\frac{1}{16}k_1^2$ very closely. Introducing this approximation and the identity of q_1 and $q^{2\eta+1}$ in (70), assuming $e^{2\alpha_a}$ large compared with unity, and solving for α_a gives

$$\alpha_a = [10 \log_{10} (e^{2\alpha_p} - 1) - 10(2\eta + 1) \log_{10} (q) - 12.04] \text{ db} \quad (71)$$

This equation is extremely useful in choosing the primary design parameters. The modular constant q is a measure of k, which is merely the ratio of the cutoff frequency f_1 to the attenuation band limit f_2. Thus the equation relates the sharpness of cutoff, the maximum pass band distortion α_p, the minimum attenuation band loss α_a and the number of ladder sections η. Aside from this necessary relationship these parameters are all arbitrary. The function $\log_{10} (q)$ is tabulated against $\sin^{-1} k$ in most elliptic function tables and in addition there are rapidly convergent series for computing q from k or vice versa.[89] Values of α_a determined by (71) are plotted in Fig. 23 against $\sin^{-1} k$ for various values of η, assuming α_p to be 0.1 db, together with a curve of the changes in α_a corresponding to changes in α_p.

Determination of the Real Roots of the Power Ratio

When k_1 is as small as (70) requires in ordinary design problems, $\text{sn}\left[(2\eta + 1)u\frac{K_1}{K}, k_1\right]$ approximates $\sin\left[(2\eta + 1)u\frac{\pi}{2K}\right]$ for real values of u and neighboring complex values. In other words for values of Ω not too remote from the pass band, (67) can be replaced by

$$\begin{aligned} e^{2\alpha} &= 1 + (e^{2\alpha_p} - 1) \sin^2\left[(2\eta + 1)u\frac{\pi}{2K}\right] \\ \Omega &= \sqrt{k}\, \text{sn}\, (u, k). \end{aligned} \quad (72)$$

The function $\text{sn}\,(u, k)$ cannot normally be replaced by $\sin\left[u\frac{\pi}{2K}\right]$ since the modulus k represents the ratio of the cutoff frequency to the attenuation band limit and is rarely small.

The above equations can be used to compute the real roots $\pm a_0$ of the power ratio, which must be determined before the formulae (65)

[89] For a tabulation of $\log_{10} (q)$ see, for instance, Silberstein's "Synopsis of Applicable Mathematics" (21).

76

for the complex zeros can be used. If u is replaced by iu', the above power ratio will be zero at the real value of u' determined by

$$\sinh\left[(2\eta + 1)u' \frac{\pi}{2K}\right] = \frac{1}{\sqrt{e^{2\alpha_p} - 1}} \tag{73}$$

The real zero a_0 can be calculated from u' by the following formula derived by replacing sn (iu', k) in $i\sqrt{k}$ sn (iu', k) by Jacobi's well known equivalent:

$$a_0 = \frac{\sqrt{k}\,\mathrm{sn}\,(u', \sqrt{1 - k^2})}{\mathrm{cn}\,(u', \sqrt{1 - k^2})} \tag{74}$$

While this procedure uses the approximate formula (72) for the power ratio at an imaginary value of u, this value turns out to be such that the approximation is reasonably close in ordinary design problems. Any small error in a_0, moreover, represents merely a small change in the primary design constant α_p or α_a. A rigorous formula for u' can readily be derived from (67) but is relatively complicated in numerical applications.

Other Special Design Formulae

A large variety of additional rigorous and approximate special design formulae can be derived by similar methods. These include not only formulae for such quantities as insertion phase, impedances, etc., but also equations for determining the constants Ω_σ, a_0, etc., in terms of well-known Θ function expansions rather than from elliptic function tables. For the particular elliptic function computations required the Θ function expansions are usually easier to use than the tables.[90]

Inverse Impedance Filters with Attenuation Band and Pass Band Tchebycheff Parameters

Inverse impedance filters with attenuation band and pass band Tchebycheff parameters call for exactly the same sort of analysis as symmetrical filters. The only real difference is in the specific periodic substitution and elliptic function transformation involved. The simplest formulae are obtained by requiring zero loss at zero frequency and at the pass band minima rather than at zero frequency, at the cutoff,

[90] These Θ function expansions are the Fourier series representations of the functions. They are particularly convenient in that the arguments of the trigonometric functions involved depend only on η. The expansions are listed in various tables of formulae such as Silberstein's "Synopsis of Applicable Mathematics" (21).

and at the pass band maxima. While this leads to a somewhat less efficient use of the elements it at least permits the terminations to be equal. The corresponding power ratio turns out to be determined by the following set of equations

$$
\begin{aligned}
e^{2\alpha} &= 1 + (e^{2\alpha_p} - 1)\,\mathrm{sn}^2\left[(2\eta)u\frac{K_1}{K}, k_1\right] \\
\Omega^2 &= k\,\mathrm{sn}\,[u, k]\,\mathrm{sn}\left[\left(u + \frac{K}{\eta}\right), k\right] \\
q_1 &= q^{2\eta} \\
f_1/f_2 &= k\,\mathrm{sn}^2\left[\frac{2\eta - 1}{2\eta}K, k\right]
\end{aligned}
\tag{75}
$$

In these equations, Ω and f_1, f_2 are defined as in the case of symmetrical filters but k no longer represents f_1/f_2, being now defined by the last of these equations.

The more efficient requirement of zero loss at zero frequency and at the pass band maxima can be satisfied by making linear transformations on $e^{2\alpha}$ and Ω^2 in these equations. The increased efficiency is indicated by the corresponding change in the formula for f_1/f_2, which becomes

$$
f_1/f_2 = k\,\mathrm{sn}\left[\frac{2\eta - 1}{2\eta}K, k\right] \tag{76}
$$

Other Types of Filters

Other types of filters with attenuation band and pass band Tchebycheff parameters besides the low-pass types described above can be obtained by means of ordinary frequency transformations. These include high-pass filters and also band-pass and band-elimination filters with symmetrical loss characteristics on a logarithmic frequency scale. More general types of characteristics can be obtained by making rational transformations on the power ratio and on the square of the frequency, but this usually leads to power ratios not realizable with networks of the symmetrical or inverse impedence types. Still more general characteristics can be obtained by combining similar transformations with the use of Tchebycheff pass band parameters but general attenuation band parameters.

Instead of transforming the power ratios described previously, the analysis used in driving them can itself be modified in such a way as to lead to new types of loss characteristics. Differences between the transfer constants of reference image parameter filters with matched impedances can be used, for instance, in the design of filters on the Tchebycheff pass band parameter basis.[91] Similarly, new types of loss characteristics can be obtained by using different elliptic functions in (67). Tchebycheff attenuation band parameters can be obtained with general pass band parameters by merely replacing cosh (Θ_I) by sech (Θ_I) in (46).

Even when the power ratio is appropriate for symmetrical or inverse impedance filters, the terminations can both be chosen arbitrarily if the requirement of networks of these particular types is abandoned.[92] It is always possible, for instance, to make one termination zero or infinite.

PART IV. DISSIPATIVE REACTANCE NETWORKS

The parasitic dissipation which must be present in actual reactance networks is referred to in Parts I, II and III only in connection with methods of estimating the effects of the dissipation in networks designed on a non-dissipative basis. These effects, however, frequently represent substantial differences between the characteristics actually realized and those which the non-dissipative design procedure assumes to be obtained. This makes it desirable to include in the insertion loss theory a modification applying directly to the design of dissipative reactance networks.

The logical procedure is to parallel the theory of non-dissipative networks with a synthesis theory applying to the design of networks which are made up of dissipative reactance elements and which produce prescribed insertion loss characteristics when terminated in prescribed resistances. The filter theory of Part III can then be applied to the choice of loss characteristics provided this can be done without violating the conditions necessary for the realizability of dissipative networks of the type required. Even when the configurations are required to be those which would be used in the absence of dissipation there frequently turn out to be realizable loss functions at least very closely approxi-

91 This amounts to saying that sections described by negative "m's" can be included in the reference filters.

92 Provided, of course, $\frac{4R_1R_2}{(R_1+R_2)^2}$ is equal to or less than the smallest value assumed by $e^{2\alpha}$ at real frequencies.

mating the desired special functions of Part III except for added constant losses.[93]

The best method to use in designing a dissipative reactance network depends upon what distribution of the dissipation is required—i.e., upon the relative dissipativeness of the various elements to be used. In the networks encountered in ordinary design problems the dissipation approximates closely certain very simple distributions for which special design procedures can be developed. This situation makes it unnecessary to develop a perfectly general theory permitting the dissipation constants of all the elements to be prescribed independently. Any such general theory must obviously be undesirably complicated since it must permit the introduction of a large number of additional arbitrary constants besides those determining the insertion loss function.

Equally Dissipative Inductances and Capacities

The simplest theory is obtained by assuming a dissipative reactance network meeting the following very special requirements. Each actual inductance is required to be equivalent to an ideal inductance combined with a constant series resistance. Similarly, each actual capacity must be equivalent to an ideal capacity combined with a constant shunt resistance. Finally, all the inductance-resistance and capacity-conductance ratios are required to be equal, which amounts to requiring the phase angles of the impedances of all the reactance elements to be of equal magnitude.

Fundamental Transformation Principle

To understand the design of networks involving the uniform type of dissipation described above it is simplest to begin by examining the effect of adding such dissipation to all the reactance elements in any known network. The determination of the function of p representing any voltage ratio, impedance, or similar complex quantity describing the known network is accomplished by properly combining the im-

[93] Bode (22) discovered ways of adding resistances to image parameter filters in such a manner as to reduce the distortion due to dissipation. This procedure increases the number of reactance elements, however, by requiring certain elements to be split in two, and in addition it only partially compensates for the effects of dissipation. Bode also discovered certain general types of equalizers which can actually duplicate the discrimination characteristics of filters and in which all inductances can be in series with resistances and all capacities in parallel with resistances and can therefore be permitted to be dissipative. These equalizers are of the constant resistance type, however, and require approximately twice the number of elements appearing in non-dissipative filters.

pedances of the various elements. On a non-dissipative basis, the reactance elements produce impedances typified by Lp and $\frac{1}{Cp}$. Upon the addition of the uniform dissipation these impedances become $L(p + d)$ and $\frac{1}{C(p + d)}$, where d is a constant representing the value assumed by the equal resistance-reactance ratios at unit ω. It follows that the modification of the complex function corresponding to the addition of the dissipation amounts to the transformation represented by the substitution of $(p + d)$ for p.[94]

It follows from the above principle that the *removal* of uniform dissipation from all the reactance elements of a network will change each corresponding complex function by the transformation represented by the substitution of $(p - d)$ for p. Thus if a network in which the elements are equally dissipative is to be designed to produce a prescribed complex function it is only necessary to design any network producing the "predistorted" function obtained by replacing p by $(p - d)$ in the prescribed function. The prescribed function will then be obtained upon the addition of the required dissipation.

When the insertion loss function of a dissipative reactance network is prescribed, the first step is to find the corresponding insertion voltage ratio exactly as in the design of non-dissipative networks. The predistortion of this voltage ratio and the computation of the corresponding predistorted insertion power ratio is then followed by the design of a pure reactance network, to which the required dissipation is finally added. When the prescribed insertion loss is to be realized exactly, the corresponding voltage ratio of the form $\frac{A + pB}{P}$ yielded by the method of Part I must normally be modified by the reversal of the signs of the poles with negative real parts before predistortion, in order to obtain a predistorted voltage ratio realizable with a pure reactance network.[95] In most actual design problems, however, a more efficient use of elements is obtained by retaining $\frac{A + pB}{P}$ and only approximating the rigorous procedure in a manner to be described later, which

[94] This transformation, which was originally discovered by Bode, is described in Guillemin's "Communication Networks" (18)—Chap. VI, footnote on pp. 245, 246.

[95] It can be shown that this change in the original voltage ratio does not change the corresponding insertion loss but only the insertion phase.

usually leads to a very good approximation to the prescribed insertion loss.

Arrangement of Design Procedure to Meet Physical Conditions

In accordance with Part I, the realization of a predistorted voltage ratio with a pure reactance network calls for the following conditions. First, the predistorted voltage ratio must be expressible in the form $\frac{A + pB}{P}$, in which A, B, and P are even polynomials with real coefficients such that $(A + pB)$ has no roots with positive real parts. Second, the corresponding predistorted power ratio must be positive and no less than $\frac{4R_1R_2}{(R_1 + R_2)^2}$ at all real frequencies. Third, in order to avoid inefficiency in the use of elements it is also best to require that $(A + pB)$ and P shall have no common factors except possibly p.[96]

The usual design procedure is to choose an insertion loss function without regard to dissipation and then to require the dissipation to be small enough to yield a predistorted voltage ratio with roots with negative real parts. Replacing p by $(p - d)$ in the original voltage ratio amounts to adding d to each of the roots and poles. Thus the restriction on the dissipation requires the dissipation constant d to be smaller than the magnitude of the smallest real part of any root of the original voltage ratio, so that the addition of d can not change a negative real part into a positive real part.

In order to meet the limitation on the magnitude of the predistorted power ratio an initially unknown constant factor can be included in the original power ratio. It turns out that this constant also appears in the predistorted power ratio as nothing more than a simple multiplier and can thus easily be chosen in such a way that the predistorted power ratio is no less than $\frac{4R_1R_2}{(R_1 + R_2)^2}$.

If a prescribed insertion loss function is such that the corresponding actual voltage ratio takes the form $\frac{A + pB}{P}$ without use of common

[96] This is necessary for the exclusion of common factors from the expression $\frac{N}{P^2}$ for the corresponding power ratio. Recall that inefficient networks are usually encountered when common factors other than constants or p^2 must be combined with numerator and denominator of the simplest rational fraction expression for $e^{2\alpha}$ in order to obtain the form $\frac{N}{P^2}$.

factors other than p, the predistorted voltage ratio will not take that form.[97] Since this condition requires the finite poles of a voltage ratio to occur in pairs differing only in sign, it cannot be satisfied both before and after the addition of d to each of the poles. Modified insertion loss functions can readily be found such that the predistorted voltage ratios meet this condition rather than the actual voltage ratios. It is normally more satisfactory, however, to start with a prescribed insertion loss function of the previous type and then to realize it only to a good approximation by means of the procedure outlined below.

A power ratio is first chosen which takes the form $\frac{N}{P^2}$ without use of identical roots of N and P^2 except possibly p^2. The theory of Part I is then used to determine the corresponding voltage ratio of the form $\frac{A + pB}{P}$. The predistorted voltage ratio is next obtained by replacing p by $(p - d)$ in the numerator of this function only, the denominator P being left unchanged. A corresponding pure reactance network is then designed, after which the required dissipation is added as before. The voltage ratio actually obtained is a third function derived from $\frac{A + pB}{P}$ by replacing p by $(p + d)$ in the denominator P.

It follows from the fact that P is real at real frequencies that the actual power ratio obtained by the above procedure will approximate the original power ratio $\frac{N}{P^2}$ very closely except near the poles. This is because the first order effect of the substitution of $(p + d)$ for p upon the magnitude of a function of ω is proportional to the derivative of the phase, while the phase of P is zero at real frequencies.[98]

Special Case of Filters

The effectiveness of the design procedure described above is illustrated by the special case in which the original loss function is formed by adding a constant loss to one of the functions described in Part III as appropriate for filter purposes. In this special case, the maximum

[97] Except in the very special case in which P is a constant, i.e., in which all frequencies of infinite loss occur at infinity.

[98] The relation between the derivative of the phase and the effect of the substitution of $(p + d)$ for p upon the magnitude of a function was discovered by Bode, who used it in the derivation of his formulae for the estimation of the effect of adding dissipation to a prescribed network. It is used in exactly the same way by Guillemin (18) in deriving Mayer's solution (10) of the same Problem.

value of the dissipation constant d for which predistortion is physically possible unusually corresponds to an amount of dissipation which would produce a pass band distortion of about 6 decibels if the desired voltage ratio were not predistorted. To avoid practical difficulties it is usually necessary to keep d well below this physical limit, a margin of perhaps 30 to 50 per cent being usually needed.[99]

When d is small enough to satisfy the physical restriction, the predistortion of only the numerator of the voltage ratio frequently leads to a final loss characteristic which differs appreciably from that originally prescribed only by the substitution of rounded finite peaks for the infinite peaks in the attenuation band. It is possible, however, for the loss to be reduced substantially below the original minima between the peaks over a portion of the attenuation band.

Serious reductions in attenuation band losses can be partly compensated for by starting with complex frequencies of infinite loss so chosen that one of each conjugate pair becomes real in the final characteristic. If Tchebycheff pass band parameters are used, for instance, complex m's can be assigned to the reference image parameter filter. It turns out that these m's must be so chosen that the addition of the required dissipation to the reference filter would produce infinite attenuation peaks. This permits known methods to be used in their determination.[100] It is also possible to obtain better approximations to non-dissipative filter characteristics by including so-called compensating resistances in addition to dissipative reactances in the networks producing prescribed insertion losses, but this requires substantial changes in design procedure.

Unequally Dissipative Inductances and Capacities

Suppose now that the previous requirement of uniform dissipation is relaxed to the extent of permitting the capacities to include a different amount of dissipation than the inductances. Except in the design of such circuits as narrow-band filters, the situation immediately becomes much more complicated.[101] The predistorted voltage ratio correspond-

[99] As d approaches the physical limit the constant loss which must be added to the loss function chosen on a non-dissipative basis approaches infinity.

[100] The possibility of using complex m's to obtain infinite attenuation peaks in dissipative image parameter filters was introduced by Bode (4) and is described by Guillemin (18).

[101] A frequency transformation can be used to transform a low-pass filter of equally dissipative inductances and capacities into a band-pass filter of series and parallel resonant circuits resonating at mid-band and with associated series

ing to the removal of all dissipation from the final network can no longer be found directly from the actual voltage ratio. In addition, it now depends upon the assumption of a 4-pole including no resistances other than those produced by the equally dissipative inductances and equally dissipative capacities. It turns out, however, that a design procedure can be obtained by combining the predistortion principle appropriate for equally dissipative inductances and capacities with an additional modification of the theory of Part I.

The simplest procedure is to replace the dissipation constants of the inductances and capacities by constants d_0 and δ representing their average and one half their difference. The typical inductive and capacitive impedances then become $L(p + d_0 + \delta)$ and $\frac{1}{C(p + d_0 - \delta)}$. The predistortion method described above can be used to determine the voltage ratio that would be obtained upon the removal of the dissipation represented by d_0. The design problem is therefore solved if a method can be found for designing a network of impedances of the types $L(p + \delta)$ and $\frac{1}{C(p - \delta)}$ producing a prescribed voltage ratio. Such a design method has been found and turns out to be very similar to that described in Part I for the non-dissipative case, in which δ is zero.

Design of Networks of Oppositely Dissipative Inductances and Capacities

The fact that the above design problem can be solved depends upon the following properties of networks of impedances of the type $L(p + \delta)$ and $\frac{1}{C(p - \delta)}$. In the first place, the elements of the determinant of any network of this type can be expressed in the form.

$$Z_{jk} = \frac{L_{jk}(p^2 - \delta^2) + \dfrac{1}{C_{jk}}}{p - \delta} \tag{77}$$

It follows that the determinant must be equivalent to the product of $(p - \delta)^{-n}$ and an even polynomial in p, where n is the degree of the determinant. The same statement is also true of any cofactor of the

---

and shunt constant resistances. If the band is narrow, the resistances approximate the total effect of dissipation no matter how it is divided between the inductances and capacities.

determinant provided n is modified to take account of the change in degree. Thus each open- and short-circuit impedance of a network of this particular type is equivalent to the product of $(p - \delta)^{-1}$ and an even rational function of p.[102] In other words, the impedances differ in form from those of non-dissipative networks only by the substitution of the factor $(p - \delta)$ for p.

The important design formulae are those for determining the open- and short-circuit impedances as functions of frequency. After these have been determined the impedances obtained upon the removal of all dissipation can easily be computed. In accordance with (77), all that is necessary is to replace the factor $(p - \delta)^{-1}$ by p^{-1} in each impedance and to replace p^2 by $(p^2 + \delta^2)$ in the even rational function multiplying this factor. Formulae for computing the impedances of the dissipative network are listed below. The derivation will be omitted as in the case of the non-dissipative networks of Part I since the formulae can be checked by simple mesh computations.

Particularly simple formulae are obtained by expressing the insertion voltage ratio in the form

$$\frac{V_{20}}{V_2} = \frac{p}{p - \delta} \frac{A + pB}{P} \tag{78}$$

in which A, B and P are even polynomials as before. The additional polynomials A' and B' are then found exactly as in the determination of non-dissipative networks producing the voltage ratio $\frac{A + pB}{P}$. In other words, A' and B' are determined from the following pair of equations formed by combining (12) and (13) of Part I:

$$\begin{aligned} A^2 - p^2B^2 - \frac{4R_1R_2}{(R_1 + R_2)^2} P^2 &= J'^2(p_1'^2 - p^2) \cdots (p_n'^2 - p^2) \\ A' + pB' &= J'(p_1' - p) \cdots (p_n' - p) \end{aligned} \tag{79}$$

[102] This theorem can also be derived from the following theorem included in Guillemin's "Communication Networks" (18)—Vol. II, Chap. X, p. 445. Suppose each element in a network produces an impedance proportional to $(p + 2\alpha)$ or $\frac{1}{(p + 2\beta)}$. Then any impedance of the network takes the form of the product of $\sqrt{\frac{p + 2\alpha}{p + 2\beta}}$ and an odd rational function of $\sqrt{(p + 2\alpha)(p + 2\beta)}$. This theorem was preceded by a still more general theorem due to Bode which states that if each element of a network produces an individual impedance proportional to Z_1 or Z_2 any impedance of the complete network will be the product of Z_2 and a rational function of $\frac{Z_1}{Z_2}$.

In terms of these polynomials, the impedances of the dissipative network are given by

$$
\begin{aligned}
Z_{O1} &= R_1 \frac{(A - A') + \delta(B + B')}{(p - \delta)(B + B')} \\
Z_{O2} &= R_2 \frac{(A + A') + \delta(B + B')}{(p - \delta)(B + B')} \\
Z_{O12} &= \frac{-2R_1R_2}{(R_1 + R_2)} \frac{P}{(p - \delta)(B + B')} \\
Z_{S1} &= R_1 \frac{p^2(B - B') + \delta^2(B + B') + 2\delta A}{(p - \delta)[(A + A') + \delta(B + B')]} \\
Z_{S2} &= R_2 \frac{p^2(B - B') + \delta^2(B + B') + 2\delta A}{(p - \delta)[(A - A') + \delta(B + B')]} \\
Z_{S12} &= \frac{R_1 + R_2}{2} \left(\frac{p^2(B - B') + \delta^2(B + B') + 2\delta A}{(p - \delta)P} \right)
\end{aligned}
\qquad (80)
$$

A very similar alternative set of formulae expressing the short-circuit impedances more simply rather than the open-circuit impedances can be obtained by starting with a voltage ratio derived from (78) by reversing the sign used with δ.

The specific type of voltage ratio function required depends more upon the particular configuration than in the case of non-dissipative networks. A mid-series high pass ladder, for instance, will produce an infinite loss point at $p = +\delta$ while the corresponding mid-shunt ladder will produce one at $p = -\delta$. The exact type of voltage ratio function required can ordinarily be determined by means of conventional circuit analysis. Since the dissipation associated with one type of element is negative, it may be possible for the corresponding power ratio to be less than $\frac{4R_1R_2}{(R_1 + R_2)^2}$ at certain real frequencies. In ordinary design problems, however, it is to be expected that there will be a minimum permissible value somewhere in the near neighborhood of $\frac{4R_1R_2}{(R_1 + R_2)^2}$.

The determination of $(A' + pB')$ by (79) is also more restricted than in the case of non-dissipative networks. This is because it is the multiplicity of solutions for $(A' + pB')$ which leads to the various non-equivalent configurations producing a given voltage ratio, while different configurations which can produce the same voltage ratio on a non-

87

dissipative basis may no longer have this property after the addition of the dissipation.

Useful information bearing on the choice of B' is obtained by examining the driving-point impedances of a network determined by (80). Simple circuit analysis shows the corresponding reflection coefficients to be

$$\begin{aligned} \frac{R_1 - Z_1}{R_1 + Z_1} &= \frac{p - \delta}{p}\left(\frac{A' + pB'}{A + pB}\right) - \frac{\delta}{p} \\ \frac{R_2 - Z_2}{R_2 + Z_2} &= \frac{p - \delta}{p}\left(\frac{-A' + pB'}{A + pB}\right) - \frac{\delta}{p} \end{aligned} \tag{81}$$

These formulae indicate that at least one of the driving-point impedances will be the negative of the corresponding termination at zero frequency unless $\frac{A'}{pB'}$ and $\frac{A}{pB}$ approach zero at that point and unless $\frac{B'}{B}$ approaches -1. It can be shown that $\frac{A'}{pB'}$ and $\frac{A}{pB}$ will approach zero and $\frac{B'}{B}$ will approach ± 1 provided zero frequency represents an even order pole of the voltage ratio (78), or else a point at which it is finite. The above conditions will then be satisfied by the proper choice of the sign of B', which (79) leaves arbitrary.

Other Types of Uniformly Dissipative Networks

Similar methods can be used in the design of dissipative reactance networks in which each element is equivalent to an inductance or capacity combined with both series and shunt constant resistances, all elements of each kind being equally dissipative at all frequencies. This permits the simulation of the variation of dissipation with frequency encountered in actual elements.

The effect of removing both series and shunt types of dissipation from both kinds of elements in like amounts can be computed by means of a bilinear predistortion transformation. The removal of the proper amount of dissipation of each kind leaves a network in which the capacitive dissipation is the negative of the inductive dissipation. The impedances are then represented by products of $(p - \delta - \gamma p^2)^{-1}$ and even rational functions of p, where δ and γ depend only upon the dissipation constants. Methods similar to those described previously show that the corresponding impedances can be determined by replacing δ by $(\delta + \gamma p^2)$ in (78) and (80).

Theoretically, networks of still more complicated elements can be designed in exactly the same way. The only requirement is the existence of a predistortion transformation producing a predistorted voltage ratio which can be realized with a network of two kinds of elements such that the ratio of their impedances is proportional to an even rational

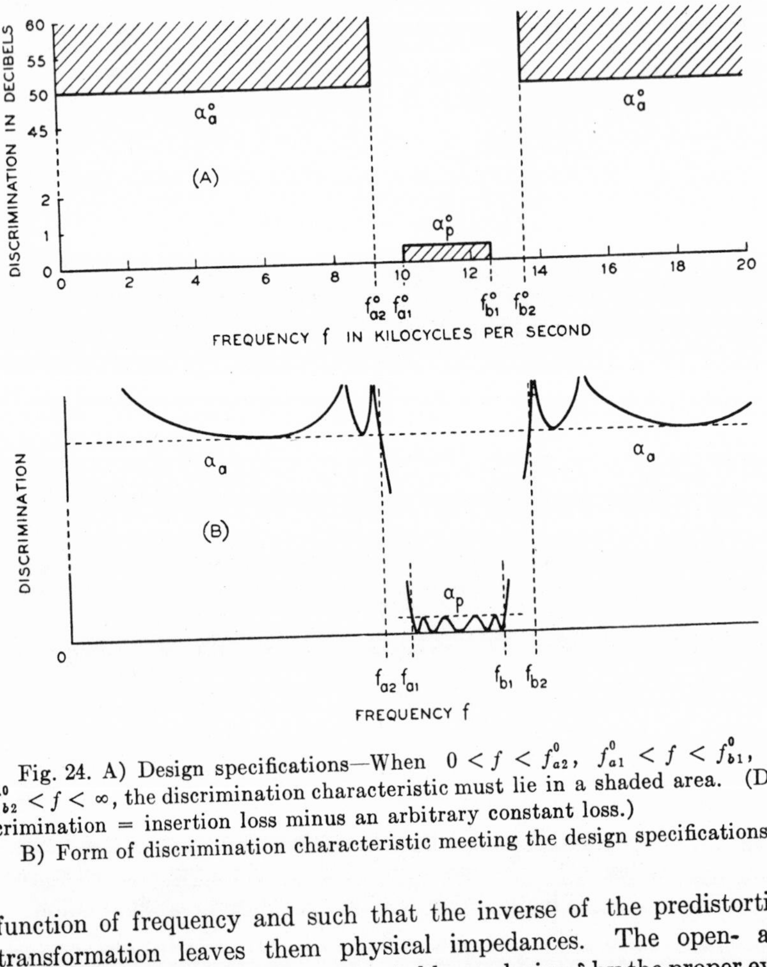

Fig. 24. A) Design specifications—When $0 < f < f^0_{a2}$, $f^0_{a1} < f < f^0_{b1}$, or $f^0_{b2} < f < \infty$, the discrimination characteristic must lie in a shaded area. (Discrimination = insertion loss minus an arbitrary constant loss.)

B) Form of discrimination characteristic meeting the design specifications.

function of frequency and such that the inverse of the predistortion transformation leaves them physical impedances. The open- and short-circuit impedances are computed by replacing δ by the proper even rational function. The principal difficulty is the determination of the required form of voltage ratio and the advantageous choice of a specific function of this form.

Concentrated Dissipation

The only non-uniformly dissipative networks which are commonly encountered approximate non-dissiptative networks to which constant resistances have been added at only a few points. Methods similar to those described above do not apply to these networks. Successive approximation methods for correcting for the effect of adding the resistances have been used successfully but are too complicated to be included here.

Illustrative Numerical Design

In the introduction, the specific design procedure appropriate for filters of a special illustrative type was outlined in order to motivate the development of the general theory. By way of conclusion, an illustrative numerical design will now be described briefly in order to clarify further the way in which the theory can be applied to actual design problems. The filters considered in the introduction were low-pass filters having the mid-series low-pass ladder configuration and satisfying the requirement of simultaneous attenuation band and pass band Tchebycheff parameters. The numerical example which will be considered now is the design of a band-pass filter which is related to one of these low-pass filters by the frequency transformation principle described in Part III.

Suppose that a band-pass filter is to be designed to meet the following specifications:

A) Frequencies to be transmitted—10.0 kc to 12.5 kc
B) Frequencies to be effectively eliminated—0 to 9.2 kc, 13.5 kc to ∞
C) Pass-band distortion—<0.4 db
D) Discrimination against unwanted frequencies—>50 db (82)
E) Dissipation compensation to be sufficient to lead to readily realizable elements
F) Final network to be a ladder involving no transformers or coupled coils

Requirements *A* through *D* are indicated graphically in Fig. 24A.

Using the notation indicated in Fig. 24A the critical frequencies f^0_{a1}, f^0_{a2}, f^0_{b1}, f^0_{b2} satisfy approximately the equation

$$f^0_{a1}f^0_{b1} = f^0_{a2}f^0_{b2} \qquad (83)$$

while the discrimination requirement in (82) is the same for both attenuation bands. This amounts to saying that the above loss requirements are symmetrical on a logarithmic frequency scale about the mid-band frequency $\sqrt{f_{a1}^0 f_{b1}^0}$. This symmetry makes it possible to simplify the design procedure by first designing a low-pass filter and then converting it to a band-pass filter by means of a frequency transformation. Due to the uniformity of the discrimination requirement at all attenuation band frequencies, the low-pass filter can be designed by the straightforward method of simultaneous attenuation band and pass band Tchebycheff parameters. The final loss characteristic then takes the form illustrated qualitatively in Fig. 24B for the special case in which the low-pass filter includes two sections.

The first operation in the design procedure is the choice of the arbitrary constants of the low-pass filter theory subject to the specificatons to be satisfied by the final band-pass filter. This choice is guided by the following considerations:

The maximum pass band distortion α_p and the minimum attenuation band discrimination α_a are the same for both the low-pass and the band-pass filters since their loss characteristics differ only as to the frequency scales. The number of sections η is also the same for the two filters provided sections of the 6-element type of configuration are considered in the band-pass case. Finally, it can be shown that the constant k measuring the sharpness of cut-off obeys the relation

$$k = \frac{f_{b1} - f_{a1}}{f_{b2} - f_{a2}} \tag{84}$$

in which f_{a1}, f_{b1}, etc., are defined by Fig. 24B and are themselves related by

$$f_{b1} f_{a1} = f_{b2} f_{a2} . \tag{85}$$

In accordance with Part III, the constants α_p, α_a, k, η of the low-pass theory are not all arbitrary but must be related by (71)—i.e., by

$$\alpha_a = [10 \log_{10}(e^{2\alpha_p} - 1) - 10(2\eta + 1) \log_{10}(q) - 12.04] \text{ db} \tag{86}$$

in which $\log_{10}(q)$ is a quantity which is tabulated against $\sin^{-1} k$ in most elliptic function tables.

A choice of constants which is consistent with (85) and (86) and with the original specifications in (82) is as follows:

$$\begin{aligned}
k &= 0.62 \\
\eta &= 2 \\
\alpha_p &= 0.30 \text{ db} \\
\alpha_a &= 52.4 \text{ db} \\
f_{a1} &= 9.96 \text{ kc} \\
f_{b1} &= 12.54 \text{ kc} \\
f_{a2} &= 9.2872 \text{ kc} \\
f_{b2} &= 13.4484 \text{ kc}
\end{aligned} \tag{87}$$

These constants fix the final loss characteristic except for changes due to the requirement of dissipative elements.

After the constants of the Tchebycheff theory have been chosen, the corresponding voltage ratio $\frac{A + pB}{P}$ is evaluated. This is easily accomplished by means of the special formulae for the roots and poles which are included in the Tchebycheff theory. If p is used to represent $i\Omega$ rather than $i\omega$, the voltage ratio corresponding to the choice of constants indicated in (87) turns out to be

$$\frac{V_{20}}{V_2} = \frac{A + pB}{P} = \frac{(a_0 + p)(\rho_1^2 + 2a_1 p + p^2)(\rho_2^2 + 2a_2 p + p^2)}{a_0 \rho_1^2 \rho_2^2 (1 + \Omega_1^2 p^2)(1 + \Omega_2^2 p^2)} \tag{88}$$

in which the constants have the following values:

$$\begin{aligned}
a_0 &= 0.37766 \\
a_1 &= 0.25943 \\
a_2 &= 0.077333 \\
\Omega_1^2 &= 0.24902 \\
\Omega_2^2 &= 0.57282 \\
\rho_1^2 &= 0.37822 \\
\rho_2^2 &= 0.66141
\end{aligned} \tag{89}$$

The low-pass type of voltage ratio is obtained by defining Ω as $\frac{\omega}{\sqrt{\omega_1 \omega_2}}$

92

as in equation (60) of Part III. The corresponding band-pass type of voltage ratio is obtained by redefining Ω in accordance with the relation

$$\Omega = \frac{\omega^2 - \omega_{b1}\omega_{a1}}{\omega\sqrt{(\omega_{b1} - \omega_{a1})(\omega_{b2} - \omega_{a2})}} \tag{90}$$

in which ω_{a1}, ω_{b1}, etc. are the values assumed by ω at the frequencies f_{a1}, f_{b1}, etc., defined by Fig. (24B).

The next operation is the predistortion of the voltage ratio (88) in order to compensate for the dissipation required in the elements of the final network. The predistortion is accomplished by replacing p by $(p - d)$ in the numerator of (88). The constant d appearing in this transformation is related to the dissipativeness of the final band-pass filter by

$$d = \frac{\sqrt{f_{a1}f_{b1}}}{\sqrt{(f_{b1} - f_{a1})(f_{b2} - f_{a2})}}(d_L + d_C) \tag{91}$$

in which d_L represents the mean of the resistance-reactance ratios of the coils, evaluated at mid-band frequency, and d_C represents the corresponding mean of the conductance-susceptance ratios of the condensers. In order to obtain a physically realizable design, d_L and d_C must be sufficiently small to render d substantially smaller than a_2 of (89). A suitable pair of values is

$$\begin{aligned} d_L &= 0.0100 \\ d_C &= 0.0025 \end{aligned} \tag{92}$$

which yields

$$d = 0.04263 \tag{93}$$

In addition to the modification of the voltage ratio (88) by the predistortion transformation described above, a temporarily unknown constant factor is added to represent an added constant loss. The general theories of Parts I and II are then applied to the design of a corresponding pure reactance network, using the frequency variable Ω in place of ω.[103] By choosing the added constant loss and the ratio of terminations in the proper way, a ladder type of network is obtained

[103] If Ω is defined as $\frac{\omega}{\sqrt{\omega_1\omega_2}}$, as for a low-pass filter, it is proportional to ω and can obviously be used in place of it in the formulae of Parts I and II. If it is defined by (90), as for a band-pass filter, the resulting network is related to the low-pass network by the frequency transformation principle.

which includes no transformers or negative impedance branches and which produces a minimum transducer loss.[104] The addition of the

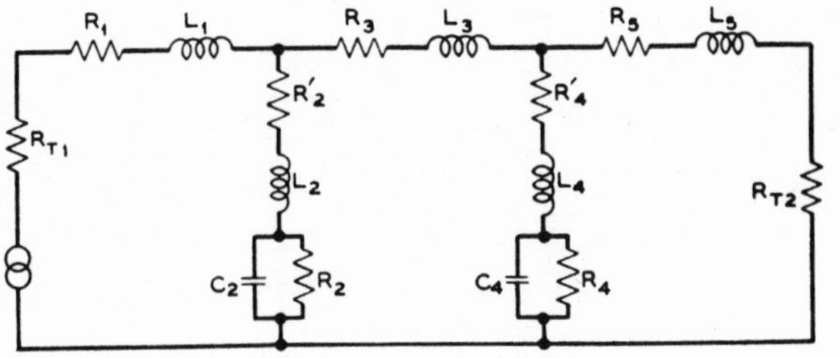

$$R_{T2} = 0.084427\, R_{T1}$$

$$L_1 = 1.1834 \frac{R_{T1}}{\sqrt{\omega_1 \omega_2}} \qquad R_1 = 0.050448\, R_{T1}$$

$$C_2 = \frac{1.8849}{R_{T1}\sqrt{\omega_1 \omega_2}} \qquad R_2 = 12.445\, R_{T1}$$

$$L_2 = 0.13211 \frac{R_{T1}}{\sqrt{\omega_1 \omega_2}} \qquad R_2' = 0.0056318\, R_{T1}$$

$$L_3 = 2.3227 \frac{R_{T1}}{\sqrt{\omega_1 \omega_2}} \qquad R_3 = 0.099017\, R_{T1}$$

$$C_4 = \frac{1.7650}{R_{T1}\sqrt{\omega_1 \omega_2}} \qquad R_4 = 13.290\, R_{T1}$$

$$L_4 = 0.32454 \frac{R_{T1}}{\sqrt{\omega_1 \omega_2}} \qquad R_4' = 0.013835\, R_{T_1}$$

$$L_5 = 0.85255 \frac{R_{T1}}{\sqrt{\omega_1 \omega_2}} \qquad R_5 = 0.036344\, R_{T1}$$

Provided R_{T1} is in ohms and ω_1, ω_2 are in radians per second:
Inductances are in henries
Capacities are in farads
Resistances are in ohms

Fig. 25. Preliminary low-pass filter

proper resistances to this network, corresponding to the substitution of $(p + d)$ for p in the impedance functions representing the different

[104] Recall that transducer loss was defined in the introduction as the difference in level between the received power and the maximum power obtainable from the generator with any passive network.

94

branches, leads to a good approximation to the loss determined by (88) plus an added constant loss.

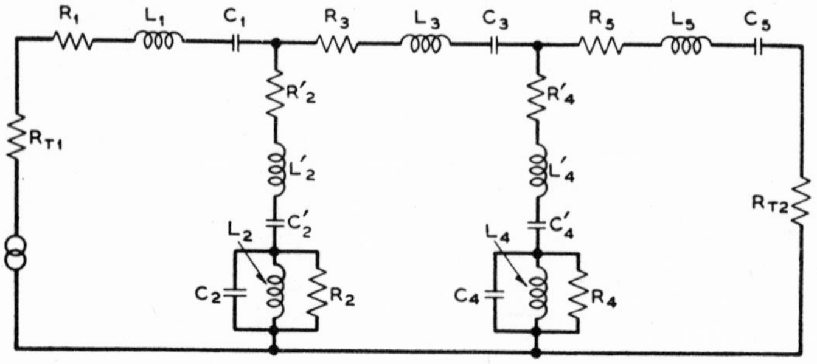

$$R_{T2} = 0.084427\ R_{T1}$$

$L_1 = 0.057481\ R_{T1}$	$C_1 = \dfrac{3.5282}{R_{T1}}$	$R_1 = 0.050448\ R_{T1}$
$L_2 = 0.0022151\ R_{T1}$	$C_2 = \dfrac{91.556}{R_{T1}}$	$R_2 = 12.445\ R_{T1}$
$L_2' = 0.0064170\ R_{T1}$	$C_2' = \dfrac{31.605}{R_{T1}}$	$R_2' = 0.0056318\ R_{T1}$
$L_3 = 0.11282\ R_{T1}$	$C_3 = \dfrac{1.7976}{R_{T1}}$	$R_3 = 0.099017\ R_{T1}$
$L_4 = 0.0023656\ R_{T1}$	$C_4 = \dfrac{85.732}{R_{T1}}$	$R_4 = 13.290\ R_{T1}$
$L_4' = 0.015764\ R_{T1}$	$C_4' = \dfrac{12.865}{R_{T1}}$	$R_4' = 0.013835\ R_{T1}$
$L_5 = 0.041411\ R_{T1}$	$C_5 = \dfrac{4.8974}{R_{T1}}$	$R_5 = 0.036344\ R_{T1}$

Provided R_{T1} is in ohms:
Inductances are in millihenries
Capacities are in microfarads
Resistances are in ohms

Fig. 26. Band-pass filter derived from the preliminary low-pass filter by a frequency transformation

If a mid-series type of ladder configuration is assumed and if Ω is defined as $\dfrac{\omega}{\sqrt{\omega_1 \omega_2}}$, the above procedure leads to the low-pass filter indicated in Fig. 25. Transforming this network by redefining Ω in terms of (90), with f_{a1}, f_{a2}, etc., fixed by (87), yields the band-pass design indicated in Fig. 26. The computed loss characteristic of this network is plotted in Fig. 27. The resistances associated with the resonant circuits can be approximated with dissipation resistances. Better approximations can be obtained, however, by replacing the shunt branches by configurations of the type indicated in Fig. 28, which can be made very nearly equivalent to them.

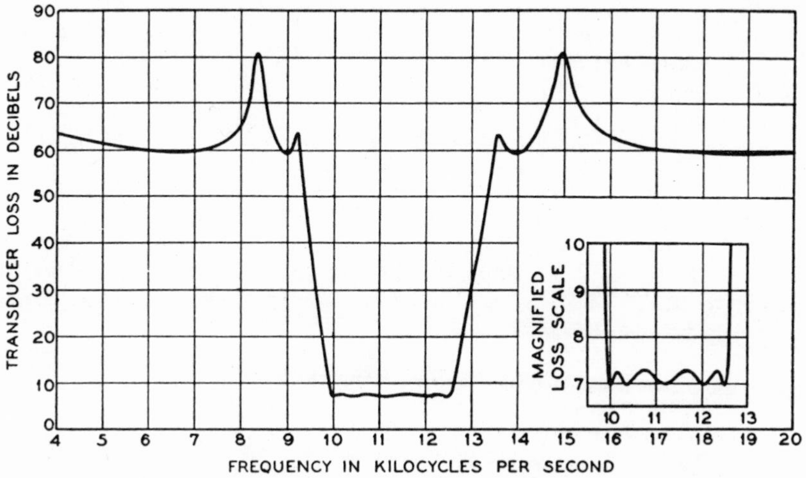

Fig. 27. Transducer loss of the network of Fig. 26 (i.e., insertion loss plus reflection loss corresponding to inequality of terminations)

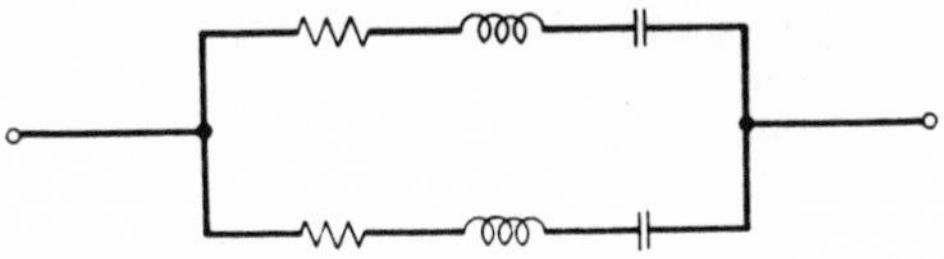

Fig. 28. Alternative shunt branch configuration

BIBLIOGRAPHY

1. Pupin, M. I.: "Wave Propagation over Non-Uniform Cables and Long Distance Air-Lines," *Trans. A. I. E. E.*, *17*, 445–507 (1900).
2. Campbell, G. A.: "On Loaded Lines in Telephonic Transmission," *Phil. Mag.*, *5*, 313–330 (1903).
3. Zobel, O. J.: "Theory and Design of Uniform and Composite Electric Wave Filters," *Bell System Tech. Journal*, *2*, 1–46 (1923), and "Transmission Characteristics of Electric Wave Filters," *Bell System Tech. Journal*, *3*, 567–620 (1924).

4. Bode, H. W.: "A General Theory of Electric Wave Filters," *Journal of Math. and Physics, 13*, 275–362 (1934).
5. Cauer, W.: "Ein Reaktanztheorem," *Preuss. Akad. d. Wissenschaften, Phys.-Math. Kl., Sitzber., Nos. 30–32*, 673–681 (1931).
6. Gewertz, C. M.: "Synthesis of a Finite Four-Terminal Network from its Prescribed Driving-Point Functions and Transfer Function," *Journal of Math. and Physics, 12*, 1–257 (1933).
7. Norton, E. L.: "Constant Resistance Networks with Applications to Filter Groups," *Bell System Tech. Journal, 16*, 178–193 (April, 1937).
8. Brune, O.: "Synthesis of a Finite Two-Terminal Network Whose Driving-Point Impedance is a Prescribed Function of Frequency," *Journal of Math. and Physics., 10*, 191–235 (1931).
9. Dietzold, R. L.: "A Mechanical Root Finder," *Bell Labs. Record, 16*, Dec., 1937.
10. Mayer, H. F.: "Ueber die Daempfung von Siebketten in Durchlaissigkeitsbereich," *E. N. T., 2*, 335–338 (1925).
11. Bode, H. W. and Dietzold, R. L.: "Ideal Wave Filters," *Bell System Tech. Journal, 14*, 215–252 (1935).
12. Fry, T. C.: "The Use of Continued Fractions in the Design of Electrical Networks," *Am. Math. Soc., Bull., 35*, 463–498 (1929).
13. Scott, R. F. and Mathews, G. B.: "The Theory of Determinants," Second Edition, 1904.
14. Norton, E. L.: U. S. Patent No. 1681554.
15. Campbell, G. A.: "Physical Theory of the Electric Wave-Filter," *Bell System Tech. Journal, 1*, 1–32 (1922).
16. Cauer, W.: "Ein Interpolationsproblem mit Funktionen mit Positiven Realteil," *Mathematische Zeitschrift, 38*, 1–44 (1933).
17. Shea, T. E.: "Transmission Networks and Wave Filters," D. Van Nostrand Co., New York, 1929—p. 253, Fig. 138.
18. Guillemin, E. A.: "Communication Networks," John Wiley and Sons, New York, N. Y., 1935—Vol. II.
19. Cayley, A.: "An Elementary Treatise on Elliptic Functions," second edition, G. Bell and Sons, London, 1895.
20. Jacobi, C. G. J.: "Fundamenta Nova Theoriae Functionum Ellipticarum," Königsberg, 1829.
21. Silberstein, L.: "Synopsis of Applicable Mathematics," G. Bell and Sons, London, 1923—Section on Elliptic Functions, pp. 152–170.
22. Bode, H. W.: U. S. Patents Nos. 1955788, 2002216, 2029014.

8

Reprinted from *IRE Trans. Circuit Theory*, **CT-2**, 320–325 (Dec. 1955)

Realizability Theorem for Mid-series or Mid-shunt Low-pass Ladders Without Mutual Induction*

T. FUJISAWA†

Introduction

In the design of nondissipative filters, it is well known that the ladder structure without mutual induction is desirable in practice.

Brune[1] showed that the terms "positive real function" and "two-terminal-impedance function" are synonymous and that the network can be realized as a ladder structure. Darlington[2] gave a different proof of this theorem and showed that any dissipative impedance can be realized by means of a nondissipative two-terminal-pair network terminated in a pure resistance. Thus Darlington established the insertion loss method of filter design. Darlington's method has been thoroughly discussed by Piloty,[3] Kiyasu,[4] Cauer,[5] and Guillemin[6] in various ways. Synthesis without mutual induction has been studied in various papers and recently Weinberg[7] gave an approach to Darlington's method.

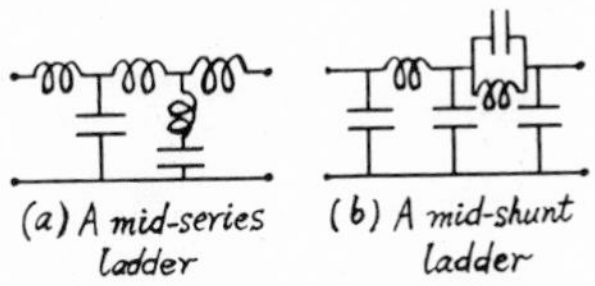

Fig. 1—Basic ladder configurations.

In the case of the low-pass filter, the basic circuit configuration is a mid-series or mid-shunt low-pass ladder[8] shown in Fig. 1. Negative series inductances or negative shunt capacities often appear. In such a case, it is inevitable to use mutual induction.[9] Darlington,[10] Kiyasu,[11] etc., gave sufficient conditions for physical realizability of the above-mentioned circuits without mutual induction. In this paper, the author gives realizability conditions, which are both necessary and sufficient, for mid-series or mid-shunt low-pass ladders without mutual induction.

The aim of this paper is mainly to present necessary and sufficient conditions in order that a nondissipative two-terminal-pair network can be realized by means of a mid-series or mid-shunt nondissipative low-pass ladder without mutual induction. The conditions are imposed upon the input impedance of the network when terminated in a pure resistance. Conditions are also necessary and sufficient so that a dissipative two-terminal impedance can be realized by means of a mid-series or mid-shunt nondissipative low-pass ladder without mutual induction terminated in a pure resistance.

Necessary Conditions

We will limit our consideration to mid-series low-pass ladders, because a mid-shunt circuit is the dual of a mid-series circuit.[12]

Let us introduce, as usual, a complex frequency variable $p=\sigma+j\omega$ and let us consider the input impedance of the network terminated in an R-ohm resistance. The input impedance $Z(p)$ is written in the form

$$Z(p) = \frac{u_1(p) + pf_1(p)}{u_2(p) + pf_2(p)}, \tag{1}$$

where u_1, u_2, f_1, and f_2 are even polynomials, and where the denominator and numerator are prime to each other. $Z(p)$ is a positive real function and its properties are well known.

Needless to say, we have

$$\frac{u_1(0)}{u_2(0)} = R. \tag{2}$$

In the first place, if the mid-series low-pass ladder has no negative series inductances, it is clear that $Z(p)$ has a pole or a zero at infinity except in trivial cases where there are no series inductances and no shunt capacities. We will not consider the trivial cases. Then we have the following necessary condition.

Condition (a). $Z(p)$ has a pole or a zero at infinity.

Secondly, we have a relation of the form

$$u_1u_2 - p^2f_1f_2 = K\left[\left(1+\frac{p^2}{\omega_1^2}\right)\cdots\left(1+\frac{p^2}{\omega_n^2}\right)\right]^2, \tag{3}$$

* Original manuscript received by PGCT November 11, 1954; revised manuscript received May 12, 1955.
† Kinki University, Osaka, Japan.

1 O. Brune, "Synthesis of a finite two-terminal network," *Jour. Math. Phys.*, vol. 10, pp. 191–236; April, 1931.
2 S. Darlington, "Synthesis of reactance 4-poles," *Jour. Math. Phys.*, vol. 18, pp. 257–353; September, 1939.
3 H. Piloty, "Kanonische kettenschaltungen für reaktanzvierpole," *T.F.T.*, vol. 29, pp. 249–258, 279–290, 320–325; September, October, November, 1940.
4 Z. Kiyasu, "Synthesis of reactance 4-poles," *Bull. Elect. Lab.*, vol. 4, pp. 429–599; June, 1940. (In Japanese)
5 W. Cauer, "Theorie der linearen Wechselstromschaltungen," Becker und Erler, Leipzig, Germany, vol. 1, 1941.
6 E. A. Guillemin, "Modern methods of network synthesis," *Advances in Electronics*, Academic Press, New York, vol. III, pp. 261–303; 1951.
7 L. Weinberg, "The Darlington problem," *Jour. Appl. Phys.*, vol. 24, pp. 776–779; June, 1953. L. Weinberg, "A general RLC synthesis procedure," Proc. IRE, vol. 42, pp. 427–437; February, 1954. L. Weinberg, "Unbalanced RLC networks," Proc. IRE, vol. 42, pp. 467–475; February, 1954.
8 Darlington, *loc. cit.*
9 See the above cited references.
10 Darlington, *loc. cit.*
11 Z. Kiyasu, "Filters without mutual induction," *J. Inst. Elect. Commun. Eng. Japan*, vol. 26, p. 400; May, 1943. (In Japanese)
12 Darlington, *loc. cit.*

where $\omega_1, \omega_2, \cdots, \omega_n$ are positive finite frequencies at which the loss is infinite,[13] and where K is a positive real constant. This is easily seen in the cases shown in Fig. 2 and moreover we obtain the following condition in these simple cases.[14]

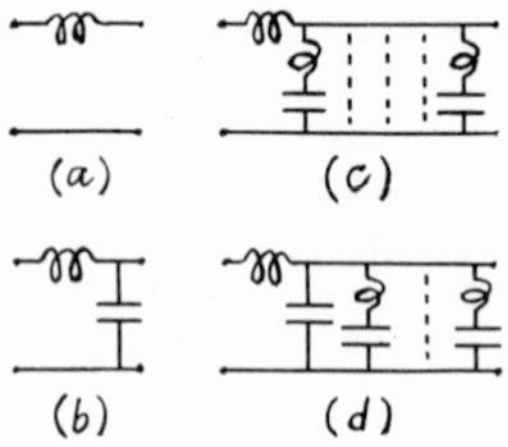

Fig. 2—Some simple special cases.

Condition (b). If the relation $\omega_1 \leqq \omega_2 \leqq \cdots \leqq \omega_n$ is assumed, then the polynomial $f_1(j\omega)$ of the variable ω has at least k roots, which are positive real frequencies, not larger than ω_k for any k $(1 \leqq k \leqq n)$.

Let us show that the relation (3) and the above condition (*b*) are true in general. In order to prove this by mathematical induction, let us suppose that circuit A in Fig. 3(a) satisfies relation (3) and condition (*b*).

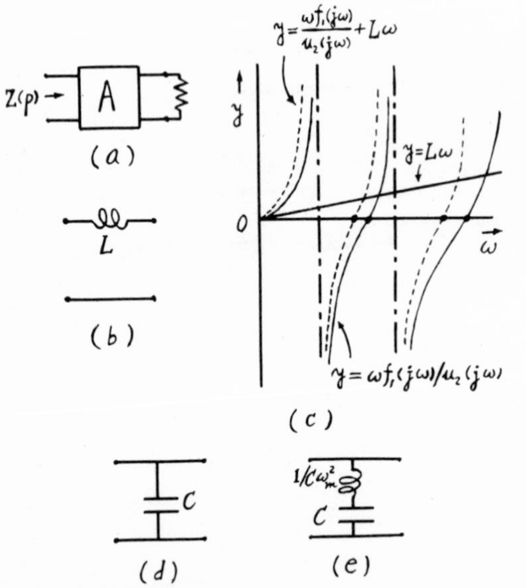

Fig. 3—Circuits used in a proof by induction: *a*, a circuit satisfying conditions (*b*); *b*, a series inductance L to be added to A; *c*, reactances before and after the addition of $L\omega$; *d*, a shunt capacity to be added to A; *e*, a resonance circuit to be added to A.

If the circuit shown in Fig. 3(*b*) is cascade-connected to the input terminals of the network A, then the input impedance of the whole network is written in the form

[13] Darlington, *loc. cit.*

[14] We will omit again the trivial case where there is no series inductance.

$$Z'(p) = Z(p) + Lp = \frac{(u_1 + Lp^2 f_2) + p(f_1 + Lu_2)}{u_2 + pf_2} = \frac{u_1' + pf_1'}{u_2 + pf_2}$$

and we have

$$u_1'u_2 - p^2 f_1' f_2 = u_1 u_2 - p^2 f_1 f_2.$$

It follows that $Z'(p)$ satisfies the relation (3). Since L is a positive number and since $pf_1(p)/u_2(p)$ is a reactance, though the two polynomials f_1, u_2 may not be prime to each other, we see from Fig. 3(c) that the condition (*b*) is true for $Z'(p)$.

If the two-terminal-pair network shown in Fig. 3(d) is cascade-connected to the input terminals of the network A, then the input impedance of the whole network is written in the form

$$Z'(p) = \frac{1}{\dfrac{1}{Z(p)} + Cp} = \frac{u_1 + pf_1}{(u_2 + Cp^2 f_1) + p(f_2 + Cu_1)} = \frac{u_1 + pf_1}{u_2' + pf_2'}$$

and we have

$$u_1 u_2' - p^2 f_1 f_2' = u_1 u_2 - p^2 f_1 f_2.$$

It follows immediately that $Z'(p)$ satisfies the relation (3) and the condition (*b*).

If the two-terminal-pair network shown in Fig. 3(e) is cascade-connected to the input terminals of the network A, then the input impedance of the whole network is written in the form

$$Z'(p) = \frac{1}{\dfrac{1}{Z(p)} + \dfrac{Cp}{1 + p^2/\omega_m^2}} = \frac{u_1(1 + p^2/\omega_m^2) + pf_1(1 + p^2/\omega_m^2)}{[u_2(1 + p^2/\omega_m^2) + Cp^2 f_1] + p[f_2(1 + p^2/\omega_m^2) + Cu_1]} = \frac{u_1' + pf_1'}{u_2' + pf_2'}$$

and we have

$$u_1'u_2' - p^2 f_1' f_2' = (1 + p^2/\omega_m^2)^2 (u_1 u_2 - p^2 f_1 f_2).$$

If the polynomials $(u_1' + pf_1')$, $(u_2' + pf_2')$ are prime to each other, it is evident that $Z'(p)$ satisfies the relation (3) and the condition (*b*). Otherwise, the polynomials u_1', u_2', f_1', f_2', must have the factor $(1 + p^2/\omega_m^2)$ in common and we have the relations

$$Z'(p) = \frac{\dfrac{u_1'}{(1 + p^2/\omega_m^2)} + p\dfrac{f_1'}{(1 + p^2/\omega_m^2)}}{\dfrac{u_2'}{(1 + p^2/\omega_m^2)} + p\dfrac{f_2'}{(1 + p^2/\omega_m^2)}} = \frac{u_1 + pf_1}{u_2'' + pf_2''}$$

and

$$u_1u_2'' - p^2f_1f_2'' = u_1u_2 - p^2f_1f_2.$$

It follows that $Z'(p)$ satisfies the relation (3) and the condition (b) even in this case. Thus the proof that the relation (3) and the condition (b) are necessary in general is complete by mathematical induction.

There are several parameters specifying a two-terminal-pair nondissipative network; for example, general circuit parameters $ABCD$, open-circuit impedances, short-circuit admittances, open-circuit and short-circuit reactances measured at one pair of terminals, etc. Relations among these parameters and the input impedance are well established. It is well known[15] that pf_1/u_2 and u_1/pf_2 are respectively a short-circuit reactance and an open-circuit reactance measured at the input terminals. The roots of $f_1(p)$, however, are not necessarily zeroes of the short-circuit reactance pf_1/u_2, because the two polynomials, f_1, u_2 may not be prime to each other. Therefore, in general, condition (b) can not be specified completely in terms of the zeros of the short-circuit reactance measured at one pair of terminals.

It is well known[16] that the open-circuit transfer impedance function $Z_{12}(p)$ is of the form

$$Z_{12}(p) = \frac{(1 + p^2/\omega_1^2) \cdots (1 + p^2/\omega_n^2)}{pf_2(p)}.$$

In order that a nondissipative two-terminal-pair network can be realized by means of a mid-series low-pass ladder without mutual induction, zeros of the transfer impedance function of the network must all be purely imaginary. Hereafter, understand the term "nondissipative two-terminal-pair network" in the sense that zeros of the transfer impedance function of the network are pure imaginary values.

Darlington[17] showed that the input impedance of the two-terminal-pair nondissipative network terminated in an R-ohm resistance specifies uniquely the two-terminal-pair network except for the existence of equivalent two-terminal-pair networks. Therefore, if we can realize the input impedance of the given network terminated in an R-ohm resistance by means of a mid-series low-pass ladder terminated in an R-ohm resistance, then given nondissipative two-terminal-pair network is equal to mid-series low-pass ladder so obtained.

SYNTHESIS

In the preceding section, we have seen that the input impedance of the nondissipative two-terminal-pair network terminated in an R-ohm resistance, $Z(p)$, necessarily satisfies the relations (2), (3) and the conditions (a), (b), so that network can be realized by means of a mid-series low-pass ladder without mutual induction.

[15] Darlington, *loc. cit.*, and Guillemin, *loc. cit.*
[16] Darlington, *loc. cit.*, and Guillemin, *loc. cit.*
[17] Darlington, *loc. cit.* The polarity of the terminals, which is considered in Darlington's theory, is not considered here.

In this section, we will show that any two-terminal impedance $Z(p)$, which satisfies the relations (2), (3) and the conditions (a), (b), can be realized by means of a mid-series low-pass ladder without mutual induction terminated in an R-ohm resistance. We will prove that the relations (2), (3) and the conditions (a), (b) are sufficient as well as necessary in order that the network may be realized by means of a mid-series low-pass ladder without mutual induction. There are several cases to consider.

If $Z(p)$ has a zero at infinity, apply the following step I. Remove a shunt capacity C from $Z(p)$ as shown in Fig. 4. $1/Z(p)$ has a pole at infinity and C is the residue.

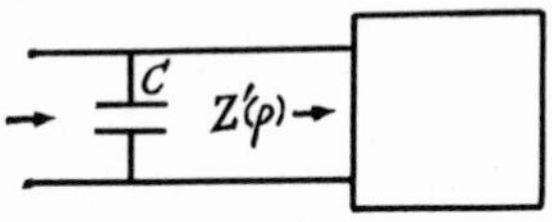

Fig. 4—Removal of a shunt capacity.

We have the relations

$$\frac{1}{Z'(p)} = \frac{1}{Z(p)} - Cp = \frac{(u_2 - Cp^2f_1) + p(f_2 - Cu_1)}{u_1 + pf_1}$$
$$= \frac{u_2' + pf_2'}{u_1 + pf_1}$$

and

$$u_1u_2' - p^2f_1f_2' = u_1u_2 - p^2f_1f_2.$$

It is evident that Z' is of lower degree than Z, and that Z' is a positive real function satisfying the relations (2), and (3) in the same manner.

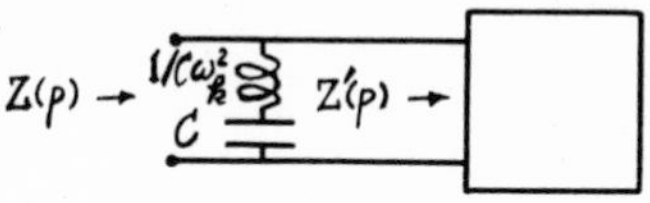

Fig. 5—Removal of a resonance circuit.

In order to prove that $Z'(p)$ satisfies the conditions (a), (b), it is sufficient to prove that $Z'(p)$ has a pole at infinity.

The condition (b) shows that the degree of the polynomial

$$u_1u_2' - p^2f_1f_2' = u_1u_2 - p^2f_1f_2$$

does not exceed that of f_1^2. It follows that the limit

$$\lim_{p\to\infty} \frac{1}{Z'(p)} = \lim_{p\to\infty} \frac{u_1u_2' - p^2f_1f_2'}{u_1^2 - p^2f_1^2}$$

must vanish. Thus the result follows.

If $Z(p)$ has a zero at $p = j\omega_k$, apply the following step II. Remove a shunt arm which is a series resonance circuit as shown in Fig. 5. $1/Z(p)$ has a pole at $p = j\omega_k$ and $C/2\omega_k^2$ is the residue. We have the relation

$$\frac{1}{Z'(p)} = \frac{1}{Z(p)} - \frac{Cp}{1+p^2/\omega_k^2} = \frac{[u_2(1+p^2/\omega_k^2) - Cp^2f_1]/(1+p^2/\omega_k^2) + p[f_2(1+p^2/\omega_k^2) - Cu_1]/(1+p^2/\omega_k^2)}{u_1/(1+p^2/\omega_k^2) + pf_1/(1+p^2/\omega_k^2)}$$

$$= \frac{u_2' + pf_2'}{u_1' + pf_1'}$$

and

$$u_1'u_2' - p^2f_1'f_2' = \frac{u_1u_2 - p^2f_1f_2}{(1+p^2/\omega_k^2)^2}.$$

It follows immediately that $Z'(p)$ satisfies the conditions (*a*), (*b*), and $Z'(p)$ is of lower degree than $Z(p)$.

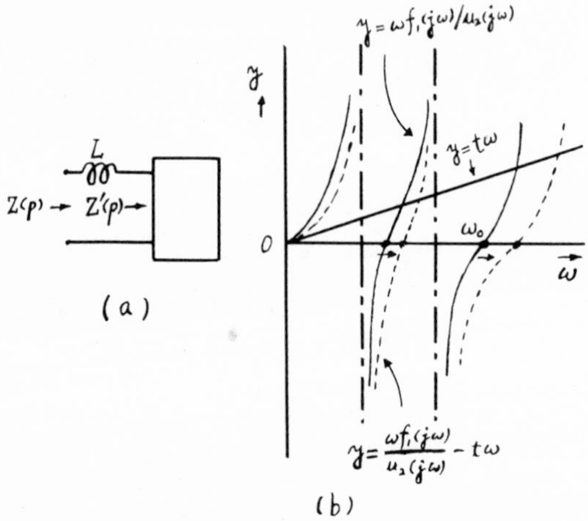

Fig. 6—Removal of a series inductance: *a*, circuit; *b*, reactance.

When we cannot apply the above two steps, we apply the following zero-shifting: step III. Remove a series inductance L from $Z(p)$ as shown in Fig. 6(a). L is determined by the relation

$$L = \text{minimum}\left(\frac{Z(j\omega_1)}{j\omega_1}, \cdots, \frac{Z(j\omega_n)}{j\omega_n}, \lim_{p\to\infty}\frac{Z(p)}{p}\right),$$

where all negative or undetermined values of $Z(j\omega_s)/j\omega_s$ are not taken into consideration. We have the relations

$$Z'(p) = Z(p) - Lp = \frac{(u_1 - Lp^2f_2) + p(f_1 - Lu_2)}{u_2 + pf_2}$$

$$= \frac{u_1' + pf_1'}{u_2 + pf_2}$$

and

$$u_1'u_2 - p^2f_1'f_2 = u_1u_2 - p^2f_1f_2.$$

If $Z(p)$ has a pole at $p = j\omega_k$, then the value $Z(j\omega_k)/j\omega_k$ is undetermined. In this case, we see[18] that the polynomial u_2 as well as f_2 has simple zeros at $p = j\omega_k$ and that the polynomial u_1 as well as f_1 has no zeros at $p = j\omega_k$. Thus we see that the reactance of $pf_1(p)/u_2(p)$ has a pole at $p = j\omega_k$, if $Z(p)$ has a pole at $p = j\omega_k$.

If $Z(p)$ does not have a pole at $p = j\omega_k$, we have the relation

$$Z(j\omega_k) = j\omega_k f_1(j\omega_k)/u_2(j\omega_k).$$

This follows from the fact that

$$Z(p) - \frac{pf_1(p)}{u_2(p)} = \frac{u_1u_2 - p^2f_1f_2}{u_2(u_2 + pf_2)}.$$

Thus L is a positive real number which does not exceed $\lim_{p\to\infty} Z(p)/p$ and hence $Z'(p)$ is a positive real function. Now we are in a position to prove that $Z'(p)$ satisfies the conditions (*a*), (*b*).

Let us consider the function

$$pf_1''(p) = p[f_1(p) - tu_2(p)],$$

where t is a nonegative number. The common roots of f_1 and u_2 are conserved in $pf_1''(p)$. Except for these common roots, the roots of $pf_1''(p)$ are the zeros of $[pf_1(p)/u_2(p) - tp]$. When t increases from zero, each zero of the function $(pf_1/u_2 - tp)$ moves to the right as shown in Fig. 6(b). The maximum root ω_0 shown in Fig. 6(b) vanishes if t teaches the value $\lim_{p\to\infty} [f_1(p)/u_2(p)]$.

The definition of L and the above consideration imply that, if $f_1(p)$ has m_k roots, which are real positive frequencies, not larger than ω_k for any k, then the same is true for $f_1'(p)$. We see $Z'(p)$ satisfies condition (*b*).

In order to show that $Z'(p)$ satisfies condition (*a*), let us first consider the case where u_1 is of higher degree than pf_1. Denote the degrees of u_1, pf_1, u_2, pf_2 by $2n+2$, $2n+1$, $2n$, $2n+1$ respectively, where n is a nonnegative integer. The degree of u_1u_2 is equal to $4n+2$ as is that of $p^2f_1f_2$. However, it follows from the condition (*b*) that the degree of $(u_1u_2 - p^2f_1f_2)$ does not exceed $4n$. Thus the coefficient of the highest power of p in this expression must vanish, or,

$$u_{10}u_{20} - f_{10}f_{20} = 0,$$

where u_{10} is the leading coefficient of u_1, and so on, and hence

$$\frac{u_{10}}{f_{20}} = \frac{f_{10}}{u_{20}} = \lim_{p\to\infty}\frac{Z(p)}{p}.$$

$Z'(p)$ has a pole at infinity if $L < \lim_{p\to\infty} Z(p)/p$. $Z'(p)$ has a zero at infinity if $L = \lim_{p\to\infty} Z(p)/p$, for the degree u_1' or f_1' is not larger than $2n$ or $2n-1$ respectively. In this case, $Z'(p)$ satisfies the condition (*a*).

[18] This is easily seen if we bear in mind the following fact: The denominator and numerator of $Z(p)$ are prime to each other; u_1/pf_1, u_1/pf_2, u_2/pf_1, u_2/pf_2 are all reactances; and the multiplicity of the zero of the right-hand side of the equality (3) is even at $p = j\omega_k$.

Let us next consider the case where u_1 is of lower degree than pf_1. Denote the degrees of u_1, pf_1, u_2, pf_2 by $2n$, $2n+1$, $2n$, $2n-1$ respectively, where n is a positive integer. $Z'(p)$ has a pole at infinity if $L<\lim_{p\to\infty} Z(p)/p$. If $L=\lim_{p\to\infty} Z(p)/p$ we cannot help concluding that there is no ω_s which is not smaller than ω_0 shown in Fig. 6(b). Consequently, the degree of $(u_1u_2-p^2f_1f_2)$ is not larger than $4n-4$ from the condition (b). We have

$$L = \lim_{p\to\infty} \frac{Z(p)}{p} = \frac{u_{10}}{f_{20}} = \frac{f_{10}}{u_{20}}$$

as before. It follows that $Z'(p)$ has a zero at infinity and that $Z'(p)$ satisfies the condition (a).

Consequently, $Z'(p)$ satisfies the relations (2), (3) and the conditions (a), (b). Since this step is a zero-shifting process, $Z'(p)$ has a zero at $p=j\omega_k$ for some $k(1\leq k\leq n)$ or else a zero at infinity. Thus we can apply step I or II to $Z'(p)$.

In order to terminate the proof of sufficiency, we must take into consideration some special cases which may occur. The first such case arises when $f_1(p)$ vanishes. This may occur at the end of step III. In this case, we have the relations

$$Z(p) = \frac{u_1}{u_2 + pf_2}$$

and

$$u_1u_2 = [(1 + p^2/\omega_1^2) \cdots (1 + p^2/\omega_n^2)]^2.$$

Since $Z(p)$ is a positive real function where denominator and numerator are prime to each other, we have

$$u_1 = Ru_2 = K'(1 + p^2/\omega_1^2) \cdots (1 + p^2/\omega_n^2)$$

where K' is a positive constant, and hence we have

$$\frac{1}{Z(p)} = \frac{1}{R} + \frac{pf_2}{u_1}.$$

The corresponding circuit is a mid-series circuit terminated in an R-ohm resistance, in which no series inductance exists.

The second special case arises when $f_2(p)$ vanishes. This can occur at the ends of steps I and II. In this case, we have as before

$$Z(p) = R + \frac{pf_1}{u_2}$$

and $u_1=Ru_2=K''(1+p^2/\omega_1^2) \cdots (1+p^2/\omega_n^2)$, where K'' is a constant. Since pf_1/u_2 is a reactance and since f_1 satisfies the condition (b), it follows that the ratio $f_1(p)/u_2(p)$ is a positive real constant. The corresponding circuit is a series connection of an inductance and an R-ohm resistance.

By repeated applications of steps I, II, and III, $Z(p)$ reduces to a constant resistance of R-ohms and the circuit obtained is a mid-series low-pass ladder which contains no negative element. Thus the proof of sufficiency is complete.

Example

The mid-series low-pass ladder shown in Fig. 7(a) contains a negative element and hence we require a coupled coil for its physical realization. If we calculate the input impedance of the circuit terminated in a 1-ohm resistance at the 2–2′ terminals, we have the relations

$$Z(p) = \frac{(1 + 5p^2) + p(2 + 5p^2)}{(1 + 3p^2 + p^4) + p(2 + p^2)} = \frac{u_1 + pf_1}{u^2 + pf_2}$$

and

$$u_1u_2 - p^2f_1f_2 = (1 + 2p^2)^2.$$

It follows immediately that $Z(p)$ satisfies the conditions (a), (b). Therefore the given circuit has an equivalent mid-series low-pass ladder without mutual induction.

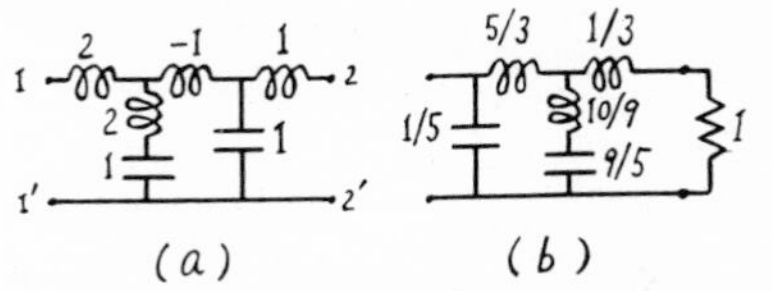

Fig. 7—An example of the elimination of a negative element: a, original circuit; b, equivalent circuit.

Since $Z(p)$ has a zero at infinity, we apply step I. The removal of a shunt capacity having the value 1/5 reduces the original impedance to

$$Z'(p) = \frac{1 + 2p + 5p^2 + 5p^3}{1 + 9p/5 + 13p^2/5}.$$

Now, it is observed that $Z'(p)$ does not have a zero at a frequency specified by (3). Hence we advance to step III and compute

$$\frac{Z'(j/\sqrt{2})}{j/\sqrt{2}} = \frac{5}{3}$$

$$\lim_{p\to\infty} \frac{Z'(p)}{p} = \frac{25}{13}.$$

We choose the minimum value and set $L=5/3$. This value of inductance is subtracted from $Z'(p)$ and we obtain

$$Z''(p) = \frac{(1 + 2p^2)(1 + p/3)}{1 + 9p/5 + 13p^2/5}.$$

Evidently this expression has a zero at the appropriate frequency so that step II may be applied. After this series resonant branch is removed, we are left with

$$Z'''(p) = 1 + p/3.$$

It is of interest to note that this is the special case where f_2 vanishes. The final circuit configuration is shown in Fig. 7(b).

Discussion

According to the preceding sections, we have the following Realizability Theorem. In order that a nondissipative two-terminal-pair network can be realized by means of a mid-series low-pass ladder without mutual induction, the relations (2), (3) and the conditions (*a*), (*b*) are necessary and sufficient. In order that a dissipative two-terminal impedance $Z(p)$ can be realized by means of a mid-series low-pass ladder without mutual induction terminated in a pure resistance, the relation (3) and the condition (*a*), (*b*) are necessary and sufficient. If we replace $f_1(p)$ by $f_2(p)$ in the condition (*b*), the theorem is true for mid-shunt low-pass ladders.

In the trivial cases which we omitted before, the realizability conditions are obvious and are not described here.

If we do not limit our consideration to pure mid-series or mid-shunt configurations, the new condition (*c*) and the following corollary are obtained. The proof is accomplished by a little modification of the method in the preceding section.

Condition (*c*). The number of ω_s in the equality (3) does not exceed the number of positive real frequency roots of $f_1(p)$ (or $f_2(p)$), and $u_1(p)$ (or $u_2(p)$) has at least k roots, which are positive real frequencies, not larger than ω_k for any k $(1 \leqq k \leqq n)$.

Corollary. In order that a nondissipative two-terminal-pair network can be realized by means of a low-pass ladder without mutual induction, the relations (2), (3) and the conditions (*a*), (*c*) are sufficient. In order that a dissipative impedance $Z(p)$ can be realized by means of a low-pass ladder without mutual induction, the relation (3) and the conditions (*a*), (*c*) are sufficient.

Darlington's sufficient conditions[19] are included in the above theorem. Kiyasu's sufficient conditions,[20] which are more precise than Darlington's conditions, are also included in it.

The synthesis of networks containing no ideal transformers is one of the central problems with which we are confronted at present. The result of this paper is a contribution to this problem.

Acknowledgment

This work was done in the author's academic years in Prof. Y. Kasahara's laboratory at Osaka University. The author expresses his gratitude to Dr. H. Ozaki for his valuable suggestions.

[19] Darlington, *loc. cit.*
[20] Kiyasu, *loc. cit.*

9

Reprinted from *Wireless Engr.*, **7**, 536–541 (1930)

On the Theory of Filter Amplifiers.*

By S. Butterworth, M.Sc.

(*Admiralty Research Laboratory*).

THE orthodox theory of electrical wave filters has been admirably presented by Mr. M. Reed in recent numbers of *E.W. & W.E.* (p. 122, March, 1930 *et seq*), and it is not proposed in the present Paper to add to or to repeat any of that theory. In this work the problem of electrical filtering is attacked from a new angle in which use is made of systems of simple filter units separated by valves so that we combine in one amplifier the property of filtering with that of amplification. The simple units employed can, in the case of low pass filters, be so designed that they take up little more space than the anode resistance employed in the ordinary straight resistance capacity amplifier. The writer has constructed filter units in which the resistances and inductances are wound round a cylinder of length 3in. and diameter 1¼in., while the necessary condensers are contained within the core of the cylinder. Units so constructed can be made of the plug-in type to admit of ready replacement. In the case of band pass filters it should be possible to design the intervalve system so as to have all the elements on two such cylindrical units. It is therefore clear that it is possible to obtain compact filter amplifier systems even if the degree of amplification is pushed to many stages.

The theory given below has been embodied in a set of design tables so that the task of the designer is reduced to as small proportions as possible.

Apart from the compactness of the system the filter amplifier has an advantage over the orthodox filter systems in that the effect of resistance is under complete control so that we may construct filters in which the sensitivity is uniform in the pass region.

1. General Scheme.

An ideal electrical filter should not only completely reject the unwanted frequencies but should also have uniform sensitivity for the wanted frequencies. In the usual type of filter circuit, the first condition is generally approximately fulfilled, but the second condition is usually either not obtained or is approximately arrived at by an empirical adjustment of the resistances of the elements.

The following theory was developed primarily in order to arrive at a logical scheme of design for low pass filters, but it will be shown that it is possible to make use of the theory for band pass, band stop, and high pass filters.

The theory of the general filter-circuit of the Campbell type including resistance is not attempted, but it is shown how to obtain the best results from a two element filter and then how to combine any number of elementary pairs, separated from each other by valves, so as to approach closer and closer to the ideal filter as the number of stages are increased. In this way we can combine amplification and filtering properties in one unit, a combination that is often required in the applications of filter circuits.

This procedure also frees us from the necessity of considering the impedances of the circuits with which the filter is to be associated so that the filter amplifier may be used for a variety of circuits provided suitable input and output transformers are employed.

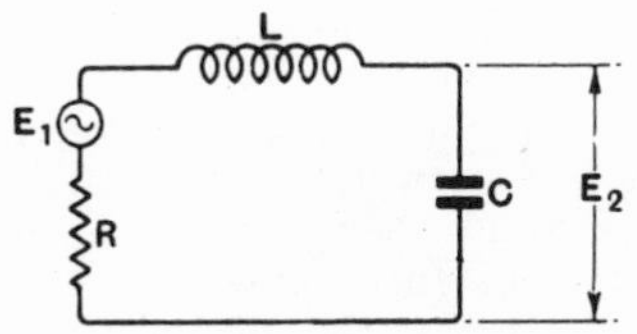

Fig. 1.

Further, in the Campbell filters, the elements (except the first and last) are similar, but in the present design we have assumed at first that we have perfect freedom in regard to the electrical constants of the elements and then these have been chosen with a view to obtaining the nearest approximation to the condition of uniform sensitivity in the "pass" region, and zero sensitivity in the "stop" region.

* The author is indebted to the Admiralty for permission to publish this paper.

MS. received by the Editor, May, 1930.

In the case of the low pass filter, if f_0 is the "cut off" frequency and f (xf_0) is any other frequency, the aim is to obtain a filter factor F, that is, the ratio of the output e.m.f. to the input e.m.f., of the form

$$F = (1 + x^m)^{-\frac{1}{2}} \quad .. \quad .. \quad .. \quad (1),$$

where m increases with the number of elements employed. It is clear that as m increases,

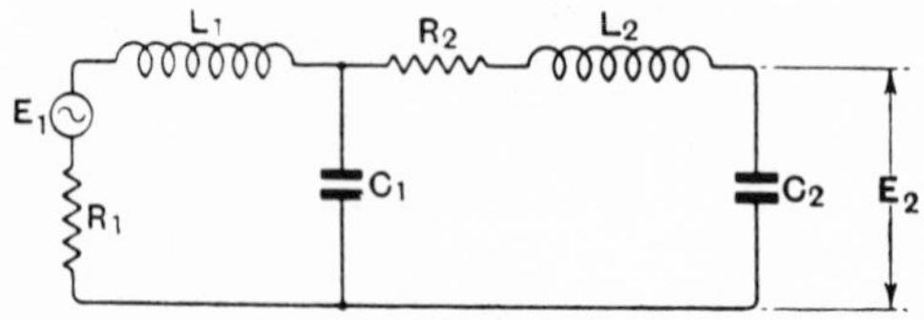

Fig. 2.

F will approximate more and more closely to the value unity when x is less than unity, and to zero when x is greater than unity.

For the case of a one element filter (Fig. 1), it is shown that by making $1/2\pi\sqrt{LC}$ equal to the cut off frequency and $L/C = \frac{1}{2}R^2$ we obtain a filter factor

$$F = (1 + x^4)^{-\frac{1}{2}} \quad .. \quad .. \quad .. \quad (2).$$

If this element follows a valve of known impedance, R is fixed, so that L/C is fixed. The value of LC follows from the required cut off frequency so that the numerical design is completely fixed.

For the case of a two element filter (Fig. 2) a filter factor

$$F = (1 + x^8)^{-\frac{1}{2}} \quad .. \quad .. \quad .. \quad (3)$$

may be obtained by satisfying four equations involving the seven quantities f_0, R_1, L_1, C_1, R_2, L_2, C_2.

For a given cut off frequency (f_0) and input resistance (R_1) we have thus four equations for five unknowns so that we are left with one further choice. It is convenient to choose a practically realisable value for $R_2/2\pi f_0 L_2$ and then the values of all the electrical constants follow.

For a series of two element filters interposed between valves we make use of the equation (*see* Todhunter's Trigonometry, p. 307)

$$1 + y^{2m} = (1 + 2y\cos\pi/2m + y^2)$$
$$(1 + 2y\cos 3\pi/2m + y^2)..(1 + 2y\cos(2m-1)\pi/2m + y^2)$$

Thus if we employ n intervalve elements and design the sth member so as to have a filter factor

$$F_s = (1 + 2x^4\cos\theta + x^8)^{-\frac{1}{2}} \quad .. \quad (4)$$

in which $\theta = \cos(2s - 1)\pi/2n$

then from the above equation the whole filter factor will be

$$F = (1 + x^{8n})^{-\frac{1}{2}} \quad .. \quad .. \quad .. \quad (5).$$

The characteristics to be expected on this scheme are shown by the curves of Fig. 3. Curve A refers to a single element and curve B to a pair of elements, while curves C, D and E are for two three and six pairs respectively, each pair being an intervalve element. By adding the usual resistance capacity combination between the filter element and the grid as in Fig. 5, we thus obtain a filter amplifier having practically uniform pass sensitivity.

2. Low Pass Filter. One Element. Fig. 1.

A source of e.m.f. is applied to a circuit of inductance L, resistance R and capacity C. This, of course, is the ordinary resonating circuit and if the applied e.m.f. (E_1) has

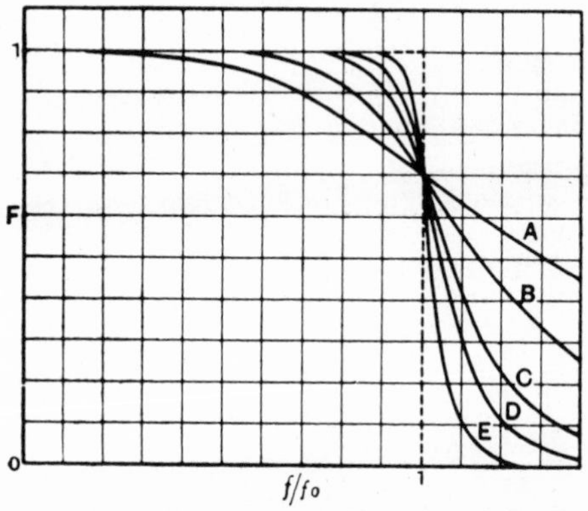

Fig. 3.

pulsatance ω, the output e.m.f. (E_2) is such that

$$(E_1/E_2)^2 = R^2\omega^2C^2 + (1 - \omega^2LC)^2 =$$
$$1 + \omega^2(R^2C^2 - 2LC) + \omega^4L^2C^2 \quad .. \quad (6).$$

Let $LC = 1/\omega_0^2$ and make $R^2 = 2L/C$ (7).

Then if $\omega/\omega_0 = x$, $(E_1/E_2)^2 = 1 + x^4$

the filter factor

$$F = (1 + x^4)^{-\frac{1}{2}} \quad .. \quad .. \quad .. \quad (8)$$

when condition (7) is satisfied.

3. Low Pass Filter. Pair of Elements. Fig. 2.

Generalise the circuit as in Fig. 4 for the purpose of carrying out the necessary algebra.

Z_1, Z_2, z_1, z_2 are the vector impedances of the various arms and on applying Ohm's Law we have

$$E_1/E_2 = (1 + Z_1/z_1)(1 + Z_2/z_2) + Z_1/z_2 \quad (9)$$

where E_1 and E_2 have now vector values.

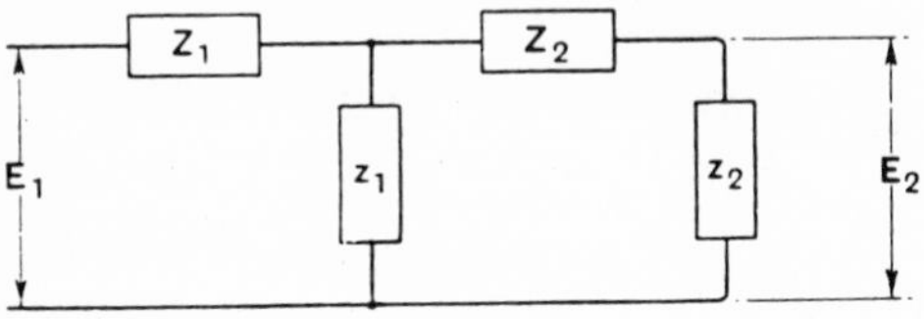

Fig. 4.

Let

$$L_1C_1 = 1/\omega_1^2,\ L_2C_2 = 1/\omega_2^2,$$
$$R_1/\omega_1L_1 = P_1,\ R_2/\omega_2L_2 = P_2 \quad .. (10).$$

Then since

$$\left.\begin{aligned} Z_1 &= R_1 + j\omega L_1,\ z_1 = 1/j\omega C_1 \\ Z_2 &= R_2 + j\omega L_2,\ z_2 = 1/j\omega C_2 \end{aligned}\right\} \quad .. (11)$$

we have

$$\left.\begin{aligned} 1 + Z_1/z_1 &= 1 - \omega^2/\omega_1^2 + jP_1\omega/\omega_1 \\ 1 + Z_2/z_2 &= 1 - \omega^2/\omega_2^2 + jP_2\omega/\omega_2 \end{aligned}\right\} \quad (12).$$

Also put

$$\omega/\omega_1 = x/a,\ \omega/\omega_2 = ax,\ C_2/C_1 = \beta \quad .. (13)$$

so that

$$x^2 = \omega^2/\omega_1\omega_2,\ a^2 = \omega_1/\omega_2 \quad .. \quad .. (14).$$

Then on substitution in (9)

$$E_1/E_2 = (1 - x^2/a^2 + jP_1x/a)$$
$$(1 - a^2x^2 + jP_1ax) + \beta(-x^2/a^2 + jP_1x/a) \quad .. (15).$$

Separate real and imaginary parts, square and add, and we obtain

$$1/F^2 = 1 + (B^2 - 2A)x^2 + (2 + A^2 - 2BC)$$
$$x^4 + (C^2 - 2A)x^6 + x^8 \quad .. \quad .. (16)$$

in which

$$\left.\begin{aligned} A &= (1 + \beta)/a^2 + a^2 + P_1P_2 \\ B &= P_1(1 + \beta)/a + P_2a \\ C &= P_1a + P_2/a \end{aligned}\right\} \quad .. (17).$$

If we can so choose the circuit constants that the coefficients of x^2, x^4, x^6 vanish then the filter factor will become

$$F = (1 + x^8)^{-\frac{1}{2}} \quad .. \quad .. \quad .. (18).$$

The conditions that these coefficients vanish are

$$B^2 = C^2 = 2A,\ 2 + A^2 = 2BC \quad .. (19).$$

On eliminating B and C we have

$$A^2 - 4A + 2 = 0 \quad .. \quad .. (20)$$

that is $\quad A = 2 \pm \sqrt{2}.$

But examination of (17) shows that A must be greater than 2 so that the only realisable root is

$$A = 2 + \sqrt{2} = 3.414 \quad .. (21)$$

and then $\quad B = C \quad = 2.613 \quad .. (22).$

Using these in (17) we have three equations to find a, β, P_1, P_2. One of these may be fixed arbitrarily. It is convenient to fix P_2/a, that is, by (14) $R_2/\sqrt{\omega_1\omega_2}L_2$, and since also by (14) $\sqrt{\omega_1\omega_2}$ is the cut off pulsatance (ω_0 say) we choose R_2/ω_0L_2 arbitrarily.

Then (17) fixes a, β, P_1, P_2. But a^2 is the ratio ω_1/ω_2 and $\omega_1\omega_2$ is fixed by the required cut off pulsatance, so that ω_1, ω_2 are fixed. The design is now completely fixed when the input resistance R_1 is specified, for P_1 determines L_1, then ω_1 determines C_1, β determines C_2, ω_2 determines L_2 and P_2 determines R_2.

4. Calculation of P_1, P_2, a, β.

The solution of equations (17) is facilitated by making use of the following substitutions.

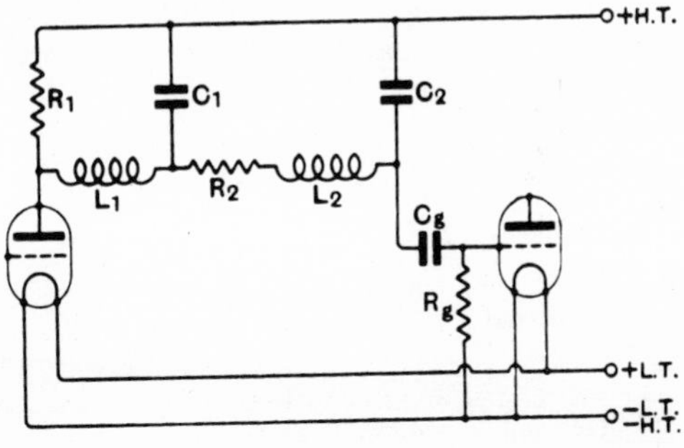

Fig. 5.

Put

$$a^2 - (1 + \beta)/a^2 = v,\ a^2 + (1 + \beta)/a^2 =$$
$$2 + kv \quad .. \quad .. (24)$$

so that

$$a^2 - 1 = \tfrac{1}{2}(1 + k)v,\ 1 - (1 + \beta)/a^2 =$$
$$\tfrac{1}{2}(1 - k)v \quad .. (25).$$

Then, solving the last two of equations (17)

for P_1 and P_2 and putting $B = C = \sqrt{2A}$ we find with the help of (25)

$$P_1\alpha = (1 + k)\sqrt{A/2},\ P_2/\alpha = (1 - k)\sqrt{A/2} \quad .. \quad (26).$$

Using these in the first of equations (17) we obtain

$$2 + kv = (1 + k^2)A/2 \quad .. \quad .. \quad (27).$$

The numerical calculations can now be carried out thus :—

P_2/α is fixed arbitrarily, then, using the value of A in (21), k follows from (26). $P_1\alpha$ follows then from (26).

The value of v is next found from (27) and then α^2 and β are obtained from (25).

5. Design Tables.

The designer is given the values of R_1 and the cut off pulsatance ω_0 and requires the correct values of L_1, L_2, C_1, C_2 and R_2.

TABLE I.

SINGLE INTERVALVE ELEMENT. FIGS. 2, 6, 7 and 8.

$R_2/\omega_0 L_2$.	X_1.	X_2.	Y_1.	Y_2.	R_2/R_1.
0.00	0.3827	1.584	0.9277	0.6560	0.0000
0.05	0.3902	1.664	0.9019	0.7201	0.0823
0.10	0.3979	1.761	0.8798	0.7966	0.1761
0.15	0.4060	1.869	0.8577	0.8846	0.2803
0.20	0.4144	1.990	0.8353	0.9873	0.3980
0.25	0.4232	2.128	0.8128	1.108	0.5320

For this purpose it is convenient to construct Tables giving

$$X_1 = \omega_0 L_1/R_1,\ X_2 = \omega_0 L_2/R_1, \text{ and } R_2/R_1$$

$$Y_1 = 1/\omega_0 C_1 R_1,\ Y_2 = 1/\omega_0 C_2 R_1.$$

These quantities can be calculated as follows when $P_1\alpha$, P_2/α α and β are known.

$$X_1 = 1/P_1\alpha$$

$$X_1/Y_1 = \omega_0^2 L_1 C_1 = \omega_1^2 L_1 C_1 + \omega_2/\omega_1 = 1/\alpha^2$$

$$Y_1/Y_2 = C_2/C_1 = \beta$$

$$X_2/Y_2 = \omega_0^2 L_2 C_2 = \omega_2^2 L_2 C_2 + \omega_1/\omega_2 = \alpha^2$$

$$R_2/R_1 = X_2 P_2/\alpha$$

giving in turn the quantities required.

Table I is the Design Table for a single pair of elements the values of $R_2/\omega_0 L_2$ in the first column being assumed.

6. Multistage Filter Amplifiers. Fig. 5.

In a filter amplifier having n intervalve stages the filter for the sth stage must be designed so that

$$A^2 - 4A + 2 = 2\cos(2s - 1)\pi/2n \quad .. \quad (28)$$

{see equations (4) and (16)} while

$$B^2 = C^2 = 2A \quad .. \quad .. \quad .. \quad (29).$$

as before.

The solution of (28) is

$$A = 4\cos^2(2s - 1)\pi/8n \quad .. \quad .. \quad (30)$$

so that the design is exactly as in section (4) except that we use (30) for A instead of (21).

TABLE II.

TWO INTERVALVE STAGES.

$R_2/\omega_0 L_2$.	X_1.	X_2.	Y_1.	Y_2.	R_2/R_1.	Stage.
0.00	0.3605	1.582	1.027	0.5556	0.000	1
	0.4252	1.749	0.7507	0.9808	0.000	2
0.05	0.3671	1.669	1.007	0.6082	0.0834	1
	0.4345	1.879	0.7233	1.129	0.0940	2
0.10	0.3740	1.762	0.9877	0.6672	0.1762	1
	0.4441	2.069	0.6951	1.322	0.2069	2
0.15	0.3811	1.869	0.9688	0.7350	0.2803	1
	0.4540	2.318	0.6670	1.578	0.3478	2
0.20	0.3885	1.984	0.9450	0.8115	0.3968	1
	0.4647	2.704	0.6364	1.974	0.5407	2
0.25	0.3962	2.113	0.9312	0.8988	0.5281	1
	0.4758	3.333	0.6054	2.620	0.8332	2

Tables II and III are the Design Tables for two stage and three stage systems respectively while Table IV holds for four, five and six stage systems.

Fig. 5 shows the nature of the intervalve system, R_g and C_g being the usual resistance capacity elements.

7. Band Pass Filters. Fig. 6.

Starting with a low pass filter system suppose we put in series with L_1, L_2 the capacities C'_1, C'_2 and in parallel with the capacities C_1, C_2 the inductances L'_1, L'_2.

TABLE III.

THREE INTERVALVE STAGES.

Stage 2 as in Table I.

$R_2/\omega_0 L_2$.	X_1.	X_2.	Y_1.	Y_2.	R_2/R_1.	Stage.
0.00	0.3566	1.587	1.046	0.5412	0.0000	1
	0.4456	1.983	0.6763	1.307	0.0000	3
0.05	0.3631	1.674	1.026	0.5921	0.0837	1
	0.4558	2.267	0.6461	1.599	0.1133	3
0.10	0.3698	1.770	1.008	0.6497	0.1770	1
	0.4664	2.715	0.6148	2.059	0.2715	3
0.15	0.3768	1.876	0.9889	0.7147	0.2814	1
	0.4776	3.548	0.5823	2.910	0.5322	3
0.20	0.3840	1.995	0.9706	0.7884	0.3985	1
	0.4893	5.720	0.5483	5.104	1.144	3
0.25	0.3915	2.123	0.9524	0.8725	0.5307	1
	0.5015	27.43	0.5125	26.84	6.858	3

Let these new capacities and inductances be such that

$$L_1 C'_1 = L_2 C'_2 = L'_1 C_1 = L'_2 C_2 = 1/\omega^2_a \quad (31).$$

Then the generalised circuit (Fig. 4) will have for its vector impedances

$$\left.\begin{aligned} Z_1 &= R_1 + j\zeta L_1,\ z_1 = 1/j\zeta C_1, \\ Z_2 &= R_2 + j\zeta L_2,\ z_2 = 1/j\zeta C_2 \end{aligned}\right\} \quad \ldots (32)$$

in which $\zeta = \omega - \omega^2{}_a/\omega$.

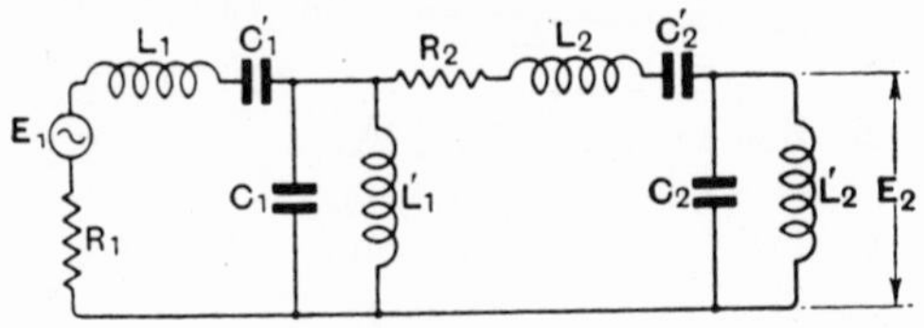

Fig. 6.

Now (32) is of the form (11) so that the whole of the low pass filter theory will hold. The interpretation of x however is now

$$x = \zeta/\sqrt{\omega_1\omega_2} = (\omega - \omega^2{}_a/\omega)/\sqrt{\omega_1\omega_2} \quad (33).$$

The form of the filter factor shows that the cut off pulsatances are such as to satisfy $x = \pm 1$ or

$$\omega - \omega^2{}_a/\omega = \pm\sqrt{\omega_1\omega_2} \quad \ldots \quad \ldots (34).$$

TABLE IV.

FOUR INTERVALVE STAGES.

$R_2/\omega_0 L_2$.	X_1.	X_2.	Y_1.	Y_2.	R_2/R_1.	Stage.
0.00	0.3553	1.588	1.052	0.5364	—	1
	0.3695	1.576	0.9838	0.5916	—	2
	0.4009	1.608	0.8463	0.7617	—	3
	0.4574	2.265	0.6358	1.630	—	4
0.05	0.3617	1.676	1.033	0.5868	0.0838	1
	0.3764	1.661	0.9635	0.6490	0.0831	2
	0.4091	1.706	0.8221	0.8491	0.0853	3
	0.4681	2.769	0.6039	2.146	0.1384	4

FIVE INTERVALVE STAGES.

$R_2/\omega_0 L_2$.	X_1.	X_2.	Y_1.	Y_2.	R_2/R_1.	Stage.
0.00	0.3546	1.589	1.055	0.5342	—	1
	0·3636	1.579	1.012	0.5676	—	2
	0.4147	1.662	0.7912	0.8713	—	4
	0.4650	2.562	0.6104	1.951	—	5
0.05	0.3610	1.677	1.036	0.5843	0.0839	1
	0.3703	1.665	0.9917	0.6219	0.0833	2
	0.4234	1.782	0.7652	0.9859	0.0891	4
	0.4760	3.387	0.5773	2.793	0.1694	5

For Stage 3, see Table I.

SIX INTERVALVE STAGES.

$R_2/\omega_0 L_2$.	X_1.	X_2.	Y_1.	Y_2.	R_2/R_1.	Stage.
0.00	0.3543	1.590	1.057	0.5330	—	1
	0.3734	1.574	0.9658	0.6087	—	3
	0.3942	1.592	0.8742	0.7180	—	4
	0.4702	2.865	0.5930	2.272	—	6
0.05	0.3607	1.678	1.038	0.5830	0.0839	1
	0.3805	1.660	0.9449	0.6685	0.0830	3
	0.4021	1.685	0.8508	0.7965	0.0843	4
	0.4816	4.146	0.5592	3.571	0.2073	6

For Stages 2 and 5, see Table II.

Putting $\sqrt{\omega_1\omega_2} = \omega_0$ as before we have now two cut off pulsatances (ω_α, ω_β) given by

$$\left.\begin{aligned} \omega_\alpha &= \tfrac{1}{2}\omega_0 + \sqrt{\omega^2{}_a + \tfrac{1}{4}\omega_0{}^2} \\ \omega_\beta &= -\tfrac{1}{2}\omega_0 + \sqrt{\omega^2{}_a + \tfrac{1}{4}\omega_0{}^2} \end{aligned}\right\} \quad \ldots (35)$$

(33) shows that if $\omega a > \omega > \omega_\beta$, x is less than unity so that the filter is now a band pass filter having cut off pulsatances given by (35). The width of the band is $\omega_\alpha - \omega_\beta = \omega_0$ and the centre of the band is

$$\omega_r = \tfrac{1}{2}(\omega_\alpha + \omega_\beta) = \sqrt{\omega^2{}_a + \tfrac{1}{4}\omega_0{}^2} \quad \ldots (36).$$

Hence for a given centre ω_r and given band width ω_0 we use the low pass filter Tables to calculate the circuit constants together with equations (31) and (36).

It is interesting to notice that for a given band width the values of L_1, L_2, C_1, C_2 and R_2 are fixed and adjustment of L'_1, L'_2, C'_1 and C'_2 simultaneously shifts the band bodily along the frequency scale.

8. High Pass Filters. Fig. 6.

In the low pass system let the inductances be replaced by capacities and the capacities by inductances. Since $j\omega L$ becomes $1/j\omega C$ and *vice versa*, terms such as $\omega^2 LC$ are replaced by $1/\omega^2 LC$ that is, if $LC = I/\omega_1{}^2$ we replace ω/ω_1 by ω_1/ω. Hence x is now interpreted as ω_0/ω and the pass region is for pulsatances *greater* than ω_0.

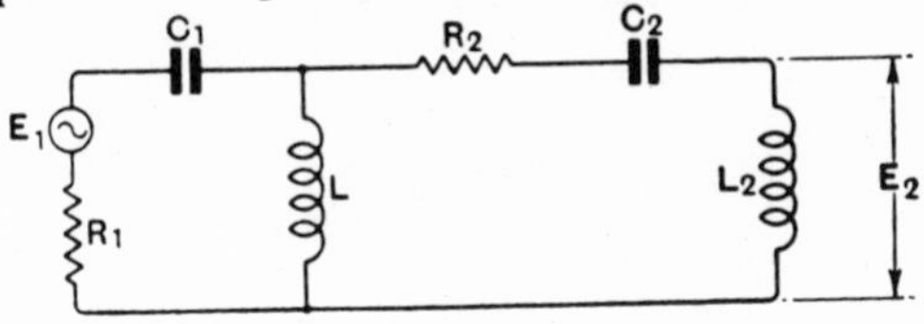

Fig. 7.

In the Design Tables we now interpret X_1, X_2, Y_1, Y_2 as follows :—

$$\begin{aligned} X_1 &= 1/\omega_0 C_1 R_1,\ X_2 = 1/\omega_0 C_2 R_1 \\ Y_1 &= \omega_0 L_1/R_1,\ Y_2 = \omega_0 L_2/R_1 \end{aligned} \quad \ldots (37)$$

9. Band Stop Filters. Fig. 8.

These are obtained from high pass filters by putting inductances L'_1, L'_2 in parallel with the capacities C_1, C_2 and capacities C'_1, C'_2 in series with the inductances L_1, L_2. The values of the new capacities and inductances are obtained from

$$L'_1 C_1 = L'_2 C_2 = L_1 C'_1 = L_2 C'_2 = 1/\omega^2 a \quad \ldots (38).$$

in which $\omega^2_a = \omega^2_r - \frac{1}{4}\omega_0^2$

ω_r = mid point of stop region.
ω_0 = width of stopped band.

DESIGN TABLES.

A. Method of use for Low Pass Filters. Fig. 2.

The given quantities are the cut off pulsatance and the input resistance R_1

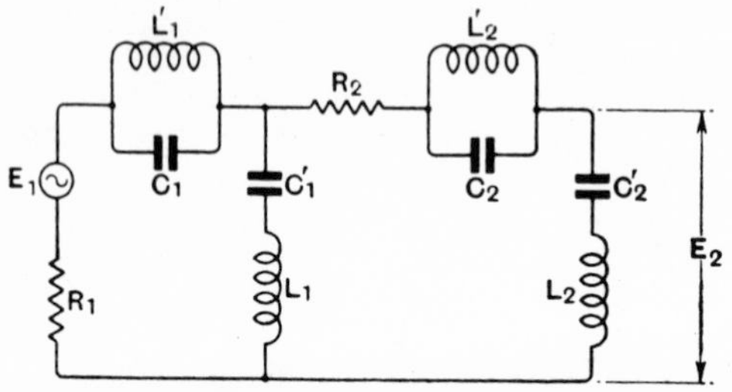

Fig. 8.

(fixed by the valve employed and its associated anode resistance).

Assume a value for $R_2/\omega_0 L_2$ and then read off X_1, X_2, Y_1, Y_2 from the appropriate Table.

Then
$L_1 = X_1 R_1/\omega_0$, $L_2 = X_2 R_1/\omega_0$, $C_1 = 1/Y_1\omega_0 R_1$, $C_1 = 1/Y_2\omega_0 R_1$

If R_1 is in ohms, then the calculated values of L_1, L_2 are in henrys and those of C_1, C_2 are in farads.

B. Use for Band Pass Filters. Fig. 6.

ω_0 is now the band width and L_1, L_2, C_1, C_2 are calculated as above. L'_1, L'_2, C'_1, C'_2 are calculated from

$$L'_1 C_1 = L'_2 C_2 = L_1 C'_1 = L_2 C'_2 = 1/\omega^2_a$$

in which $\omega^2_a = \omega^2_r - \frac{1}{4}\omega_0^2$
and ω_r is the mid point of the band.

C. Use for High Pass Filters. Fig. 7.

ω_0 is the cut off pulsatance and

$C_1 = 1/X_1\omega_0 R_1$, $C_2 = 1/X_2\omega_0 R_1$, $L_1 = R_1 Y_1/\omega_0$, $L_2 = R_1 Y_2/\omega_0$

the units being as before.

D. Use for Band Stop Filters. Fig. 8.

ω_0 is the band width and ω_r the mid point of the band. Calculate L_1, L_2, C_1, C_2 as in (*C*) and then L'_1, L'_2, C'_1, C'_2 as in (B).

III
Feedback Amplifier Theory

Editor's Comments on Papers 10 Through 13

10 Nyquist: *Regeneration Theory*

11 Bode: *Relations Between Attenuation and Phase in Feedback Amplifier Design*

12 Blackman: *Effect of Feedback on Impedance*

13 Mason: *Feedback Theory–Further Properties of Signal Flow Graphs*

The use of feedback around electronic amplifiers was introduced by engineers at the Bell Telephone Laboratories in the 1920s. Its use was motivated by problems associated with the operation of long-distance telephone transmission systems, in which large numbers of repeater amplifiers were used. Feedback made possible the reduction of changes in overall transmission resulting from tube aging, voltage variation, and the like, so that the system maintained more nearly constant characteristics. Thus feedback made it possible to achieve low sensitivity but at the price of possible instability. The pioneer paper by Nyquist arose from attempts to understand the design of feedback systems (Paper 10).

Prior to Nyquist it was believed that the feedback amplifier would be unstable if, at a frequency at which the open-loop phase shift was 180 degrees, the loop gain was greater than unity. This was the Barkhausen stability criterion, proposed in 1919, which was based on the concept of around-and-around circulations of signals in the feedback loop. In the late 1920s Black, at the Bell Telephone Laboratories, found that he could build stable feedback amplifiers that violated the Barkhausen criterion (reported in 1934). This dilemma led to Nyquist's theoretical study and the now-famous Nyquist criterion.

The understanding of feedback that followed Nyquist led to the design of wide-band highly linear amplifiers required for multichannel carrier telephony. A landmark in understanding of this design is contained in the 1940 paper of Bode, which demonstrates the relationship between attenuation and phase through integral expressions (Paper 11; see also Bode, 1945, 1960).

Feedback is one of the important concepts in circuit theory (as it is in other fields, such as process control and automation) and finds application in many ways. A modern use of feedback is to alter the impedance of elements in some desired way and then to exploit these changed characteristics by subnetwork embedding in larger networks. This use has achieved new importance with the widespread application of active elements realized in inexpensive integrated-circuit form. Early and far-sighted research relating to this principle was done by Blackman in 1943 (Paper 12).

Signal flow graphs and the rules governing their use and manipulation are due to Mason's work in 1953. Few concepts have permeated the circuit theory literature to the extent of the signal flow graph. It is especially valuable in the analysis of complex feedback systems. Mason's contribution was published in two parts; reprinted here (Paper 13) is the second part of that work, because it contains the main results.

References

Black, H. S. "Stabilized Feedback Amplifiers," *Bell System Tech. J.*, **14**, 1–18 (1934).

Bode, H. W. *Network Analysis and Feedback Amplifier Design,* Van Nostrand Reinhold Company, New York, 551 pp., 1945.

Bode, H. W. "Feedback—The History of an Idea," *Proc. Symp. Active Networks and Feedback Systems* (Polytechnic Institute of Brooklyn), **10**, 1–17 (1960).

Mason, S. J. "Feedback Theory—Some Properties of Signal Flow Graphs," *Proc. IRE,* **41**, 1144–1156 (1953).

10

Reprinted with permission from *Bell System Tech. J.*, **11**, 126–147 (Jan. 1932)

Regeneration Theory

By H. NYQUIST

American Telephone and Telegraph Company

Regeneration or feed-back is of considerable importance in many applications of vacuum tubes. The most obvious example is that of vacuum tube oscillators, where the feed-back is carried beyond the singing point. Another application is the 21-circuit test of balance, in which the current due to the unbalance between two impedances is fed back, the gain being increased until singing occurs. Still other applications are cases where portions of the output current of amplifiers are fed back to the input either unintentionally or by design. For the purpose of investigating the stability of such devices they may be looked on as amplifiers whose output is connected to the input through a transducer. This paper deals with the theory of stability of such systems.

Preliminary Discussion

WHEN the output of an amplifier is connected to the input through a transducer the resulting combination may be either stable or unstable. The circuit will be said to be stable when an impressed small disturbance, which itself dies out, results in a response which dies out. It will be said to be unstable when such a disturbance results in a response which goes on indefinitely, either staying at a relatively small value or increasing until it is limited by the non-linearity of the amplifier. When thus limited, the disturbance does not grow further. The net gain of the round trip circuit is then zero. Otherwise stated, the more the response increases the more does the non-linearity decrease the gain until at the point of operation the gain of the amplifier is just equal to the loss in the feed-back admittance. An oscillator under these conditions would ordinarily be called stable but it will simplify the present paper to use the definitions above and call it unstable. Now, this fact as to equality of gain and loss appears to be an accident connected with the non-linearity of the circuit and far from throwing light on the conditions for stability actually diverts attention from the essential facts. In the present discussion this difficulty will be avoided by the use of a strictly linear amplifier, which implies an amplifier of unlimited power carrying capacity. The attention will then be centered on whether an initial impulse dies out or results in a runaway condition. If a runaway condition takes place in such an amplifier, it follows that a non-linear amplifier having the same gain for small current and decreasing gain with increasing current will be unstable as well.

Steady-State Theories and Experience

First, a discussion will be made of certain steady-state theories; and reasons why they are unsatisfactory will be pointed out. The most obvious method may be referred to as the series treatment. Let the complex quantity $AJ(i\omega)$ represent the ratio by which the amplifier and feed-back circuit modify the current in one round trip, that is, let the magnitude of AJ represent the ratio numerically and let the angle of AJ represent the phase shift. It will be convenient to refer to AJ as an admittance, although it does not have the dimensions of the quantity usually so called. Let the current

$$I_0 = \cos \omega t = \text{real part of } e^{i\omega t} \tag{a}$$

be impressed on the circuit. The first round trip is then represented by

$$I_1 = \text{real part of } AJe^{i\omega t} \tag{b}$$

and the nth by

$$I_m = \text{real part of } A^n J^n e^{i\omega t}. \tag{c}$$

The total current of the original impressed current and the first n round trips is

$$I_n = \text{real part of } (1 + AJ + A^2J^2 + \cdots A^nJ^n)e^{i\omega t}. \tag{d}$$

If the expression in parentheses converges as n increases indefinitely, the conclusion is that the total current equals the limit of (d) as n increases indefinitely. Now

$$1 + AJ + \cdots A^nJ^n = \frac{1 - A^{n+1}J^{n+1}}{1 - AJ}. \tag{e}$$

If $|AJ| < 1$ this converges to $1/(1 - AJ)$ which leads to an answer which accords with experiment. When $|AJ| > 1$ an examination of the numerator in (e) shows that the expression does not converge but can be made as great as desired by taking n sufficiently large. The most obvious conclusion is that when $|AJ| > 1$ for some frequenc there is a runaway condition. This disagrees with experiment, for instance, in the case where AJ is a negative quantity numerically greater than one. The next suggestion is to assume that somehow the expression $1/(1 - AJ)$ may be used instead of the limit of (e). This, however, in addition to being arbitrary, disagrees with experimental results in the case where AJ is positive and greater than 1, where the expression $1/(1 - AJ)$ leads to a finite current but where experiment indicates an unstable condition.

The fundamental difficulty with this method can be made apparent by considering the nature of the current expressed by (a) above. Does the expression cos ωt indicate a current which has been going on for all time or was the current zero up to a certain time and cos ωt thereafter? In the former case we introduce infinities into our expressions and make the equations invalid; in the latter case there will be transients or building-up processes whose importance may increase as n increases but which are tacitly neglected in equations (b) – (e). Briefly then, the difficulty with this method is that it neglects the building-up processes.

Another method is as follows: Let the voltage (or current) at any point be made up of two components

$$V = V_1 + V_2, \tag{f}$$

where V is the total voltage, V_1 is the part due directly to the impressed voltage, that is to say, without the feed-back, and V_2 is the component due to feed-back alone. We have

$$V_2 = AJV. \tag{g}$$

Eliminating V_2 between (f) and (g)

$$V = V_1/(1 - AJ). \tag{h}$$

This result agrees with experiment when $|AJ| < 1$ but does not generally agree when AJ is positive and greater than unity. The difficulty with this method is that it does not investigate whether or not a steady state exists. It simply assumes tacitly that a steady state exists and if so it gives the correct value. When a steady state does not exist this method yields no information, nor does it give any information as to whether or not a steady state exists, which is the important point.

The experimental facts do not appear to have been formulated precisely but appear to be well known to those working with these circuits. They may be stated loosely as follows: There is an unstable condition whenever there is at least one frequency for which AJ is positive and greater than unity. On the other hand, when AJ is negative it may be very much greater than unity and the condition is nevertheless stable. There are instances of $|AJ|$ being about 100 without the conditions being unstable. This, as will appear, accords closely with the rule deduced below.

Notation and Restrictions

The following notation will be used in connection with integrals:

$$\int_I \phi(z)dz = \lim_{M \to \infty} \int_{-iM}^{+iM} \phi(z)dz, \tag{1}$$

the path of integration being along the imaginary axis (see equation 9), i.e., the straight line joining $-iM$ and $+iM$;

$$\int_{s+} \phi(z)dz = \lim_{M \to \infty} \int_{-iM}^{iM} \phi(z)dz, \tag{2}$$

the path of integration being along a semicircle [1] having the origin for center and passing through the points $-iM$, M, iM;

$$\int_C \phi(z)dz = \lim_{M \to \infty} \int_{-iM}^{-iM} \phi(z)dz, \tag{3}$$

the path of integration being first along the semicircle referred to and then along a straight line from iM to $-iM$. Referring to Fig. 1 it

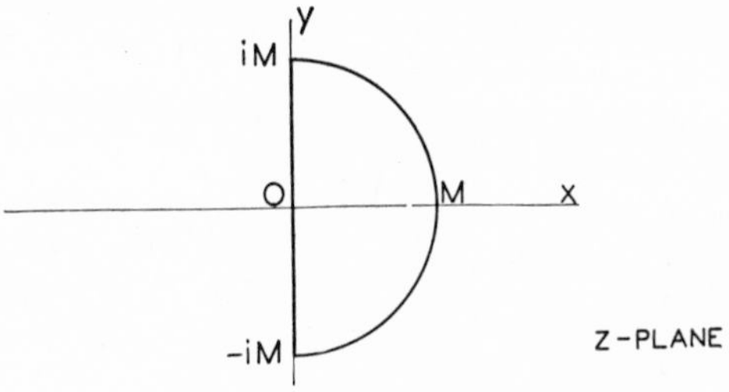

Fig. 1—Paths of integration in the z-plane.

will be seen that

$$\int_{s+} - \int_I = \int_C . \tag{4}$$

The total feed-back circuit is made up of an amplifier in tandem with a network. The amplifier is characterized by the amplifying ratio A which is independent of frequency. The network is characterized by the ratio $J(i\omega)$ which is a function of frequency but does not depend on the gain. The total effect of the amplifier and the network is to multiply the wave by the ratio $AJ(i\omega)$. An alternative way of characterizing the amplifier and network is to say that the amplifier is

[1] For physical interpretation of paths of integration for which $x > 0$ reference is made to a paper by J. R. Carson, "Notes on the Heaviside Operational Calculus," *B. S. T. J.*, Jan. 1930. For purposes of the present discussion the semicircle is preferable to the path there discussed.

characterized by the amplifying factor A which is independent of time, and the network by the real function $G(t)$ which is the response caused by a unit impulse applied at time $t = 0$. The combined effect of the amplifier and network is to convert a unit impulse to the function $AG(t)$. Both these characterizations will be used.

The restrictions which are imposed on the functions in order that the subsequent reasoning may be valid will now be stated. There is no restriction on A other than that it should be real and independent of time and frequency. In stating the restrictions on the network it is convenient to begin with the expression G. They are

$$G(t) \text{ has bounded variation}, \quad -\infty < t < \infty. \tag{AI}$$

$$G(t) = 0, \quad -\infty < t < 0. \tag{AII}$$

$$\int_{-\infty}^{\infty} |G(t)|\, dt \text{ exists.} \tag{AIII}$$

It may be shown [2] that under these conditions $G(t)$ may be expressed by the equation

$$G(t) = \frac{1}{2\pi i}\int_I J(i\omega)e^{i\omega t}d(i\omega), \tag{5}$$

where

$$J(i\omega) = \int_{-\infty}^{\infty} G(t)e^{-i\omega t}dt. \tag{6}$$

These expressions may be taken to define J. The function may, however, be obtained directly from computations or measurements; in the latter case the function is not defined for negative values of ω. It must be defined as follows to be consistent with the definition in (6):

$$J(-i\omega) = \text{complex conjugate of } J(i\omega). \tag{7}$$

While the final results will be expressed in terms of $AJ(i\omega)$ it will be convenient for the purpose of the intervening mathematics to define an auxiliary and closely related function

$$w(z) = \frac{1}{2\pi i}\int_I \frac{AJ(i\omega)}{i\omega - z}d(i\omega), \qquad 0 < x < \infty, \tag{8}$$

where

$$z = x + iy \tag{9}$$

and where x and y are real. Further, we shall define

$$w(iy) = \lim_{x \to 0} w(z). \tag{10}$$

[2] See Appendix II for fuller discussion.

The function will not be defined for $x < 0$ nor for $|z| = \infty$. As defined it is analytic[3] for $0 < x < \infty$ and at least continuous for $x = 0$.

The following restrictions on the network may be deduced:

$$\lim_{y \to \infty} y|J(iy)| \text{ exists.} \tag{BI}$$

$$J(iy) \text{ is continuous.} \tag{BII}$$

$$w(iy) = AJ(iy). \tag{BIII}$$

Equation (5) may now be written

$$AG(t) = \frac{1}{2\pi i}\int_I w(z)e^{zt}dz = \frac{1}{2\pi i}\int_{s+} w(z)e^{zt}dz. \tag{11}$$

From a physical standpoint these restrictions are not of consequence. Any network made up of positive resistances, conductances, inductances, and capacitances meets them. Restriction (AII) says that the response must not precede the cause and is obviously fulfilled physically. Restriction (AIII) is fulfilled if the response dies out at least exponentially, which is also assured. Restriction (AI) says that the transmission must fall off with frequency. Physically there are always enough distributed constants present to insure this. This effect will be illustrated in example 8 below. Every physical network falls off in transmission sooner or later and it is ample for our purposes if it begins to fall off, say, at optical frequencies. We may say then that the reasoning applies to all linear networks which occur in nature. It also applies to other linear networks which are not physically producible but which may be specified mathematically. See example 7 below.

A temporary wave $f_0(t)$ is to be introduced into the system and an investigation will be made of whether the resultant disturbance in the system dies out. It has associated with it a function $F(z)$ defined by

$$f_0(t) = \frac{1}{2\pi i}\int_I F(z)e^{zt}dz = \frac{1}{2\pi i}\int_{s+} F(z)e^{zt}dz. \tag{12}$$

$F(z)$ and $f_0(t)$ are to be made subject to the same restrictions as $w(z)$ and $G(t)$ respectively.

Derivation of a Series for the Total Current

Let the amplifier be linear and of infinite power-carrying capacity. Let the output be connected to the input in such a way that the

[3] W. F. Osgood, "Lehrbuch der Funktionentheorie," 5th ed., Kap. 7, § 1, Hauptsatz. For definition of "analytic" see Kap. 6, § 5.

amplification ratio for one round trip is equal to the complex quantity AJ, where A is a function of the gain only and J is a function of ω only, being defined for all values of frequency from 0 to ∞.

Let the disturbing wave $f_0(t)$ be applied anywhere in the circuit. We have

$$f_0(t) = \frac{1}{2\pi}\int_{-\infty}^{+\infty} F(i\omega)e^{i\omega t}d\omega \tag{13}$$

or

$$f_0(t) = \frac{1}{2\pi i}\int_{s+} F(z)e^{zt}dz. \tag{13'}$$

The wave traverses the circuit and on completing the first trip it becomes

$$f_1(t) = \frac{1}{2\pi}\int_{-\infty}^{\infty} w(i\omega)F(i\omega)e^{i\omega t}d\omega \tag{14}$$

$$= \frac{1}{2\pi i}\int_{s+} w(z)F(z)e^{zt}dz. \tag{14'}$$

After traversing the circuit a second time it becomes

$$f_2(t) = \frac{1}{2\pi i}\int_{s+} Fw^2e^{zt}dz, \tag{15}$$

and after traversing the circuit n times

$$f_n(t) = \frac{1}{2\pi i}\int_{s+} Fw^ne^{zt}dz. \tag{16}$$

Adding the voltage of the original impulse and the first n round trips we have a total of

$$s_n(t) = \sum_{k=0}^{n} f_k(t) = \frac{1}{2\pi i}\int_{s+} F(1 + w + \cdots w^n)e^{zt}dz. \tag{17}$$

The total voltage at the point in question at the time t is given by the limiting value which (17) approaches as n is increased indefinitely [4]

$$s(t) = \sum_{k=0}^{\infty} f_k(t) = \lim_{n\to\infty}\frac{1}{2\pi i}\int_{s+} S_n(z)e^{zt}dz, \tag{18}$$

where

$$S_n = F + Fw + Fw^2 + \cdots Fw^n = \frac{F(1 - w^{n+1})}{1 - w}. \tag{19}$$

[4] Mr. Carson has called my attention to the fact that this series can also be derived from Theorem IX, p. 49, of his Electric Circuit Theory. Whereas the present derivation is analogous to the theory expressed in equations (*a*)–(*e*) above, the alternative derivation would be analogous to that in equations (*f*)–(*h*).

Convergence of Series

We shall next prove that the limit $s(t)$ exists for all finite values of t. It may be stated as of incidental interest that the limit

$$\int_{s+} S_\infty(z)e^{izt}dz \tag{20}$$

does not necessarily exist although the limit $s(t)$ does. Choose M_0 and N such that

$$|f_0(\lambda)| \leq M_0. \qquad 0 \leq \lambda \leq t. \tag{21}$$

$$|G(t-\lambda)| \leq N. \quad 0 \leq \lambda \leq t. \tag{22}$$

We may write [5]

$$f_1(t) = \int_{-\infty}^{\infty} G(t-\lambda)f_0(\lambda)d\lambda. \tag{23}$$

$$|f_1(t)| \leq \int_0^t M_0 N d\lambda = M_0 N t. \tag{24}$$

$$f_2(t) = \int_{-\infty}^{\infty} G(t-\lambda)f_1(\lambda)d\lambda. \tag{25}$$

$$|f_2(t)| \leq \int_0^t M_0 N^2 t dt = M_0 N^2 t^2/2! \tag{26}$$

Similarly

$$|f_n(t)| \leq M_0 N^n t^n/n! \tag{27}$$

$$|s_n(t)| \leq M_0(1 + Nt + \cdots N^n t^n/n!). \tag{28}$$

It is shown in almost any text [6] dealing with the convergence of series that the series in parentheses converges to e^{Nt} as n increases indefinitely. Consequently, $s_n(t)$ converges absolutely as n increases indefinitely.

Relation Between $s(t)$ and w

Next consider what happens to $s(t)$ as t increases. As t increases indefinitely $s(t)$ may converge to zero, indicating a condition of stability, or it may go beyond any value however large, indicating a runaway condition. The question which presents itself is: *Referring to (18) and (19), what properties of $w(z)$ and further what properties of $AJ(i\omega)$ determine whether $s(t)$ converges to zero or diverges as t increases*

[5] G. A. Campbell, "Fourier Integral," *B. S. T. J.*, Oct. 1928, Pair 202.
[6] E.g., Whittaker and Watson, "Modern Analysis," 2d ed., p. 531.

indefinitely? From (18) and (19)

$$s(t) = \lim_{n\to\infty} \frac{1}{2\pi i} \int_{s+} F\left(\frac{1}{1-w} - \frac{w^{n+1}}{1-w}\right) e^{zt}dz. \tag{29}$$

We may write

$$s(t) = \frac{1}{2\pi i}\int_{s+} [F/(1-w)]e^{zt}dz - \lim_{n\to\infty}\frac{1}{2\pi i}\int_{s+} [Fw^{n+1}/(1-w)]e^{zt}dz \tag{30}$$

provided these functions exist. Let them be called $q_0(t)$ and $\lim_{n\to\infty} q_n(t)$ respectively. Then

$$q_n(t) = \int_{-\infty}^{\infty} q_0(t-\lambda)\phi(\lambda)d\lambda. \tag{31}$$

where

$$\phi(\lambda) = \frac{1}{2\pi i}\int_{s+} w^{n+1}e^{z\lambda}dz. \tag{32}$$

By the methods used under the discussion of convergence above it can then be shown that this expression exists and approaches zero as n increases indefinitely provided $q_0(t)$ exists and is equal to zero for $t < 0$. Equation (29) may therefore be written, subject to these conditions

$$s(t) = \frac{1}{2\pi i}\int_{s+} [F/(1-w)]e^{zt}dz. \tag{33}$$

In the first place the integral is zero for negative values of t because the integrand approaches zero faster than the path of integration increases. Moreover,

$$\int_I [F/(1-w)]e^{zt}dz \tag{34}$$

exists for all values of t and approaches zero for large values of t if $1 - w$ does not equal zero on the imaginary axis. Moreover, the integral

$$\int_C [F/(1-w)]e^{zt}dz \tag{35}$$

exists because

1. Since F and w are both analytic within the curve the integrand does not have any essential singularity there,
2. The poles, if any, lie within a finite distance of the origin because $w \to 0$ as $|z|$ increases, and
3. These two statements insure that the total number of poles is finite.

We shall next evaluate the integral for a very large value of t. It will suffice to take the C integral since the I integral approaches zero. Assume originally that $1 - w$ does not have a root on the imaginary axis and that $F(z)$ has the special value $w'(z)$. The integral may be written

$$\frac{1}{2\pi i}\int_C [w'/(1 - w)]e^{zt}dz. \tag{36}$$

Changing variables it becomes

$$\frac{1}{2\pi i}\int_D [1/(1 - w)]e^{zt}dw, \tag{37}$$

where z is a function of w and D is the curve in the w plane which corresponds to the curve C in the z plane. More specifically the imaginary axis becomes the locus $x = 0$ and the semicircle becomes a small curve which spirals around the origin. See Fig. 2. The function

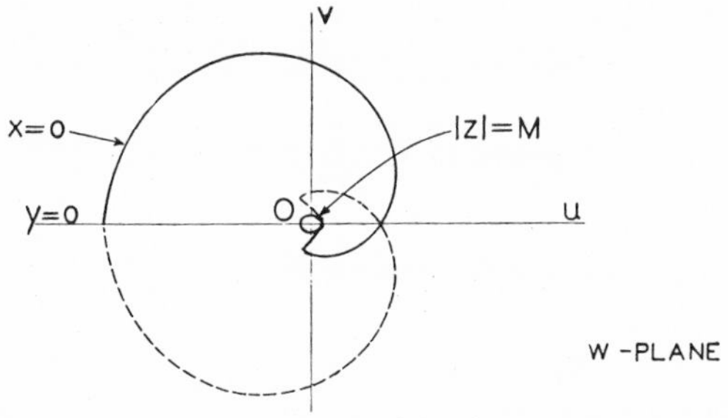

Fig. 2—Representative paths of integration in the w-plane corresponding to paths in Fig. 1.

z and, therefore, the integrand is, in general, multivalued and the curve of integration must be considered as carried out over the appropriate Riemann surface.[7]

Now let the path of integration shrink, taking care that it does not shrink across the pole at $w = 1$ and initially that it does not shrink across such branch points as interfere with its passage, if any. This shrinking does not alter the integral [8] because the integrand is analytic at all other points. At branch points which interfere with the passage of the path the branches stopped may be severed, transposed and connected in such a way that the shrinking may be continued past the branch point. This can be done without altering the value of the integral. Thus the curve can be shrunk until it becomes one or more very small circles surrounding the pole. The value of the total integral

[7] Osgood, loc. cit., Kap. 8.
[8] Osgood, loc. cit., Kap. 7, § 3, Satz 1.

(for very large values of t) is by the method of residues [9]

$$\sum_{j=1}^{n} r_j e^{z_j t}, \tag{38}$$

where z_j ($j = 1, 2 \cdots n$) is a root of $1 - w = 0$ and r_j is its order. The real part of z_j is positive because the curve in Fig. 1 encloses points with $x > 0$ only. The system is therefore stable or unstable according to whether

$$\sum_{j=1}^{n} r_j$$

is equal to zero or not. But the latter expression is seen from the procedure just gone through to equal the number of times that the locus $x = 0$ encircles the point $w = 1$.

If F does not equal w' the calculation is somewhat longer but not essentially different. The integral then equals

$$\sum_{j=1}^{n} \frac{F(z_j)}{w(z_j)} e^{z_j t} \tag{39}$$

if all the roots of $1 - w = 0$ are distinct. If the roots are not distinct the expression becomes

$$\sum_{j=1}^{n} \sum_{k=1}^{r_j} A_{jk} t^{k-1} e^{z_j t}, \tag{40}$$

where A_{jr_j}, at least, is finite and different from zero for general values of F. It appears then that unless F is specially chosen the result is essentially the same as for $F = w'$. The circuit is stable if the point lies wholly outside the locus $x = 0$. It is unstable if the point is within the curve. It can also be shown that if the point is on the curve conditions are unstable. We may now enunciate the following

Rule: Plot plus and minus the imaginary part of $AJ(i\omega)$ against the real part for all frequencies from 0 to ∞. If the point $1 + i0$ lies completely outside this curve the system is stable; if not it is unstable.

In case of doubt as to whether a point is inside or outside the curve the following criterion may be used: Draw a line from the point ($u = 1$, $v = 0$) to the point $z = -i\infty$. Keep one end of the line fixed at ($u = 1$, $v = 0$) and let the other end describe the curve from $z = -i\infty$ to $z = i\infty$, these two points being the same in the w plane. If the net angle through which the line turns is zero the point ($u = 1$, $v = 0$) is on the outside, otherwise it is on the inside.

If AJ be written $|AJ|(\cos\theta + i \sin\theta)$ and if the angle always

[9] Osgood, loc. cit., Kap. 7, § 11, Satz 1.

changes in the same direction with increasing ω, where ω is real, the rule can be stated as follows: The system is stable or unstable according to whether or not a real frequency exists for which the feed-back ratio is real and equal to or greater than unity.

In case $d\theta/d\omega$ changes sign we may have the case illustrated in Figs. 3 and 4. In these cases there are frequencies for which w is real and

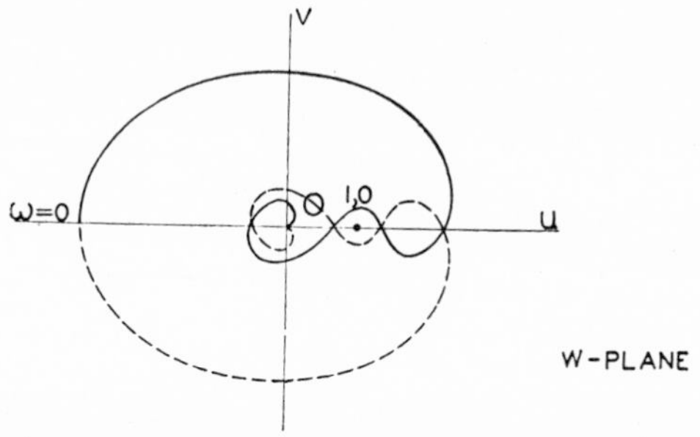

Fig. 3—Illustrating case where amplifying ratio is real and greater than unity for two frequencies, but where nevertheless the path of integration does not include the point 1, 0.

greater than 1. On the other hand, the point (1, 0) is outside of the locus $x = 0$ and, therefore, according to the rule there is a stable condition.

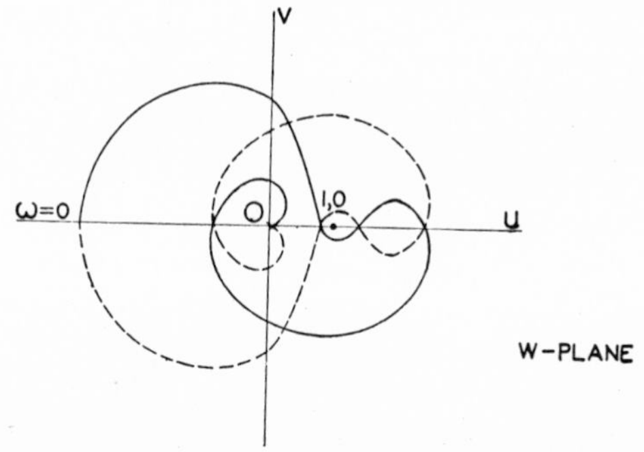

Fig. 4—Illustrating case where amplifying ratio is real and greater than unity for two frequencies, but where nevertheless the path of integration does not include the point 1, 0.

If networks of this type were used we should have the following interesting sequence of events: For low values of A the system is in a stable condition. Then as the gain is increased gradually, the system becomes unstable. Then as the gain is increased gradually still further, the system again becomes stable. As the gain is still further increased the system may again become unstable.

Examples

The following examples are intended to give a more detailed picture of certain rather simple special cases. They serve to illustrate the previous discussion. In all the cases F is taken equal to AJ so that f_0 is equal to AG. This simplifies the discussion but does not detract from the illustrative value.

1. Let the network be pure resistance except for the distortionless amplifier and a single bridged condenser, and let the amplifier be such that there is no reversal. We have

$$AJ(i\omega) = \frac{B}{\alpha + i\omega}, \tag{41}$$

where A and α are real positive constants. In (18)[10]

$$\begin{aligned} f_n &= \frac{1}{2\pi i}\int_I A^{n+1}J^{n+1}(i\omega)e^{i\omega t}di\omega \\ &= Be^{-\alpha t}(B^n t^n/n!). \end{aligned} \tag{42}$$

$$s(t) = Be^{-\alpha t}(1 + Bt + B^2t^2/2! + \cdots). \tag{43}$$

The successive terms f_0, f_1, etc., represent the impressed wave and the successive round trips. The whole series is the total current.

It is suggested that the reader should sketch the first few terms graphically for $B = \alpha$, and sketch the admittance diagrams for $B < \alpha$, and $B > \alpha$.

The expression in parentheses equals e^{Bt} and

$$s(t) = Be^{(B-\alpha)t}. \tag{44}$$

This expression will be seen to converge to 0 as t increases or fail to do so according to whether $B < \alpha$ or $B \geq \alpha$. This will be found to check the rule as applied to the admittance diagram.

2. Let the network be as in 1 except that the amplifier is so arranged that there is a reversal. Then

$$AJ(i\omega) = \frac{-B}{\alpha + i\omega}. \tag{45}$$

$$f_n = (-1)^{n+1}Be^{-\alpha t}(B^n t^n/n!). \tag{46}$$

The solution is the same as in 1 except that every other term in the series has its sign reversed:

$$\begin{aligned} s(t) &= -Be^{-\alpha t}(1 - Bt + B^2t^2/2! + \cdots) \\ &= -Be^{(-\alpha-B)t}. \end{aligned} \tag{47}$$

[10] Campbell, loc. cit., Pair 105.

This converges to 0 as t increases regardless of how great B may be taken. If the admittance diagram is drawn this is again found to check the rule.

3. Let the network be as in 1 except that there are two separated condensers bridged across resistance circuits. Then

$$AJ(i\omega) = \frac{B^2}{(\alpha + i\omega)^2}. \tag{48}$$

The solution for $s(t)$ is obtained most simply by taking every other term in the series obtained in 1.

$$\begin{aligned} s(t) &= Be^{-\alpha t}(Bt + B^3t^3/3! + \cdots) \\ &= Be^{-\alpha t} \sinh Bt. \end{aligned} \tag{49}$$

4. Let the network be as in 3 except that there is a reversal. Then

$$AJ(i\omega) = \frac{-B^2}{(\alpha + i\omega)^2}. \tag{50}$$

The solution is obtained most directly by reversing the sign of every other term in the series obtained in 3.

$$\begin{aligned} s(t) &= -Be^{-\alpha t}(Bt - B^3t^3/3! + \cdots) \\ &= -Be^{-\alpha t} \sin Bt. \end{aligned} \tag{51}$$

This is a most instructive example. An approximate diagram has been made in Fig. 5, which shows that as the gain is increased the

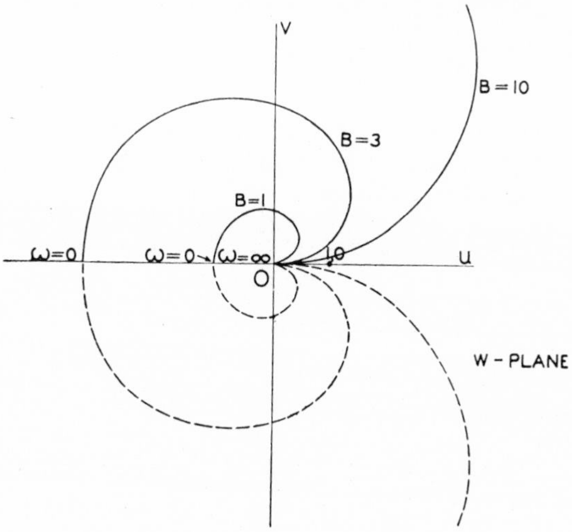

Fig. 5—Illustrating Example 4, with three values for B.

14

feed-back ratio may be made arbitrarily great and the angle arbitrarily small without the condition being unstable. This agrees with the expression just obtained, which shows that the only effect of increasing the gain is to increase the frequency of the resulting transient.

5. Let the conditions be as in 1 and 3 except for the fact that four separated condensers are used. Then

$$AJ(i\omega) = \frac{B^4}{(\alpha + i\omega)^4}. \tag{52}$$

The solution is most readily obtained by selecting every fourth term in the series obtained in 1.

$$\begin{aligned} s(t) &= Be^{-\alpha t}(B^3t^3/3! + B^7t^7/7! + \cdots) \\ &= \tfrac{1}{2}Be^{-\alpha t}\,(\sinh Bt - \sin Bt). \end{aligned} \tag{53}$$

This indicates a condition of instability when $B \geq \alpha$, agreeing with the result deducible from the admittance diagram.

6. Let the conditions be as in 5 except that there is a reversal. Then

$$Y = \frac{-B^4}{(\alpha + i\omega)^4}. \tag{54}$$

The solution is most readily obtained by changing the sign of every other term in the series obtained in 5.

$$s(t) = Be^{-\alpha t}(-B^3t^3/3! + B^7t^7/7! - \cdots). \tag{55}$$

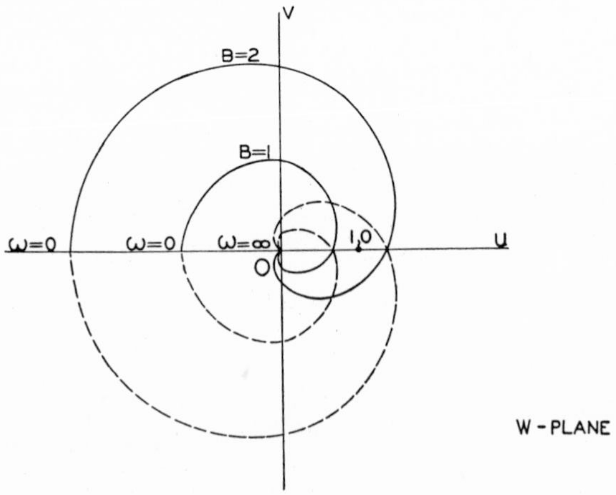

Fig. 6—Illustrating Example 6, with two values for B.

For large values of t this approaches

$$s(t) = -\tfrac{1}{2}Be^{(B/\sqrt{2}-\alpha)t}\sin(Bt/\sqrt{2} - \pi/4). \tag{56}$$

This example is interesting because it shows a case of instability although there is a reversal. Fig. 6 shows the admittance diagram for $B\sqrt{2} - \alpha < 0$ and for $B\sqrt{2} - \alpha > 0$.

7. Let

$$AG(t) = f_0(t) = A(1 - t), \qquad 0 \leq t \leq 1. \tag{57}$$

$$AG(t) = f_0(t) = 0, \qquad -\infty < t < 0, \qquad 1 < t < \infty. \tag{57'}$$

We have

$$\begin{aligned} AJ(i\omega) &= A\int_0^1 (1 - t)e^{-i\omega t}dt \\ &= A\left(\frac{1 - e^{-i\omega}}{\omega^2} + \frac{1}{i\omega}\right). \end{aligned} \tag{58}$$

Fig. 7 is a plot of this case for $A = 1$.

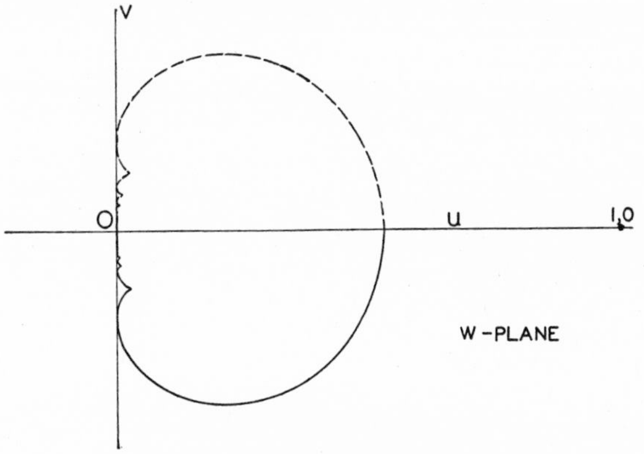

Fig. 7—Illustrating Example 7.

8. Let

$$AJ(i\omega) = \frac{A(1 + i\omega)}{(1 + i2\omega)}. \tag{59}$$

This is plotted on Fig. 8 for $A = 3$. It will be seen that the point 1 lies outside of the locus and for that reason we should expect that the system would be stable. We should expect from inspecting the diagram that the system would be stable for $A < 1$ and $A > 2$ and that it would be unstable for $1 \leq A \leq 2$. We have overlooked one fact, however; the expression for $AJ(i\omega)$ does not approach zero as ω

16

increases indefinitely. Therefore, it does not come within restriction (BI) and consequently the reasoning leading up to the rule does not apply.

The admittance in question can be made up by bridging a capacity in series with a resistance across a resistance line. This admittance

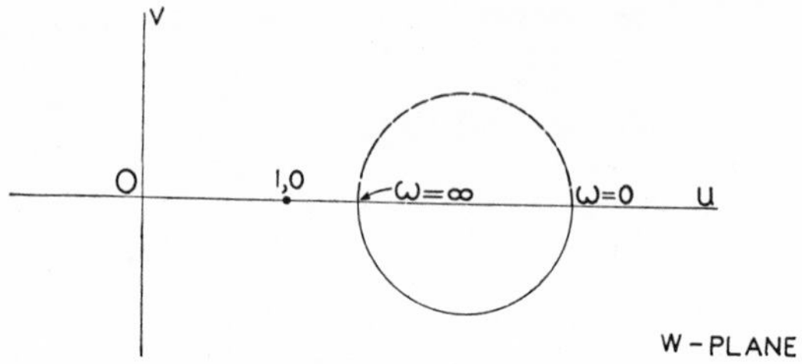

Fig. 8—Illustrating Example 8, without distributed constants.

obviously does not approach zero as the frequency increases. In any actual network there would, however, be a small amount of distributed capacity which, as the frequency is increased indefinitely, would cause the transmission through the network to approach zero. This is shown graphically in Fig. 9. The effect of the distributed capacity is

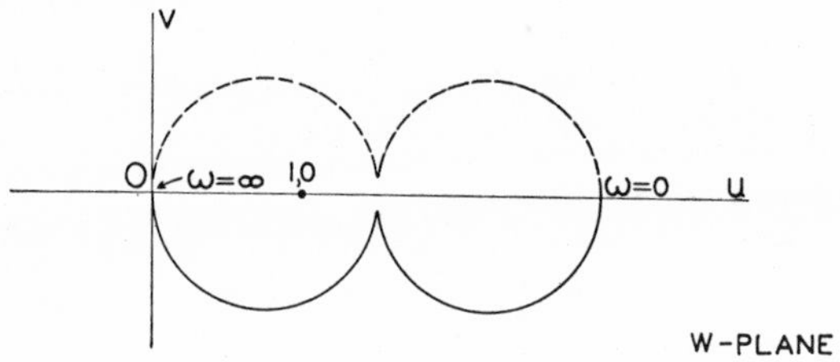

Fig. 9—Illustrating Example 8, with distributed constants.

essentially to cut a corridor from the circle in Fig. 8 to the origin, which insures that the point lies inside the locus.

Appendix I

Alternative Procedure

In some cases $AJ(i\omega)$ may be given as an analytic expression in $(i\omega)$. In that case the analytic expression may be used to define w for all values of z for which it exists. If the value for $AJ(i\omega)$ satisfies all the restrictions the value thus defined equals the w defined above for $0 \leq x < \infty$ only. For $-\infty < x < 0$ it equals the analytic continuation of the function w defined above. If there are no essential singularities anywhere including at ∞, the integral in (33) may be evaluated by the theory of residues by completing the path of integration so that all the poles of the integrand are included. We then have

$$s(t) = \sum_{j=1}^{j=n} \sum_{k=1}^{r_j} A_{jk} t^{k-1} e^{z_j t}. \tag{60}$$

If the network is made up of a finite number of lumped constants there is no essential singularity and the preceding expression converges because it has only a finite number of terms. In other cases there is an infinite number of terms, but the expression may still be expected to converge, at least, in the usual case. Then the system is stable if all the roots of $1 - w = 0$ have $x < 0$. If some of the roots have $x \geqq 0$ the system is unstable.

The calculation then divides into three parts:

1. The recognition that the impedance function is $1 - w$.[11]

2. The determination of whether the impedance function has zeros for which $x \geqq 0$.[12]

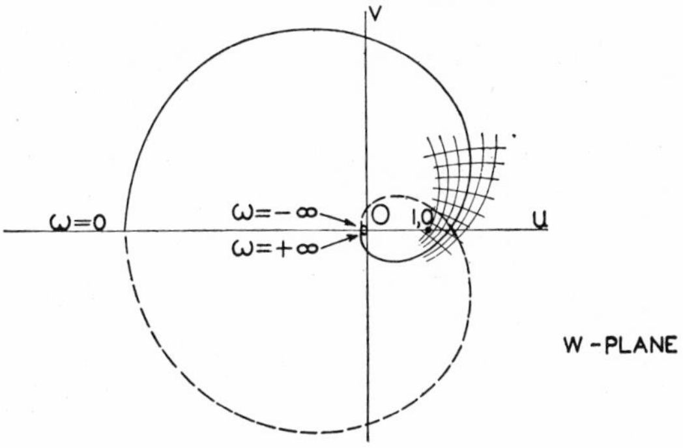

Fig. 10—Network of loci x = const., and y = const.

3. A deduction of a rule for determining whether there are roots for which $x \geqq 0$. The actual solution of the equation is usually too laborious.

To proceed with the third step, plot the locus $x = 0$ in the w plane, i.e., plot the imaginary part of w against the real part for all the values of y, $-\infty < y < \infty$. See Fig. 10. Other loci representing

$$x = \text{const.} \tag{61}$$

and

$$y = \text{const.} \tag{62}$$

[11] Cf. H. W. Nichols, *Phys. Rev.*, vol. 10, pp. 171–193, 1917.
[12] Cf. Thompson and Tait, "Natural Philosophy," vol. I, § 344

may be considered and are indicated by the network shown in the figure in fine lines. On one side of the curve x is positive and on the other it is negative. Consider the equation

$$w(z) - 1 = 0$$

and what happens to it as A increases from a very small to a very large value. At first the locus $x = 0$ lies wholly to the left of the point. For this case the roots must have $x < 0$. As A increases there may come a time when the curve or successive convolutions of it will sweep over the point $w = 1$. For every such crossing at least one of the roots changes the sign of its x. We conclude that if the point $w = 1$ lies inside the curve the system is unstable. It is now possible to enunciate the rule as given in the main part of the paper but there deduced with what appears to be a more general method.

Appendix II

Discussion of Restrictions

The purpose of this appendix is to discuss more fully the restrictions which are placed on the functions defining the network. A full discussion in the main text would have interrupted the main argument too much.

Define an additional function

$$n(z) = \frac{1}{2\pi i}\int_I \frac{AJ(i\lambda)}{i\lambda - z} d(i\lambda), \qquad -\infty < x < 0. \tag{63}$$

$$n(iy) = \lim_{x \to 0} n(z).$$

This definition is similar to that for $w(z)$ given previously. It is shown in the theorem [13] referred to that these functions are analytic for $x \neq 0$ if $AJ(i\omega)$ is continuous. We have not proved, as yet, that the restrictions placed on $G(t)$ necessarily imply that $J(i\omega)$ is continuous. For the time being we shall assume that $J(i\omega)$ may have finite discontinuities. The theorem need not be restricted to the case where $J(i\omega)$ is continuous. From an examination of the second proof it will be seen to be sufficient that $\int_I J(i\omega)d(i\omega)$ exist. Moreover, that proof can be slightly modified to include all cases where conditions (AI)–(AIII) are satisfied.

[13] Osgood, loc. cit.

For, from the equation at top of page 298 [13]

$$\left|\frac{w(z_0 - \Delta z) - w(z_0)}{\Delta z} - \frac{1}{2\pi i}\int_I \frac{AJ(i\lambda)}{(i\lambda - z_0)^2} d(i\lambda)\right|$$

$$\leq |\Delta z| \left|\frac{1}{2\pi i}\int_I \frac{AJ(i\lambda)d(i\lambda)}{(i\lambda - z_0 - \Delta z)(i\lambda - z_0)^2}\right|, \qquad x_0 > 0. \quad (64)$$

It is required to show that the integral exists. Now

$$\int_I \frac{AJ(i\lambda)d(i\lambda)}{(i\lambda - z_0 - \Delta z)(i\lambda - z_0)^2}$$

$$= \int_I \frac{AJ(i\lambda)d(i\lambda)}{(i\lambda - z_0)^3}\left(1 + \frac{\Delta z}{i\lambda - z_0} + \frac{\Delta z^2}{i\lambda - z_0} + \text{etc.}\right) \quad (65)$$

if Δz is taken small enough so the series converges. It will be sufficient to confine attention to the first term. Divide the path of integration into three parts,

$$-\infty < \lambda < -|z_0| - 1, \qquad -|z_0| - 1 < \lambda < |z_0| + 1, \qquad |z_0| + 1 < \lambda < \infty.$$

In the middle part the integral exists because both the integrand and the range of integration are finite. In the other ranges the integral exists if the integrand falls off sufficiently rapidly with increasing λ. It is sufficient for this purpose that condition (BI) be satisfied. The same proof applies to $n(z)$.

Next, consider $\lim_{x \to 0} w(z) = w(iy)$. If iy is a point where $J(iy)$ is continuous, a straightforward calculation yields

$$w(iy) = AJ(iy)/2 + P(iy). \quad (66a)$$

Likewise,

$$n(iy) = -AJ(iy)/2 + P(iy) \quad (66b)$$

where $P(iy)$ is the principal value [14] of the integral

$$\frac{1}{2\pi i}\int_I \frac{AJ(i\lambda)}{i\lambda - iy} d(i\lambda).$$

Subtracting

$$w(iy) - n(iy) = AJ(iy) \quad (67)$$

If (iy) is a point of discontinuity of $J(iy)$

$$|w| \text{ and } |n| \text{ increase indefinitely as } x \to 0. \quad (68)$$

Next, evaluate the integral

$$\frac{1}{2\pi i}\int_{x+I} w(z)e^{zt}dz,$$

[14] E. W. Hobson, "Functions of a Real Variable," vol. I, 3d edition, § 352.

where the path of integration is from $x - i\infty$ to $x + i\infty$ along the line $x =$ const. On account of the analytic nature of the integrand this integral is independent of x (for $x > 0$). It may be written then

$$\lim_{x\to 0} \frac{1}{2\pi i}\int_{x+I} w(z)e^{zt}dz = \lim_{x\to 0}\frac{1}{2\pi i}\int_{x+I}\frac{1}{2\pi i}\int_I \frac{AJ(i\lambda)}{i\lambda - z}e^{zt}d(i\lambda)dz$$

$$= \lim_{x\to 0}\frac{1}{2\pi i}\int_{x+I}\frac{1}{2\pi i}\lim_{M\to\infty}\left[\int_{-iM}^{iy-i\delta} + \int_{iy-i\delta}^{iy+i\delta} + \int_{iy+i\delta}^{iM}\right]\frac{AJ(i\lambda)}{i\lambda - z}e^{zt}d(i\lambda)dz$$

$$= \lim_{x\to 0}\left[\frac{1}{2\pi i}\int_{x+I}\frac{1}{2\pi i}\int_{iy-i\delta}^{iy+i\delta}\frac{AJ(i\lambda)}{i\lambda - z}e^{zt}d(i\lambda)dz + Q(t,\delta)\right],\ x>0, \quad (69)$$

where δ is real and positive. The function Q defined by this equation exists for all values of t and for all values of δ. Similarly,

$$\lim_{x\to 0}\frac{1}{2\pi i}\int_{x+I} n(z)e^{zt}dz$$

$$= \left[\lim_{x\to 0}\frac{1}{2\pi i}\int_{x+I}\frac{1}{2\pi i}\int_{iy+i\delta}^{iy+i\delta}\frac{AJ(i\lambda)}{i\lambda - z}e^{zt}d(i\lambda)dz + Q(t,\delta)\right],\ x<0, \quad (70)$$

Subtracting and dropping the limit designations

$$\frac{1}{2\pi i}\int_{x+I} w(z)e^{zt}dz - \frac{1}{2\pi i}\int_{x+I} n(z)e^{zt}dz = \frac{1}{2\pi i}\int_I AJ(i\lambda)e^{i\lambda t}d(i\lambda). \quad (71)$$

The first integral is zero for $t < 0$ as can be seen by taking x sufficiently large. Likewise, the second is equal to zero for $t > 0$. Therefore,

$$\frac{1}{2\pi i}\int_{x+I} w(z)e^{zt}dz = \frac{1}{2\pi i}\int_I AJ(i\omega)e^{i\omega t}d(i\omega) = AG(t),\quad 0 < t < \infty \quad (72)$$

$$-\frac{1}{2\pi i}\int_{x+I} n(z)e^{zt}dz$$

$$= \frac{1}{2\pi i}\int_I AJ(i\omega)e^{i\omega t}d(i\omega) = AG(t) \quad -\infty < t < 0. \quad (73)$$

We may now conclude that

$$\int_I n(iy)e^{iyt}d(iy) = 0, \qquad -\infty < t < \infty \quad (74)$$

provided

$$G(t) = 0, \qquad -\infty < t < 0. \quad (\text{AII})$$

But (74) is equivalent to

$$n(z) = 0, \quad (74')$$

which taken with (67) gives

$$w(iy) = AJ(iy). \tag{BIII}$$

(BIII) is, therefore, a necessary consequence of (AII). (74′) taken with (68) shows that

$$J(iy) \text{ is continuous.} \tag{BII}$$

It may be shown [15] that (BI) is a consequence of (AI). Consequently all the B conditions are deducible from the A conditions.

Conversely, it may be inquired whether the A conditions are deducible from the B conditions. This is of interest if $AJ(i\omega)$ is given and is known to satisfy the B conditions, whereas nothing is known about G.

Condition AII is a consequence of BIII as may be seen from (67) and (74). On the other hand AI and AIII cannot be inferred from the B conditions. It can be shown by examining (5), however, that if the slightly more severe condition

$$\lim_{y \to \infty} y^{\gamma} J(iy) \text{ exists,} \qquad (\gamma > 1), \tag{BIa}$$

is satisfied then

$$G(t) \text{ exists,} \qquad -\infty < t < \infty, \tag{AIa}$$

which, together with AII, insures the validity of the reasoning.

It remains to show that the measured value of $J(i\omega)$ is equal to that defined by (6). The measurement consists essentially in applying a sinusoidal wave and determining the response after a long period. Let the impressed wave be

$$E = \text{real part of } e^{i\omega t}, \qquad t \geq 0. \tag{75}$$

$$E = 0, \qquad t < 0. \tag{75′}$$

The response is

$$\text{real part of} \int_0^t AG(\lambda) e^{i\omega(t-\lambda)} d\lambda$$

$$= \text{real part of } Ae^{i\omega t} \int_0^t G(\lambda) e^{-i\omega\lambda} d\lambda. \tag{76}$$

For large values of t this approaches

$$\text{real part of } Ae^{i\omega t} J(i\omega). \tag{77}$$

Consequently, the measurements yield the value $AJ(i\omega)$.

[15] See Hobson, loc. cit., vol. II, 2d edition, § 335. It will be apparent that K depends on the total variation but is independent of the limits of integration.

11

Reprinted with permission from *Bell System Tech. J.*, **19**, 421–454 (July 1940)

Relations Between Attenuation and Phase in Feedback Amplifier Design

By H. W. BODE
Bell Telephone Laboratories

Introduction

THE engineer who embarks upon the design of a feedback amplifier must be a creature of mixed emotions. On the one hand, he can rejoice in the improvements in the characteristics of the structure which feedback promises to secure him.[1] On the other hand, he knows that unless he can finally adjust the phase and attenuation characteristics around the feedback loop so the amplifier will not spontaneously burst into uncontrollable singing, none of these advantages can actually be realized. The emotional situation is much like that of an impecunious young man who has impetuously invited the lady of his heart to see a play, unmindful, for the moment, of the limitations of the $2.65 in his pockets. The rapturous comments of the girl on the way to the theater would be very pleasant if they were not shadowed by his private speculation about the cost of the tickets.

In many designs, particularly those requiring only moderate amounts of feedback, the bogy of instability turns out not to be serious after all. In others, however, the situation is like that of the young man who has just arrived at the box office and finds that his worst fears are realized. But the young man at least knows where he stands. The engineer's experience is more tantalizing. In typical designs the loop characteristic is always satisfactory—except for one little point. When the engineer changes the circuit to correct that point, however, difficulties appear somewhere else, and so on ad infinitum. The solution is always just around the corner.

Although the engineer absorbed in chasing this rainbow may not realize it, such an experience is almost as strong an indication of the existence of some fundamental physical limitation as the census which the young man takes of his pockets. It reminds one of the experience of the inventor of a perpetual motion machine. The perpetual motion machine, likewise, always works—except for one little factor. Evidently, this sort of frustration and lost motion is inevitable in

[1] A general acquaintance with feedback circuits and the uses of feedback is assumed in this paper. As a broad reference, see H. S. Black, "Stabilized Feedback Amplifiers," *B. S. T. J.*, January, 1934.

1

feedback amplifier design as long as the problem is attacked blindly. To avoid it, we must have some way of determining in advance when we are either attempting something which is beyond our resources, like the young man on the way to the theater, or something which is literally impossible, like the perpetual motion enthusiast.

This paper is written to call attention to several simple relations between the gain around an amplifier loop, and the phase change around the loop, which impose limits to what can and cannot be done in a feedback design. The relations are mathematical laws, which in their sphere have the same inviolable character as the physical law which forbids the building of a perpetual motion machine. They show that the attempt to build amplifiers with certain types of loop characteristics *must* fail. They permit other types of characteristic, but only at the cost of certain consequences which can be calculated. In particular, they show that the loop gain cannot be reduced too abruptly outside the frequency range which is to be transmitted if we wish to secure an unconditionally stable amplifier. It is necessary to allow at least a certain minimum interval before the loop gain can be reduced to zero.

The question of the rate at which the loop gain is reduced is an important one, because it measures the actual magnitude of the problem confronting both the designer and the manufacturer of the feedback structure. Until the loop gain is zero, the amplifier will sing unless the loop phase shift is of a prescribed type. The cutoff interval as well as the useful transmission band is therefore a region in which the characteristics of the apparatus must be controlled. The interval represents, in engineering terms, the price of the ticket.

The price turns out to be surprisingly high. It can be minimized by accepting an amplifier which is only conditionally stable.[2] For the customary absolutely stable amplifier, with ordinary margins against singing, however, the price in terms of cutoff interval is roughly one octave for each ten db of feedback in the useful band. In practice, an additional allowance of an octave or so, which can perhaps be regarded as the tip to the hat check girl, must be made to insure that the amplifier, having once cut off, will stay put. Thus in an amplifier with 30 db feedback, the frequency interval over which effective control of the loop transmission characteristics is necessary is at least four octaves, or sixteen times, broader than the useful band. If we raise the feedback to 60 db, the effective range must be more than a hundred times the useful range. If the useful band is itself large these factors

[2] Definitions of conditionally and unconditionally stable amplifiers are given on page 432.

may lead to enormous effective ranges. For example, in a 4 megacycle amplifier they indicate an effective range of about 60 megacycles for 30 db feedback, or of more than 400 megacycles if the feedback is 60 db.

The general engineering implications of this result are obvious. It evidently places a burden upon the designer far in excess of that which one might anticipate from a consideration of the useful band alone. In fact, if the required total range exceeds the band over which effective control of the amplifier loop characteristics is physically possible, because of parasitic effects, he is helpless. Like the young man, he simply can't pay for his ticket. The manufacturer, who must construct and test the apparatus to realize a prescribed characteristic over such wide bands, has perhaps a still more difficult problem. Unfortunately, the situation appears to be an inevitable one. The mathematical laws are inexorable.

Aside from sounding this warning, the relations between loop gain and loop phase can also be used to establish a definite method of design. The method depends upon the development of overall loop characteristics which give the optimum result, in a certain sense, consistent with the general laws. This reduces actual design procedure to the simulation of these characteristics by processes which are essentially equivalent to routine equalizer design. The laws may also be used to show how the characteristics should be modified when the cutoff interval approaches the limiting band width established by the parasitic elements of the circuit, and to determine how the maximum realizable feedback in any given situation can be calculated. These methods are developed at some length in the writer's U. S. Patent No. 2,123,178 and are explained in somewhat briefer terms here.

Relations Between Attenuation and Phase in Physical Networks [3]

The amplifier design theory advanced here depends upon a study of the transmission around the feedback loop in terms of a number of general laws relating the attenuation and phase characteristics of physical networks. In attacking this problem an immediate difficulty presents itself. It is apparent that no entirely definite and universal

[3] Network literature includes a long list of relations between attenuation and phase discovered by a variety of authors. They are derived typically from a Fourier analysis of the transient response of assumed structures and are frequently ambiguous, because of failure to recognize the minimum phase shift condition. No attempt is made to review this work here, although special mention should be made of Y. W. Lee's paper in the *Journal for Mathematics and Physics* for June, 1932. The proof of the relations given in the present paper depends upon a contour integration in the complex frequency plane and can be understood from the disclosure in the patent referred to previously.

relation between the attenuation and the phase shift of a physical structure can exist. For example, we can always change the phase shift of a circuit without affecting its loss by adding either an ideal transmission line or an all-pass section. Any attenuation characteristic can thus correspond to a vast variety of phase characteristics.

For the purposes of amplifier design this ambiguity is fortunately unimportant. While no unique relation between attenuation and phase can be stated for a general circuit, a unique relation does exist between any given loss characteristic and the *minimum* phase shift which must be associated with it. In other words, we can always add a line or all-pass network to the circuit but we can never subtract such a structure, unless, of course, it happens to be part of the circuit originally. If the circuit includes no surplus lines or all-pass sections, it will have at every frequency the least phase shift (algebraically) which can be obtained from any physical structure having the given attenuation characteristic. The least condition, since it is the most favorable one, is, of course, of particular interest in feedback amplifier design.

For the sake of precision it may be desirable to restate the situations in which this minimum condition fails to occur. The first situation is found when the circuit includes an all-pass network either as an individual structure or as a portion of a network which can be replaced by an all-pass section in combination with some other physical structure.[4] The second situation is found when the circuit includes a transmission line. The third situation occurs when the frequency is so high that the tubes, network elements and wiring cannot be considered to obey a lumped constant analysis. This situation may be found, for example, at frequencies for which the transit time of the tubes is important or for which the distance around the feedback loop is an appreciable part of a wave-length. The third situation is, in many respects, substantially the same as the second, but it is mentioned separately here as a matter of emphasis. Since the effective band of a feedback amplifier is much greater than its useful band, as the introduction pointed out, the considerations it reflects may be worth taking into account even when they would be trivial in the useful band alone.

It will be assumed here that none of these exceptional situations is found. For the minimum phase condition, then, it is possible to derive

[4] Analytically this condition can be stated as follows: Let it be supposed that the transmission takes place between mesh 1 and mesh 2. The circuit will include an all-pass network, explicit or concealed, if any of the roots of the minor Δ_{12} of the principal circuit determinant lie below the real axis in the complex frequency plane. This can happen in bridge configurations, but not in series-shunt configurations, so that all ladder networks are automatically of minimum phase type.

a large number of relations between the attenuation and phase characteristics of a physical network. One of the simplest is

$$\int_{-\infty}^{\infty} B du = \frac{\pi}{2}(A_\infty - A_0), \tag{1}$$

where u represents $\log f/f_0$, f_0 being an arbitrary reference frequency, B is the phase shift in radians, and A_0 and A_∞ are the attenuations in nepers at zero and infinite frequency, respectively. The theorem states, in effect, that the total area under the phase characteristic plotted on a logarithmic frequency scale depends only upon the difference between the attenuations at zero and infinite frequency, and not upon the course of the attenuation between these limits. Nor does it depend upon the physical configuration of the network unless a non-minimum phase structure is chosen, in which case the area is necessarily increased. The equality of phase areas for attenuation characteristics of different types is illustrated by the sketches of Fig. 1.

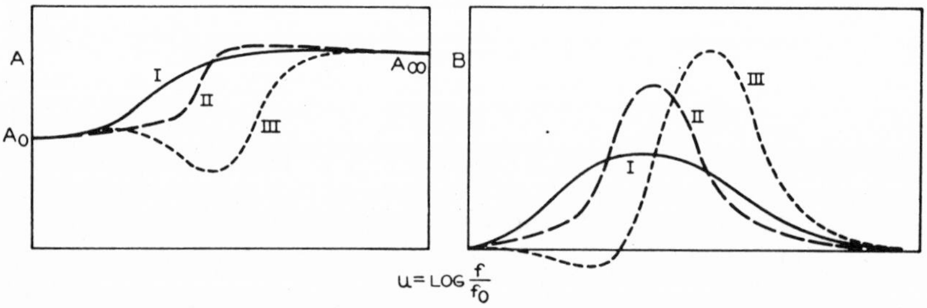

Fig. 1—Diagram to illustrate relation between phase area and change in attenuation.

The significance of the phase area relation for feedback amplifier design can be understood by supposing that the practical transmission range of the amplifier extends from zero to some given finite frequency. The quantity $A_0 - A_\infty$ can then be identified with the change in gain around the feedback loop required to secure a cut-off. Associated with it must be a certain definite phase area. If we suppose that the maximum phase shift at any frequency is limited to some rather low value the total area must be spread out over a proportionately broad interval on the frequency scale. This must correspond roughly to the cut-off region, although the possibility that some of the area may be found above or below the cut-off range prevents us from determining the necessary interval with precision.

A more detailed statement of the relationship between phase shift and change in attenuation can be obtained by turning to a second theorem. It reads as follows:

$$B(f_c) = \frac{1}{\pi}\int_{-\infty}^{\infty} \frac{dA}{du} \log \coth \frac{|u|}{2}\, du, \tag{2}$$

where $B(f_c)$ represents the phase shift at any arbitrarily chosen frequency f_c and $u = \log f/f_c$. This equation, like (1), holds only for the minimum phase shift case.

Although equation (2) is somewhat more complicated than its predecessor, it lends itself to an equally simple physical interpretation. It is clear, to begin with, that the equation implies broadly that the

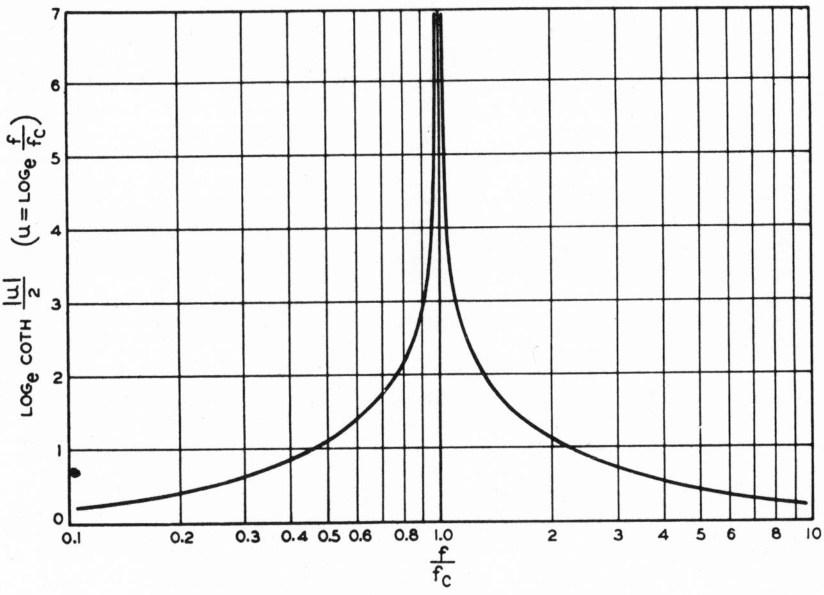

Fig. 2—Weighting function in loss-phase formula.

phase shift at any frequency is proportional to the derivative of the attenuation on a logarithmic frequency scale. For example, if dA/du is doubled B will also be doubled. The phase shift at any particular frequency, however, does not depend upon the derivative of attenuation at that frequency alone, but upon the derivative at all frequencies, since it involves a summing up, or integration, of contributions from the complete frequency spectrum. Finally, we notice that the contributions to the total phase shift from the various portions of the frequency spectrum do not add up equally, but rather in accordance with the function $\log \coth |u|/2$. This quantity, therefore, acts as a weighting function. It is plotted in Fig. 2. As we might expect physically, it is much larger near the point $u = 0$ than it is in other regions. We can, therefore, conclude that while the derivative of attenuation at all frequencies enters into the phase shift at any particular frequency $f = f_c$ the derivative in the neighborhood of f_c is relatively much more important than the derivative in remote parts of the spectrum.

As an illustration of (2), let it be supposed that $A = ku$, which corresponds to an attenuation having a constant slope of 6 k db per octave. The associated phase shift is easily evaluated. It turns out, as we might expect, to be constant, and is equal numerically to $k\pi/2$ radians. This is illustrated by Fig. 3. As a second example, we may consider

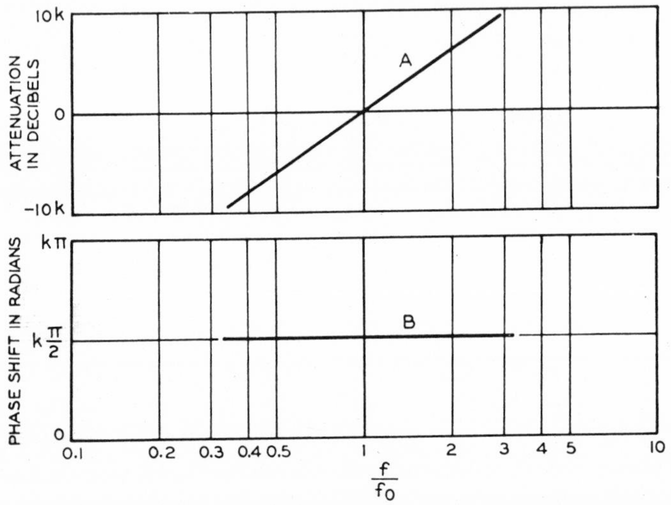

Fig. 3—Phase characteristic corresponding to a constant slope attenuation.

a discontinuous attenuation characteristic such as that shown in Fig. 4. The associated phase characteristic, also shown in Fig. 4, is proportional to the weighting function of Fig. 2.

The final example is shown by Fig. 5. It consists of an attenuation characteristic which is constant below a specified frequency f_b and has a constant slope of 6 k db per octave above f_b. The associated phase characteristic is symmetrical about the transition point between the two ranges. At sufficiently high frequencies, the phase shift approaches the limiting $k\pi/2$ radians which would be realized if the constant slope were maintained over the complete spectrum. At low frequencies the phase shift is substantially proportional to frequency and is given by the equation

$$B = \frac{2k}{\pi} \frac{f}{f_b} \cdot \tag{3}$$

7

Solutions developed in this way can be added together, since it is apparent from the general relation upon which they are based that the phase characteristic corresponding to the sum of two attenuation

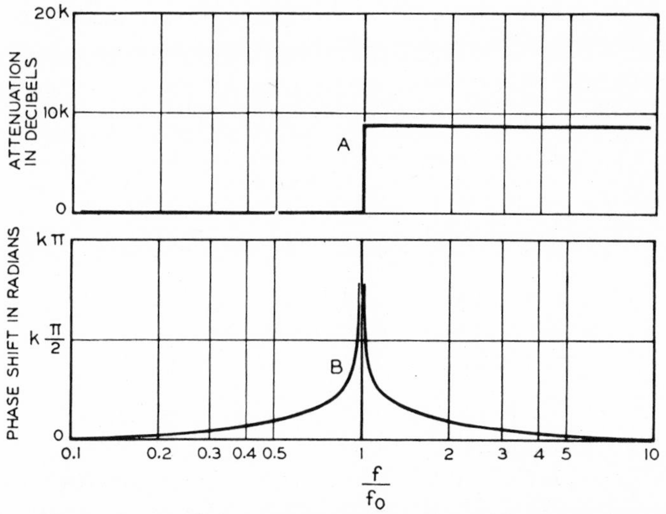

Fig. 4—Phase characteristic corresponding to a discontinuity in attenuation.

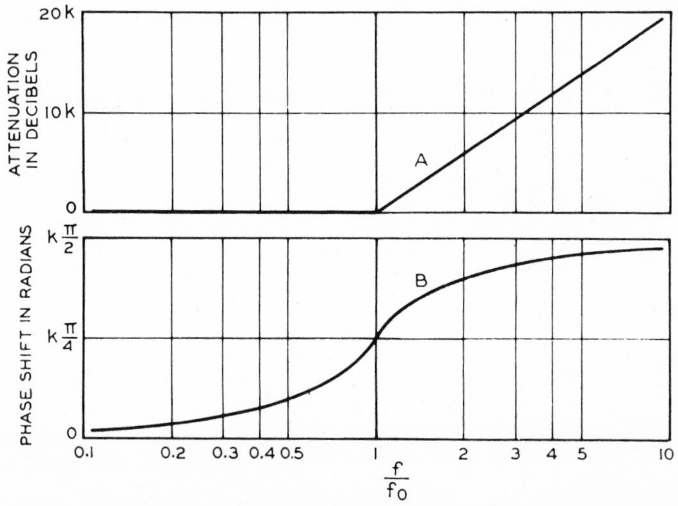

Fig. 5—Phase characteristic corresponding to an attenuation which is constant below a prescribed frequency and has a constant slope above it.

characteristics will be equal to the sum of the phase characteristics corresponding to the two attenuation characteristics separately. We can therefore combine elementary solutions to secure more complicated

8

characteristics. An example is furnished by Fig. 6, which is built up from three solutions of the type shown by Fig. 5. By proceeding sufficiently far in this way, an approximate computation of the phase characteristic associated with almost any attenuation characteristic can be made, without the labor of actually performing the integration in (2).

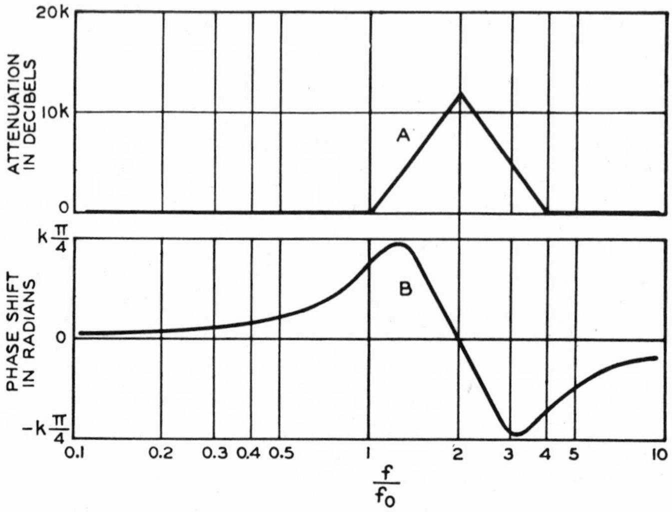

Fig. 6—Diagram to illustrate addition of elementary attenuation and phase characteristics to produce more elaborate solutions of the loss-phase formula.

Equations (1) and (2) are the most satisfactory expressions to use in studying the relation between loss and phase in a broad physical sense. The mechanics of constructing detailed loop cut-off characteristics, however, are simplified by the inclusion of one other, somewhat more complicated, formula. It appears as

$$\int_0^{f_0} \frac{A\,df}{\sqrt{f_0^2 - f^2}\,(f^2 - f_c^2)} + \int_{f_0}^{\infty} \frac{B\,df}{\sqrt{f^2 - f_0^2}\,(f^2 - f_c^2)}$$

$$= \frac{\pi}{2} \frac{B(f_c)}{f_c\sqrt{f_0^2 - f_c^2}}, \qquad f_c < f_0$$

$$= -\frac{\pi}{2} \frac{A(f_c)}{f_c\sqrt{f_c^2 - f_0^2}}, \qquad f_c > f_0, \quad (4)$$

where f_0 is some arbitrarily chosen frequency and the other symbols have their previous significance.

The meaning of (4) can be understood if it is recalled that (2) implies that the minimum phase shift at any frequency can be computed if the attenuation is prescribed at all frequencies. In the same way (4) shows how the complete attenuation and phase characteristics can be determined if we begin by prescribing the attenuation below f_0 and the phase shift above f_0. Since f_0 can be chosen arbitrarily large or small this is evidently a more general formula than either (1) or (2), while it can itself be generalized, by the introduction of additional irrational factors, to provide for more elaborate patterns of bands in which A and B are specified alternately.

As an example of this formula, let it be assumed that $A = K$ for $f < f_0$ and that $B = k\pi/2$ for $f > f_0$. These are shown by the solid lines in Fig. 7. Substitution in (4) gives the A and B characteristics in the rest of the spectrum as

$$B = k \sin^{-1}\frac{f}{f_0}, \qquad f < f_0$$

$$A = K + k \log\left[\sqrt{\frac{f^2}{f_0^2} - 1} + \frac{f}{f_0}\right], \qquad f > f_0. \tag{5}$$

These are indicated by broken lines in Fig. 7. In this particularly

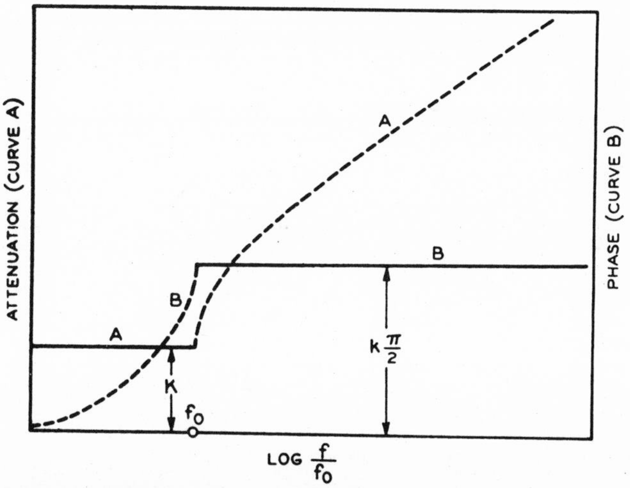

Fig. 7—Construction of complete characteristics from an attenuation characteristic specified below a certain frequency and a phase characteristic above it. The solid lines represent the specified attenuation and phase characteristics, and the broken lines their computed extensions to the rest of the spectrum.

simple case all four fragments can be combined into the single analytic formula

$$A + iB = K + k \log\left[\sqrt{1 - \frac{f^2}{f_0^2}} + i\frac{f}{f_0}\right]. \tag{6}$$

This expression will be used as the fundamental formula for the loop cut-off characteristic in the next section.

Overall Feedback Loop Characteristics

The survey just concluded shows what combinations of attenuation and phase characteristics are physically possible. We have next to determine which of the available combinations is to be regarded as representing the transmission around the overall feedback loop. The choice will naturally depend somewhat upon exactly what we assume that the amplifier ought to do, but with any given set of assumptions it is possible, at least in theory, to determine what combination is most appropriate.

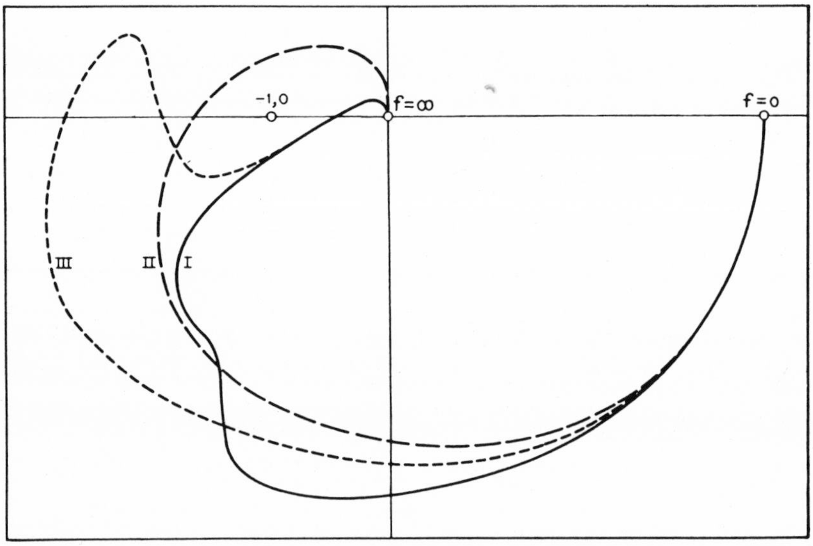

Fig. 8—Nyquist stability diagrams for various amplifiers. Curve I represents "absolute" stability, Curve II instability, and Curve III "conditional" stability. In accordance with the convention used in this paper the diagram is rotated through 180° from its normal position so that the critical point occurs at $-1, 0$ rather than $+1, 0$.

The situation is conveniently investigated by means of the Nyquist stability diagram [5] illustrated by Fig. 8. The diagram gives the path

[5] *Bell System Technical Journal*, July, 1932. See also Peterson, Kreer, and Ware, *Bell System Technical Journal*, October, 1934. The Nyquist diagrams in the present paper are rotated through 180° from the positions in which they are usually drawn, turning the diagrams in reality into plots of $-\mu\beta$. In a normal amplifier there is one net phase reversal due to the tubes in addition to any phase shifts chargeable directly to the passive networks in the circuit. The rotation of the diagram allows this phase reversal to be ignored, so that the phase shifts actually shown are the same as those which are directly of design interest.

traced by the vector representing the transmission around the feedback loop as the frequency is assigned all possible real values. In accordance with Nyquist's results a path such as II, which encircles the point $-1, 0$, indicates an unstable circuit and must be avoided. A stable amplifier is obtained if the path resembles either I or III, neither of which encircles $-1, 0$. The stability represented by Curve III, however, is only "Nyquist" or "conditional." The path will enclose the critical point if it is merely reduced in scale, which may correspond physically to a reduction in tube gain. Thus the circuit may sing when the tubes begin to lose their gain because of age, and it may also sing, instead of behaving as it should, when the tube gain increases from zero as power is first applied to the circuit. Because of these possibilities conditional stability is usually regarded as undesirable and the present discussion will consequently be restricted to "absolutely" or "unconditionally" stable amplifiers having Nyquist diagrams of the type resembling Curve I.

The condition that the amplifier be absolutely stable is evidently that the loop phase shift should not exceed 180° until the gain around the loop has been reduced to zero or less. A theoretical characteristic which just met this requirement, however, would be unsatisfactory, since it is inevitable that the limiting phase would be exceeded in fact by minor deviations introduced either in the detailed design of the amplifier or in its construction. It will therefore be assumed that the limiting phase is taken as 180° less some definite margin. This is illustrated by Fig. 9, the phase margin being indicated as $y\pi$ radians. At frequencies remote from the band it is physically impossible, in most circuits, to restrict the phase within these limits. As a supplement, therefore, it will be assumed that larger phase shifts are permissible if the loop gain is x db below zero. This is illustrated by the broken circular arc in Fig. 9. A theoretical loop characteristic meeting both requirements will be developed for an amplifier transmitting between zero and some prescribed limiting frequency with a constant feedback, and cutting off thereafter as rapidly as possible. This basic characteristic can be adapted to amplifiers with varying feedback in the useful range or with useful ranges lying in other parts of the spectrum by comparatively simple modifications which are described at a later point. It is, of course, contemplated that the gain and phase margins x and y will be chosen arbitrarily in advance. If we choose large values we can permit correspondingly large tolerances in the detailed design and construction of the apparatus without risk of instability. It turns out, however, that with a prescribed width of cutoff interval the amount of feedback which can be realized in the useful range is decreased as the assumed margins are increased, so that it is generally desirable to choose as small margins as is safe.

The essential feature in this situation is the requirement that the diminution of the loop gain in the cutoff region should not be accompanied by a phase shift exceeding some prescribed amount. In view of the close connection between phase shift and the slope of the attenuation characteristic evidenced by (2) this evidently demands that the amplifier should cut off, on the whole, at a well defined rate which is not too fast. As a first approximation, in fact, we can choose the cutoff

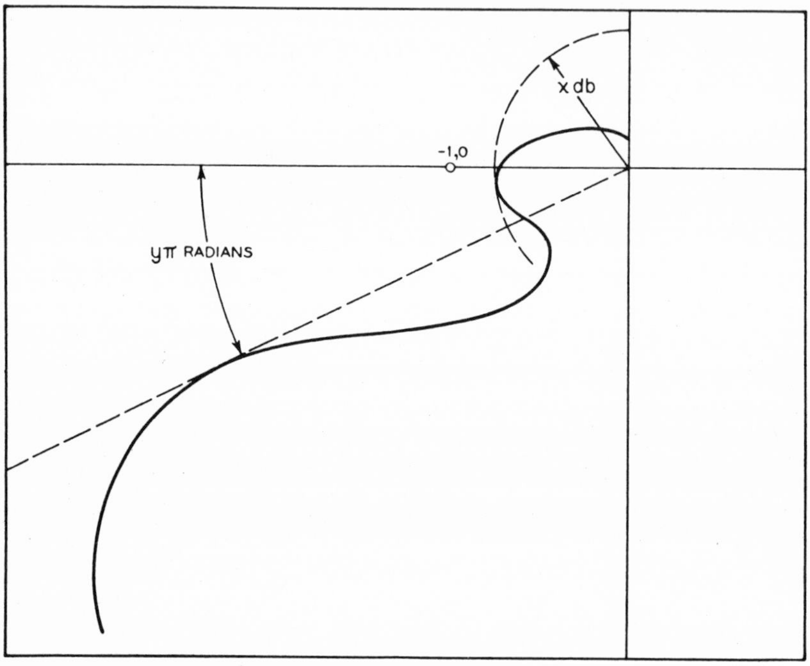

Fig. 9—Diagram to illustrate definitions of phase and gain margins for the feedback loop.

characteristic as an exactly constant slope from the edge of the useful band outward. Such a characteristic has already been illustrated by Fig. 5 and is shown, replotted,[6] by the broken lines in Fig. 10. If we choose the parameter corresponding to k in Fig. 5 as 2 the cutoff rate is 12 db per octave and the phase shift is substantially 180° at high frequencies. This choice thus leads to zero phase margin. By choosing a somewhat smaller k on the other hand, we can provide a definite

[6] To prevent confusion it should be noticed that the general attenuation-phase diagrams are plotted in terms of relative loss while loop cutoff characteristics, here and at later points, are plotted in terms of relative gain.

margin against singing, at the cost of a less rapid cutoff. For example, if we choose $k = 1.5$ the limiting phase shift in the $\mu\beta$ loop becomes 135°, which provides a margin of 45° against instability, while the rate of cutoff is reduced to 9 db per octave. The value $k = 1.67$, which corresponds to a cutoff rate of 10 db per octave and a phase margin of 30°, has been chosen for illustrative purposes in preparing Fig. 10. The loss margin depends upon considerations which will appear at a later point.

Although characteristics of the type shown by Fig. 5 are reasonably satisfactory as amplifier cutoffs they evidently provide a greater phase

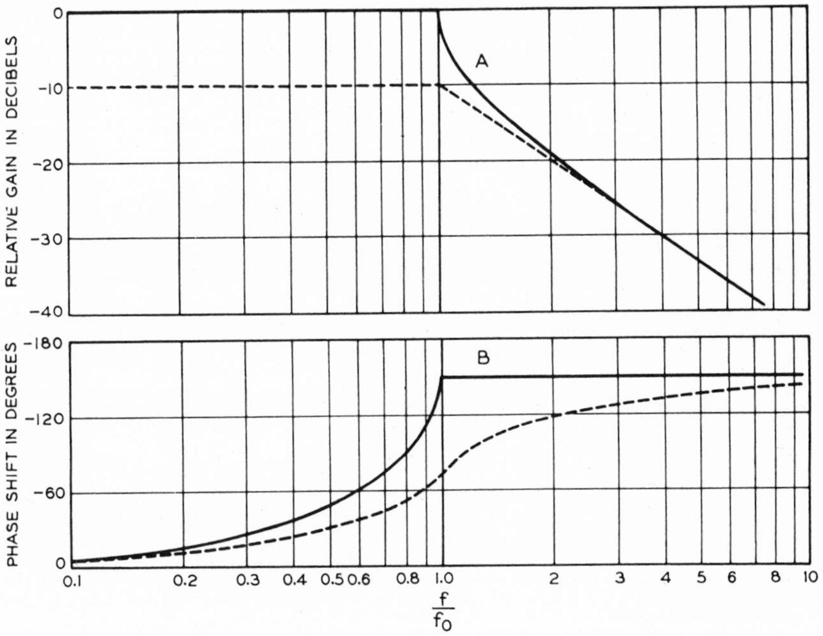

Fig. 10—Ideal loop cutoff characteristics. Drawn for a 30° phase margin.

margin against instability in the region just beyond the useful band than they do at high frequencies. In virtue of the phase area law this must be inefficient if, as is supposed here, the optimum characteristic is one which would provide a constant margin throughout the cutoff interval. The relation between the phase and the slope of the attenuation suggests that a constant phase margin can be obtained by increasing the slope of the cutoff characteristic near the edge of the band, leaving its slope at more remote frequencies unchanged, as shown by the solid lines in Fig. 10. The exact expression for the required curve can be found from (6), where the problem of determining such a characteristic appeared as an example of the use of the general formula (4).

At high frequencies the new phase and attenuation characteristics merge with those obtained from the preceding straight line cutoff, as Fig. 10 indicates. In this region the relation between phase margin and cutoff slope is fixed by the k in the equation (6) in the manner already described for the more elementary cutoff. At low frequencies, however, the increased slope near the edge of the band permits 6 k db more feedback.

It is worth while to pause here to consider what may be said, on the basis of these characteristics, concerning the breadth of cutoff interval required for a given feedback, or the "price of the ticket," as it was expressed in the introduction. If we adopt the straight line cutoff and assume the k used in Fig. 10 the interval between the edge of the useful band and the intersection of the characteristic with the zero gain axis is evidently exactly 1 octave for each 10 db of low frequency feedback. The increased efficiency of the solid line characteristic saves one octave of this total if the feedback is reasonably large to begin with. This apparently leads to a net interval one or two octaves narrower than the estimates made in the introduction. The additional interval is required to bridge the gap between a purely mathematical formula such as (6), which implies that the loop characteristics follow a prescribed law up to indefinitely high frequencies, and a physical amplifier, whose ultimate loop characteristics vary in some uncontrollable way. This will be discussed later. It is evident, of course, that the cutoff interval will depend slightly upon the margins assumed. For example, if the phase margin is allowed to vanish the cutoff rate can be increased from 10 to 12 db per octave. This, however, is not sufficient to affect the order of magnitude of the result. Since the diminished margin is accompanied by a corresponding increase in the precision with which the apparatus must be manufactured such an economy is, in fact, a Pyrrhic victory unless it is dictated by some such compelling consideration as that described in the next section.

Maximum Obtainable Feedback

A particularly interesting consequence of the relation between feedback and cutoff interval is the fact that it shows why we cannot obtain unconditionally stable amplifiers with as much feedback as we please. So far as the purely theoretical construction of curves such as those in Fig. 10 is concerned, there is clearly no limit to the feedback which can be postulated. As the feedback is increased, however, the cutoff interval extends to higher and higher frequencies. The process reaches a physical limit when the frequency becomes so high that parasitic effects in the circuit are controlling and do not permit the prescribed cutoff characteristic to be simulated with sufficient precision. For example, we are obviously in physical difficulties if the cutoff characteristic specifies a net gain around the loop at a frequency so high that the tubes themselves working into their own parasitic capacitances do not give a gain.

This limitation is studied most easily if the effects of the parasitic elements are lumped together by representing them in terms of the asymptotic characteristic of the loop as a whole at extremely high frequencies. An example is shown by Fig. 11. The structure is a

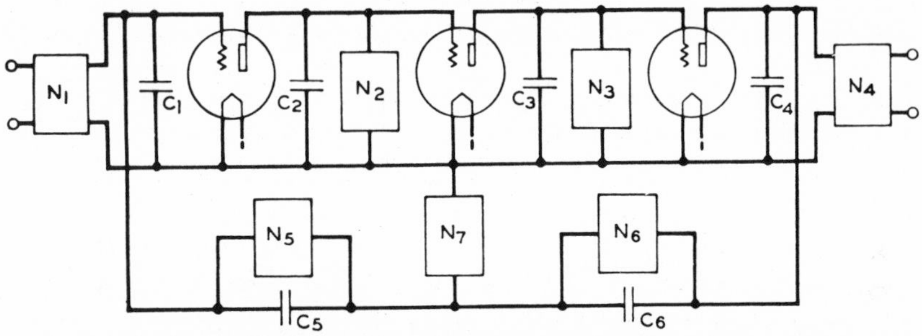

Fig. 11—Elements which determine the asymptotic loop transmission characteristic in a typical amplifier.

shunt feedback amplifier. The β circuit is represented by the T composed of networks N_5, N_6 and N_7. The input and output circuits are represented by N_1 and N_4 and the interstage impedances by N_2 and N_3. The C's are parasitic capacitances with the exception of C_5 and C_6, which may be regarded as design elements added deliberately to N_5 and N_6 to obtain an efficient high frequency transmission path from output to input. At sufficiently high frequencies the loop transmission will depend only upon these various capacitances, without regard to the N's. Thus, if the transconductances of the tubes are represented by G_1, G_2, and G_3 the asymptotic gains of the first two tubes are $G_1/\omega C_1$ and $G_2/\omega C_3$. The rest of the loop includes the third tube and the potentiometer formed by the capacitances C_1, C_4, C_5 and C_6. Its asymptotic transmission can be written as $G_3/\omega C$, where

$$C = C_1 + C_4 + \frac{C_1 C_4}{C_5 C_6}(C_5 + C_6).$$

Each of these terms diminishes at a rate of 6 db per octave. The complete asymptote is $G_1 G_2 G_3/\omega^3 C C_2 C_3$. It appears as a straight line with a slope of 18 db per octave when plotted on logarithmic frequency paper.

A similar analysis can evidently be made for any amplifier. In the particular circuit shown by Fig. 11 the slope of the asymptote, in units of 6 db per octave, is the same as the number of tubes in the circuit. The slope can evidently not be less than the number of tubes but it may be greater in some circuits. For example if C_5 and C_6 were omitted in Fig. 11 and N_5 and N_6 were regarded as degenerating into resistances the asymptote would have a slope of 24 db per octave and would lie below the present asymptote at any reasonably high frequency. In any event the asymptote will depend only upon the parasitic elements of the circuit and perhaps a few of the most significant design elements. It can thus be determined from a skeletonized version of the final structure. If waste of time in false starts is to be avoided such a determination should be made as early as possible, and certainly in advance of any detailed design.

The effect of the asymptote on the overall feedback characteristic is illustrated by Fig. 12. The curve *ABEF* is a reproduction of the ideal

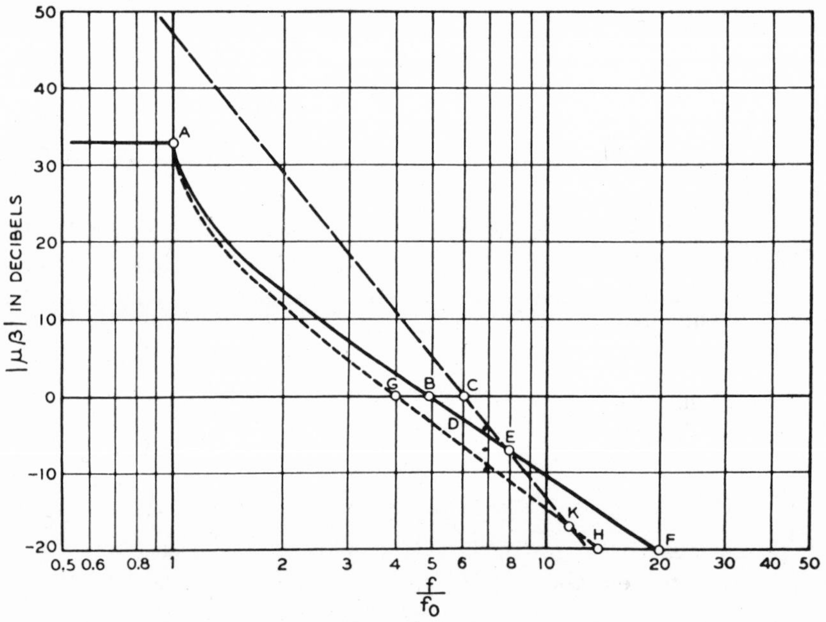

Fig. 12—Combination of asymptotic characteristic and ideal cutoff characteristic.

cutoff characteristic originally given by the solid lines in Fig. 10. It will be recalled that the curve was drawn for the choice $k = 5/3$, which corresponds to a phase margin of 30° and an almost constant slope, for the portion *DEF* of the characteristic, of about 10 db per octave. The straight line *CEK* represents an asymptote of the type just described, with a slope of 18 db per octave. Since the asymptote may be assumed to represent the practical upper limit of gain in the high-frequency region, the effect of the parasitic elements can be obtained by replacing the theoretical cutoff by the broken line characteristic *ABDEK*. In an actual circuit the corner at *E* would, of course, be rounded off, but this is of negligible quantitative importance. Since *EF* and *EK* diverge by 8 db per octave the effect can be studied by adding curves of the type shown by Fig. 5 to the original cutoff characteristic.

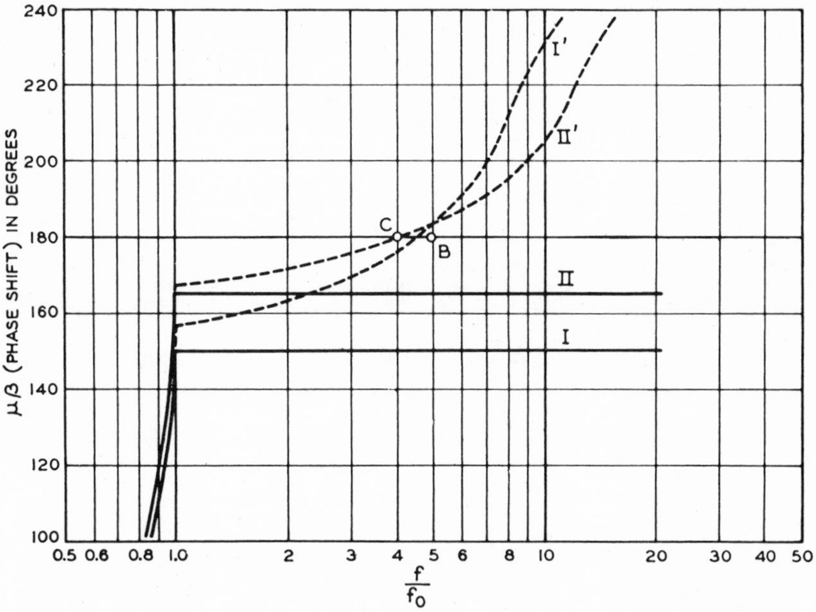

Fig. 13—Phase characteristics corresponding to gain characteristics of Fig. 12.

The phase shift in the ideal case is shown by Curve I of Fig. 13. The addition of the phase corresponding to the extra slope of 8 db per octave at high frequencies produces the total phase characteristic shown by Curve I′. At the point *B* where $|\mu\beta| = 1$, the additional phase shift amounts to 35 degrees. Since this is greater than the original phase margin of 30 degrees the amplifier is unstable when parasitic elements are considered. In the present instance stability can be regained by increasing the coefficient *k* to 1-5/6, which leads to the broken line characteristic *AGKH* in Fig. 12. This reduces the nominal phase margin to 15 degrees, but the frequency interval between *G* and *K* is so much greater than that between *B* and *E* that the added phase is reduced still more and is just less than 15° at the new

18

cross over point G. This is illustrated by II and II′ in Fig. 13. On the other hand, if the zero gain intercept of the asymptote CEK had occurred at a slightly lower frequency, no change in k alone would have been sufficient. It would have been necessary to reduce the amount of feedback in the transmitted range in order to secure stability.

The final characteristic in Fig. 13 reaches the limiting phase shift of 180° only at the crossover point. It is evident that a somewhat more efficient solution for the extreme case is obtained if the limiting 180° is approximated throughout the cutoff interval. This result is attained by the cutoff characteristic shown in Fig. 14. The characteristic con-

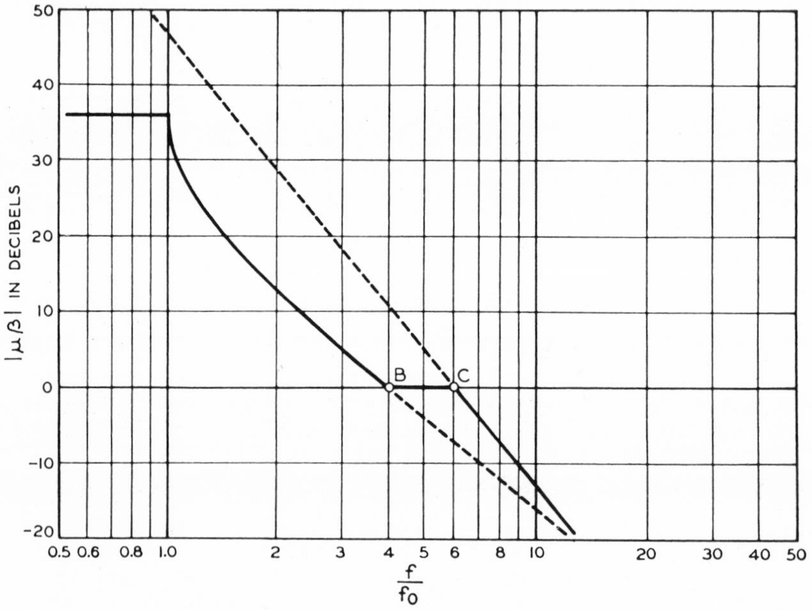

Fig. 14—Ideal cutoff modified to take account of asymptotic characteristic. Drawn for zero gain and phase margins.

sists of the original theoretical characteristic, drawn for $k = 2$, from the edge of the useful band to its intercept with the zero gain axis, the zero gain axis from this frequency to the intercept with the high-frequency asymptote, and the asymptote thereafter. It can be regarded as a combination of the ideal cutoff characteristic and two characteristics of the type shown by Fig. 5. One of the added characteristics starts at B and has a positive slope of 12 db per octave, since the ideal cutoff was drawn for the limiting value of k. The other starts at C and has the negative slope, $-$ 18 db per octave, of the asymptote itself. As (3) shows, the added slopes correspond at lower frequencies to approximately linear phase characteristics of opposite sign. If the frequencies B and C at which the slopes begin are in the same ratio, 12 : 18, as the slopes themselves the contributions of the added slopes will substantially cancel each other and the net phase shift throughout the cutoff interval will be almost the same as that of the ideal curve alone. The exact phase characteristic is shown by Fig. 15. It dips

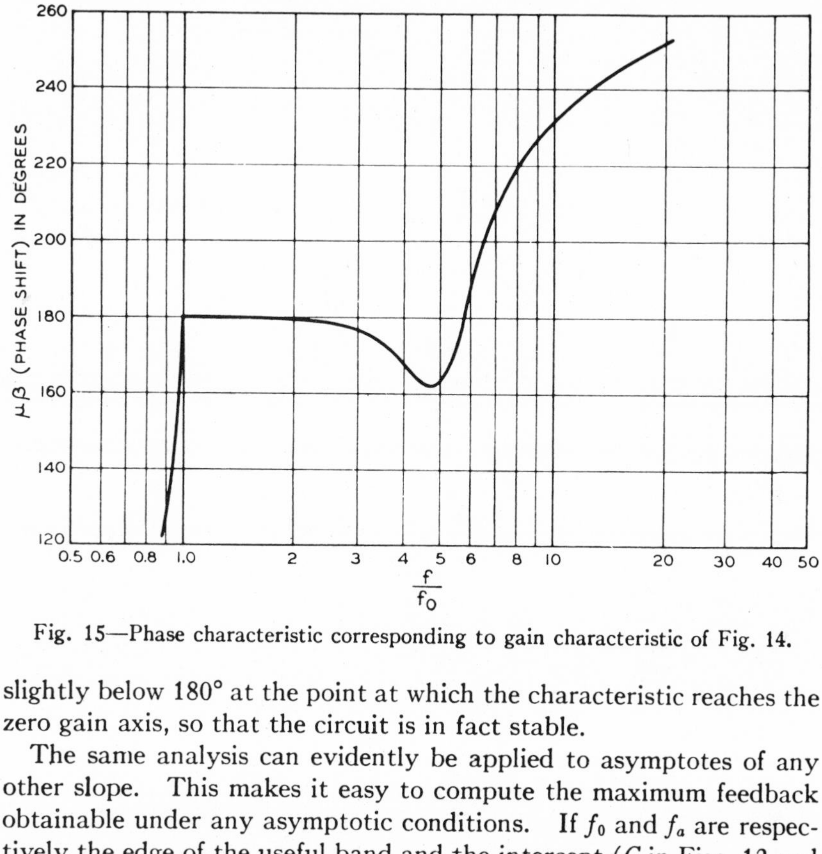

Fig. 15—Phase characteristic corresponding to gain characteristic of Fig. 14.

slightly below 180° at the point at which the characteristic reaches the zero gain axis, so that the circuit is in fact stable.

The same analysis can evidently be applied to asymptotes of any other slope. This makes it easy to compute the maximum feedback obtainable under any asymptotic conditions. If f_0 and f_a are respectively the edge of the useful band and the intercept (C in Figs. 12 and 14) of the asymptote with the zero gain axis, and n is the asymptotic slope, in units of 6 db per octave, the result appears as

$$A_m = 40 \log_{10} \frac{4f_a}{f_0}, \tag{7}$$

where A_m is the maximum feedback in db.[7]

[7] The formulæ for maximum feedback given here and in the later equation (8) are slightly conservative. It follows from the phase area law that more feedback should be obtained if the phase shift were exactly 180° below the crossover and rose

For the sake of generality it is convenient to extend this formula to include also situations in which there exists some further linear phase characteristic in addition to those already taken into account. In exceptional circuits, the final asymptotic characteristic may not be completely established by the time the curve reaches the zero gain axis and the additional phase characteristic may be used to represent the effect of subsequent changes in the asymptotic slope. Such a situation might occur in the circuit of Fig. 11, for example, if C_5 or C_6 were made extremely small. The additional term may also be used to represent departures from a lumped constant analysis in high-frequency amplifiers, as discussed earlier, If we specify the added phase characteristic, from whatever source, by means of the frequency f_d at which it would equal $2n/\pi$ radians, if extrapolated, the general formula corresponding to (7) becomes

$$A_m = 40 \log_{10} \frac{4}{nf_0} \frac{f_a f_d}{f_a + f_d} \cdot \qquad (8)$$

It is interesting to notice that equations (7) and (8) take no explicit account of the final external gain of the amplifier. Naturally, if the external gain is too high the available μ circuit gain may not be sufficient to provide it and also the feedback which these formulæ promise. This, however, is an elementary question which requires no further discussion. In other circumstances, the external gain may enter the situation indirectly, by affecting the asymptotic characteristics of the β path, but in a well chosen β circuit this is usually a minor consideration. The external gain does, however, affect the parts of the circuit upon which reliance must be placed in controlling the overall loop characteristic. For example, if the external gain is high the μ circuit will ordinarily be sharply tuned and will drop off rapidly in gain beyond the useful band. The β circuit must therefore provide a decreasing loss to bring the overall cutoff rate within the required limit. Since the β circuit must have initially a high loss to correspond to the high final gain of the complete amplifier, this is possible. Conversely, if the gain of the amplifier is low the μ circuit will be relatively flexible and the β circuit relatively inflexible.

rapidly to its ultimate value thereafter. These possibilities can be exploited approximately by various slight changes in the slope of the cutoff characteristic in the neighborhood of the crossover region, or a theoretical solution can be obtained by introducing a prescribed phase shift of this type in the general formula (4). The theoretical solution gives a Nyquist path which, after dropping below the critical point with a phase shift slightly less than 180°, rises again with a phase shift slightly greater than 180° and continues for some time with a large amplitude and increasing phase before it finally approaches the origin. These possibilities are not considered seriously here because they lead to only a few db increase in feedback, at least for moderate n's, and the degree of design control which they envisage is scarcely feasible in a frequency region where, by definition, parasitic effects are almost controlling.

In setting up (7) and (8) it has been assumed that the amplifier will, if necessary, be built with zero margins against singing. Any surplus which the equations indicate over the actual feedback required can, of course, be used to provide a cutoff characteristic having definite phase and gain margins. For example, if we begin with a lower feedback in the useful band the derivative of the attenuation between this region and the crossover can be proportionately reduced, with a corresponding decrease in phase shift. We can also carry the flat portion of the characteristic below the zero gain axis, thus providing a gain margin when the phase characteristic crosses 180°. In reproportioning the characteristic to suit these conditions, use may be made of the approximate formula

$$A_m - A = (A_m + 17.4)y + \frac{n-2}{n}x + \frac{2}{n}xy, \tag{9}$$

where A_m is the maximum obtainable feedback (in db), A is the actual feedback, and x and y are the gain and phase margins in the notation of Fig. 9. Once the available margin has been divided between the x and

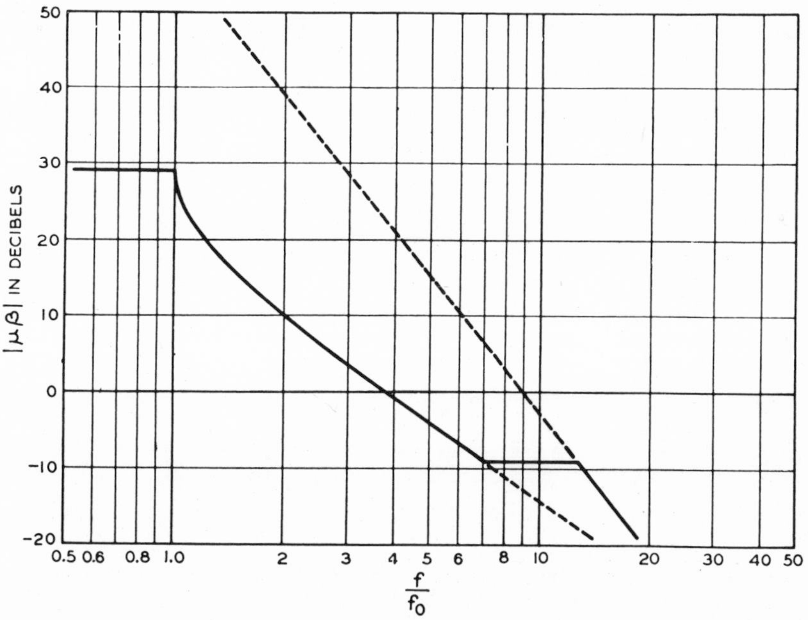

Fig. 16—Modified cutoff permitting 30° phase margin and 9 db gain margin.

y components by means of this formula the cutoff characteristic is, of course, readily drawn in. An example is furnished by Figs. 16 and 17, where it is assumed that $A_m = 43$ db, $A = 29$ db, $x = 9$ db, $n = 3$ and $y = 1/6$. The Nyquist diagram for the structure is shown by Fig. 18. It evidently coincides almost exactly with the diagram postulated originally in Fig. 9.

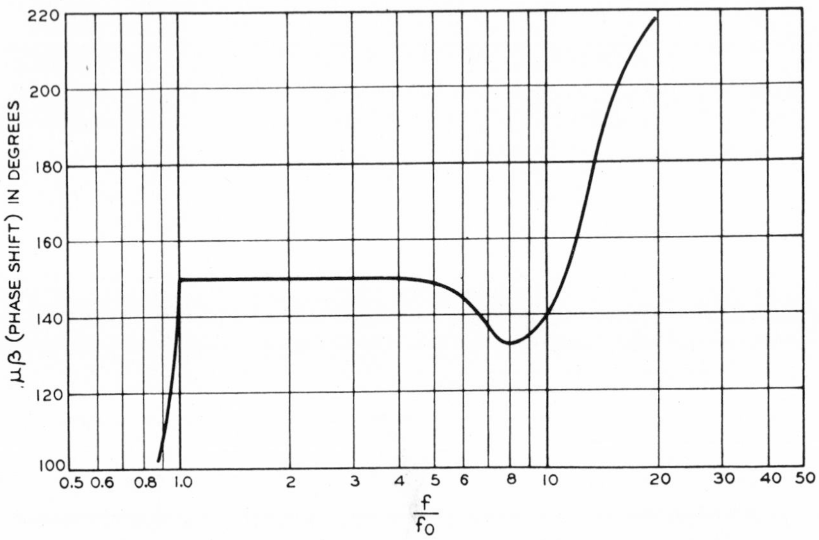

Fig. 17—Phase characteristic corresponding to gain characteristic of Fig. 16.

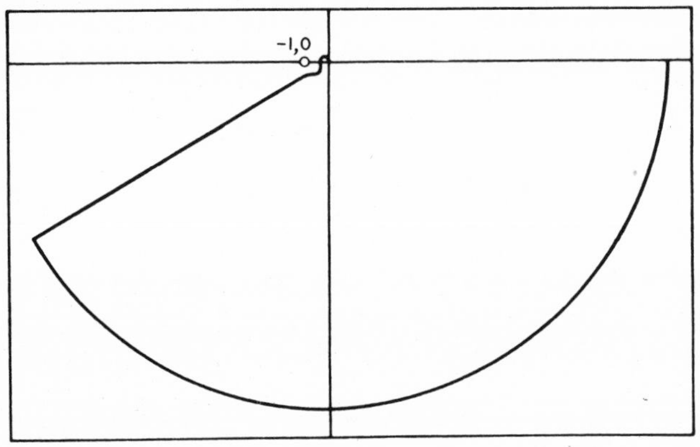

Fig. 18—Nyquist diagram corresponding to gain and phase characteristics of Figs. 16 and 17. As in Fig. 8 the diagram is rotated to place the critical point at $-1, 0$ rather than $+1, 0$.

With the characteristic of Fig. 16 at hand, we can return once more to the calculation of the total design range corresponding to any given feedback. From the useful band to the intersection of the cutoff characteristic with the zero gain axis the calculation is the same as that made previously in connection with Fig. 10. From the zero gain intercept to the junction with the asymptote, where we can say that design control is finally relaxed, there is, however, an additional interval of nearly two octaves. Although Fig. 16 is fairly typical, the exact breadth of the additional interval will depend somewhat on circumstances. It is increased by an increase in the asymptotic slope and reduced by decreasing the gain margin.

Relative Importance of Tubes and Circuit in Limiting Feedback [8]

The discussion just finished leads to the general conclusion that the feedback which can be obtained in any given amplifier depends ultimately upon the high-frequency asymptote of the feedback loop. It is a matter of some importance, then, to determine what fixes the asymptote and how it can be improved. Evidently, the asymptote is finally restricted by the gains of the tubes alone. We can scarcely improve upon the result secured by connecting the output plate directly to the input grid. Within this limit, however, the actual asymptotic characteristic will depend upon the configuration and type of feedback employed, since a given distribution of parasitic elements may evidently affect one arrangement more than another. The salient circuit problem is therefore that of choosing a general configuration for the feedback circuit which will allow the maximum efficiency of transmission at high frequencies.

The relative importance of tube limitations and circuit limitations is most easily studied if we replace (7) by

$$A_m = 40 \log_{10} \frac{4f_t}{nf_0} - \frac{2A_t}{n}, \tag{10}$$

where f_t is the frequency at which the tubes themselves working into their own parasitic capacitances have zero gain [9] and A_t is the asymptotic loss of the complete feedback loop in db at $f = f_t$. The first term

[8] The material of this section was largely inspired by comments due to Messrs. G. H. Stevenson and J. M. West.

[9] I.e., $f_t = \frac{G_m}{2\pi C}$, where G_m and C are respectively the transconductance and capacitance of a representative tube. The ratio $\frac{G_m}{C}$ is the so-called "figure of merit" of the tube. The analysis assumes that the interstage network is a simple shunt impedance, so that the parasitic capacitance does correctly represent its asymptotic behavior. More complicated four-terminal interstage networks, such as transformer coupling circuits and the like, are generally inadmissible in a feedback amplifier because of the high asymptotic losses and consequent high-phase shifts which they introduce.

of (10) shows how the feedback depends upon the intrinsic band width of the available tubes. In low-power tubes especially designed for the purpose f_t may be 50 mc or more, but if f_0 is small the first term will be substantial even if tubes with much lower values of f_t are selected. The second term gives the loss in feedback which can be ascribed to the rest of the circuit. It is evidently not possible to provide input and output circuits and a β-path without making some contribution to the asymptotic loss, so that A_t cannot be zero. In an amplifier designed with particular attention to this question, however, it is frequently possible to assign A_t a comparatively low value, of the order of 20 to 30 db or less. Without such special attention, on the other hand, A_t is likely to be very much larger, with a consequent diminution in available feedback.

In addition to f_t and A_t, (10) includes the quantity n, which represents the final asymptotic slope in multiples of 6 db per octave. Since the tubes make no contribution to the asymptotic loss at $f = f_t$ we can vary n without affecting A_t by changing the number of tubes in the circuit. This makes it possible to compute the optimum number of tubes which should be used in any given situation in order to provide the maximum possible feedback. If A_t is small the first term of (10) will be the dominant one and it is evidently desirable to have a small number of stages. The limit may be taken as $n = 2$ since with only one stage the feedback is restricted by the available forward gain, which is not taken into account in this analysis. On the other hand since the second term varies more rapidly than the first with n, the optimum number of stages will increase as A_t is increased. It is given generally by

$$n = \frac{A_t}{8.68} \tag{11}$$

or in other words the optimum n is equal to the asymptotic loss at the tube crossover in nepers.

This relation is of particular interest for high-power circuits, such as radio transmitters, where circuit limitations are usually severe but the cost of additional tubes, at least in low-power stages, is relatively unimportant. As an extreme example, we may consider the problem of providing envelope feedback around a transmitter. With the relatively sharp tuning ordinarily used in the high-frequency circuits of a transmitter the asymptotic characteristics of the feedback path will be comparatively unfavorable. For illustrative purposes we may assume that $f_a = 40$ kc. and $n = 6$. In accordance with (7) this would provide a maximum available feedback over a 10 kc. voice band of 17 db. It will also be assumed that the additional tubes for the low-power portions of the circuit have an f_t of 10 mc.[10] The corresponding A_t is 33 nepers [11] so that equation (11) would say that the feedback would be increased by the addition of as many as 27 tubes to the circuit. Naturally in such an extreme case this result can be looked upon only as a qualitative indication of the direction in which to proceed. If we add only 4 tubes, however, the available feedback becomes 46 db while if we add 10 tubes it reaches 60 db. It is to be observed that only a small part of the available gain of the added tubes is used in directly increasing the feedback. The remainder is consumed in compensating for the unfortunate phase shifts introduced by the rest of the circuit.

Amplifiers of Other Types

The amplifier considered thus far is of a rather special type. It has a useful band extending from zero up to some prescribed frequency f_0, constant feedback in the useful band, and it is absolutely stable. Departures from absolute stability are rather unusual in practical amplifiers and will not be considered here. It is apparent from the phase area relation that a conditionally stable amplifier may be expected to have a greater feedback for a cut-off interval of given breadth than a structure which is unconditionally stable, but a detailed discussion of the problem is beyond the scope of this paper.

Departures from the other assumptions are easily treated. For example, if a varying feedback in the useful band is desired, as it may be in occasional amplifiers, an appropriate cut-off characteristic can be constructed by returning to the general formula (4), performing the integrations graphically, if necessary. If the phase requirement in the cut-off region is left unchanged only the first integral need be modified. The most important question, for ordinary purposes, is that of determining how high the varying feedback can be, in comparison with a corresponding constant feedback characteristic, for any given asymptote. This can be answered by observing the form to which the first integral in (4) reduces when f_c is made very large. It is easily seen that the asymptotic conditions will remain the same provided the

[10] In tubes operating at a high-power level f_t may, of course, be quite low. It is evident, however, that only the tubes added to the circuit are significant in interpreting (11). The additional tubes may be inserted directly in the feedback path if they are made substantially linear in the voice range by subsidiary feedback of their own. This will not affect the essential result of the present analysis.

[11] It is, of course, not to be expected that the actual asymptotic slope will be constant from 40 kc. to 10 mc. Since only the region extending a few octaves above 40 kc. is of interest in the final design, however, the apparent A_t can be obtained by extrapolating the slope in this region.

feedback in the useful band satisfies a relation of the form

$$\int_0^{\pi/2} A\,d\phi = \text{constant}, \tag{12}$$

where $\phi = \sin^{-1} f/f_0$. Thus the area under the varying characteristic, when plotted against ϕ, should be the same as that under a corresponding constant characteristic having the same phase and gain margins and the same final asymptote. This is exemplified by Fig. 19, the

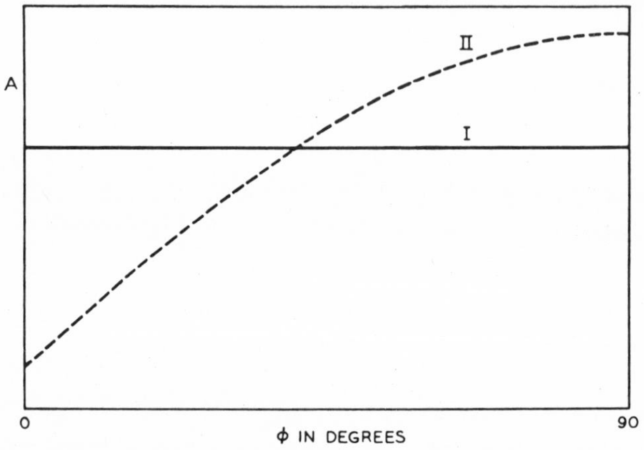

Fig. 19—Diagram to illustrate the computation of available feedback when the required feedback in the useful band is not the same at all frequencies.

varying characteristic being chosen for illustrative purposes as a straight line on an arithmetic frequency scale.

The most important question has to do with the assumption that the useful transmission band extends down to zero frequency. In most amplifiers, of course, this is not true. It is consequently necessary to provide a cut-off characteristic on the lower as well as the upper side of the band. The requisite characteristics are easily obtained from the ones which have been described by means of frequency transformations of a type familiar in filter theory. Thus if the cut-off characteristics studied thus far are regarded as being of the "low-pass" type the characteristics obtained from them by replacing f/f_0 by its reciprocal may be regarded as being of the "high-pass" type. If the band width of the amplifier is relatively broad it is usually simplest to treat the upper and lower cut-offs as independent characteristics of low-pass and high-pass types. In this event, the asymptote for the lower cut-off is furnished by such elements as blocking condensers and choke coils in the plate supply leads. The low-frequency asymptote is usually not so serious a problem as the high-frequency asymptote since it can be placed as far from the band as we need by using large enough elements in the power supply circuits. The superposition of a low-frequency cutoff on the idealized loop gain and phase characteristics of a "low-pass" circuit is illustrated by the broken lines in Fig. 20.

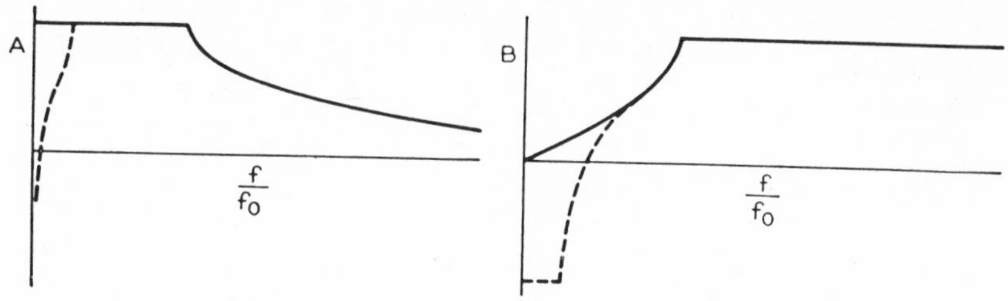

Fig. 20—Modification of loop characteristics to provide a lower cutoff in a broad-band amplifier.

If the band width is relatively narrow it is more efficient to use the transformation in filter theory which relates a low-pass to a symmetrical band-pass structure. The transformation is obtained by replacing f/f_0 in the low-pass case by $(f^2 - f_1f_2/f(f_2 - f_1))$, where f_1 and f_2 are the edges of the prescribed band. It substitutes resonant and anti-resonant circuits tuned to the center of the band for the coils and condensers in the low-pass circuit. In particular each parasitic inductance is tuned by the addition of a series condenser and each parasitic capacity is tuned by a shunt coil. The parameters of the transformation must, of course, be so chosen that the parasitic elements have the correct values for use in the new branches.

This leads to a simple but important result. If the inductance of a series resonant circuit is fixed, the interval represented by $f_b - f_a$ in Fig. 21, between the frequencies at which the absolute value of the reactance reaches some prescribed limit X_0, is always constant and equal to the frequency at which the untuned inductance would exhibit the reactance X_0, whatever the tuning frequency may be. The same relation holds for the capacity in an anti-resonant circuit. Thus the frequency range over which the branches containing parasitic elements exhibit comparable impedance variations is the same in the band-pass structure and in the prototype low-pass structure. But since the transformation does not affect the relative impedance levels of the various branches in the circuit, this result can be extended to the complete $\mu\beta$ characteristic. We can therefore conclude that *the feedback which is obtainable in an amplifier of given general configuration and with given parasitic elements and given margins depends only upon the breadth*

28

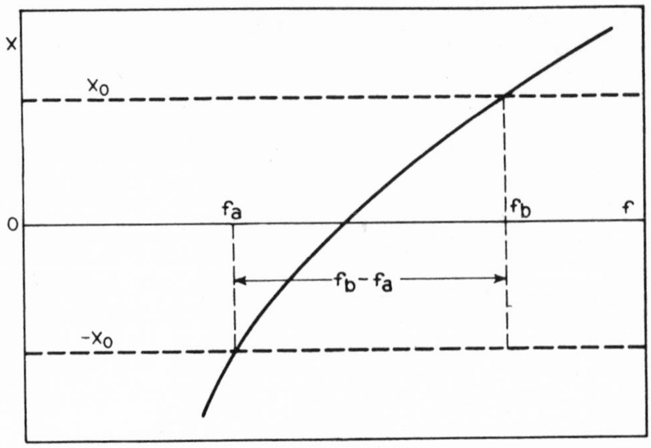

Fig. 21—Frequency interval between prescribed reactances of opposite sign in a resonant circuit with fixed inductance.

of the band in cycles and is independent of the location of the band in the frequency spectrum.

These relations are exemplified by the plots of a low-pass cutoff characteristic and the equivalent band-pass characteristic shown by Fig. 22. The equality of corresponding frequency intervals is indicated by the horizontal lines A, B and C.

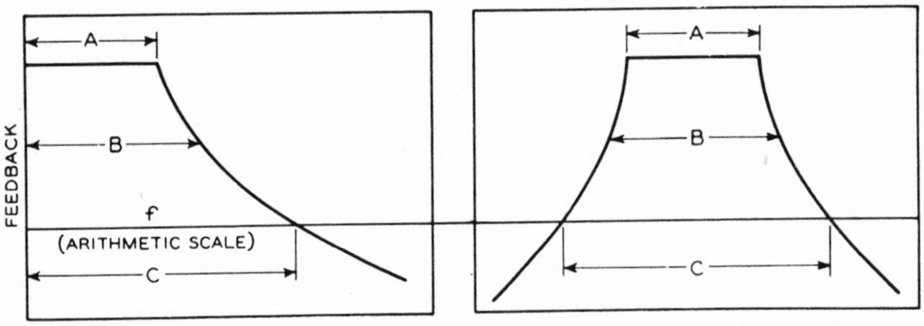

Fig. 22—Diagram to illustrate the conservation of band width in the low-pass to band-pass transformation with fixed parasitic elements. A, B and C represent typical corresponding intervals of equal breadth.

Example

An example showing the application of the method in an actual design is furnished by Fig. 23. The structure is a feedback amplifier intended to serve as a repeater in a 72-ohm coaxial line.[12] The useful frequency range extends from 60 to 2,000 kc. Coupling to the line is

[12] The author's personal contact with this amplifier was limited to the evolution of a paper schematic for the high frequency design. The other aspects of the problem are the work of Messrs. K. C. Black, J. M. West and C. H. Elmendorf.

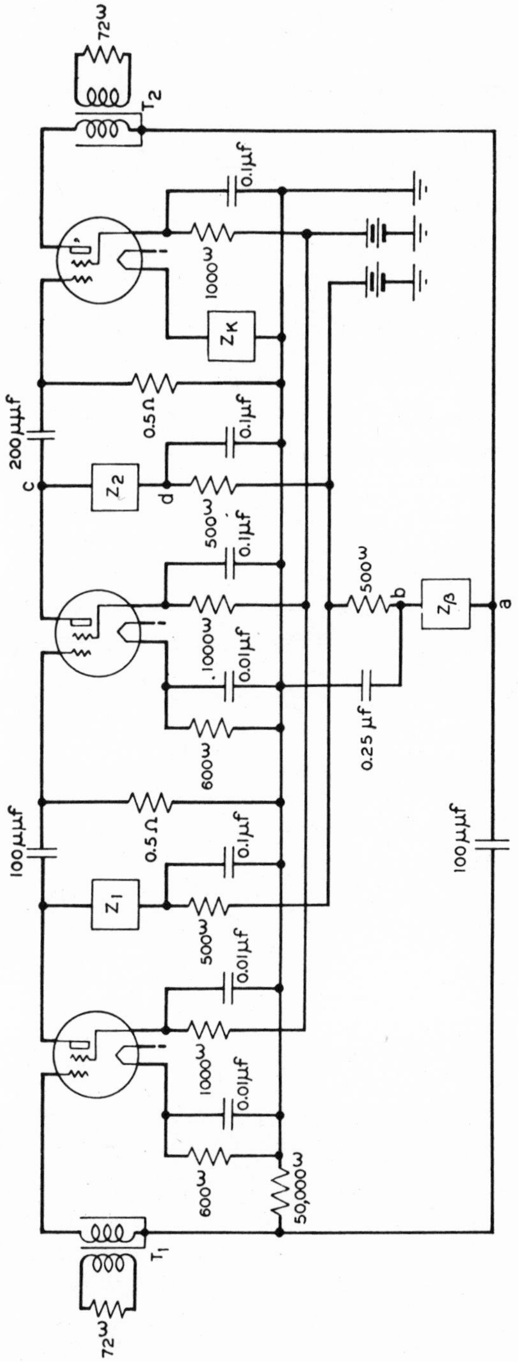

Fig. 23—General schematic of a 2 mc. feedback amplifier.

obtained through the shielded input and output transformers T_1 and T_2. The three stages in the μ circuit are represented in Fig. 23 as single tubes. Physically, however, each stage employes two tubes in parallel, the transconductances of the individual tubes being about 2000 micromhos. The principal feedback is obtained through the impedance Z_β. There is in addition a subsidiary local feedback on the power stage through the impedance Z_K. This is advantageous in producing a further reduction in the effects of modulation in this stage but it does not materially affect the feedback available around the principal loop.

The elements shown explicitly include resistance-capacitance filters in the power supply leads to the plates and screens, cathode resistances and by-pass condensers to provide grid bias potentials, and blocking condenser-grid-leak combinations for the several tubes. In addition to serving these functions, the various resistance-capacitance combinations are also used to provide the cutoff characteristic below the useful band. The low-frequency asymptote is established by the grid leak resistances and the associated coupling condensers and the approach of the feedback characteristic to the asymptote is controlled mainly by the cathode impedances and the resistance-capacitance filters in the power supply leads to the plates. The principal parts of the circuit entering into the $\mu\beta$ characteristic at high frequencies are the interstage impedances Z_1 and Z_2, the feedback impedance Z_β,[13] the cathode impedance Z_K, and the two transformers. The four network designs are shown in detail in Figs. 24, 25, 26, and 27.

The joint transconductance, 4000 micromhos, of two tubes in parallel operating into an average interstage capacity of 14 mmf, as indicated by Figs. 24 and 25, gives an f_t of about 50 mc. The parasitic capacities (chiefly transformer high side and ground capacities) in the other parts of the feedback loop provide a net loss, A_t, of about 18 db at this frequency. Since the asymptotic slope is 18 db per octave the intercept of the complete asymptote with the zero gain axis occurs about one octave lower, at slightly less than 25 mc. This is a relatively high intercept and may be attributed in part to the high gain of the vacuum tubes. The care used in minimizing parasitic capacities in the construction of the amplifier and the general circuit arrangement, including in particular the use of single shunt impedances for the coupling and feedback networks, are also helpful.

[13] The relative complexity of this network is explained by the fact that it actually serves as a regulator to compensate for the effects of changes in the line temperature. (See H. W. Bode, "Variable Equalizers," *Bell System Technical Journal*, April, 1938.) The present discussion assumes that the controlling element is at its normal setting. For this setting the network is approximately equal to a resistance in series with a small inductance. The fact that the amplifier must remain stable over a regulation range may serve to explain why the design includes such large stability margins.

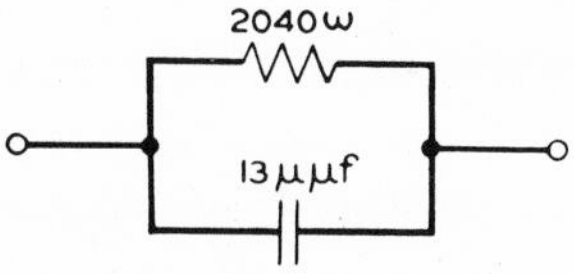

Fig. 24—First interstage for the amplifier of Fig. 23.

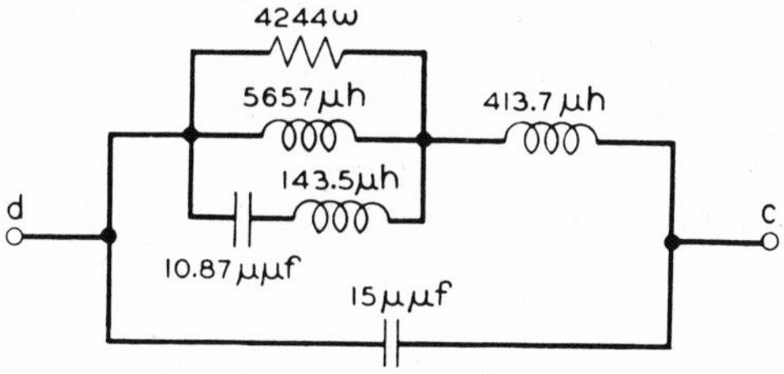

Fig. 25—Second interstage for the amplifier of Fig. 23.

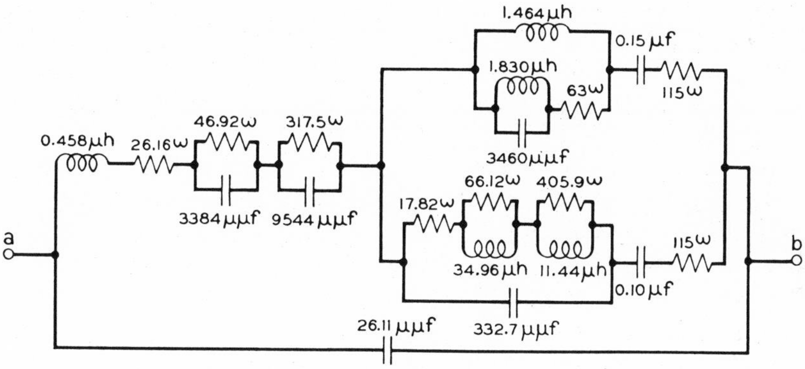

Fig. 26—β circuit impedance for the amplifier of Fig. 23.

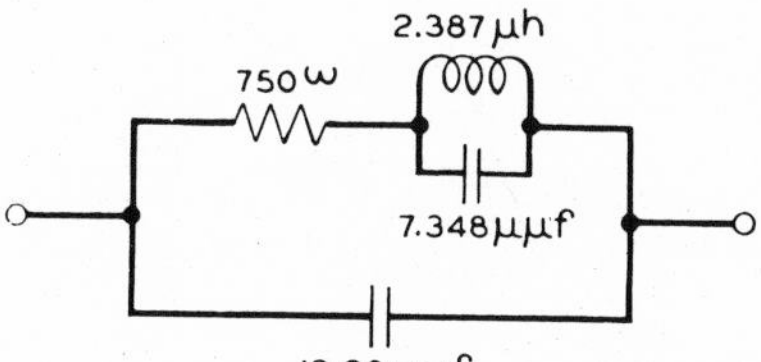

Fig. 27—Cathode impedance for the amplifier of Fig. 23.

In accordance with (7) the maximum available feedback A_m is 48 db. For design purposes, however, x and y in (9) were chosen as 15 db and 1/5 respectively. This reduces the actual feedback A to about 28 db. The theoretical cutoff characteristic corresponding to

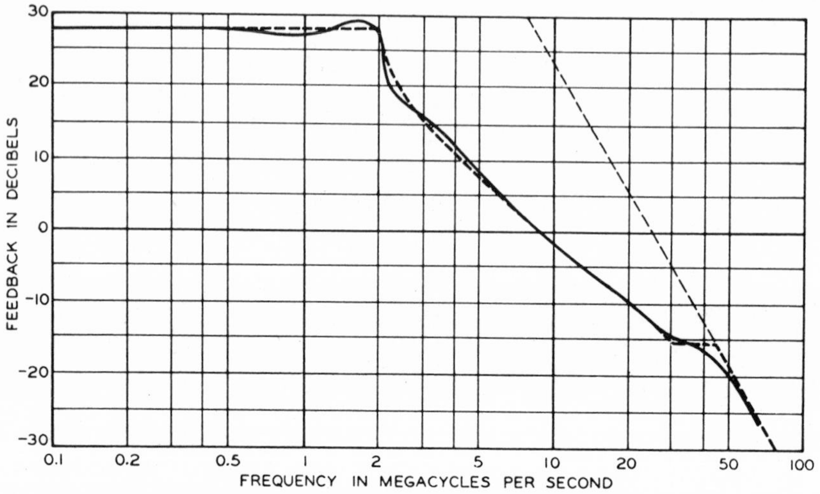

Fig. 28—Loop gain characteristic for the amplifier of Fig. 23.

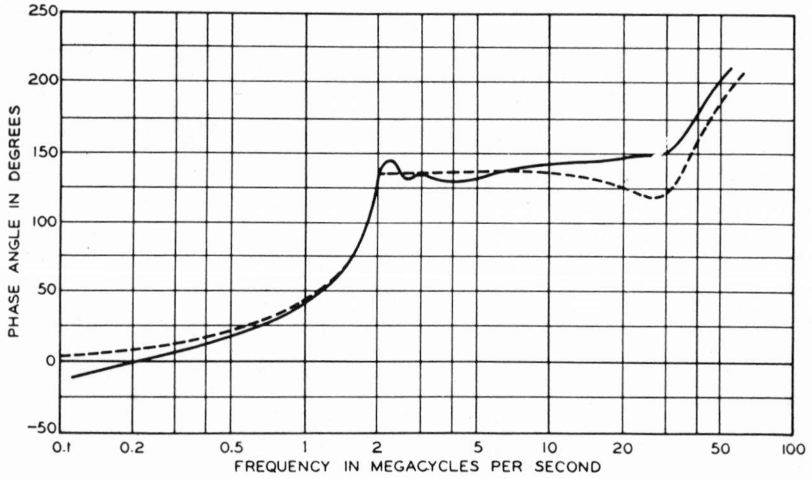

Fig. 29—Loop phase characteristic for the amplifier of Fig. 23.

these parameters is shown by the broken lines in Figs. 28 and 29, and the actual design characteristic by the solid lines. Since this is a structure in which the required forward gain is considerably less than the maximum available gain, the general course of the cutoff characteristic is controlled, in accordance with the procedure outlined previously, by the elements in the μ circuit. The sharp slope just beyond the edge of the useful band is obtained from a transformer anti-resonance. The relatively flat portion of the characteristic near its intersection with the asymptote is due partly to an anti-resonance of the β circuit with its distributed capacitance and partly to an increase in the gain of the third tube because of the filter-like action of the elements of Z_K in cutting out the local feedback on the tube in this region.

The large margins in the design made it possible to secure a substantial increase in feedback without instability. For example, with a loss margin as great as 15 db the feedback can be increased by adjusting the screen and plate voltages to increase the tube gains. A higher feedback can also be obtained by adjusting the resistance in the first interstage. As this interstage was designed, an increase in the resistance results in an increased amplifier gain and a correspondingly increased feedback which follows a new theoretical characteristic with a somewhat reduced phase margin. The adjustment, in effect, produces a change in the value of the constant k in equation (6). With this adjustment the feedback can be increased to about 40 db before the amplifier sings.

12

Reprinted with permission from *Bell System Tech. J.*, **22**(3), 269–277 (1943)

Effect of Feedback on Impedance

By R. B. BLACKMAN

THE impedance of a network is defined as the complex ratio of the alternating potential difference maintained across its terminals by an external source of electromotive force, to the resulting current flowing into these terminals. If the network contains active elements such as vacuum tubes, the resulting current (or potential difference if the input current is taken as the independent variable) may be due in part to the excitation of the active elements. The definition of impedance does not discriminate between the part of the current (or potential difference) due directly to the external source of electromotive force and the part due to the excitation of the active elements by the external source. Hence the impedance will in general depend upon the degree of activity of the active elements.

These observations were made early in the development of feedback amplifiers by H. S. Black[1] who made two important uses of the effect of feedback on impedance. In the first place it afforded a method of measuring feedback which has some advantages over the method which involves opening the feedback loop, providing proper terminations for it and measuring the transmission around it. In the second place the effect of feedback on impedance was used to control the impedances presented by a feedback amplifier to the external circuits connected to it.

Relations between impedance and feedback were derived by Black and others for a number of specific feedback amplifier configurations. In some cases these relations turned out to be very simple. For the most part, however, these relations were so complicated that they defied reduction to a common form.[2] The difficulty seems to have been due, in part at least, to the attempt to formulate the relationship, in each case, in terms of the normal feedback of the amplifier. In some cases the difficulty seems to have been due partly also to the valid, but, as it turns out, irrelevant observation that the feedback is affected by the impedance of the measuring circuit as

[1] H. S. Black, "Stabilized Feedback Amplifiers", *B.S.T.J.*, January, 1934.

[2] Shortly after the general relationship between feedback and impedance was derived, it was independently established by H. W. Bode and J. M. West by examination of a variety of feedback amplifier designs. The generality of the relationship was also independently proved for amplifiers with a single feedback path by J. G. Kreer and by C. H. Elmendorf.

well as by the removal of any impedance elements or circuits which are normally connected to the amplifier.

These difficulties are avoided by the method of derivation adopted in this paper. Illustrative examples are then given of some of the uses to which the general relationship between feedback and impedance may be put.

Derivation

The derivation of the general relationship between feedback and impedance will be made here with reference to the diagram shown in Fig. 1.

One of the vacuum tubes in the network, namely that one to which the feedback is to be referred, is shown explicitly at the top of the box in the diagram. The grid lead to this tube is broken at terminals 2, 2′. In practice, the break in the grid lead would leave the grid still coupled to some

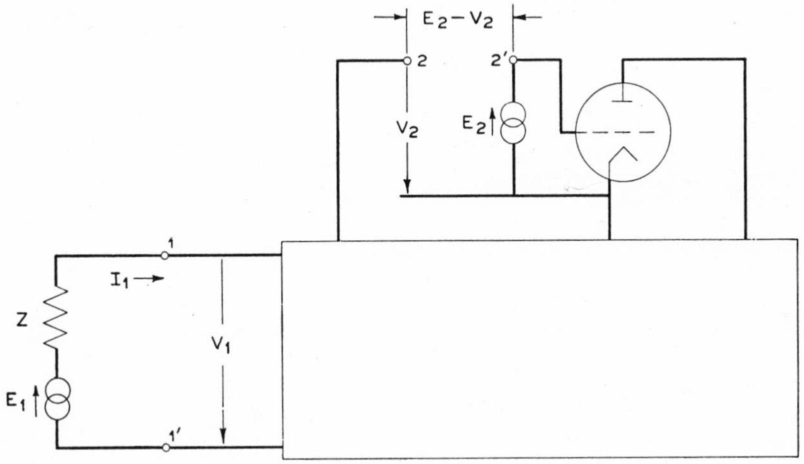

Fig. 1—Relation between feedback and impedance.

degree to the other electrodes of the tube through parasitic interelectrode admittance. For analytical purposes, however, it may be assumed that the parasitic admittances between the grid and the other electrodes of the tubes are connected not directly to the grid within the tube but to some point farther out along the grid lead. Under this assumption the break in the grid lead not only removes the feedback to the tube completely, but also leaves the parasitic admittances connected in the network in such a way that their contribution to the feedback is implicitly taken into account. Furthermore, the impedance looking into the grid of the tube is now infinite so that if a voltage is applied to the grid no current will be drawn from the source of the voltage.

At the left-hand side of the box in the diagram, terminals 1, 1′ are brought out. These are the terminals to which the impedance is to be referred. In the normal condition of the network these terminals may be connected through an external impedance branch. This is the case, for example, when terminals 1, 1′ are the input terminals of a feedback amplifier whose input impedance is under investigation. However, this external impedance may also be zero or infinite according as terminals 1, 1′ are "mesh-terminals" obtained by breaking open a mesh of the network, or "junction-terminals" obtained by bringing out two junctions of the network.

It is assumed that the network, including all of the vacuum tubes, is a linear system in which, therefore, the Superposition Principle holds. Hence, if an e.m.f. E_1 is applied in series with terminals 1, 1′ and a second e.m.f. E_2 is applied between the grid and the cathode of the tube, the potential difference V_1 developed across the input terminals 1, 1′ and the potential difference V_2 developed between the terminal 2 and the cathode of the tube will be linearly related to E_1 and E_2. If the source of E_1 has internal impedance the coefficients in these relations will depend upon this impedance. However, if the input current I_1 is used as an independent variable in place of the e.m.f. E_1 the coefficients will not depend upon the impedance of the source of the current I_1. It is also convenient to consider the potential difference $E_2 - V_2$ developed across the terminals 2, 2′ as one of the dependent variables in place of V_2. Therefore,

$$\left.\begin{aligned} V_1 &= AI_1 + BE_2 \\ E_2 - V_2 &= CI_1 + DE_2 \end{aligned}\right\} \tag{1}$$

where the coefficients are independent of Z.

From these equations we obtain

$$\left(\frac{V_1}{I_1}\right)_{E_2=V_2} = \frac{AD - BC}{D}$$

$$\left(\frac{V_1}{I_1}\right)_{E_2=0} = A$$

$$\left(\frac{E_2 - V_2}{E_2}\right)_{V_1=0} = \frac{AD - BC}{A}$$

$$\left(\frac{E_2 - V_2}{E_2}\right)_{I_1=0} = D$$

Hence

$$\frac{\left(\dfrac{V_1}{I_1}\right)_{E_2=V_2}}{\left(\dfrac{V_1}{I_1}\right)_{E_2=0}} = \frac{1 - \left(\dfrac{V_2}{E_2}\right)_{V_1=0}}{1 - \left(\dfrac{V_2}{E_2}\right)_{I_1=0}} \tag{2}$$

This equation expresses the relationship between feedback and impedance. To make this more apparent the physical significance of each of the factors in this equation will be examined and suitable symbols will be substituted for them.

In equations (1) E_2 and I_1 were regarded as independent variables. However, the ratio $\left(\frac{V_1}{I_1}\right)_{E_2=V_2}$ implies that E_2 is adjusted to be equal to V_2. This means that E_2 is dependent upon I_1. The reason for the imposition of this dependence is that with E_2 equal to V_2 the terminals 2, 2′ may be connected together and the source of E_2 may be removed without affecting, in particular, the potential difference V_1 across terminals 1, 1′ and the current I_1 into these terminals.

Obviously, therefore, the ratio $\left(\frac{V_1}{I_1}\right)_{E_2=V_2}$ is the impedance which will be seen at the terminals 1, 1′ when terminals 2, 2′ are connected together and the only source of e.m.f. acting on the network is the external circuit connected to the terminals 1, 1′. This ratio will be symbolized by Z_A.

The ratio $\left(\frac{V_1}{I_1}\right)_{E_2=0}$ implies that no voltage is applied between the grid and the cathode of the tube. However, it is immaterial whether or not a voltage is applied to the grid of the tube if the amplification of the tube is nullified. Obviously, therefore, this ratio is the impedance which will be seen at the terminals 1, 1′ when terminals 2, 2′ are connected together and the amplification of the tube is nullified. This ratio will be symbolized by Z_P.

Finally, the ratios $\left(\frac{V_2}{E_2}\right)_{V_1=0}$ and $\left(\frac{V_2}{E_2}\right)_{I_1=0}$ are readily recognized from the definition of feedback to be the feedback to the vacuum tube with the terminals 1, 1′ connected together in the first case, and left open in the second. These ratios will be symbolized by F_{Sh} and F_{Op} respectively.

Hence, equation (2) may be written in the more significant form

$$\frac{Z_A}{Z_P} = \frac{1 - F_{Sh}}{1 - F_{Op}} \tag{3}$$

Determination of Feedback

One of the uses to which the relationship (3) may be put is in the determination of feedback by impedance measurement. However, since this relationship involves two feedbacks, only one of which may be identified with the feedback to be determined, one of these feedbacks must be known.

In the most common types of feedback amplifiers it is possible to choose terminals 1, 1′ so that either F_{Sh} or F_{Op} is zero. If $F_{Op} = 0$ and $F_{Sh} = F_N$ where F_N is the normal feedback, then

$$F_N = 1 - \frac{Z_A}{Z_P} \tag{4}$$

On the other hand, if $F_{Sh} = 0$ and $F_{Op} = F_N$ then

$$F_N = 1 - \frac{Z_P}{Z_A} \tag{5}$$

Fig. 2 shows a feedback amplifier in which the μ-circuit and the β-network are connected in series at one end and in parallel at the other end. At terminals 1, 1′ in this figure the conditions for formula (4) are obviously fulfilled. Hence, if the impedance measurements are made at these ter-

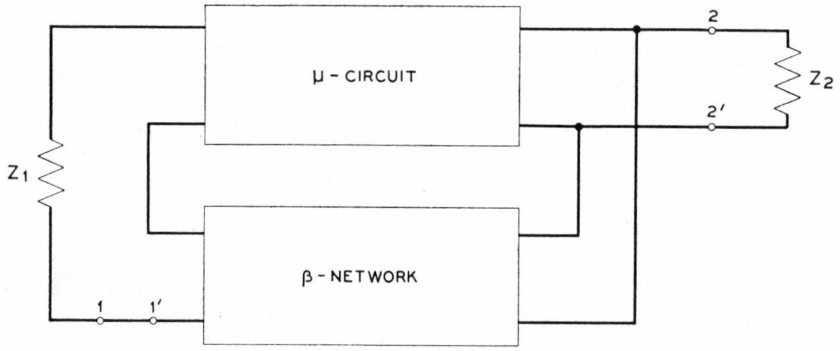

FIG. 2—Feedback amplifier with series feedback at one end and shunt feedback at the other end.

minals, the feedback is given by formula (4). On the other hand, at terminals 2, 2′ in Fig. 2 the conditions for formula (5) are obviously fulfilled. Hence, if the impedance measurements are made at these terminals, the feedback is given by formula (5).

If the grid-plate parasitic admittance of a tube in a feedback amplifier is not negligible it is not possible to open any physical mesh in the amplifier so that $F_{Op} = 0$ for that tube. In such a case, therefore, (4) is not applicable. However, if the impedance measurements are made between the grid and the cathode of that tube the conditions for formula (5) are obviously fulfilled, and the feedback is given by formula (5). Hence, of the two particular forms (4) and (5) of the general relationship (3), only (5) enjoys complete generality in the determination of feedback by impedance measurements.

Feedback during Impedance Measurements

While the feedback computed from impedance measurements by formula (4) or (5) is the normal feedback, the feedback during the impedance measurements may be quite different, due to the impedance of the impedance measuring circuit. Referring to Fig. 1 we see that the feedback during measurement is by definition

$$F_Z = \left(\frac{V_2}{E_2}\right)_{V_1=-ZI_1}$$

where Z is the impedance of the impedance measuring circuit. By equations (1) this is easily reduced to

$$F_Z = \frac{Z_P F_{Sh} + ZF_{Op}}{Z_P + Z} \tag{6}$$

Under the conditions to which formula (4) applies

$$F_Z = \frac{F_N}{1 + \dfrac{Z}{Z_P}} \tag{7}$$

It is clear therefore that even if F_N satisfies Nyquist's Stability Criterion, Z may be of such a character that F_Z violates that criterion. In that case it will be impossible to make the impedance measurements.

Contrariwise, if F_N violates Nyquist's Stability Criterion, it is possible to choose Z so that F_Z satisfies that criterion and make it possible to measure the impedance. Substituting (4) into (7) we find that a sufficient but not necessary condition in order that $|F_Z| < 1$ is that

$$|Z| > |Z_A| + 2|Z_P|$$

Under the conditions to which formula (5) applies

$$F_Z = \frac{F_N}{1 + \dfrac{Z_P}{Z}} \tag{8}$$

Similar observations may be made with respect to (8) as were made with respect to (7). Substituting (5) into (8) we find that a sufficient but not necessary condition in order that $|F_Z| < 1$ is that

$$\frac{1}{|Z|} > \frac{1}{|Z_A|} + \frac{2}{|Z_P|}$$

Feedback Control of Impedance

The application of the relationship (3) to the feedback control of impedance may be illustrated by a few concrete examples.

Let us assume that we are interested in the impedance faced by the line impedance Z_1 in Fig. 2. If the terminals 1, 1′ in Fig. 3 are left open the feedback is obviously zero. Let the feedback when the terminals are shorted together be denoted by F_{Sh}. If the impedances of the μ-circuit and the β-network are denoted by Z_μ and Z_β, respectively, then

$$Z_A = Z_P(1 - F_{Sh}) \tag{9}$$

where

$$Z_P = Z_\mu + Z_\beta$$

This shows the now well-known fact that series feedback may be used to magnify impedance.

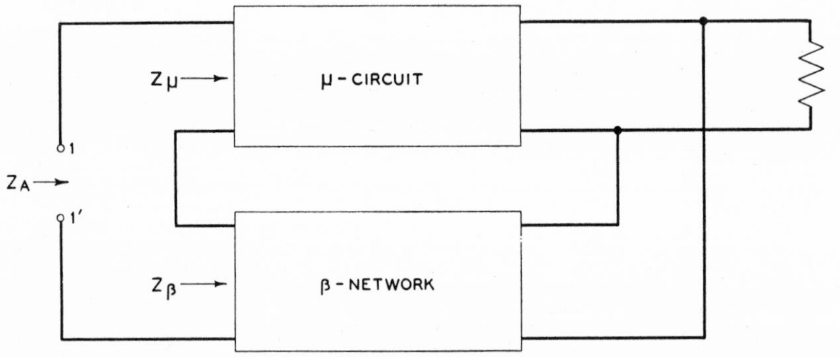

FIG. 3—Impedance faced by the line at the series feedback end of a feedback amplifier.

However, it should be noted that the feedback F_{Sh} involved in (9) is not now equal to the normal feedback F_N as it was when the terminals 1, 1′ were taken as in Fig. 2. The relation between F_N and F_{Sh} may be obtained from (6) by identifying F_N with F_Z, and Z_1 with Z. Hence

$$F_N = \frac{F_{Sh}}{1 + \dfrac{Z_1}{Z_P}} \tag{10}$$

From (9) and (10) it follows that even with a very modest amount of normal feedback the magnification of the impedance may be very large. For example, if $Z_P = 1000$ ohms, $Z_1 = 1$ megohm and $F_{Sh} = -1000$, then Z_A is better than 1000 times as large as Z_P although F_N is not quite unity in magnitude.

Similarly, the impedance faced by the line impedance Z_2 in Fig. 2, as shown in Fig. 4, is

$$Z_A = \frac{Z_P}{1 - F_{Op}} \tag{11}$$

where

$$Z_P = \frac{Z_\mu Z_\beta}{Z_\mu + Z_\beta}$$

This shows the now well-known fact that shunt feedback may be used to reduce impedance.

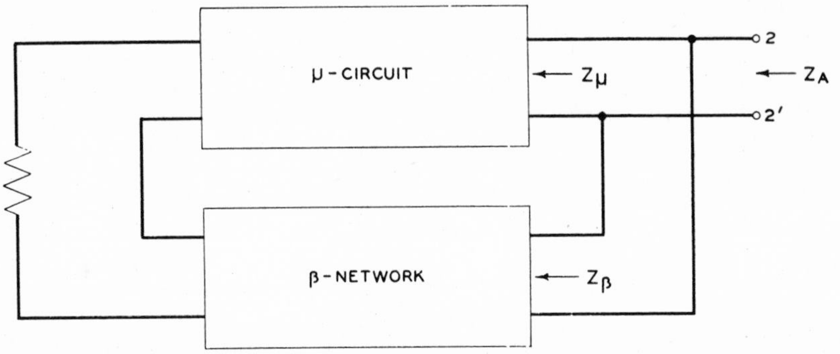

FIG. 4—Impedance faced by the line at the shunt feedback end of a feedback amplifier.

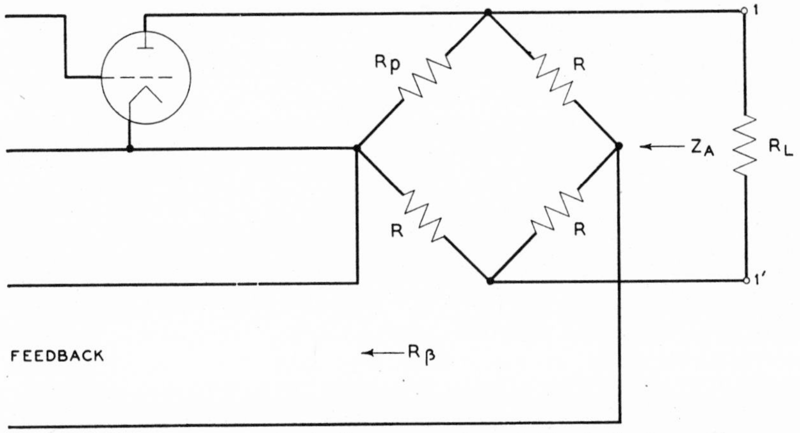

FIG. 5—Impedance faced by the line connected to a bridge feedback amplifier.

The relation between the normal feedback F_N and the feedback F_{Op} involved in (11) is, by (6)

$$F_N = \frac{F_{Op}}{1 + \dfrac{Z_P}{Z_2}} \tag{12}$$

From (11) and (12) it follows that even with a very modest amount of normal feedback the reduction in impedance may be very large. For example, if $Z_P = 100{,}000$ ohms, $Z_2 = 100$ ohms and $F_{Op} = -1000$, then Z_A is less than 100 ohms although F_N is not quite unity in magnitude.

The two examples given above illustrate the use of feedback to magnify or to reduce the impedance of a network. This impedance, however, will be correspondingly sensitive to changes in the characteristics of the vacuum tubes. A third example of the use of the relationship (3) will show that feedback may also be used to make the impedance of a network less sensitive to changes in the characteristics of the vacuum tubes.

In the case of the bridge-type feedback network shown in Fig. 5 we have, with respect to the terminals 1, 1′

$$Z_P = R(1 + Q)$$

$$F_{Sh} = A\,\frac{1 + \dfrac{2R + 3R_\beta}{R + R_\beta}Q}{1 + Q}$$

$$F_{Op} = A\left(1 + \frac{2R + 3R_\beta}{R + R_\beta}Q\right) = (1 + Q)F_{Sh}$$

where A is the feedback designed for the condition $R_p = R$, and

$$Q = \frac{(R_p - R)(R + R_\beta)}{(R + R_p)(2R + R_\beta) + 2RR_\beta}$$

Then, by (3)

$$Z_A = R\left(1 + \frac{Q}{1 - F_{Op}}\right)$$

Hence, if the feedback F_{Op} is very large the effect of bridge unbalance on the impedance presented to the line will be very small. If, for example, the design feedback is 40 db the output impedance cannot change more than 1 per cent however severely the bridge might be unbalanced by R_p being larger than R.

The feedback when the line impedance R_L is connected may be obtained by identifying R_L with Z in formula (6). It is

$$F = A\,\frac{1 + \dfrac{2R + 3R_\beta}{R + R_\beta}Q}{1 + \dfrac{R}{R + R_L}Q}$$

whence

$$\frac{\partial \log F}{\partial \log R_L} = \frac{RR_L}{R + R_L}\,\frac{Q}{R_L + Z_P}$$

The effect of bridge unbalance is to make the feedback sensitive to changes in the line impedance R_L.

13

Reprinted from *Proc. IRE,* **44**(7), 920–926 (1956)

Feedback Theory—Further Properties of Signal Flow Graphs*

SAMUEL J. MASON†

Summary—**A way to enhance**
Writing gain at a glance.
Dr. Tustin extended
Proof appended.
Examples illustrative
Pray not frustrative.

Background

THERE ARE many different paths to the solution of a set of linear equations. The formal method involves inversion of a matrix. We know, however, that there are many different ways of inverting a matrix: determinantal expansion in minors, systematic reduction of a matrix to diagonal form, partitioning into submatrices, and so forth, each of which has its particular interpretation as a sequence of algebraic manipulations within the original equations. A determinantal expansion of special interest is

$$D = \sum a_{1i}a_{2j}a_{3k} \cdots a_{nx} \tag{1}$$

where a_{mp} is the element in the mth row and pth column of a determinant having n rows, and the summation is taken over all possible permutations of the column subscripts. (The sign of each term is positive or negative in accord with an even or odd number of successive adjacent column-subscript interchanges required to produce a given permutation.) Since the solution of a set of linear equations involves ratios of determinantal quantities, (1) suggests the general idea that a linear system analysis problem should be interpretable as a search for all possible combinations of something or other, and that the solution should take the form of a sum of products of the somethings, whatever they are, divided by another such sum of products. Hence, instead of undertaking a sequence of operations, we can find the solution by looking for certain combinations of things. The method will be especially useful if these combinations have a simple interpretation in the context of the problem.

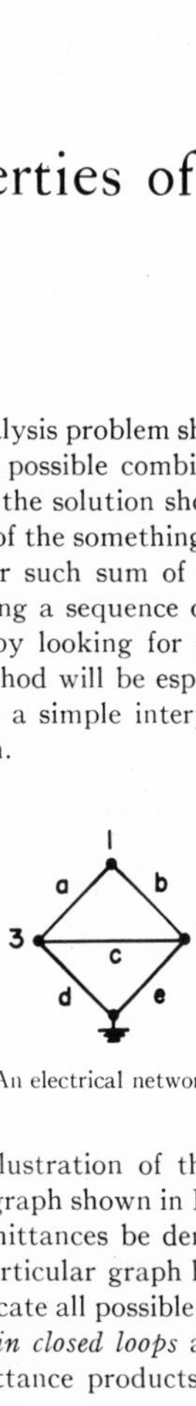

Fig. 1—An electrical network graph.

As a concrete illustration of the idea, consider the electrical network graph shown in Fig. 1. For simplicity, let the branch admittances be denoted by letters *a, b, c, d,* and *e.* This particular graph has *three* independent node-pairs. First locate all possible sets of *three* branches which *do not contain closed loops* and write the sum of their branch admittance products as the denominator of (2).

$$Z_{12} = \frac{ab + ac + bc + bd}{abd + abe + acd + ace + ade + bcd + bce + bde} \tag{2}$$

Now locate all sets of *two* branches which *do not form*

* Original manuscript received by the IRE, August 16, 1955; revised manuscript received, February 27, 1956. This work was supported in part by the Army (Signal Corps), the Air Force (Office of Scientific Research, Air Research and Development Command), and the Navy (Office of Naval Research).

† Dept. of Elect. Engrg. and Res. Lab. of Electronics, M.I.T., Cambridge, Mass.

closed loops and which also *do not contain any paths from node 1 to ground or from node 2 to ground.* Write the sum of their branch admittance products as the numerator of expression (2). The result is the transfer impedance between nodes 1 and 2, that is, the voltage at node 2 when a unit current is injected at node 1. Any impedance of a branch network can be found by this process.[1]

So much for electrical network graphs. Our main concern in this paper is with signal flow graphs,[2] whose branches are directed. Tustin[3] has suggested that the feedback factor for a flow graph of the form shown in Fig. 2 can be formulated by combining the feedback loop

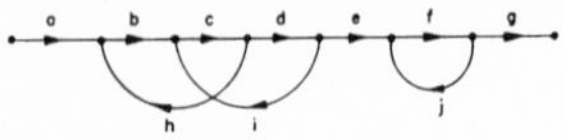

Fig. 2—The flow graph of an automatic control system.

gains in a certain way. The three loop gains are

$$T_1 = bch \tag{3a}$$

$$T_2 = cdi \tag{3b}$$

$$T_3 = fj \tag{3c}$$

and the forward path gain is

$$G_0 = abcdefg. \tag{4}$$

The gain of the complete system is found to be

$$G = \frac{G_0}{[1 - (T_1 + T_2)](1 - T_3)} \tag{5}$$

and expansion of the denominator yields

$$G = \frac{G_0}{1 - (T_1 + T_2 + T_3) + (T_1T_3 + T_2T_3)}\cdot \tag{6}$$

Tustin recognized the denominator as unity plus the sum of all possible products of loop gains taken one at a time $(T_1+T_2+T_3)$, two at a time, $(T_1T_3+T_2T_3)$, three at a time, and so forth, excluding products of loops that touch or partially coincide. The products T_1T_2 and $T_1T_2T_3$ are properly and accordingly missing in this particular example. The algebraic sign alternates, as shown, with each succeeding group of products.

Tustin did not take up the general case but gave a hint that a graph having several different forward paths could be handled by considering each path separately and superposing the effects. Detailed examination of the general problem shows, in fact, that the form of (6) must be modified to include possible feedback factors in the numerator. Otherwise (6) applies only to those graphs in which each loop touches all forward paths.

[1] Y. H. Ku, "Resume of Maxewll's and Kirchoff's rules for network analysis," *J. Frank. Inst.*, vol. 253, pp. 211–224; March, 1952.

[2] S. J. Mason, "Feedback theory—some properties of signal flow graphs," PROC. IRE, vol. 41, pp. 1144–1156; September, 1953.

[3] A. Tustin, "Direct Current Machines for Control Systems," The Macmillan Company, New York, pp. 45–46, 1952.

The purposes of this paper are: to extend the method to a general form applicable to any flow graph; to present a proof of the general result; and to illustrate the usefulness of such flow graph techniques by application to practical linear analysis problems. The proof will be given last. It is tempting to add, at this point, that a better understanding of linear analysis is a great aid in problems of nonlinear analysis and linear or nonlinear design.

A Brief Statement of Some Elementary Properties of Linear Signal Flow Graphs

A signal flow graph is a network of directed branches which connect at nodes. Branch *jk* originates at node *j* and terminates upon node *k*, the direction from *j* to *k* being indicated by an arrowhead on the branch. Each branch *jk* has associated with it a quantity called the branch gain g_{jk} and each node *j* has an associated quantity called the node signal x_j. The various node signals are related by the associated equations

$$\sum_j x_j g_{jk} = x_k, \qquad k = 1, 2, 3, \cdots \tag{7}$$

The graph shown in Fig. 3, for example, has equations

$$ax_1 + dx_3 = x_2 \tag{8a}$$

$$bx_2 + fx_4 = x_3 \tag{8b}$$

$$ex_2 + cx_3 = x_4 \tag{8c}$$

$$gx_3 + hx_4 = x_5. \tag{8d}$$

We shall need certain definitions. A *source* is a node having only outgoing branches (node 1 in Fig. 3). A *sink* is a node having only incoming branches. A *path* is any continuous succession of branches traversed in the indicated branch directions. A *forward path* is a path from source to sink along which no node is encountered more than once (*abch, aeh, aefg, abg,* in Fig. 3).

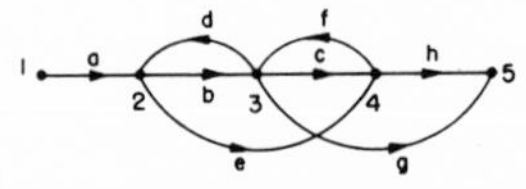

Fig. 3—A simple signal flow graph.

A *feedback loop* is a path that forms a closed cycle along which each node is encountered once per cycle (*bd, cf, def,* but not *bcfd*, in Fig. 3). A *path gain* is the product of the branch gains along that path. The *loop gain* of a feedback loop is the product of the gains of the branches forming that loop. The *gain of a flow graph* is the signal appearing at the sink per unit signal applied at the source. Only one source and one sink need be considered, since sources are superposable and sinks are independent of each other.

Additional terminology will be introduced as needed.

General Formulation of Flow Graph Gain

To begin with an example, consider the graph shown in Fig. 4. This graph exhibits three feedback loops, whose gains are

$$T_1 = h \tag{9a}$$

$$T_2 = fg \tag{9b}$$

$$T_3 = de \tag{9c}$$

and two forward paths, whose gains are

$$G_1 = ab \tag{10a}$$

$$G_2 = ceb. \tag{10b}$$

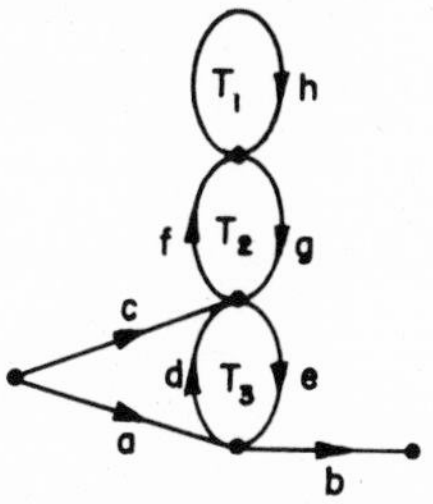

Fig. 4—A flow graph with three feedback loops.

To find the graph gain, first locate all possible *sets of nontouching loops* and write the algebraic sum of their gain products as the denominator of (11).

$$G = \frac{G_1(1 - T_1 - T_2) + G_2(1 - T_1)}{1 - T_1 - T_2 - T_3 + T_1T_3} \tag{11}$$

Each term of the denominator is the gain product of a set of nontouching loops. The algebraic sign of the term is plus (or minus) for an even (or odd) number of loops in the set. The graph of Fig. 4 has no sets of three or more nontouching loops. Taking the loops two at a time we find only one permissible set, T_1T_3. When the loops are taken one at a time the question of touching does not arise, so that each loop in the graph is itself an admissible "set." For completeness of form we may also consider the set of loops taken "none at a time" and, by analogy with the zeroth power of a number, interpret its gain product as the unity term in the denominator of (11). The numerator contains the sum of all forward path gains, each multiplied by a factor. The factor for a given forward path is made up of all possible sets of loops which *do not touch each other* and which also do *not touch that forward path*. The first forward path ($G_1 = ab$) touches the third loop, and T_3 is therefore absent from the first numerator factor. Since the second path ($G_2 = ceb$) touches both T_2 and T_3, only T_1 enters the second factor.

The general expression for graph gain may be written as

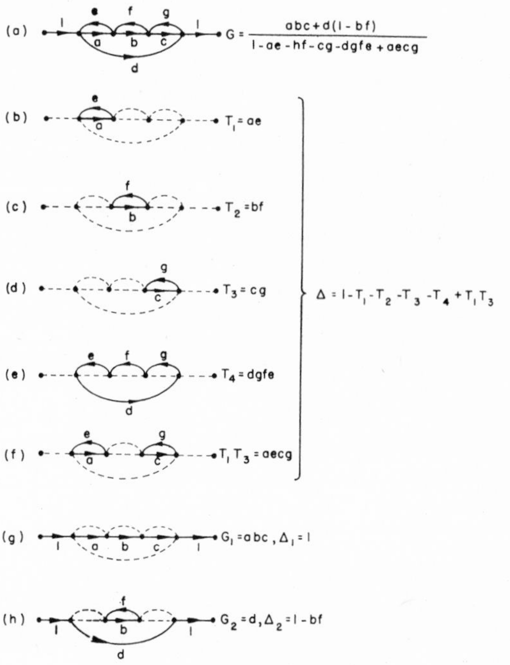

Fig. 5—Identification of paths and loop sets.

$$G = \frac{\sum_k G_k\Delta_k}{\Delta} \tag{12a}$$

wherein

$$G_k = \text{gain of the } k\text{th forward path} \tag{12b}$$

$$\Delta = 1 - \sum_m P_{m1} + \sum_m P_{m2} - \sum_m P_{m3} + \cdots \tag{12c}$$

P_{mr} = gain product of the mth possible combination of r nontouching loops (12d)

Δ_k = the value of Δ for that part of the graph not touching the kth forward path. (12e)

The form of (12a) suggests that we call Δ the *determinant* of the graph, and call Δ_k the *cofactor* of forward path k.

A subsidiary result of some interest has to do with graphs whose feedback loops form nontouching subgraphs. To find the *loop subgraph* of any flow graph, simply remove all of those branches *not* lying in feedback loops, leaving all of the feedback loops, and nothing but the feedback loops. In general, the loop subgraph may have a number of nontouching parts. The useful fact is that the *determinant of a complete flow graph is equal to the product of the determinants of each of the nontouching parts in its loop subgraph.*

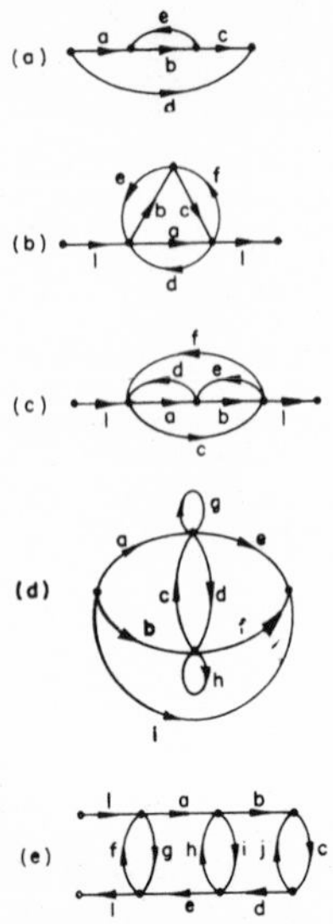

(a) $\dfrac{d(1-be)+abc}{1-be}$

(b) $\dfrac{a+bc}{1-ad-be-cf-bcd-afe}$

(c) $\dfrac{c+ab}{1-ad-be-abf-ced-cf}$

(d) $\dfrac{i(1-g-h-cd+gh)+ae(1-h)+bf(1-g)+adf+bce}{1-g-h-cd+gh}$

(e) $\dfrac{g(1-hi-jc-hbcd+hijc)+aie(1-jc)+abcde}{1-fg-hi-jc-faie-hbcd-fabcde+fghi+fgjc+hijc+faiejc+fghbcd-fghijc}$

Fig. 6—Sample flow graphs and their gain expressions.

Illustrative Examples of Gain Evaluation by Inspection of Paths and Loop Sets

Eq. (12) is formidable at first sight but the idea is simple. More examples will help illustrate its simplicity. Fig. 5 (on the previous page) shows the first of these displayed in minute detail: (a) the graph to be solved; (b)–(f) the loop sets contributing to Δ; (g) and (h) the forward paths and their cofactors. Fig. 6 gives several additional examples on which you may wish to practice evaluating gains by inspection.

Illustrative Applications of Flow Graph Techniques to Practical Analysis Problems

The study of flow graphs is a fascinating topological game and therefore, from one viewpoint, worthwhile in its own right. Since the associated equations of a linear flow graph are in cause-and-effect form, each variable expressed explicitly in terms of others, and since physical problems are often very conveniently formulated in just this form, the study of flow graphs assumes practical significance.

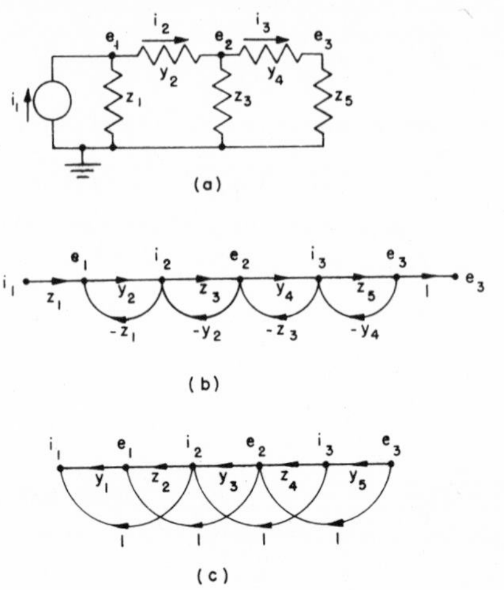

Fig. 7—The transfer impedance of a ladder.

Consider the ladder network shown in Fig. 7(a). The problem is to find the transfer impedance e_3/i_1. One possible formulation of the problem is indicated by the flow graph Fig. 7(b). The associated equations state that $e_1 = z_1(i_1 - i_2)$, $i_2 = y_2(e_1 - e_2)$, and so forth. By inspection of the graph,

$$\frac{e_3}{i_1} = \frac{z_1y_2z_3y_4z_5}{1 + z_1y_2 + y_2z_3 + z_3y_4 + y_4z_5 + z_1y_2z_3y_4 + z_1y_2y_4z_5 + y_2z_3y_4z_5} \tag{13a}$$

or, with numerator and denominator multiplied by $y_1y_3y_5 = 1/z_1z_3z_5$,

$$\frac{e_3}{i_1} = \frac{y_2y_4}{y_1y_3y_5 + y_2y_3y_5 + y_1y_2y_5 + y_1y_4y_5 + y_1y_3y_4 + y_2y_4y_5 + y_2y_3y_4 + y_1y_2y_4}. \tag{13b}$$

This result can be checked by the branch-combination method mentioned at the beginning of this paper.

A different formulation of the problem is indicated by the graph of Fig. 7(c), whose equations state that $i_3 = y_5e_3$, $e_2 = e_3 + z_4i_3$, $i_2 = i_3 + y_3e_2$, and so forth. In the physical problem i_1 is the primary cause and e_3 the final effect. We may, however, *choose* a value of e_3 and then calculate the value of i_1 *required* to produce that e_3. The resulting equations will, from the analysis viewpoint, treat e_3 as a primary cause (source) and i_1 as the final effect (sink) *produced* by the chain of calculations. *This does not in any way alter the physical role of* i_1. The new graph (c) may appear simpler to solve than that of (b). Since graph (c) contains no feedback loops, the determinant and path cofactors are all equal to unity. There are many forward paths, however, and careful inspection is required to identify the sum of their gains as

$$\frac{i_1}{e_3} = y_1z_2y_3z_4y_5 + y_1z_2y_3 + y_1z_2y_5 + y_1z_4y_5 + y_3z_4y_5 + y_1 + y_3 + y_5 \qquad (13c)$$

which proves to be, as it should, the reciprocal of (13b). Incidentally, graph (c) is obtainable directly from graph (b), as are all other possible cause-and-effect formulations involving the same variables, by the process of path inversion discussed in a previous paper.[2] This example points out the two very important facts: 1) the primary *physical* source does not *necessarily* appear as a source node in the graph, and 2) of two possible flow graph formulations of a problem, the one having fewer feedback loops is not necessarily simpler to solve by inspection, since it may also have a much more complicated set of forward paths.

Fig. 8(a) offers another sample analysis problem, determination of the voltage gain of a feedback amplifier. One possible chain of cause-and-effect reasoning, which leads from the circuit model, Fig. 8(b), to the flow graph formulation, Fig. 8(c), is the following. First notice that e_{g1} is the difference of e_1 and e_k. Next express i_1 as an effect due to causes e_{g1} and e_k, using superposition to write the gains of the two branches entering node i_1. The dependency of e_{g2} upon i_1 follows directly. Now, e_2 would be easy to evaluate in terms of either e_{g2} or i_f if the other were zero, so superpose the two effects as indicated by the two branches entering node e_2. At this point in the formulation e_k and i_f are as yet not explicitly specified in terms of other variables. It is a simple matter, however, to visualize e_k as the superposition of the voltages in R_k caused by i_1 and i_f, and to identify i_f as the superposition of two currents in R_f caused by e_k and e_2. This completes the graph.

The path from e_k to e_{g1} to i_1 may be lumped in parallel with the branch entering i_1 from e_k. This simplification, convenient but not necessary, yields the graph shown in Fig. 9. We could, of course, have expressed i_1 in terms of e_1 and e_k at the outset and arrived at Fig. 9 directly. All simplifications of a graph are themselves

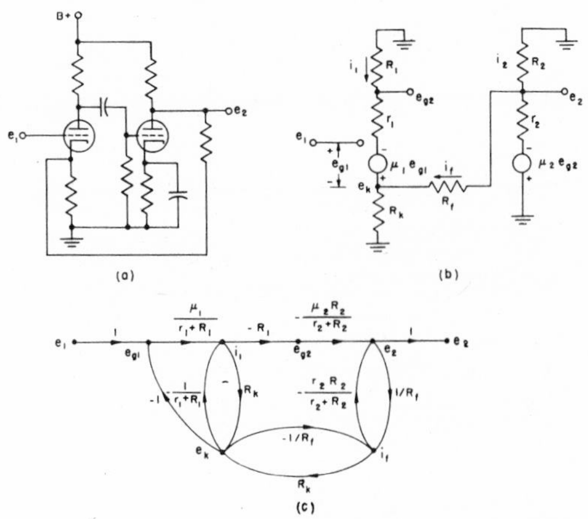

Fig. 8—Voltage gain of a feedback amplifier. (a) A feedback amplifier; (b) The midband linear incremental circuit model; (c) A possible flow graph.

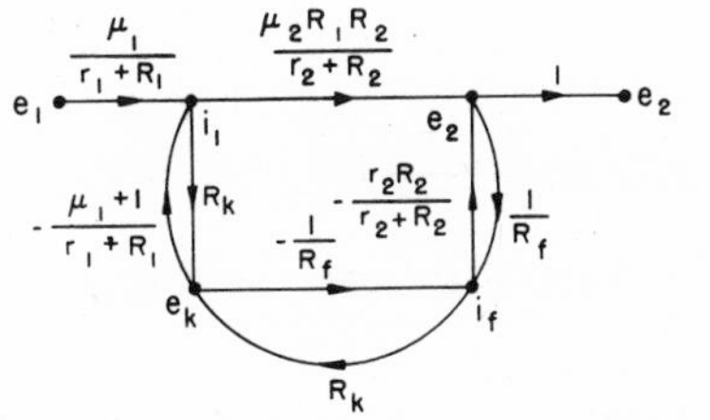

Fig. 9—Elimination of superfluous nodes e_{g1} and e_{g2}.

possible formulations. The better our perception of the workings of a circuit, the fewer variables will we need to introduce at the outset and the simpler will be the resulting flow graph structure.

In discussing the feedback amplifier of Fig. 8(a) it is common practice to neglect the loading effect of the feedback resistor R_f in parallel with R_k, the loading effect of R_f in parallel with R_2, and the leakage transmission from e_k to e_2 through R_f. Such an approximation is equivalent to the removal of the branches from e_k to i_f and i_f to e_2 in Fig. 9. It is sometimes dangerous to make early approximations, however, and in this case no appreciable labor is saved, since we can write the exact answer by inspection of Fig. 9:

$$\frac{e_2}{e_1} = \frac{\dfrac{\mu_1\mu_2R_1R_2}{(r_1+R_1)(r_2+R_2)}\left[1+\dfrac{R_k}{R_f}\right] + \dfrac{\mu_1R_kr_2R_2}{(r_1+R_1)(R_f)(r_2+R_2)}}{1+\dfrac{(\mu_1+1)R_k}{r_1+R_1}+\dfrac{R_k}{R_f}+\dfrac{r_2R_2}{R_f(r_2+R_2)}+\dfrac{(\mu_1+1)\mu_2R_1R_kR_2}{R_f(r_1+R_1)(r_2+R_2)}+\dfrac{(\mu_1+1)R_kr_2R_2}{R_f(r_1+R_1)(r_2+R_2)}} \qquad (14)$$

The two forward paths are $e_1i_1e_2$ and $e_1i_1e_ki_je_2$, the first having a cofactor due to loop e_ki_j. The principal feedback loop is $i_1e_2i_je_k$ and its gain is the fifth term of the denominator. Physical interpretations of the various paths and loops could be discussed but our main purpose, to illustrate the formulation of a graph and the evaluation of its gain by inspection, has been covered.

As a final example, consider the calculation of microwave reflection from a triple-layered dielectric sandwich. Fig. 10(a) shows the incident wave A, the reflection B, and the four interfaces between adjacent regions of different material. The first and fourth interfaces, of course, are those between air and solid. Let r_1 be the reflection coefficient of the first interface, relating the incident and reflected components of tangential electric field. It follows from the continuity of tangential E that the interface transmission coefficient is $1+r_1$, and from symmetry that the reflection coefficient from the opposite side of the interface is the negative of r_1. A suitable flow graph is sketched in Fig. 10(b). Node signals along the upper row are right-going waves just to the left or right of each interface, those on the lower row are left-going waves, and quantities d are exponential phase shift factors accounting for the delay in traversing each layer.

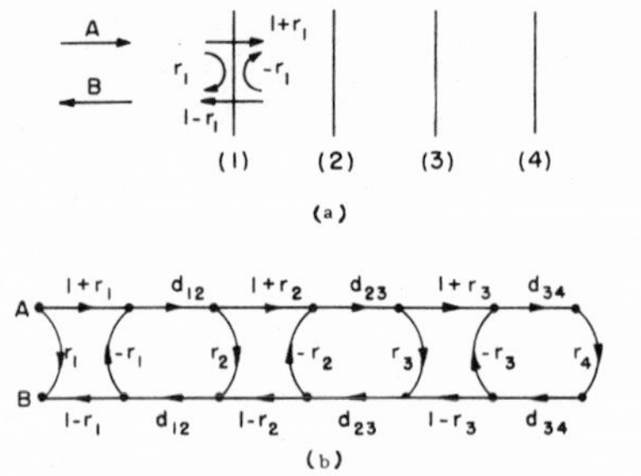

Fig. 10—A wave reflection problem. (a) Reflection of waves from a triple-layer; (b) A possible flow graph.

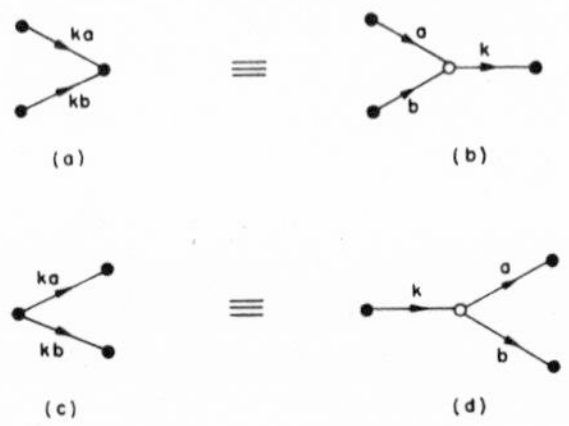

Fig. 11—Two touching paths.

Apart from the first branch r_1, the graph has the same structure as that of Fig. 6(e). Hence the reflectivity of the triple layer will be

$$\frac{B}{A} = r_1 + (1 + r_1)(1 - r_1)G \tag{15}$$

where G is in the same form as the gain of Fig. 6(e). We shall not expand it in detail. The point is that the answer can be written by inspection of the paths and loops in the graph.

Proof of the General Gain Expression

In an earlier paper[2] a quantity Δ was defined as

$$\Delta = (1 - T_1')(1 - T_2') \cdots (1 - T_n') \tag{16}$$

for a graph having n nodes, where

T_k' = loop gain of the kth node as computed with all higher-numbered nodes split.

Splitting a node divides that node into a new source and a new sink, all branches entering that node going with the new sink and all branches leaving that node going with the new source. The loop gain of a node was defined as the gain from the new source to the new sink, when that node is split. *It was also shown that Δ, as computed according to (16), is independent of the order in which the nodes are numbered, and that consequently Δ is a linear function of each branch gain in the graph. It follows that Δ is equal to unity plus the algebraic sum of various branch-gain products.*

We shall first show that each term of Δ, other than the unity term, is a product of the gains of nontouching feedback loops. This can be done by contradiction. Consider two branches which either enter the same node or leave the same node, as shown in Fig. 11(a) and (c). Imagine these branches imbedded in a larger graph, the remainder of which is not shown. Call the branch gains ka and kb. Now consider the equivalent replacements (b) and (d). The new node may be numbered zero, whence $T_0' = 0$, the other T' quantities in (16) are unchanged, and Δ is therefore unaltered. If both branches ka and kb appear in a term of the Δ of graph (a) then the square of k must appear in a term of the Δ of graph (b). This is impossible since Δ must be a linear function of branch gain k. Hence no term of Δ can contain the gains of two touching paths.

Now suppose that of the several nontouching paths appearing in a given term of Δ, some are feedback loops and some are open paths. Destruction of all other branches eliminates some terms from Δ but leaves the given term unchanged. It follows from (16) and the definitions of T_k', however, that the Δ for the subgraph containing only these nontouching paths is just

$$\Delta = (1 - T_1)(1 - T_2) \cdots (1 - T_m) \tag{17}$$

where T_k is the gain of the kth feedback loop in the subgraph. Hence the open path gains cannot appear in the given term and it follows that each term of Δ is the product of gains of nontouching feedback loops. Moreover, it is clear from the structure of Δ that a term in any subgraph Δ must also appear as a term in the Δ of the complete graph, and conversely, every term of Δ is a term of some subgraph Δ. Hence, to identify all possible terms in Δ we must look for all possible subgraphs comprising sets of nontouching loops. Eq. (17) also shows that the algebraic sign of a term is plus or minus in accord with an even or odd number of loops in that term. This verifies the form of Δ as given in (12c) and (12d).

We shall next establish the general expression for graph gain (12a). The following notation will prove convenient. Consider the graph shown schematically in Fig. 12, with node $n+1$ given special attention. Let

Δ' = the Δ for the complete graph of $n+1$ nodes.

Δ = the value of Δ with node $n+1$ split or removed.

T = the loop gain of node $n+1$.

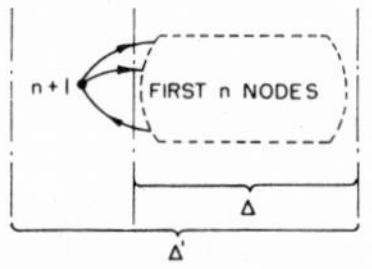

Fig. 12—A flow graph with one node placed strongly in evidence.

There will in general be several different feedback loops containing node $n+1$. Let

T_k = gain of the kth feedback loop containing node $n+1$,

Δ_k = the value of Δ for that part of the graph not touching loop T_k.

With the above notation, we have from (16) that

$$1 - T = \frac{\Delta'}{\Delta}. \tag{18}$$

Remembering that any Δ is the algebraic sum of gain products of nontouching loops, we find it possible to write

$$\Delta' = \Delta - \sum_k T_k \Delta_k. \tag{19}$$

Eq. (19) represents the count of all possible nontouching loop sets in Δ'. The addition of node $n+1$ creates new loops T_k but the only new loop sets of Δ' not already in Δ are the nontouching sets $T_k\Delta_k$. The negative sign in (19) suffices to preserve the sign rule, since the product of T_k and a positive term of Δ_k will contain an odd number of loops.

Substitution of (19) into (18) yields the general result:

$$T = \frac{\sum_k T_k \Delta_k}{\Delta}. \tag{20}$$

With node $n+1$ permanently split, T is just the source-to-sink gain of the graph and T_k is the kth forward path. This verifies (12a).

Acknowledgment

The writer is indebted to Prof. K. Wildes and also to J. Cruz, L. Boffi, G. Amster, and other students in his 1955 summer-term class in subject 6.633, Electronic Circuit Theory, M.I.T.; for helpful suggestions.

IV
Active Networks

Editor's Comments on Papers 14, 15, and 16

14 Llewellyn: *Some Fundamental Properties of Transmission Systems*

15 Sallen and Key: *A Practical Method of Designing RC Active Filters*

16 Kuh: *Regenerative Modes of Active Networks*

The need for a theory of active networks was felt early in the evolution of circuit theory, because of the early use of vacuum tubes in combination with passive elements. Progress has been relatively slow compared with that for passive networks, for a number of reasons. Models for electronic devices are difficult to construct with accuracy and there is often poor agreement between theory and practice. Added to this is the generality required to treat the combined active-and-passive case. One of the first contributions was that of Llewellyn in 1952, who gave the conditions for absolute stability in active two-port networks (Paper 14).

Even though a theory for active networks has evolved slowly, procedures for the realization of linear active networks developed rapidly, relatively unrestricted compared with the restrictive passive-network synthesis procedures. Given this flexibility it has been possible to concentrate on such things as design for minimum sensitivity. A survey of techniques available is provided in the book edited by Mitra (1971). One of the first methods to appear was that of Sallen and Key in 1955 (Paper 15). The practical aspects of this method were not fully recognized until operational amplifiers became inexpensive and readily available in integrated-circuit form. Thus recognition of this classic has come only recently.

One problem to receive the attention of circuit theorists is the distinction of an active network as compared with a passive network. One of the earliest papers on this subject, which provided a definition of measure of activity and, in particular, a condition for two-port activity, was that of Mason (1954). Representative of the approach to the study of active networks since that time was the contribution of Kuh in 1960 (Paper 16). In a related work, Desoer and Kuh (1960) define activity in terms of the ability of the linear, time-invariant network to support an unstable mode under passive loading. Comparable studies were made by Youla, Castriota, and Carlin in 1959 (Paper 18), and a study that is of a fundamental nature and related is that of Raisbeck in 1954 (Paper 22). Both of these papers appear elsewhere in this volume.

References

Desoer, C. A., and E. S. Kuh. "Bounds on the Natural Frequencies of Linear Active Networks," *Proc. Symp. Active Networks and Feedback Systems* (Polytechnic Institute of Brooklyn), **10**, 415–436 (1960).

Mason, S. J. "Power Gain in Feedback Amplifiers," *IRE Trans. Circuit Theory,* **CT-1**, 20–25 (1954).

Mitra, S. K. *Active Inductorless Filters,* IEEE Press, New York, 232 pp., 1971.

14

Reprinted from *Proc. IRE*, **40**, 271–283 (Mar. 1952)

Some Fundamental Properties of Transmission Systems*

F. B. LLEWELLYN†, FELLOW, IRE

Summary—**The problem of the minimum loss in relation to the singing point is investigated for generalized transmission systems that must be stable for any combination of passive terminating impedances. It is concluded that the loss may approach zero db only in those cases where the image impedances seen at the ends of the system are purely resistive. Moreover, in such cases, the method of overcoming the transmission loss, whether by conventional repeaters or by series and shunt negative impedance loading, or otherwise, is quite immaterial to the external behavior of the system as long as the image impedances are not changed. The use of impedance-correcting networks provides one means of insuring that phase of the image impedance of the over-all system approaches zero.**

General relations are derived which connect the image impedance and the image gain of an active system with its over-all performance properties.

SINCE THE TIME when amplifiers first were introduced into the telephone plant, the properties of two-way repeaters have been subjected to extensive analysis. From this it might be inferred that further study is likely to uncover very little that is not already known. Nevertheless, it frequently happens that new types and permutations of repeater and loading circuits are proposed, and current methods of analysis are found to be quite difficult.

In the face of this situation, the present paper is intended to review the underlying fundamentals, and to present them in what is hoped to be a form that will allow them to be simply and easily applied in determining the over-all performance. In a wider sense, what is attempted is to state certain basic physical properties and limitations in a way that allows one to say, "Regardless of detail, if these rules are violated, it follows that the circuit cannot perform as predicted," or, on the other hand, to say, "The ideal performance of such-and-such a system is so-and-so. If the proposed plan does not approximate this ideal, it must be possible to find a better one."

As sometimes happens, this review of the properties of transmission systems has led to several concepts which are thought to be new. Their importance becomes more pronounced in connection with the current tendency to reduce the net operating loss of telephone systems to lower values than were customary in the past.

In the case of the telephone repeater, the extent of the various combinations and permutations that are encountered in practice has made difficult the statement of generalizations in simple terms. The present attempt is based on the development of linear network theory in respect to active four-poles that has been progressing perhaps quietly but nonetheless steadily in the past years. Like most mathematical generalizations, the solution of one problem is really the solution of a class of problems, and it will be found that in their broadest form, the generalizations which are now presented are just as applicable to the case of four-wire telephone and radio systems as they are to the conventional two-way repeater.

The system to be considered may contain repeaters of the 22-type, such as is illustrated in Fig. 1, or it may contain any of the other varieties. Moreover, there is no restriction placed on whether the gain is the same in both directions or not, and sections of line or of other circuit networks may be included as part of the unit under consideration. Even more broadly, the unit considered may consist either of a single repeater section, or of an unlimited number of repeater sections in tandem comprising an entire system. Restrictions are placed on these broad limits only in dealing with specific applications.

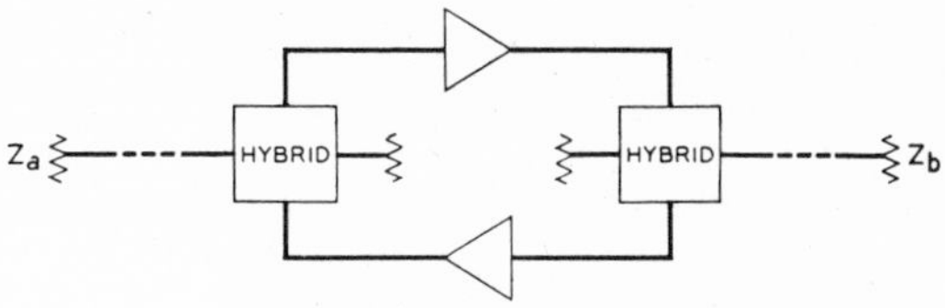

Fig. 1—Schematic of 22-type repeater.

The analysis then directs itself to the general linear four pole such as is illustrated in Fig. 2 where the rectangular box may contain as much or as little as meets the needs of the particular situation. When terminations are added, the diagram illustrates the situation.

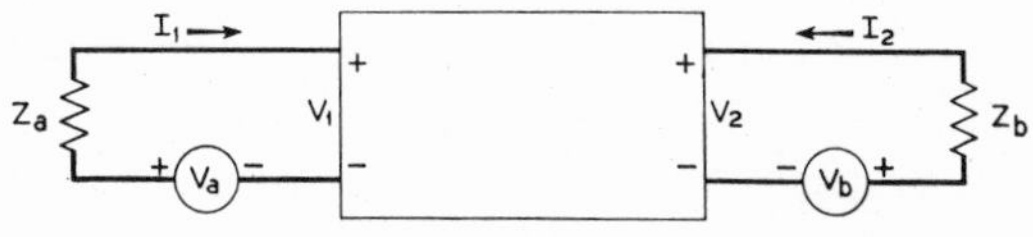

Fig. 2—Diagram of linear four-pole with terminations.

The equations describing Fig. 2 may be written:

$$\begin{aligned} Z_{11}I_1 + Z_{12}I_2 &= V_1 \\ Z_{21}I_1 + Z_{22}I_2 &= V_2 \end{aligned} \qquad (1)$$

where the Z's are characteristic of the four-pole only, and do not involve the terminations. More will be said later about their properties and how they are derived.

* Decimal classification: 621.385. Original manuscript received by the Institute, May 12, 1951; revised manuscript received, October 15, 1951.

† Bell Telephone Laboratories, 463 West Street, New York 14, N. Y.

The corresponding equations including the terminations may be written down immediately by noting that the terminations Z_a and Z_b are related to the currents and voltages by the formulas

$$V_1 = V_a - Z_a I_1$$
$$V_2 = V_b - Z_b I_2. \tag{2}$$

When combined with (1) these give

$$(Z_{11} + Z_a)I_1 + Z_{12}I_2 = V_a$$
$$Z_{21}I_1 + (Z_{22} + Z_b)I_2 = V_b, \tag{3}$$

which may be solved for the currents,

$$I_1 = \frac{V_a(Z_{22} + Z_b) - V_b Z_{12}}{\Delta} \tag{4}$$

$$I_2 = \frac{(Z_a + Z_{11})V_b - Z_{21}V_a}{\Delta} \tag{5}$$

where

$$\Delta = Z_{11}Z_{22} - Z_{12}Z_{21} + Z_a Z_{22} + Z_b Z_{11} + Z_a Z_b \tag{6}$$

is the determinant of the system of (3). It will be noted, and later use will be made of the fact, that the determinant of (1) does not depend on the terminations, and is given by

$$\Delta_0 = Z_{11}Z_{22} - Z_{12}Z_{21}, \tag{7}$$

and, consequently, that (6) may be written

$$\Delta = \Delta_0 + Z_a Z_{22} + Z_b Z_{11} + Z_a Z_b. \tag{8}$$

When the four-pole of Fig. 2 is driven from the left, V_b may be set equal to zero in (4) and (5). Under these conditions, the generator V_a sees the internal impedance Z_a in series with the impedance presented by the four-pole. From (4) we have then

$$V_a = I_1 \frac{\Delta}{Z_{22} + Z_b}. \tag{9}$$

But we can write

$$V_a = I_1(Z_a + Z_A) \tag{10}$$

where Z_A is the input impedance of the four-pole when it is terminated by Z_b. It results from (9) and (10) that

$$Z_A = \frac{\Delta}{Z_{22} + Z_b} - Z_a. \tag{11}$$

or, by (8),

$$Z_A = \frac{\Delta_0 + Z_b Z_{11}}{Z_{22} + Z_b}. \tag{12}$$

An exactly similar procedure based on driving the four-pole from the right instead of from the left, gives the impedance seen looking into that end when the left-hand termination is Z_a. The result is:

$$Z_B = \frac{\Delta_0 + Z_a Z_{22}}{Z_{11} + Z_a}. \tag{13}$$

From these last two relations, it is possible to find the impedance values for the terminations Z_a and Z_b that would simultaneously match the impedances Z_A and Z_B. These are the so-called image impedances, and are found by putting

$$Z_a = Z_A = Z_{\mathrm{I}}$$
$$Z_b = Z_B = Z_{\mathrm{II}},$$

and solving (12) and (13) simultaneously. The result is:

$$Z_{\mathrm{I}} = \sqrt{\frac{Z_{11}}{Z_{22}}\Delta_0} \tag{14}$$

$$Z_{\mathrm{II}} = \sqrt{\frac{Z_{22}}{Z_{11}}\Delta_0}. \tag{15}$$

When the terminations have these values, there are no reflections from the terminating impedances (although there may be internal reflections within the four-pole) and, in the cases where the image impedances (14) and (15) are pure resistances, the gain in power resulting from the presence of the four-pole is a maximum.

Concerning these power relationships, there is a good deal more that needs to be said. In the first place, it turns out to be more convenient to deal in terms of "virtual" power rather than real power. The difference is that the former is given by I^2Z in general, even when the currents are represented by complex numbers involving imaginaries, while the latter is equal to the product of the square of the magnitude of the current times the resistive component of the impedance. Consequently, the writing is greatly simplified by the concept of virtual power, whereas the real power may be found from it when that is required, and the phase of the terminations is known.

When the four-pole is driven from the left, so that V_b may be put equal to zero in (5), the virtual power in the output termination Z_b is given by

$$I_2^2 Z_b = V_a^2 \frac{Z_{21}^2}{\Delta^2} Z_b = V_a^2 \frac{Z_{21}}{Z_{12}} \frac{Z_{12}Z_{21}}{\Delta^2} Z_b. \tag{16}$$

The operating gain is defined as the ratio of this to the virtual power that the generator V_a would deliver directly to a matched load, Z_a. Thus, when the generator V_a is connected to a matched load, the current is $V_a/2Z_a$, and the virtual power in the load is $V_a^2/4Z_a$. The operating gain[1] is therefore

$$\Gamma_{21} = \frac{Z_{21}}{Z_{12}} \cdot \frac{Z_{12}Z_{21}}{\Delta^2} \cdot 4Z_a Z_b \tag{17}$$

where symbol Γ_{21} indicates that the gain is from left to right. In the opposite case, where the four-pole is driven from V_b on the right while the virtual output power is absorbed in Z_a on the left, the corresponding expression for operating gain is

[1] The insertion gain may be found by multiplying the operating gain by

$$\frac{(Z_a + Z_b)^2}{4Z_a Z_b}.$$

$$\Gamma_{12} = \frac{Z_{12}}{Z_{21}} \frac{Z_{12}Z_{21}}{\Delta^2} 4Z_aZ_b. \tag{18}$$

It is obvious, therefore, that the ratio of the gains in the two directions is $(Z_{21}/Z_{12})^2$. When Z_{12} is equal to Z_{21}, the gain (or loss, which is the reciprocal of the gain) in the two directions is likewise the same.

The expressions (17) and (18) are not particularly complicated, but for physical interpretation they may be put into very much better shape. This requires a little algebra, but, to make our proofs complete, it is worth outlining the procedure in some detail, rather than merely stating the final result.

The first two steps may probably be combined into one without impossing undue difficulties. Thus, from (7) the expression Z_{12}/Z_{21} may be replaced by $Z_{11}Z_{22}-\Delta_0$. This is the first step. The next one is a matter of definition, and merely eliminates Z_a and Z_b by introducing the ratios

$$a = Z_a/Z_I \qquad b = Z_b/Z_{II}. \tag{19}$$

When these substitutions are made in (17), remembering that Δ is given by (8), we have, with the help of (14) and (15),

$$\Gamma_{21} = \frac{Z_{21}}{Z_{12}} \frac{[Z_{11}Z_{22}-\Delta_0]4ab\Delta_0}{[\Delta_0(1+ab)+(a+b)\sqrt{Z_{11}Z_{22}\Delta_0}]^2}$$

$$= \frac{Z_{21}}{Z_{12}} \frac{\left[1-\frac{\Delta_0}{Z_{11}Z_{22}}\right]4ab}{\left[a+b+(1+ab)\sqrt{\frac{\Delta_0}{Z_{11}Z_{22}}}\right]^2}$$

$$= \frac{Z_{21}}{Z_{12}} \frac{1-\sqrt{\frac{\Delta_0}{Z_{11}Z_{22}}}}{1+\sqrt{\frac{\Delta_0}{Z_{11}Z_{22}}}} \left[\frac{1+\sqrt{\frac{\Delta_0}{Z_{11}Z_{22}}}}{a+b+(1+ab)\sqrt{\frac{\Delta_0}{Z_{11}Z_{22}}}}\right]^2 4ab. \tag{20}$$

In the event that the terminations on input and output sides are matched to the image impedances, so that a and b are both equal to unity, the gain from (20) is given by

$$\Gamma_{21}' = \frac{Z_{21}}{Z_{12}} \frac{1-\sqrt{\frac{\Delta_0}{Z_{11}Z_{22}}}}{1+\sqrt{\frac{\Delta_0}{Z_{11}Z_{22}}}}, \tag{21}$$

which is often written in the alternative form

$$\Gamma_{21}' = \frac{Z_{21}}{Z_{12}} \frac{1-\tanh\theta}{1+\tanh\theta}$$

where θ is the propagation constant of the four-pole. It is convenient to write this matched gain in the more condensed form

$$\Gamma_{21}' = \frac{Z_{21}}{Z_{12}} \Gamma_0$$

where the image gain Γ_0 is defined by the relation

$$\Gamma_0 = \frac{1-\sqrt{\frac{\Delta_0}{Z_{11}Z_{22}}}}{1+\sqrt{\frac{\Delta_0}{Z_{11}Z_{22}}}}. \tag{22}$$

The quantities under the radical may then be written as follows by solving (22):

$$\sqrt{\frac{\Delta_0}{Z_{11}Z_{22}}} = \frac{1-\Gamma_0}{1+\Gamma_0}. \tag{23}$$

For its physical meaning, note that in the matched condition, Γ_0 is the geometric mean between the gains in the two directions.

Substitution of (23) into (20) gives:

$$\Gamma_{21} = \frac{Z_{21}}{Z_{12}} \Gamma_0 \frac{4a}{(1+a)^2} \cdot \frac{4b}{(1+b)^2} \cdot \frac{1}{\left(1-\Gamma_0 \frac{1-a}{1+a} \cdot \frac{1-b}{1+b}\right)^2}. \tag{24}$$

The expression (24) is now in the form which we were seeking. Its advantage is the physical interpretation which may be given to factors of the form

$$\frac{4a}{(1+a)^2} \quad \text{and} \quad \frac{1-a}{1+a}.$$

The first of these might be called the "mismatch" factor, and expresses the ratio of the virtual power which a generator puts into a load connected directly across its terminals to the virtual power it would put into a matched load similarly connected. The situation is well known for the case where a is a real number, and calculation illustrates how slowly the gain departs from its matched value as the impedance ratio departs from unity. For example, a two-to-one impedance mismatch means a loss of 0.5 db only. Even a ten-to-one mismatch gives only 4.8 db loss. Note, too, the curve is symmetrical about the value of unity for the impedance ratio.

The other factor is the ratio of the reflected to the incident current at the end of a line terminated by an impedance mismatch. Its reciprocal is thought to constitute a more precise definition of "return loss" than is usually given in current literature. Note also that the two factors are related through the equation

$$\frac{4a}{(1+a)^2} + \left(\frac{1-a}{1+a}\right)^2 = 1, \tag{25}$$

which states the physical fact that the sum of the absorbed power and the reflected power is equal to the incident power.

With these relations in mind, it is possible now to interpret the various factors in (24) in connection with the diagram of Fig. 2. Imagine the generator V_a to send a wave into the four-pole represented by the rectangle in the drawing. Disregarding the factor Z_{21}/Z_{12} for the moment, we can visualize the wave as progressing from the generator V_a toward the right until it meets the impedance discontinuity between Z_a and Z_I, the image impedance of the four-pole seen from the left. Of the virtual power in the incident wave, the fraction $4a/(1+a)^2$ progresses on into the four-pole while the remainder is reflected and lost in the generator impedance. Having entered the four-pole, the current wave is amplified by the factor $\sqrt{\Gamma_0 Z_{21}/Z_{12}}$, and emerges from the right-hand end of the rectangle. Here another impedance discontinuity is encountered and the fraction $4b/(1+b)^2$ of the power enters the load, while the fraction $(1-b)/(1+b)$ of the current is reflected and progresses back toward the left through the four-pole. The current is amplified by the amount $\sqrt{\Gamma_0 Z_{21}/Z_{12}}$, is reflected in part by the factor $(1-a)/(1+a)$ at the left-hand termination, and moves once more toward the right. Thus, within the four-pole there is set up a to-and-fro surging which, each time the wave arrives at the right, contributes a little more to the power in the output.

In a single round trip through the four-pole, the wave of current or voltage is modified by the factor

$$\sqrt{\Gamma_0 \frac{Z_{21}}{Z_{12}}}\sqrt{\frac{Z_{12}}{Z_{21}}\Gamma_0}\,\frac{1-a}{1+a}\cdot\frac{1-b}{1+b} = \Gamma_0\frac{1-a}{1+a}\cdot\frac{1-b}{1+b},$$

and the sum of an infinite number of round trips assumes the form

$$S = 1 + x + x^2 + x^3 + \cdots = \frac{1}{1-x} \tag{26}$$

where

$$x = \Gamma_0\frac{1-a}{1+a}\cdot\frac{1-b}{1+b},$$

and when $|x|<1$. The square of the sum must be taken in (24) because S represents a current, while (24) represents a power ratio. It is thus seen then that all of the factors in (24) may be accounted for on a physical basis, and the whole action may consequently be thought of in pictorial perspective. The usefulness of introducing the image gain Γ_0, which is the geometrical mean of the forward and reverse gains, has also been demonstrated in this connection.

However, its usefulness does not stop with (24), and the impedances presented by the four-pole may also be expressed in terms of Γ_0. Thus (12) and (13) may be written respectively, with the help of (23):

$$\frac{Z_A}{Z_I} = \frac{1-\Gamma_0\dfrac{1-b}{1+b}}{1+\Gamma_0\dfrac{1-b}{1+b}} \tag{27}$$

$$\frac{Z_B}{Z_{II}} = \frac{1-\Gamma_0\dfrac{1-a}{1+a}}{1+\Gamma_0\dfrac{1-a}{1+a}}. \tag{28}$$

These show immediately that the impedance presented by the four-pole becomes the same as the image impedance whenever $a=1$ in the one case, or $b=1$ in the other. This is, of course, axiomatic. A much more striking property is shown by noting that the intrinsic algebraic sign of the impedance must perforce be the same as that of the image impedance whenever the magnitude of $\Gamma_0(1-b)/(1+b)$ in the one case, and of $\Gamma_0(1-a)/(1+a)$ in the other, is less than unity. The converse is true when the magnitudes are greater than unity, so that whenever the image gain is sufficiently large, the input impedance is the negative of the image impedance unless a or b, as the case may be, is identically unity.

When $\Gamma_0=1$, it is interesting to note that $Z_A=Z_bZ_I/Z_{II}$, and when $\Gamma_0=-1$, that $Z_A=Z_IZ_{II}/Z_b$

Having dealt now with the derivation and discussion of expressions for impedances and gains, we come to the very important question of stability, that is, freedom from oscillation. This may be approached in several ways, but the most rigorous is probably to return to the general equations (3), and their solutions given by (4) and (5). From (4) and (5) it is seen that the currents I_1 and I_2 may be different from zero even in the absence of the driving sources V_a and V_b whenever $\Delta=0$. But Δ is a function of all of the internal network impedances as well as of the terminations Z_a and Z_b. In turn, all of these impedances are functions of $j\omega$. For purposes of analysis, $j\omega$ may be replaced by the more general variable $p=\alpha+j\omega$, so that currents and voltages of the form $e^{j\omega t}$ now become $e^{pt}=e^{(\alpha+j\omega)t}$. The significance of α then becomes apparent. When it is positive, the currents and voltages increase indefinitely with time. When it is zero, they are the usual sinusoids of constant amplitude, and when it is negative, the currents and voltages decrease with time and eventually die away altogether.

For stability it is evident that the relation $\Delta=0$ must be satisfied for negative values of α only, and not for positive values, as otherwise the currents in (4) and (5) would increase indefinitely with time, even in the absence of the driving sources V_a and V_b. If the equation $\Delta=0$ is satisfied only for negative values of α, the currents die away when the sources are removed and the system is stable, except in the contingency that one of the coefficients of V_a or V_b in (4) or (5) should become infinite for some positive value of α while, at the same time, Δ itself remained finite. Since Δ may be written

$$\Delta = (Z_{11}+Z_a)(Z_{22}+Z_b) - Z_{12}Z_{21},$$

and consequently involves all of the aforementioned coefficients, Δ can remain finite when one of the coefficients becomes infinite only if the coefficient with which it is paired in the above expression for Δ becomes zero simultaneously or (a more usual situation) is identically zero for all values of p. That is, either $(Z_{11}+Z_a)$ is infinite for the same value of p that causes $(Z_{22}+Z_b)$ to become zero, or vice versa, or else Z_{12} is infinite for the same value of p that causes Z_{21} to become zero. In either event, instability would require that the real part of p should be positive. This alternative contingency seldom occurs in bilateral systems, but is not infrequently encountered in unilateral cases. One particular example that is illustrative happens when the interstage coupling circuit between two unilateral amplifier stages contains negative impedances and, when isolated, is unstable. Connecting it between two vacuum tubes does not cause it to become stable, and it will be found that the four-pole equations for the system show that Z_{12} is zero for all frequencies, but that Z_{21} may become infinite for a positive value of α.

Whenever (3) is derived by first writing the mesh equations for the entire multi-mesh network, one equation for each mesh, and from these equations eliminating all currents but the two corresponding to the input and output meshes, the stability conditions are completely determined either by the vanishing of Δ, or by the simultaneous vanishing of one of a pair of factors forming Δ together with the vanishing of the reciprocal of the other.

Possibility of trouble occurs, however, when approximations are made. For example, when a vacuum tube with feedback is considered, the impedance looking into a pair of terminals may become negative in certain frequency ranges. There is then a strong inclination to simplify by replacing the complete details of the circuit which produced the negative impedance by the negative impedance itself. Actually, there is no objection to doing this providing that the negative impedance is completely and accurately specified over the whole frequency range.

This point is very important. For example, note that a negative impedance which was the exact negative of some passive impedance over the whole frequency range from zero to infinity, could not possibly be unstable on either open or short circuit. This is at once evident when it is considered that the values of p which satisfy the passive equation

$$Z(p) = 0$$

are identical with those that satisfy the active equation

$$-Z(p) = 0$$

and hence, if the α for the one is always negative, so also is the α for the other. From this it may further be concluded that a negative impedance which is unstable on either open or short circuit cannot possibly be the exact negative of any passive impedance whatever over the whole frequency range. One can go even further, however, and invoke some of the methods of complex function theory to show that such a negative impedance cannot even be the exact negative of any passive impedance over any finite frequency band, no matter how small.

The point of this discussion is to bring out the fact that stability or lack of it in systems involving negative impedances is often determined by the departure of the negative impedances from being the negatives of passive impedances, and hence that any disregard of this fundamental fact is likely to lead to trouble. These departures may, and in fact often do, exist at frequencies outside of the band that is of interest from the standpoint of normal use. Their effect reflects back into that band nonetheless.

How then should one proceed? Is the device of using the concept of negative impedances of no practical value? The answer to this is supplied in part by Crisson, who, some years ago, introduced the concept of series and shunt types of negative resistances. By definition, the series type is unstable on short circuit, and the shunt type is unstable on open circuit. Interpreted in the light of the foregoing discussion, these definitions may be rephrased somewhat as follows:

A negative resistance is one which behaves very nearly like the negative of a positive resistance over a fairly large frequency range. Outside of that range, however, a series type negative resistance departs from that approximation in such a way that the circuit element is unstable on short circuit, and a shunt type negative resistance departs in such a way that the circuit element is unstable on open circuit. Graphically, this would imply that, if the imaginary part of the negative impedance were plotted against the real part for all values of frequency, that is for all values of $p = \alpha + j\omega$ where $\alpha = 0$, the graph of a series type would look something like Fig. 3, and the graph of a shunt type would look something like Fig. 4. They both encircle the origin, but in different directions. It is obvious that this approximation, useful as it is, has certain limitations, and that the safest way of dealing with new or untried circuits is to be sure that the negative impedance is, in fact, specified to a sufficient extent over the whole pertinent

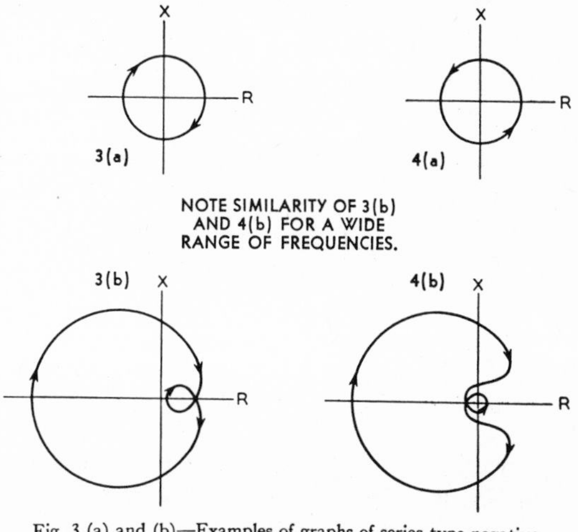

Fig. 3 (a) and (b)—Examples of graphs of series type negative impedance.

Fig. 4 (a) and (b)—Examples of graphs of shunt type negative impedance.

frequency range. Such a range would have to be sufficient to insure that the combination of the negative element with the remainder of the circuit did not have a resistive component that became negative at any higher or any lower frequency.

The practical effect of all of this is to point out that certain combinations of series and shunt type negative resistances may be quite stable, while others may not. The general stability criterion, when all things are taken into account, is the determination of the values of p that cause Δ in (4) or (5) to become zero. The alternative condition that results in instability can usually be detected by general inspection of the circuit, or may be tested for each of the four possible contingencies separately.

The investigation of Δ itself turns out to be rather cumbersome, and an easier alternative arises when it is noticed that the expression for gain Γ, given by (17), contains Δ in the denominator. It follows that Γ has an infinity whenever Δ has a zero. Also, Γ has no infinities that are not contributed by zeros of Δ. This may be verified by inserting (6) into (17) and noting that infinities of Z_a and Z_b contribute only zeros to Γ, while infinities of Z_{12} and Z_{21} contribute neither zeros nor infinities to Γ. Consequently, except for the case mentioned before where Z_{21}/Z_{12} has an infinity while $Z_{12}Z_{21}$ does not the zeros of Δ are uniquely determined by the infinities of Γ and, when Δ has no zeros with positive real parts, Γ has no infinities with positive real parts. For every zero of Δ that does have a positive real part, Γ has an infinity with a positive real part.

It may be taken then that, leaving aside the exceptions mentioned, (17) can be used as a basis for determining stability, and therefore that (24), which is merely (17) written in another form, can likewise be used. The infinities of (24) must be investigated to determine whether the real parts of any of them are positive.

The possible infinities of Γ are all determined by the equation

$$\left(1 - \Gamma_0 \frac{1-a}{1+a} \cdot \frac{1-b}{1+b}\right) = 0, \tag{29}$$

as may be seen from (24) by trying all of the other alternatives; namely $1/a=0$, $1/b=0$, $\Gamma_0=0$, $(1+a)=0$, and $(1+b)=0$. None of these others yields infinities.

There is a striking similarity between the form of (29) and the famous equation for the stability of feedback amplifiers, usually written

$$(1 - \mu\beta) = 0.$$

In fact, the similarity goes further than one of form only, and the discussion leading to (26) shows that the physical meaning of the factors involved is quite analogous. This at once suggests the possibility of applying the Nyquist stability criterion and plotting

$$\Gamma_0 \frac{1-a}{1+a} \cdot \frac{1-b}{1+b}$$

on the complex plane as a function of the frequency ω, and seeing whether the plot encircles the point $(1, j0)$. The trouble is that encirclement of the point $(1, j0)$ would indicate instability only under certain special conditions, and cannot be applied with complete generality. It happens that those conditions are fulfilled in the standard type of feedback amplifier, but very often are not in the more general cases which it is now attempted to discuss.

This fact is so important, and the appreciation of it seems so limited in extent, that a brief explanation of the fundamentals involved appears to be in order. The key to the situation is furnished by the realization that, in the conventional feedback amplifier, both μ and β are of the nature of constants multiplied by the ratio of output to input voltage across passive impedance functions (either self or transfer) and hence that neither of them has infinities whose real parts are positive. In the generalized repeater case of (29), where negative impedance elements may be involved, there is no assurance that this is so. In fact, it is readily seen that the reflection coefficient $(1-a)/(1+a)$ may become infinite when a negative resistance is connected facing a positive one, for then a is negative. This seems at first to be very discouraging to an attempt to draw simple conclusions and rules relating to the more general case. The situation is helped only by limiting the problem and being content, not with complete generality, but with an amount sufficient to cover the particular class of problem that is encountered in considering the telephone repeater. Here this analysis requires broadly, not only that the system be stable with a given pair of terminations, but that it be stable when its end terminations are either open circuited or short circuited in any possible combination of the terminations, and, moreover, that it shall be stable for any values of passive terminations in between these two extremes.

Further, it is evident in such a system that the ultimate terminations at the final terminals must consist of passive impedances. This at once implies that the image impedances of the four-pole representing the entire system must likewise have the properties of a passive impedance, as otherwise it may be shown, from (12) for example, that there always exists a value of passive termination that will result in instability. This is really a very important conclusion for it says that, in the design of systems involving negative impedances, care must be taken that the image impedances must have these passive properties at all frequencies if stability is to be guaranteed. This means that their resistive components must be positive at all frequencies from zero to infinity.

If the image impedances were entirely resistive at all frequencies, while the terminations were restricted to being passive, the greatest as well as the least magnitude that could be attained by factors of the form $(1-a)/(1+a)$ would occur when the termination approached a pure reactance, either positive or negative. The magnitude of the factor would then be unity for any value of terminating reactance. Its phase, however, could lie in any of the four quadrants of the complex impedance plane, depending upon the value of terminating reactance. Hence, when the magnitude of the image gain passed through unity, and in the event that the gain factor Γ_0 had even the smallest phase angle, it always would be possible to find values of terminating reactance that would cause the graph of

$$\Gamma_0 \frac{1-a}{1+a} \cdot \frac{1-b}{1+b}$$

to pass through the point $(1, j0)$ on the complex impedance plane. Minute changes in terminating reactance would then cause the graph to pass on one side or the other of the point $(1, j0)$ and consequently change the system from a stable one to an unstable one, or vice versa.

For this case, where the image impedances are pure resistances at all frequencies, it is clear that the system will be stable for any values of passive terminating impedances if and only if the magnitude of the image gain is less than unity. This is the situation that can be approached by the 22-type repeater of Fig. 1 when its image impedances are pure resistances, though even here departures of the hybrid coils from the ideal can introduce phase into the image impedances and create the more general situation which must now be discussed.

In this more general case, the restriction that the resistive components of the image impedances must be positive at all frequencies is still retained, as otherwise passive terminations which will cause singing can always be found. However, no restriction is now placed on the reactive component of the image impedances. Under these conditions, it can be shown from a theorem in functions of complex variables[2] that factors of the form $(1-a)/(1+a)$ attain their greatest and their least magnitudes as well as their greatest and smallest real and imaginary components when the terminations are pure reactances. For this condition, we can write:

$$\frac{1-a}{1+a} = \frac{Z_I - Z_a}{Z_I + Z_a} = \frac{R_I + j(X_I - X_a)}{R_I + j(X_I + X_a)}$$

$$= \sqrt{\frac{1 + y^2 - 2y \sin \phi_I}{1 + y^2 + 2y \sin \phi_I}}\, e^{-j\tan^{-1}(2y\cos\phi_I/(1-y^2))}, \quad (30)$$

where

$$y = X_a/|Z_I| \quad \text{and} \quad \phi_I = \tan^{-1}\frac{X_I}{R_I}.$$

This attains its greatest magnitude when $|X_a| = |Z_I|$. and the algebraic sign of X_a is opposite to that of X_I. In that event, the magnitude becomes

$$\left|\frac{1-a}{1+a}\right|_{\max} = \sqrt{\frac{1 + |\sin \phi_I|}{1 - |\sin \phi_I|}}, \quad (31)$$

and the phase is $\pm\pi/2$ depending upon the phase of Z_I.

[2] H. W. Bode, "Network Analysis and Feedback Amplifier Design," D. Van Nostrand, 1945; p. 169.

The minimum magnitude is the reciprocal of (31), or

$$\left|\frac{1-a}{1+a}\right|_{\min} = \sqrt{\frac{1 - |\sin \phi_I|}{1 + |\sin \phi_I|}}, \quad (32)$$

and occurs at an angle of π with respect to that for the maximum. The real component of (30) attains its maximum value when

$$y = -\frac{1 \mp \cos \phi_I}{\sin \phi_I}. \quad (33)$$

A graph of (30) for the four cases where the phase of Z_I is 0, 30°, 45° and 60°, respectively, is shown on Fig. 5 which illustrates the locus of the function as X_a takes on all values from $-\infty$ to $+\infty$. Further study of this figure, and the equations above relating to it, shows that the curves are true circles and that the distance from the origin to the center of a given circle is equal to $\tan \phi$, where ϕ is the phase of the corresponding image impedance.

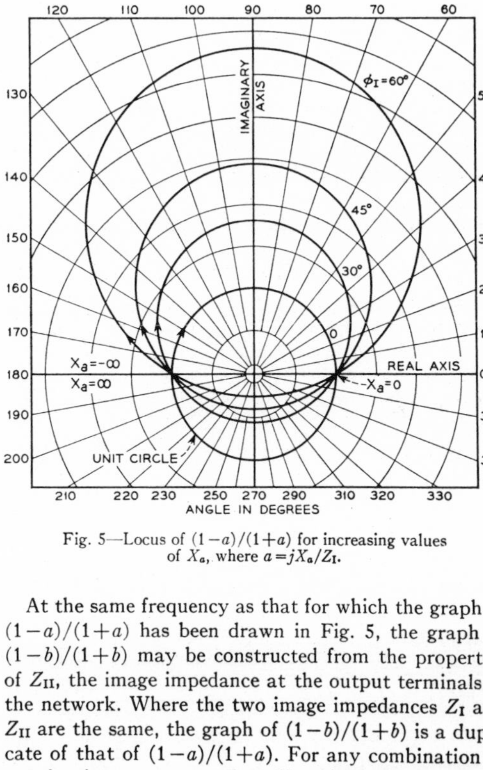

Fig. 5—Locus of $(1-a)/(1+a)$ for increasing values of X_a, where $a=jX_a/Z_I$.

At the same frequency as that for which the graph of $(1-a)/(1+a)$ has been drawn in Fig. 5, the graph of $(1-b)/(1+b)$ may be constructed from the properties of Z_{II}, the image impedance at the output terminals of the network. Where the two image impedances Z_I and Z_{II} are the same, the graph of $(1-b)/(1+b)$ is a duplicate of that of $(1-a)/(1+a)$. For any combination of terminating reactance values X_a and X_b, the product of the two graphs always falls within the envelope obtained by letting $X_a = X_b$. For this condition, Fig. 6 shows the product curve for several values of the phase of the image impedance, and it will be noted that the external envelope of the complete surface is always equal to or greater than unity.

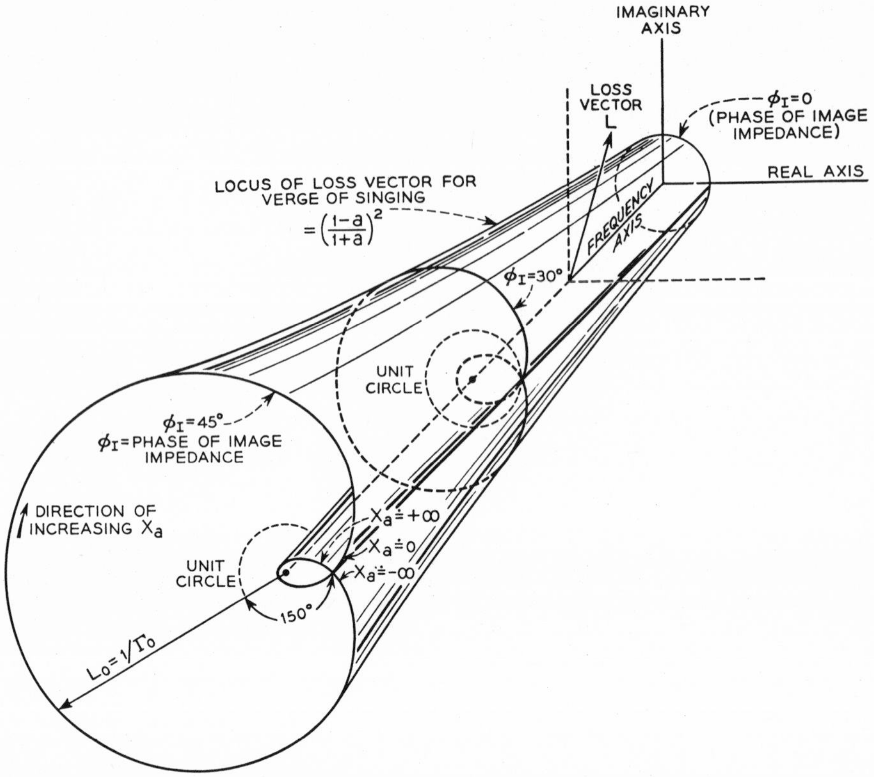

Fig. 6—Locus of stability for the loss vector for different frequencies.

Suppose, at the frequency for which the largest curve in Fig. 6 is drawn, that the gain Γ_0 has, for example, a phase of 150°. The image loss, which is the reciprocal of Γ_0, then has a phase of $-150°$. If it had a magnitude equal to that of the vector shown on the figure, the product

$$\Gamma_1 \frac{1-a}{1+a} \cdot \frac{1-a}{1+a}$$

would be exactly equal to unity, and hence, according to (29), the system would be on the verge of singing. A smaller gain at the same phase would be needed for stability or else, for this case, the same gain at a lesser phase. Fig. 6 therefore sets the relation between the allowable phase and magnitude of the gain at the particular frequency it represents, and for the symmetrical case where the image impedances at both ends of the system are the same. A curve analogous to the one considered above must be constructed for every frequency, and the three-dimensional envelope of all of them determines the allowable relationship between the maximum magnitude and phase of the gain over the frequency range. Several such curves are shown on Fig. 6 for different values of the frequency and the phase of the image impedance. The only cases in which the magnitude of the gain can approach unity are first, those for which the image impedances are both pure resistances and, second, those for which the phase of the gain is exactly zero. In all other cases, the magnitude of the gain must be less than unity to avoid singing. When the phase of the gain is 180°, its magnitude must be less than

$$\frac{1-|\sin \phi_I|}{1+|\sin \phi_I|}.$$

When the image impedances are pure resistances, the gain can approach unity regardless of its phase. The three-dimensional surface shown in Fig. 6 can then be regarded as setting the lower limit on the loss. The end of the loss vector must always fall outside of this surface at every frequency.

In many practical cases, the phase of the gain is not under control. For example, the phase changes very rapidly with frequency in a circuit several hundred miles long, and it would not be feasible to attempt to keep it within narrow limits over the speech band. For these usual cases, the curve on Fig. 7, which is plotted from the above expression for the magnitude of the gain, gives the allowable operating condition. For example, when the phase angle of the image impedance is 60°, the loss must exceed 11.3 db. The required loss is, of course, in addition to the allowance of a margin to take care of such things as changes in amplifier gains and in line losses.

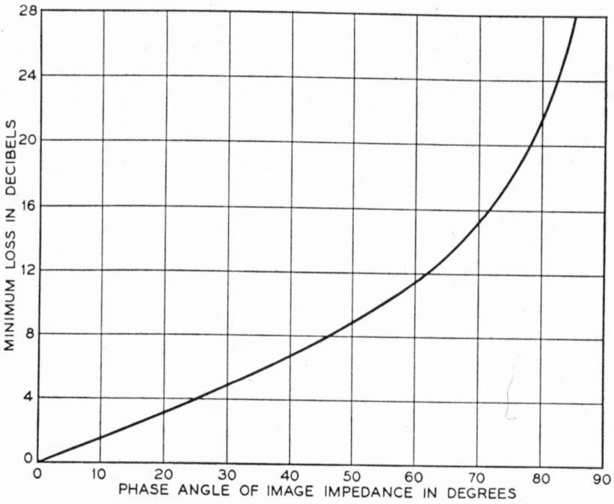

Fig. 7—Relation of minimum loss to phase angle of image impedance.

Of course, a magnitude of gain approaching unity is greatly to be desired in repeatered systems. In those which are composed of similar sections in tandem, and where, in addition, it is desired that the individual sections be stable when isolated and terminated with any combination of passive impedances, the above conditions become very important for they apply to the individual sections and severely restrict the freedom of design. Where, however, the sections need not be stable individually when subjected to all combinations of passive termination, but the system as a whole must nonetheless be stable, a good deal more freedom in design is permissible. The individual sections can actually have gains greater than unity, providing that it is removed before the final terminals are encountered.

For example, Fig. 8 shows a possible system in which repeaters or negative-resistance loading may be used quite freely in a transmission line, with the result that the system would sing if terminated at Z_I, with certain combinations of passive impedances. In general, also, the image impedances of the line will have a reactive component. At each end of the line there is placed an impedance-equalizing four-pole which matches the line on the one side, but presents a purely resistive impedance on the other. Such a network unavoidably introduces a certain loss if composed of passive elements only. An active network, however, such as the 22-type repeater circuit, can accomplish the impedance transformation without loss, or even with gain. In any event, the overall system, now having purely resistive image impedances, can be adjusted to have an overall gain that, with ideal impedance matches, approaches unity as closely as desired, and still will be completely stable for all combinations of passive termination. Any gain greater than unity, however, will result in singing, and the margin needed in practical design is a matter of how constant with time the values of the components of the system can be made, and how accurately the impedances may be matched.

In the event that the phase equalizers or converters of Fig. 8 are to be composed of passive circuit elements, the minimum loss required to consummate the impedance transformation can be found from Fig. 5. It is necessary to note that,

$$\Gamma_0 \frac{1-a}{1+a} \cdot \frac{1-b}{1+b}$$

can never pass through the point $(1, j0)$. For a reactive image impedance on the input side of the network, Fig. 5 would show the graph of the factor $(1-a)/(1+a)$ as one of the off-center circles. The graph of $(1-b)/(1+b)$ would be a unit circle, however, because of the resistive image impedance on the output side of the network, and the factor could have its phase anywhere in the four quadrants. The envelope of the product of the two factors would therefore be a circle whose radius vector was equal in magnitude to the maximum value of $(1-a)/(1+a)$, and whose phase could lie anywhere in the four quadrants. The minimum loss possible in a passive phase converter network would therefore be the

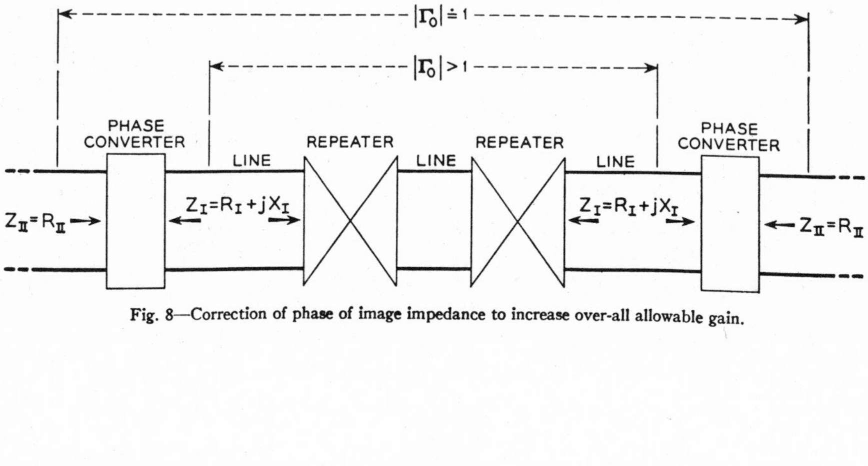

Fig. 8—Correction of phase of image impedance to increase over-all allowable gain.

reciprocal of this value, or

$$\sqrt{\frac{1+|\sin\phi_{\text{I}}|}{1-|\sin\phi_{\text{I}}|}}.$$

For systems in general, when the phase of the image impedances differ on the two ends, the stability conditions are in a sense more severe than for the symmetrical case. Here the criterion may be visualized by referring to the two curves on Fig. 5 corresponding to phases of the image impedances 30° and 60°. The first may be thought of as the reflection factor for the image impedance Z_{I}, and the second as the reflection factor for the image impedance Z_{II}. What is desired is an envelope analogous to Fig. 6, giving at each phase the maximum possible value of the product of the two factors. When a point on the envelope has a certain phase, ψ, the sum of the phases of the individual factors must be equal to ψ. Thus if ρ_a represents the magnitude of $(1-a)/(1+a)$ and ϕ_a represents its phase, and if ρ_b represents the magnitude of $(1-b)/(1+b)$ and ϕ_b represents its phase, the envelope at the angle ψ has the magnitude $\rho_a\rho_b$ for which $\phi_a+\phi_b=\psi$. The problem is to determine the maximum magnitude of this product for each value of ψ.

The easiest approach seems to be to deal, not with ρ_a and ρ_b directly, but with their logarithms, so that we have

$$\log \rho_a\rho_b = \log \rho_a + \log \rho_b,$$

and the problem is shifted from that of finding the maximum value of a product to the somewhat easier one of finding the maximum value of a sum, subject, however, to the same condition concerning phase, namely,

$$\phi_a + \phi_b = \psi.$$

Fig. 9 is constructed from Fig. 5, and shows $\log \rho_a$ or $\log \rho_b$ plotted against ϕ_a or ϕ_b, as the case may be, for different values of the phase ϕ_{I} or ϕ_{II} of the respective image impedance. The curves resemble sine waves somewhat, but are not true sinusoids although, for a rough approximation, the assumption that they are would not give large errors for moderately small value of ϕ_{I}.

$$\rho = \frac{1-a}{1+a}$$

$$a = jX_a/Z_{\text{I}}$$

$$Z_{\text{I}} = \text{image impedance}$$

$$= |Z_{\text{I}}|\, e^{j\phi_{\text{I}}}.$$

To illustrate the use of Fig. 9, assume for example that we are dealing with a system which has a phase of $\phi_{\text{I}}=30°$ for the image impedance seen from the left-hand end, and of $\phi_{\text{II}}=60°$ seen from the right-hand end. We deal then with the corresponding curves on Fig. 9, and the lower one on the left corresponds to $\log \rho_a$ and ϕ_a, and the higher to $\log \rho_b$ and ϕ_b. The envelope curve which takes the place of Fig. 6 for this case of image impedances of different phases is then constructed by finding the two ordinates $\log \rho_a$ and $\log \rho_b$, which correspond to the two angles ϕ_a and ϕ_b, such

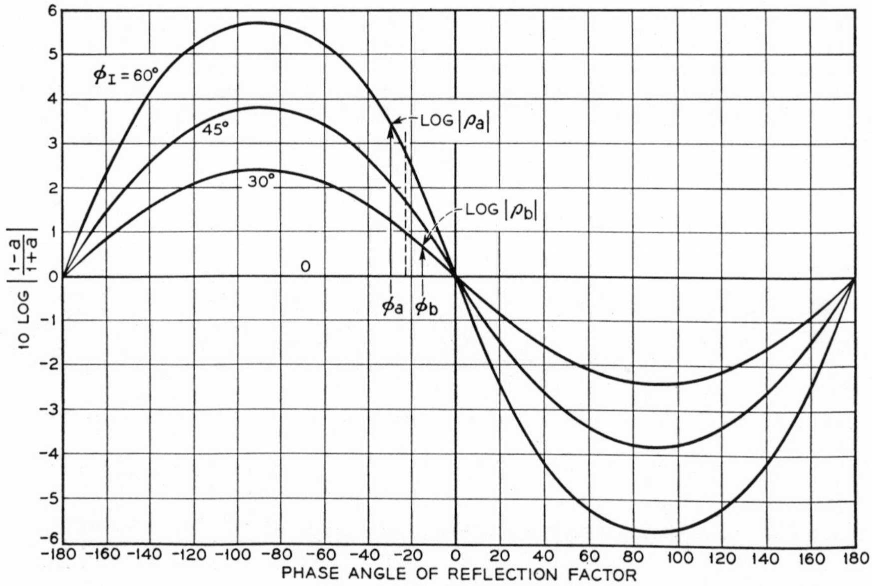

Fig. 9—Graph of logarithm of reflection factor

that the sum of the two ordinates is a maximum when $\phi_a+\phi_b$ equals the envelope angle ψ.

Algebraically this is equivalent to requiring:

$$y_a + y_b = \text{maximum when } x_a + x_b = \text{constant}.$$

Hence, for the maximum

$$\frac{dy_a}{dx_a} dx_a + \frac{dy_b}{dx_b} dx_b = 0.$$

But,

$$dx_b = - dx_a,$$

so that

$$\frac{dy_a}{dx_a} = \frac{dy_b}{dx_b}.$$

This gives the clue to the graphic solution. Suppose that $\psi = -45°$. Starting then with ϕ_a and ϕ_b, both equal to $-22.5°$, we note that the slope of the curve for $\phi_I = 60°$ is much greater than that for $\phi_I = 30°$. Consequently, we take a value for ϕ_a greater in magnitude than $-22.5°$, and for ϕ_b an equal amount less in magnitude than $-22.5°$ so that their sum again equals 45°. We note whether the slope of the $\phi_I = 60°$ curve corresponding to ϕ_a still remains greater than that of the $\phi_I = 30°$ curve corresponding to ϕ_b. If so, the departures of ϕ_a and ϕ_b from the mean of $-22.5°$ should be increased further until the two slopes are equal. When such a pair of values has been found, the envelope of the product curve for $\psi = 45°$ has the value

$$\log \rho_a + \log \rho_b$$

for its logarithm.

The process is tedious and would have to be repeated for each value ψ of the envelope phase, and besides, the whole graphical construction would have to be repeated for each frequency in order to find the entire three-dimensional contour outside of which the loss must remain if stability is to be insured. Straightforward analytical solution does not offer much hope, either, for the difficulties appear to become even greater. It may happen, of course, that some change of variable or other algorithm will be found, but the probability is not at present very favorable.

In some cases, and particularly when attention is directed toward a general philosophical approach rather than to operating criteria, an extension of the stability conditions along the line proposed by Gewertz[3] has proved useful. In this extension of the work of Gewertz, the coefficients of (1) are written in the matrix form

$$\begin{vmatrix} R_{11} + jX_{11} & R_{12} + jX_{12} \\ R_{21} + jX_{21} & R_{22} + jX_{22} \end{vmatrix}.$$

It may then be shown by the argument given above, that the system is stable for any passive termination, providing that the following conditions are satisfied at all frequencies:

$$R_{11} > 0$$

$$R_{22} > 0$$

$$4(R_{11}R_{22} + X_{12}X_{21})(R_{11}R_{22} - R_{12}R_{21}) - (R_{12}X_{21} - R_{21}X_{12})^2 > 0.$$

In the event that $Z_{12} = Z_{21}$, it will readily be recognized that the last of these three relations reduces to the form

$$(R_{11}R_{22} - R_{12}^2) > 0,$$

which has come to be known as the Gewertz condition. On the other hand, in the event that $Z_{21} = -Z_{12}$ we have the alternative

$$(R_{11}R_{22} - X_{12}^2) > 0.$$

In the symmetrical case, where $Z_{11} = Z_{22}$, and where $Z_{12} = Z_{21}$, it may be shown that the Gewertz condition becomes

$$R_{11} - R_{12} > 0,$$

whence it follows that the real part of

$$Z_I \frac{1 - \sqrt{\Gamma_0}}{1 + \sqrt{\Gamma_0}}$$

must be greater than zero. This means that the resistive component of the short-circuit impedance of a hypothetical network of half the electrical length of the actual network, should always be positive. From this it is easy to deduce the relationship

$$(1 - |\Gamma_0|) \cos \phi - 2\sqrt{|\Gamma_0|} \sin \phi \sin \beta > 0,$$

where

$$Z_I = |Z_I| e^{j\phi}$$

$$\Gamma_0 = |\Gamma_0| e^{-2j\beta}.$$

From this equation it follows that, in case $\sin \beta = 1$, we have

$$|\Gamma_0| < \frac{1 - |\sin \phi|}{1 + |\sin \phi|},$$

which agrees with the results previously attained, and shows the connection between the two methods of approach, namely the consideration of the matrix components on the one hand, and of the image parameters on the other.

It is important to point out that, in the present extended form of the Gewertz relations, all that is assured is that the network shall remain passive regardless of what passive terminations are attached. It does not follow that the network has all of the properties of a passive system, in the sense that it may be imbedded in a general network system involving passive feedback from the output to the input and still remain completely stable. A system composed entirely of passive ele-

[3] C. M. Gewertz, "Network Synthesis," The Waverly Press, 1933; pp. 45–63.

ments would, of course, be stable under these conditions.

Notwithstanding the difficulties of the general case where the image impedances at the two ends of the system are different from each other, some of the conclusions which have been pointed out are of broad validity, and in the more restricted case where the two image impedances are the same, quantitative results may be computed with moderate ease for any phase angle of the image impedances. Expressed in terms of the image gain and the image impedances, the relations are so important and general that a few examples may make their meaning clearer.

For the first one, the unilateral amplifier will be considered in order to show that, even in this case, the general principles and method of analysis apply. The schematic circuit diagram is shown in Fig. 10, and the four-pole equations are given on the figure. It is important to note that a feedback admittance Y_x is included for purposes of analysis, but that this is ultimately allowed to become so small that feedback disappears.

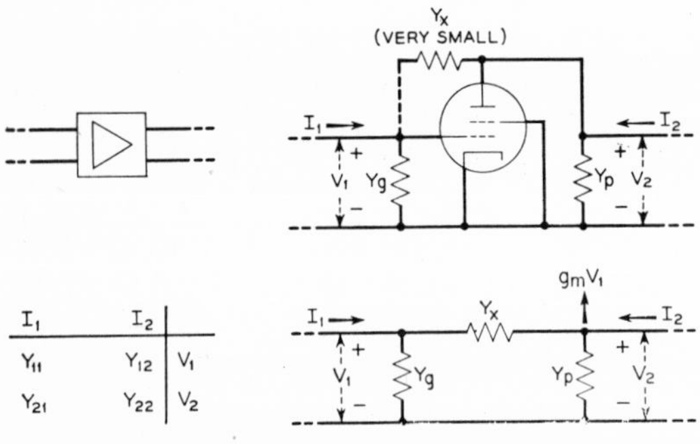

Fig. 10—Development of unilateral amplifier.

It happens that admittances are more convenient than impedances to deal with here, but the form of the various relations is not changed thereby. Application of (14), (15), and (22), gives the image parameters $Y_{\rm I}$, $Y_{\rm II}$, and Γ_0 as follows:

$$Y_{\rm I} = (Y_g + Y_x)\sqrt{1 - \frac{Y_x(g_m - Y_x)}{(Y_g + Y_x)(Y_p + Y_x)}}$$

$$Y_{\rm II} = (Y_p + Y_x)\sqrt{1 - \frac{Y_x(g_m - Y_x)}{(Y_g + Y_x)(Y_p + Y_x)}}$$

$$\Gamma_0 = \frac{1 - \sqrt{1 - \dfrac{Y_x(g_m - Y_x)}{(Y_g + Y_x)(Y_p + Y_x)}}}{1 + \sqrt{1 - \dfrac{Y_x(g_m - Y_x)}{(Y_g + Y_x)(Y_p + Y_x)}}}.$$

As the feedback admittance Y_x is allowed to become very small, (impedance very high), the image admittances easily and gracefully approach the values Y_g and Y_p, respectively. At the lower frequencies before transit-time effects enter, these are ordinary passive admittances. The image gain approaches zero, but the way it does this can best be seen by using the binomial theorem to expand the radical in the numerator. This gives

$$\Gamma_0 \to \frac{1}{4}\,\frac{Y_x(g_m - Y_x)}{(Y_g + Y_x)(Y_p + Y_x)} \to \frac{1}{4}\,\frac{g_m Y_x}{Y_g Y_p}. \qquad (34)$$

In this form, it can be seen from (24) that the operating gain from left to right with matched terminations becomes

$$\Gamma_{21}' \doteq \frac{g_m - Y_x}{Y_x}\cdot\frac{1}{4}\,\frac{g_m Y_x}{Y_g Y_p} \to \frac{1}{4}\,\frac{g_m^2}{Y_g Y_p}, \qquad (35)$$

which is recognizable as the conventional expression for this gain. On the other hand, Γ_0 itself approaches zero, and the matched gain Γ_{12}' from right to left likewise approaches zero, and in such a way that the image gain Γ_0 is the geometric mean between Γ_{12}' and Γ_{21}'. Whenever the feedback admittance Y_x is not quite zero, the circuit may yet be stable for all passive impedance terminations, but only providing that the image gain Γ_0, multiplied by the reflection coefficients, does not encircle the point $(1, j0)$.

This rather extreme illustration was chosen first to demonstrate the generality of the analysis, and to show how it applies in the unilateral case.

As an example of the bilateral case, the properties of the 22-type repeater will be considered. Fig. 1 shows the general schematic and, when the impedances seen by the hybrid coils on the network side, the transmitting side, and the receiving side are completely balanced, the image impedances are equal to the impedance of the passive balancing networks and are independent of the repeater gain. When the over-all image impedance of a system containing 22-type repeaters is a pure resistance, the repeaters may be adjusted until the gain of the system approaches unity before singing can take place. In the more usual case, the image impedances of the individual repeaters are adjusted to match that of the connecting line, which has an appreciable phase angle. Consequently the gain of the system must be held to a flat value of

$$\frac{1 - |\sin \phi_{\rm I}|}{1 + |\sin \phi_{\rm I}|},$$

or else must be tailored to fit the conditions discussed in connection with Fig. 6. However, the expedient of providing initial and terminating repeaters, whose input and output hybrids are matched to a pure resistance, will allow the system gain to be brought up to unity even in this case. With ideal impedance matches, the margin which must be allowed in practical design then depends upon the variations in repeater gains and line losses under operating conditions, and not upon the number of sections in the system or upon the over-all line loss. Extra margins are required for unavoidable impedance mismatches resulting from line irregularities.

With other types of repeaters, such as the 21-type illustrated in Fig. 11, the image gain and the image impedances are not so easy to adjust independently. However, there seems to be nothing fundamental to prevent supplying the terminals of the system with networks to provide a purely resistive image impedance. Practically, there are many cases where the construction of such networks offer excessive complexity, though in others their use may be quite feasible. The gain of the system can then be made unity before singing will occur, with any possible combination of passive impedances attached to the terminals.

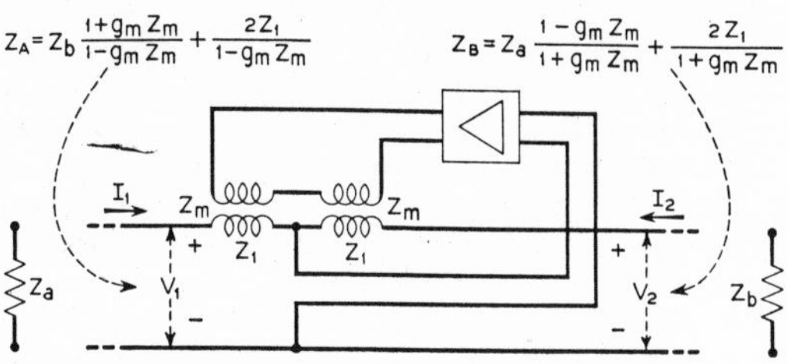

Fig. 11—Schematic of 21-type repeater.

This brings out the point that it is perfectly possible to have a system that is quite stable when terminated at mid-section points, but may be quite unstable when terminated at half-load points, and vice versa. This can happen, for example, when the image impedance is a pure resistance in the one case, but not in the other. It can happen more generally, however, whenever the phase of the image impedance is different in the two cases, and consequently the allowable gain is different.

The analysis also shows that the attempt to improve the singing margin of a system by the addition of pads of various kinds is quite futile in some cases but not in others. If a system has an image impedance that is a pure resistance, its gain may be made unity before singing can occur. The addition of a resistive pad will allow the repeater gains to be increased until the over-all gain is gain unity, but no more, and the system is back again where it started. A reactive pad added to such a system will do harm because it will produce image impedances with reactive components, and the over-all gain must remain correspondingly less than unity. However, as shown before, a phase-correcting network, or pad, applied to a system having an appreciable phase angle for its image impedance, will be helpful. We are thus led to the conclusion that an advantageous terminating four-pole is one that transforms the image impedance into a pure resistance in those cases where it initially has a reactive component.

This observation also gives the key to the best design objective for repeatered systems in general. That objective is to cause the image impedance of the system to be purely resistive to as close a degree as possible, while bringing the gain as nearly to unity as is consistent with safe singing margins.

These examples also illustrate a general conclusion that may be stated as follows:

The external stability of all systems depends only upon the phase of the image impedances and magnitude of the image gain, and not at all upon the details of the internal arrangements of the system by which these quantities are attained.

It does not follow, however, that all systems are alike in terms of the percentage change of voltage on the vacuum tubes which provide the repeater gain or the negative impedance loading, or in terms of the complexity of the equalizing and phase-correcting networks required to give the desired image-impedance terminations. The image gain of a 22-type repeater without feedback is proportional almost directly to the effective voltage of the dc supply source, while the image impedances are almost independent of this voltage. The image gain of a line with negative impedance loading may, under some conditions, vary much less rapidly with supply voltage to the tubes that furnish the negative impedance. Also, systems vary greatly in the amount of trouble resulting from line impedance irregularities.

Consequently, rather than regarding the theory here presented as saying that all systems having the same image parameters will behave alike, it may be more useful to turn the statement around and regard the theory as saying what has to be done to a given system in order that it shall be capable of operating as well as some other system. Conversely, the theory also tells how much more loss the given system perforce must have than a reference system in order to remain unconditionally stable, and it sets up specific and definite standards for the reference system. The present paper has stressed the applicability of the image parameter concept to the determination of singing conditions in telephone systems. However, the methods developed are also capable of dealing with such other properties as talker and listener echo, which are equally important in some applications. These have not been discussed in detail because the paper already is fairly long and because, with the fundamental background as presented, the reader is in a position to carry out a number of extensions for himself.

As a closing word, a few remarks concerning bibliography references are in order. It will be noticed that very few occur in the text. This is because the writer is aware of very few that have a specific and direct bearing on the mode of development of the subject which was employed. He wishes however to express appreciation of the helpful and stimulating conversations he has had with many of his colleagues on the technical staff of the Bell Telephone Laboratories. As general background to the use of image parameters in active circuit analysis, the following may be mentioned in addition to the standard modern text books:

1. H. A. Wheeler, "Wide-band amplifiers for television," PROC. I.R.E., vol. 27, pp. 429–438; July, 1939.
2. A. J. Ferguson, "Termination effects in feedback amplifier chains," *Canad. Jour. Phys.*, Section A, vol. 24, pp. 56–278; July, 1946.

Reprinted from *IRE Trans. Circuit Theory*, **CT-2**, 74–85 (Mar. 1955)

A Practical Method of Designing RC Active Filters*

R. P. SALLEN† AND E. L. KEY†

Introduction

IN THE FREQUENCY range below about 30 cps, the dissipation factors of available inductors are generally too large to permit the practical design of inductance-capacitance (LC) or resistance-inductance-capacitance (RLC) filter networks. The circuits described in the following pages were developed and collected to provide an alternative method of realizing sharp cut-off filters at very low frequencies. In many cases the active elements can be simple cathode-follower circuits that have stable gain, low output impedance and a large dynamic range.

General Method of Achieving Arbitrary Transfer Characteristics

A passive two-terminal pair network consisting of resistive and capacitive elements has an open-circuit transfer ratio of the form

$$G(s) = \frac{N(s)}{D(s)} = \frac{a_m s^m + a_{m-1}s^{m-1} + \cdots + a_1 s + a_0}{b_m s^m + b_{m-1}s^{m-1} + \cdots + b_1 s + b_0}, \quad (1)$$

where s is the complex frequency variable $(\sigma + j\omega)$, the a_i and b_i are real positive constants, and the b_i are nonzero. All the poles of $G(s)$ lie on the negative real axis of the s-plane, a property that severely limits the application of passive RC circuits to sharp cut-off filters.

The unbalanced $(n+1)$-terminal pair RC network shown in Fig. 1 can be characterized by the relation

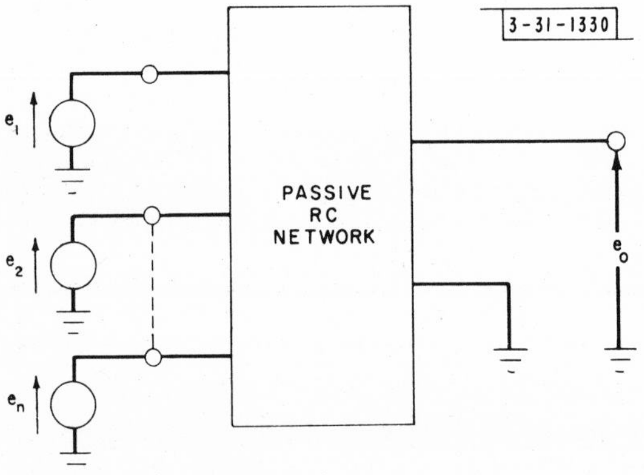

Fig. 1—The $(n+1)$-terminal pair network.

$$e_0(s) = \frac{e_1 N_1(s) + e_2 N_2(s) + \cdots + e_n N_n(s)}{D(s)}, \quad (2)$$

where the individual transfer ratios, $[N_i(s)]/[D(s)]$, have the same properties ascribed to (1).

If active elements are added to the multiterminal network in the manner shown in Fig. 2, the over-all transfer ratio, $[e_0(s)]/[e_1(s)]$, is given by

$$G(s) = \frac{K_{11}N_1(s) + K_{12}N_2(s) + \cdots + K_{1n}N_n(s)}{D(s) - [K_{01}N_1(s) + K_{02}N_2(s) + \cdots + K_{0n}N_n(s)]}. \quad (3)$$

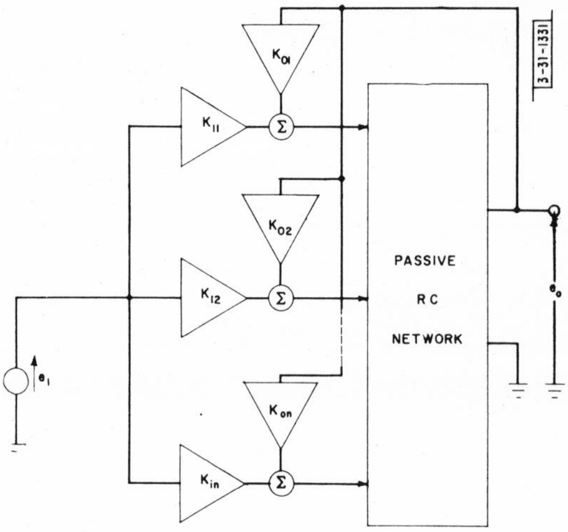

Fig. 2—Multi-terminal active network.

Generally speaking, it is possible to select an appropriate network and a series of constants K_{1i}, K_{0i} so that the poles and zeros of $G(s)$ can be placed anywhere in the complex plane. (Complex critical frequencies will occur, of course, in conjugate pairs.) Under certain conditions, all transfer functions of a given degree can be achieved with one fixed network by selection of appropriate K's.

Actually, a maximum of four active elements are required for the circuit of Fig. 2; two amplifiers for each of the two sets of K's, one with positive gain, one with negative gain. The remaining values of K_{1i} and K_{0i} can be obtained by means of passive attenuators.

Any transfer voltage ratio ordinarily realizable by means of passive RLC networks can be achieved with the circuit of Fig. 2. In addition, a variety of oscillators can also be characterized by the transfer function $G(s)$ in (3).

* This paper is an abridged version of Tech. Rep. No. 50, published by Lincoln Lab., M.I.T., Lexington, Mass., on May 6, 1954, under same title. Reference may be made to that report for useful design charts of Butterworth and Tchebycheff filters. The research in this document was supported jointly by the Army, Navy, and the Air Force under contract with M.I.T.

† Lincoln Lab., Mass. Inst. Tech., Cambridge, Mass.

In most cases it is desirable to limit the application of the general circuit of Fig. 2 to transfer ratios with only two conjugate poles. Any given transfer ratio can be achieved by a cascade of simpler circuits of this kind and one or more passive RC networks.

The second-order transfer function,

$$G(s) = \frac{a_2s^2 + a_1s + a_0}{b_2s^2 + b_1s + b_0}, \tag{4}$$

can be realized by means of the circuits of Fig. 3, which are special cases of that of Fig. 2. The arrangement of Fig. 3(b) includes two active elements that may be separate amplifiers or one amplifier with two input points. The RC passive networks generally have two capacitors and two resistors each, and a circuit-design procedure is available that affords one considerable control over the orders of magnitude of the components.

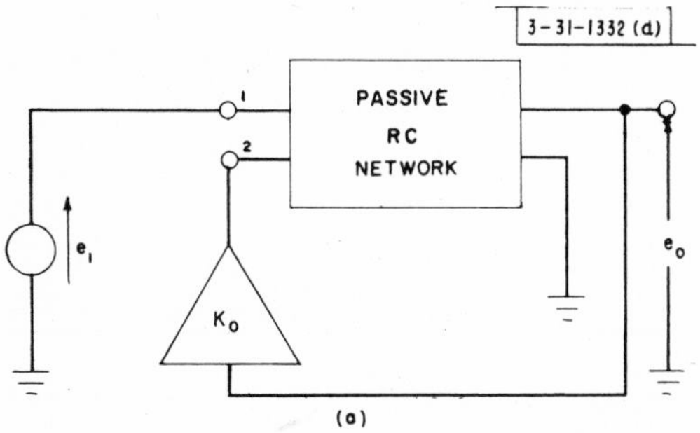

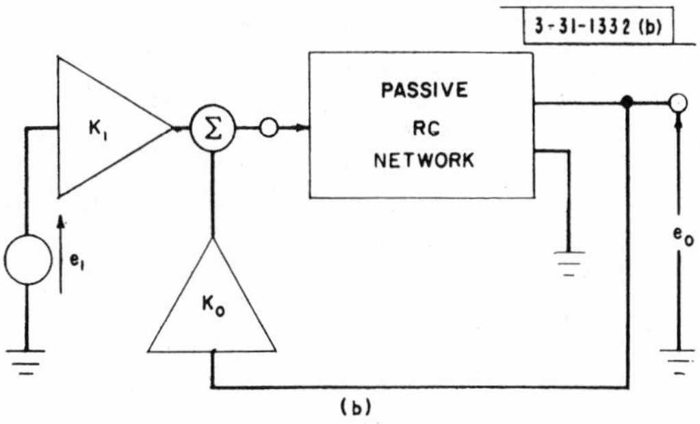

Fig. 3—Second-order active networks.

The design of circuits of the form illustrated in Fig. 3 can be facilitated by means of the catalog of possible circuit arrangements that has been compiled and is given in a later section. There are probably other useful circuits of this form that could be added to the catalog, but for most applications the present list of eighteen networks will be found adequate.

General Approach to Network Design

The basic objective of a design procedure based on the network catalog is the control over the locations of the poles of the transfer voltage ratio (4). These poles are the zeros of the denominator polynomial which, for convenience, can be normalized in the following manner. We have

$$D_0(s) = [D(s) - K_0N(s)] = b_2s^2 + b_1s + b_0 \text{ (Figs. 3 and 4)}$$

$$= b_0\left[\left(\frac{s}{\sqrt{\frac{b_0}{b_2}}}\right)^2 + \frac{s}{\sqrt{\frac{b_0}{b_2}}}\cdot\frac{b_1}{\sqrt{b_0b_2}} + 1\right]$$

$$= b_0\left[\left(\frac{s}{\omega_0}\right)^2 + \left(\frac{s}{\omega_0}\right)d + 1\right], \tag{5}$$

where

$$\omega_0 = \sqrt{\frac{b_0}{b_2}} \quad \text{and} \quad d = \frac{b_1}{\sqrt{b_0b_2}}.$$

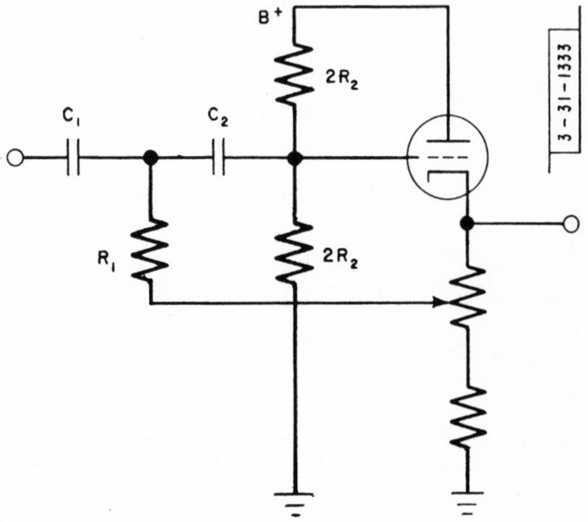

Fig. 4—High-pass filter circuit.

In the s-plane, the zeros of $D_0(s)$ lie on a circle of radius ω_0 and have a real part equal to $-d\omega_0/2$.[1] The shape of the frequency characteristics of $D_0(s)$ are dependent only on the value of the parameter d; the constant ω_0 determines their positions in the frequency domain, and b_0 determines the relative amplitude. The parameter ω_0 can be given the physical interpretations "resonant frequency," "cut-off frequency," etc., depending upon the nature of the numerator of $G(s)$.

It is convenient, in designing a circuit for a given $D_0(s)$, for one to set $\omega_0 = 1$ radian per second temporarily, and to establish the required value of d. The network response can then be shifted in frequency to ω_0 by dividing the resistive elements or the capacitive elements of the circuit by the desired value of ω_0.

In most of the networks in the catalog, there are five basic design variables: two resistances, two capacitors,

[1] This is true only for $d \leq 2$ when $d > 2$, the zeros lie on the negative real axis, a case that is not of present interest.

the gain K. The relationships between the variables that are independent of d are given with each network. Several additional parameters that have been found useful for designing a network for a given d include two products of a resistance and capacitance (designated T_1 and T_2), the ratio of the resistances (ρ), and the ratio of the capacitors (γ). The establishment of a specified value of d is accomplished by means of two of these parameters and the gain K. With each network in the catalog is a short table that specifies, for a given choice of parameters, the appropriate group of design relations for d given at the end of the catalog.

The form of the numerator of $G(s)$ is determined by the particular network chosen for the function. In some cases the numerator constants can easily be established at the desired values; in others an attempt to do this may severely limit the parameters affecting the value of d in the denominator and lead to an unsatisfactory circuit design. In this case, the numerator polynomial can be realized by means of additional passive or active networks. A method of network design is discussed later.

Catalog of Second-Order Active Networks

Definitions of Parameters:

$$T_1 = R_1C_1$$
$$T_2 = R_2C_2$$
$$\rho = \frac{R_1}{R_2}$$
$$\gamma = \frac{C_2}{C_1}$$

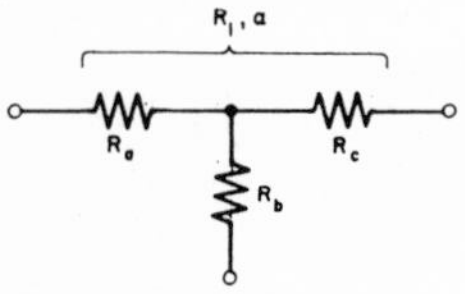

$$R_1 = R_c + \frac{R_aR_b}{R_a + R_b}$$
$$\alpha = \frac{R_a}{R_a + R_b}$$

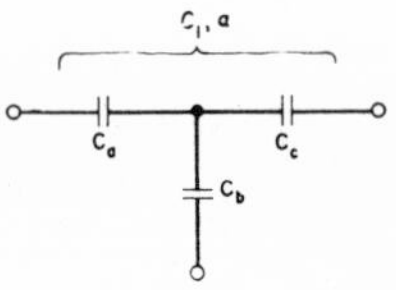

$$C_1 = \frac{C_c(C_a + C_b)}{C_a + C_b + C_c}$$
$$\alpha = \frac{C_b}{C_a + C_b}$$

FUNCTIONS OF FORM $\frac{h}{s^2 + ds + 1}$ | FUNCTIONS OF FORM $\frac{hs^2}{s^2 + ds + 1}$

1

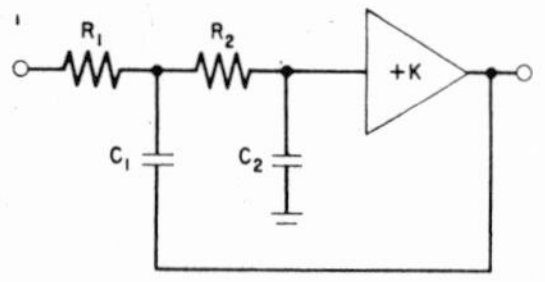

$T_1T_2 = 1$ $h = k$

Parameters	d Formulas Group
ρ, T_1	III
ρ, T_2	I
γ, T_1	IV
γ, T_2	II

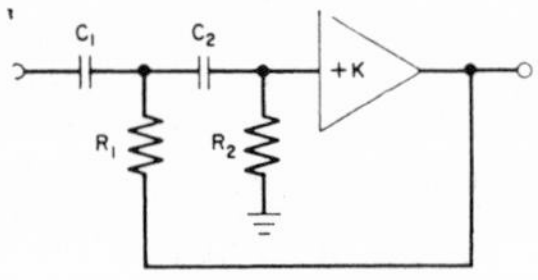

$T_1T_2 = 1$ $h = K$

Parameters	d Formulas Group
ρ, T_1	II
ρ, T_2	IV
γ, T_1	I
γ, T_2	III

2

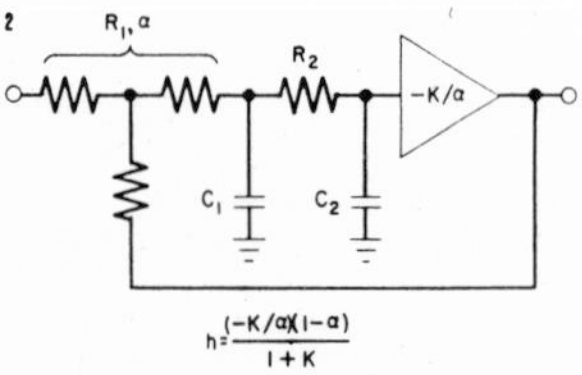

$$h = \frac{(-K/\alpha)(1-\alpha)}{1+K}$$

$$\frac{T_1T_2}{1+K} = 1$$

Alternative with 2 active inputs:

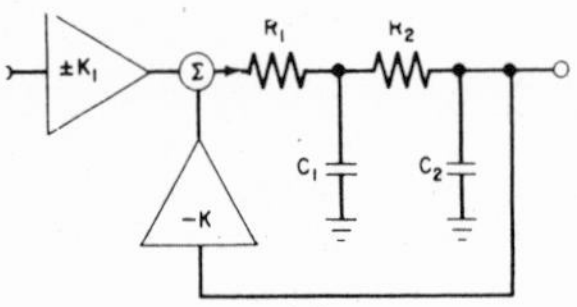

$$h = \frac{\pm K_1}{1+K}$$

Parameters	d Formulas Group
ρ, T_1	VIII
ρ, T_2	VII
γ, T_1	VII
γ, T_2	VIII

4

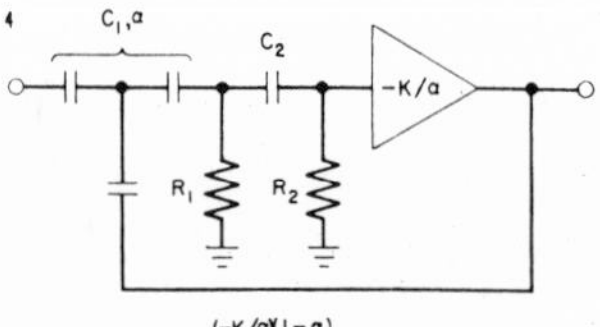

$$h = \frac{(-K/\alpha)(1-\alpha)}{1+K}$$

$$T_1T_2(1+K) = 1$$

Alternative with 2 active inputs:

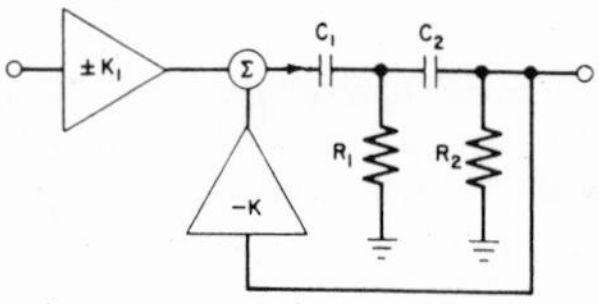

$$h = \frac{\pm K_1}{1+K}$$

Parameters	d Formulas Group
ρ, T_1	X
ρ, T_2	IX
γ, T_1	IX
γ, T_2	X

FUNCTIONS OF FORM $\frac{hs}{s^2+ds+1}$

5

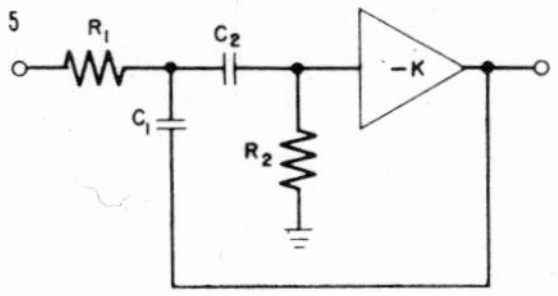

$T_1T_2(1+K)=1 \quad h=-KT_2$

Parameters	d Formulas Group
ρ, T_1	X
ρ, T_2	IX
γ, T_1	IX
γ, T_2	X

6

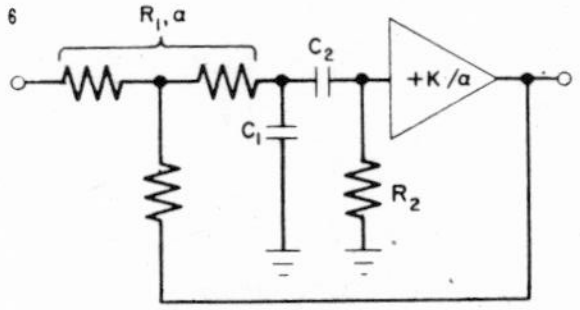

$h=(K/a)(1-a)T_2$

$T_1T_2 = 1$

Parameters	d Formulas Group
ρ, T_1	II
ρ, T_2	IV
γ, T_1	I
γ, T_2	III

Alternative with 2 active inputs:

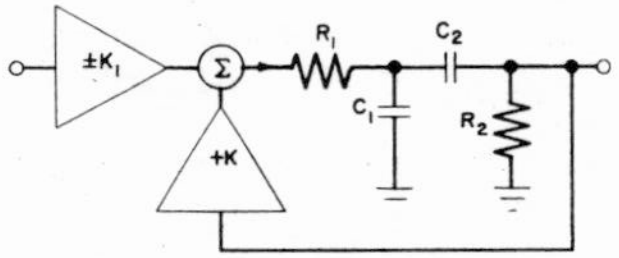

$h=\pm K_1T_2$

FUNCTIONS OF FORM $\frac{hs}{s^2+ds+1}$

7

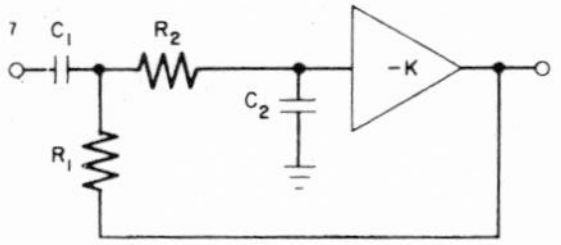

$\frac{T_1T_2}{1+K}=1 \quad h=-K/T_2$

Parameters	d Formulas Group
ρ, T_1	VIII
ρ, T_2	VII
γ, T_1	VII
γ, T_2	VIII

8

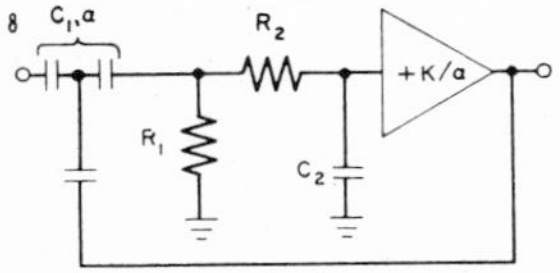

$h=(K/a)(1-a)T_1$

$T_1T_2 = 1$

Parameters	d Formulas Group
ρ, T_1	III
ρ, T_2	I
γ, T_1	IV
γ, T_2	II

Alternative with 2 active inputs:

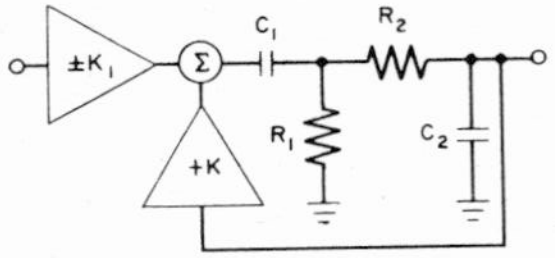

$h=\pm K_1T_1$

FUNCTIONS OF FORM $\frac{hs}{s^2 + ds + 1}$

9

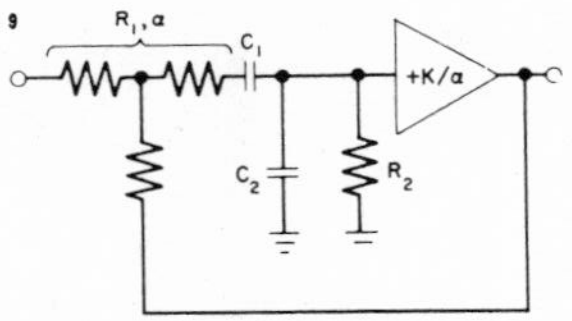

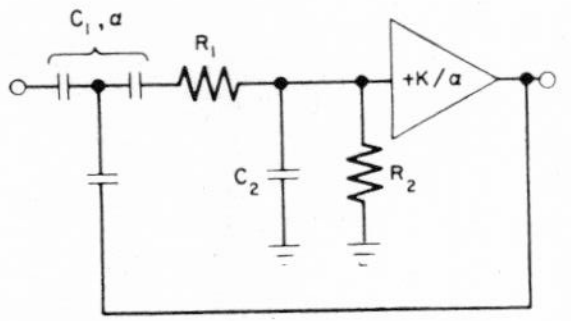

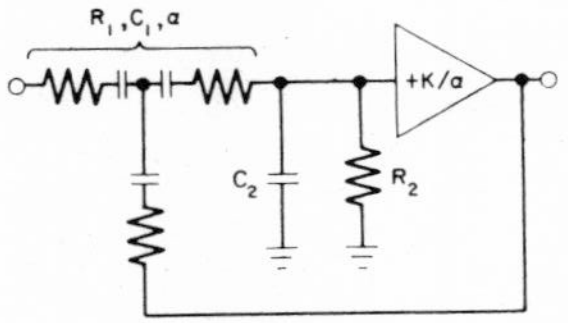

$$h = (K/\alpha)(1 - \alpha)\rho T_2$$
$$= (K/\alpha)(1 - \alpha)\gamma T_1$$
$$K > 1$$
$$T_1 T_2 = 1$$

Parameters	*d* Formulas Group
ρ, T_1	VI
ρ, T_2	V
γ, T_1	V
γ, T_2	VI

Alternative with 2 active inputs:

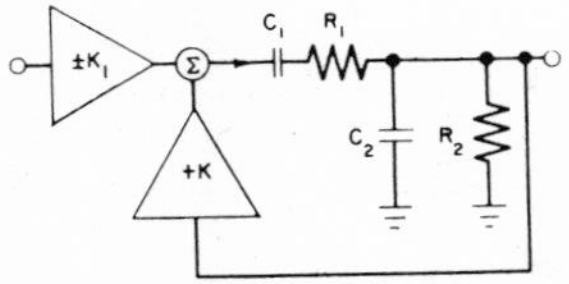

$h = \pm K_1 \rho T_2 = \pm K_1 \gamma T_1$

FUNCTIONS OF FORM $\frac{h[s + (1/\tau)]}{s^2 + ds + 1}$

10

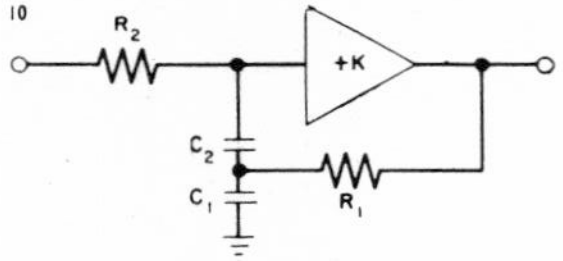

$T_1 T_2 = 1$ $\quad \tau = [(1/T_2) + \rho T_2] = T_1(1 + \gamma)$

$h = K[(1/T_2) + \rho T_2] = KT_1(1 + \gamma)$

Parameters	*d* Formulas Group
ρ, T_1	II
ρ, T_2	IV
γ, T_1	I
γ, T_2	III

1

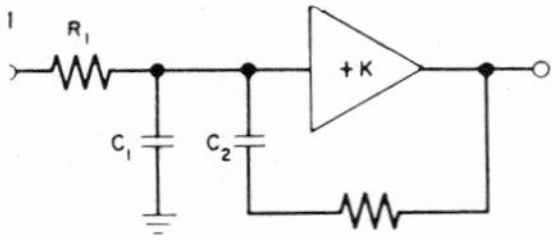

$T_1 T_2 = 1 \quad h = KT_2 \quad \tau = T_2$

Parameters	*d* Formulas Group
ρ, T_1	VI
ρ, T_2	V
γ, T_1	V
γ, T_2	VI

12

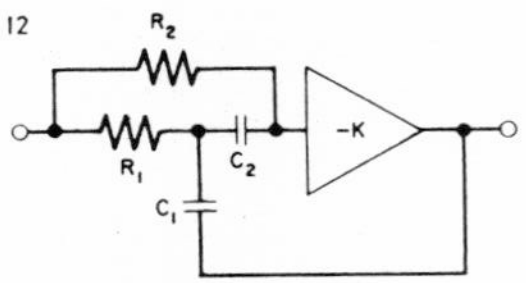

$T_1 T_2 (1 + K) = 1 \quad h = -Kd \quad \tau = d$

Parameters	*d* Formulas Group
ρ, T_1	X
ρ, T_2	IX
γ, T_1	IX
γ, T_2	X

FUNCTIONS OF FORM $\frac{hs[s + (1/\tau)]}{s^2 + ds + 1}$

13

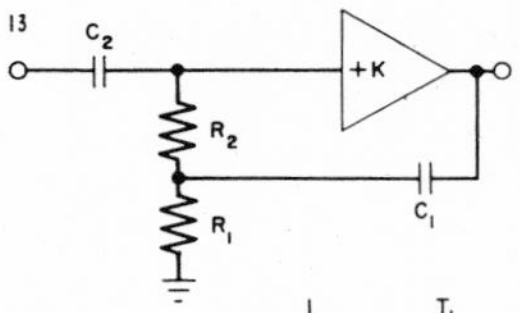

$T_1T_2 = 1 \quad h = K \quad \tau = \frac{1}{T_2(1+\rho)} = \frac{T_1}{1+\gamma T_1^2}$

Parameters	d Formulas Group
ρ, T_1	III
ρ, T_2	I
γ, T_1	IV
γ, T_2	II

14

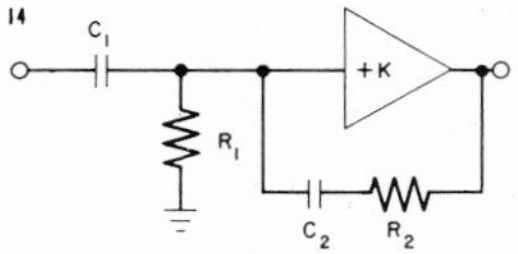

$T_1T_2 = 1 \quad h = K \quad \tau = T_2$

Parameters	d Formulas Group
ρ, T_1	II
ρ, T_2	IV
γ, T_1	I
γ, T_2	III

15

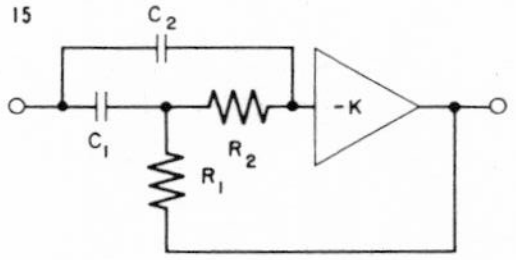

$\frac{T_1T_2}{1+K} = 1 \quad h = -K \quad \tau = 1/d$

Parameters	d Formulas Group
ρ, T_1	VIII
ρ, T_2	VII
γ, T_1	VII
γ, T_2	VIII

FUNCTIONS OF FORM $h\frac{s^2 + bs + 1}{s^2 + ds + 1}$

16

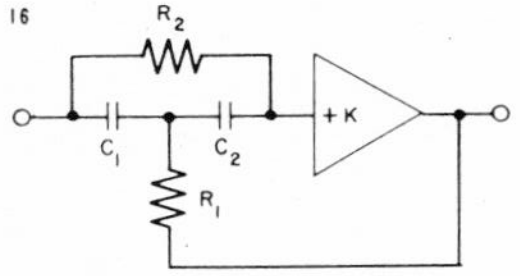

$T_1T_2 = 1 \quad h = K \quad b = (1/T_2) + \rho T_2 = T_1(1+\gamma)$

Parameters	d Formulas Group
ρ, T_1	II
ρ, T_2	IV
γ, T_1	I
γ, T_2	III

17

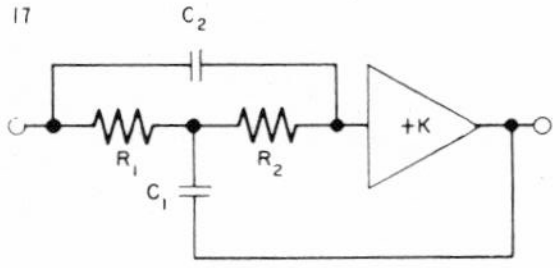

$T_1T_2 = 1 \quad h = K \quad b = (1/T_1) + \gamma T_1 = T_2(1+\rho)$

Parameters	d Formulas Group
ρ, T_1	III
ρ, T_2	I
γ, T_1	IV
γ, T_2	II

18

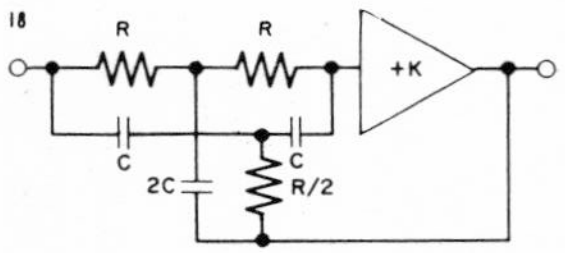

$RC = 1$

$h = K$

$b = 0$

$d = 4(1\text{-}K)$

DESIGN FORMULAS* FOR DISSIPATION FACTOR d

Group	(a) $d(x, K, T)$	(b) $T(x, K, d)$	(c) $K(x, d, T)$	(d) $K_{\min}$	(e) $x_{\min}$	(f) $T_{K_{\min}}$	(g) $K_{\max}$
I	$\frac{(1-K)}{T}+T(1+x)$	$\frac{d}{2(1+x)}\left[1 \pm \sqrt{1-\frac{4(1+x)(1-K)}{d^2}}\right]$	$T^2(1+x)+1-dT$	$\frac{4(1+x)-d^2}{4(1+x)}(<1)$	$\frac{d^2-4(1-K)}{4(1-K)}$	$\frac{d}{2(1+x)}$	$T^2(1+x)+1$
II	$\frac{(1+x-K)}{T}+T$	$\frac{d}{2}\left[1 \pm \sqrt{1-\frac{4(1+x-K)}{d^2}}\right]$	$T^2+(1+x)-dT$	$\frac{4(1+x)-d^2}{4}$	$\frac{d^2-4(1-K)}{4}$	$\frac{d}{2}$	$T^2+(1+x)$
III	$\frac{(1+x)}{T}+T(1-K)$	$\frac{d}{2(1-K)}\left[1 \pm \sqrt{1-\frac{4(1+x)(1-K)}{d^2}}\right]$	$\frac{T^2+(1+x)-dT}{T^2}$	$\frac{4(1+x)-d^2}{4(1+x)}(<1)$	$\frac{d^2-4(1-K)}{4(1-K)}$	$\frac{2(1+x)}{d}$	$\frac{T^2+(1+x)}{T^2}$
IV	$\frac{1}{T}+T(1+x-K)$	$\frac{d}{2(1+x-K)}\left[1 \pm \sqrt{1-\frac{4(1+x-K)}{d^2}}\right]$	$\frac{T^2(1+x)+1-dT}{T^2}$	$\frac{4(1+x)-d^2}{4}$	$\frac{d^2-4(1-K)}{4}$	$\frac{2}{d}$	$\frac{T^2(1+x)+1}{T^2}$
V	$\frac{1}{T}+T[1+x(1-K)]$	$\frac{d}{2[1+x(1-K)]}\left\{1 \pm \sqrt{1-\frac{4[1+x(1-K)]}{d^2}}\right\}$	$\frac{T^2(1+x)+1-dT}{xT^2}$	$\frac{4(1+x)-d^2}{4x}(>1)$	$\frac{4-d^2}{4(K-1)}$	$\frac{2}{d}$	$\frac{T^2(1+x)+1}{xT^2}$
VI	$\frac{[1+x(1-K)]}{T}+T$	$\frac{d}{2}\left[1 \pm \sqrt{1-\frac{4(1+x)}{d^2(1+K)}}\right]$	$\frac{T^2+(1+x)-dT}{x}$	$\frac{4(1+x)-d^2}{4x}(>1)$	$\frac{4-d^2}{4(K-1)}$	$\frac{d}{2}$	$\frac{T^2+(1+x)}{x}$
VII	$\frac{1}{T}+\frac{T(1+x)}{(1+K)}$	$\frac{d(1+K)}{2(1+x)}\left[1 \pm \sqrt{1-\frac{4(1+x)}{d^2(1+K)}}\right]$	$\frac{T^2(1+x)+1-dT}{(dT-1)}$	$\frac{4(1+x)-d^2}{d^2}$	$\frac{(1+K)d^2-4}{4}$	$\frac{2}{d}$	∞
VIII	$\frac{(1+x)}{T}+\frac{T}{(1+K)}$	$\frac{d(1+K)}{2}\left[1 \pm \sqrt{1-\frac{4(1+x)}{d^2(1+K)}}\right]$	$\frac{T^2+(1+x)-dT}{dT-(1+x)}$	$\frac{4(1+x)-d^2}{d^2}$	$\frac{(1+K)d^2-4}{4}$	$\frac{2(1+x)}{d}$	∞
IX	$\frac{1}{T(1+K)}+T(1+x)$	$\frac{d}{2(1+x)}\left[1 \pm \sqrt{1-\frac{4(1+x)}{d^2(1+K)}}\right]$	$\frac{T^2(1+x)+1-dT}{dT-T^2(1+x)}$	$\frac{4(1+x)-d^2}{d^2}$	$\frac{(1+K)d^2-4}{4}$	$\frac{d}{2(1+x)}$	∞
X	$\frac{(1+x)}{T(1+K)}+T$	$\frac{d}{2}\left[1 \pm \sqrt{1-\frac{4(1+x)}{d^2(1+K)}}\right]$	$\frac{T^2+(1+x)-dT}{dT-T^2}$	$\frac{4(1+x)-d^2}{d^2}$	$\frac{(1+K)d^2-4}{4}$	$\frac{d}{2}$	∞

* $T=T_1$ or T_2, as appropriate; $x=\rho$ or γ, as appropriate.

Design of Active Networks by Means of the Catalog

While anyone may start with a given transfer function or with the networks in the catalog and work out his own method for selecting parameters, one approach has been found useful and is described here for those wishing to design such circuits most directly. In any event, it is strongly recommended that one work through at least one group of relations in the foregoing section to gain insight into their meaning.

The basis of the general procedure suggested below is the necessity that the practical design of an active filter must be carried out within the limitations imposed by the available components. Restrictions on the size of capacitors, number and complexity of amplifier stages, and requirements for variability are typical factors that impose practical circuit limitations and must be controlled.

Realizing a Specified Value of d

When a particular network has been selected, one chooses a set of two parameters—(ρ, T_1), (ρ, T_2), (γ, T_1) or (γ, T_2)[2]—and locates the appropriate group of design formulas in the foregoing section. If the problem involves restrictions on the size of capacitors, then γ is a useful parameter. On the other hand, if control of the resistance values is more important, one may use ρ. In general, T_1 and T_2 are equally convenient parameters except where one of them determines a factor in the numerator of $G(s)$.

In the same section, each formula group includes:

(a) The expression for d in terms of K and the two parameters (x, T), selected above (x stands for ρ or γ, T for T_1 or T_2);

(b) The solution of the equation in (a) for T;

(c) The solution of the equation in (a) for K;

(d) The minimum value of K satisfying the equation in (a) with arbitrary x, positive T [$K_{\min}$ is obtained by solving the equation $\partial K(d, x, T)/\partial T = 0$ for T and substituting the solution, $T_{K\min}$ into the expression for $K(d, x, T)$];

(e) The minimum value of x satisfying the equation in (a) with arbitrary K, positive T (x has a minimum in the same sense as K, above);

(f) The value of expression (b) when $K = K_{\min}$ ($x = x_{\min}$ is the same condition and both values of T are the same in this case);

(g) The value of expression (c) when $d = 0$ (the significance of $K_{\max}$ is discussed below).

In establishing the values of x, T and K, one selects any two of them arbitrarily (in an algebraic sense—with more purpose in the practical sense), subject to the algebraic limitations $K \geqq K_{\min}$, $x \geqq x_{\min}$. The value of the remaining parameter is then determined from formula (b) or (c). For most purposes, a recommended procedure is the assumption of x, K, and the solution for T.

As an example, suppose that a required

$$G(s) = \frac{s^2}{s^2 + 1.414s + 1}$$

is to be realized by means of network No. 3 in the catalog, using one cathode follower as the active element. (The circuit to be used is shown in Fig. 4.) Suppose further that R_2 (parallel combination of the two biasing resistors) shall be 1 megohm and that both capacitors shall have the same value.

First of all, we shall choose (γ, T_2) as our design parameters, so that we may easily control the ratio of the capacitors and the value of R_2. According to the catalog, Formula Group III for d is indicated.

Setting $\gamma = 1$, we have

$$K_{\min} = \frac{4(1+\gamma) - d^2}{4(1+\gamma)} = \frac{4\cdot 2 - (1.414)^2}{4\cdot 2} = 0.75.$$

If we set $K = 0.9$, a reasonable value for the amplifier of Fig. 4, then

$$T_2 = \frac{d}{2(1-K)}\left[1 \pm \sqrt{1 - \frac{4(1+x)(1-K)}{d^2}}\right]$$

$$= \frac{1.414}{2(1-0.9)}\left[1 \pm \sqrt{1 = \frac{4\cdot 2.(1-0.9)}{(1.414)^2}}\right]$$

$$= 1.59, 12.5.$$

The expression for d in Formula Group III, $d = (1+\gamma)/(T_2) + T_2(1-K)$, suggests that a choice of the smaller value of T_2 above would result in a more stable circuit, in that variations in the active element will have less effect on the value of d.

Up to this point, we have $R_2 = 10^6$, $T_2 = 1.59$, and $C_1 = C_2$. Then, making use of the relation $T_1T_2 = 1$, we find $C_1 = C_2 = 1.59\ \mu f$ and $R_1 = 3.93 \times 10^5$.

It can be stated as a general rule of thumb that values of d greater than 0.5 can be realized most easily and with the simplest circuits; as d approaches 0.2, more care becomes necessary in the circuit design. Finally, values of d of the order of 0.1 or less demand active elements that are more complicated and highly stabilized, and passive elements that have been carefully adjusted within close tolerances. The latter values of d are not generally encountered in low-frequency filters.

As stated it has been found most convenient to design second-order networks on the basis of 1 radian per second, and to make a subsequent shift of their characteristics to the appropriate frequency by altering the passive elements. The basic invariants under a frequency transformation of this kind are the parameters

[2] One might also employ (ρ, γ) as design parameters, but this pair appears to be less useful than the others since it provides less control over the actual magnitudes of the components.

ρ and γ; as long as the ratio of resistors and the ratio of capacitors remain constant, the frequency characteristics of the networks will have the same shape. The necessary invariance of ρ and γ indicates the technique for making filters with variable cut-off frequencies.

Imperfections in the Active Elements

It has been assumed, heretofore, that the active elements of the networks in Figs. 2 and 3 and in the catalog possessed the ideal attributes: infinite input impedance, zero output impedance and stable gain. It is therefore important, in the design of an active network of this kind, for one to insure that the imperfections in the amplifiers used do not appreciably deteriorate the desired performance of the circuit.

With regard to finite output impedance, it can be seen that in many cases the active elements drive a portion of the passive network through a resistive element. In this case, the design can be made to incorporate the output impedance in the resistive element and effectively neutralize its effects.

On the other hand, where an amplifier drives a capacitive branch of a network, it is imperative that the output impedance be considerably smaller than any of the resistive elements of the network. This condition is most serious when the value of d is very small and the gain of the active element is close to $+1$. The limitation of the amplifier output impedance to a reasonably small value will generally prevent any significant alteration of the network characteristics in the vicinity of the cut-off frequency(ies). On the other hand, the attenuation achieved in certain networks in regions well beyond cut-off will fall short of the expected value because the output impedance, though small, is still finite. This situation has been observed in low-pass networks at high frequencies and in "notch" circuits at the null frequency. Behavior of this kind can best be investigated by a direct analysis of the particular circuit involved. Since only "very high frequencies" or "null frequencies" are of interest in this case, the analysis can be simplified by the assumption of these extreme frequency conditions. Under these circumstances, the output impedance should be negligibly small in comparison with $(1-K)$ times the value of resistive elements of the networks. Fortunately, strict requirements of this sort do not occur often in low-frequency filters.

Another interesting departure of the active elements from the ideal is the drift in their gain. With most of the networks in the catalog (those with $T_1T_2=1$), the position of the transfer characteristics in the frequency domain is independent of the active element; with a few others this is not so. In both cases, however, a drift in gain will result in a change in the actual value of d and in the shape of the frequency characteristics. It is often possible, as in the previous example, to reduce this dependence by an appropriate choice of parameters; but in any event the active gain should generally be at least $1/d$ times as stable as the expected value of d.

There is another kind of instability often characteristic of active RC networks of the kind discussed here, namely, their tendency to become oscillators. This tendency is most prevalent when the value of d is small. Even in some circuits where the active gain is ostensibly free from drift, oscillations may be sustained by an amplifier that drives itself into a region of its characteristics where the gain is far greater than expected.

A basic cure for a situation of this kind is the use of a feedback amplifier for the active element, such that the gain K is given by

$$K = \frac{A}{1+\beta A}, \tag{6}$$

where A is the gain of the amplifier without feedback, β is the feedback ratio derived from passive elements. It is easily seen that the value of K is absolutely limited to $1/\beta$, regardless of the value of A. The simple cathode-follower circuit illustrated in Fig. 4 is an example of this kind of active element. The critical value of K for a given network, K_{max}, is given at the end of the network catalog. A practical circuit design must include means for insuring that the active gain does not approach this value.

Adjustment of Physical Networks

When an active network has been constructed with physical components, minor adjustments in the latter are frequently required to achieve the performance indicated by the design. If the departure from the expected characteristics is not large, the trimming of a single capacitor or resistor may suffice to properly position the network characteristics in the frequency domain. The shapes of the characteristics are most easily altered by adjustment of the gain K.

In the event that the departure from expected characteristics is large and is not accountable to the usual tolerances in components, one may look to the following as possible sources of error: miscalculation of design parameters, excessive amplifier output impedance, poor capacitor "Q". It is unreasonable to ignore the "Q" of large paper capacitors in networks where the resistive elements are of the order of 1 megohm or more.

When the network design includes an active element whose gain is slightly less than $+1$ (e.g., the circuit of Fig. 4), it is usually difficult to measure or adjust the quantity $(1-K)$ directly with necessary accuracy. If a potentiometer is available for trimming the gain (as in Fig. 4), one may effect the adjustment in a simple manner by observing the over-all network amplitude-frequency response. The expected frequency-response characteristic for three transfer functions is illustrated in Fig. 5. Note that, although the latter are written on the basis of 1 radian per second, the frequency response is indicated at the true "resonant" frequency ω_0.

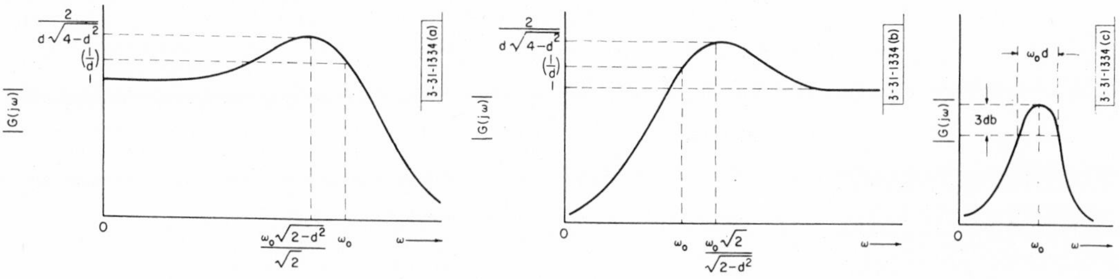

(a) $G(s) = \dfrac{1}{s^2 + ds + 1}$ (b) $G(s) = \dfrac{s^2}{s^2 + ds + 1}$ (c) $G(s) = \dfrac{s}{s^2 + ds + 1}$

Fig. 5—Amplitude-frequency response of several active networks.

It is sometimes desirable to avoid the necessity for adjusting the elements of a network, particularly the passive ones, by their prior selection within specified tolerances. Although the subject of tolerances has not been studied in detail, the expected variability of the network characteristics will generally be of the same order of magnitude as the tolerances in the passive elements, when the active elements are properly adjusted.

Factoring a High-Order $G(s)$ into Second-Order Transfer Ratios

A third- or higher-order transfer ratio can be written in the form

$$G(s) = \frac{N(s)}{D(s)} = \frac{N_1(s)}{D_1(s)} \times \frac{N_2(s)}{D_2(s)} \times \cdots \times \frac{N_n(s)}{D_n(s)}, \quad (7)$$

in a variety of ways such that the transfer ratios, $G_i(s) = [N_i(s)]/[D_i(s)]$, contain first- or second-order polynomials with real coefficients in the numerator and denominator.

All of the first-order denominators can be achieved with one passive RC network or by means of a cascade of isolated RC networks which, incidentally, can be made to absorb some of the factors of $N(s)$. The remaining second-order denominator polynomials are each identified with an individual active RC network of a form shown in the catalog. Practical considerations determine the pairing of the appropriate numerator factors with the second-order denominators; the relative properties of the various networks are the key to the pairing process.

No general rules can be given for selecting a group of second-order networks, for the choice is dependent on the particular requirements of the over-all circuit. However, an example can be used to indicate the kind of reasoning one may employ in a given situation. Consider the transfer ratio

$$G(s) = \frac{s^2(s+1)}{(s^2 + 0.5s + 1)(s^2 + s + 1)(s + 2)},$$

and assume we are interested in realizing $G(s)$ with the simplest and most easily designed circuits possible.

First of all, one passive network is indicated: either $1/(s+2)$, $s/(s+2)$ or $(s+1)/(s+2)$. The possible second-order active networks will then have functions of the form

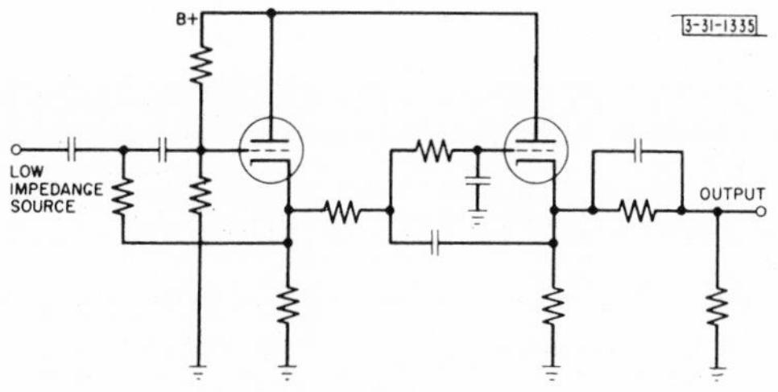

$$\frac{S^2}{S^2 + 0.5s + 1} \; \frac{1}{S^2 + S + 1} \; \frac{S+1}{S+2}$$

Fig. 6—Circuit with transfer function.

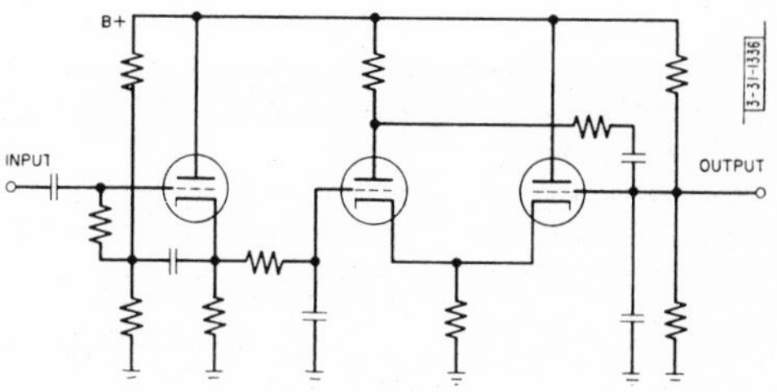

$$\frac{S(S+1)}{S^2 + 0.5s + 1} \; \frac{1}{S+2} \; \frac{S}{S^2 + S + 1}$$

Fig. 7—One alternative to circuit of Fig. 6.

$$\frac{hs^2(s+1)}{(S^2+0.5s+1)(S^2+s+1)(S+2)}$$

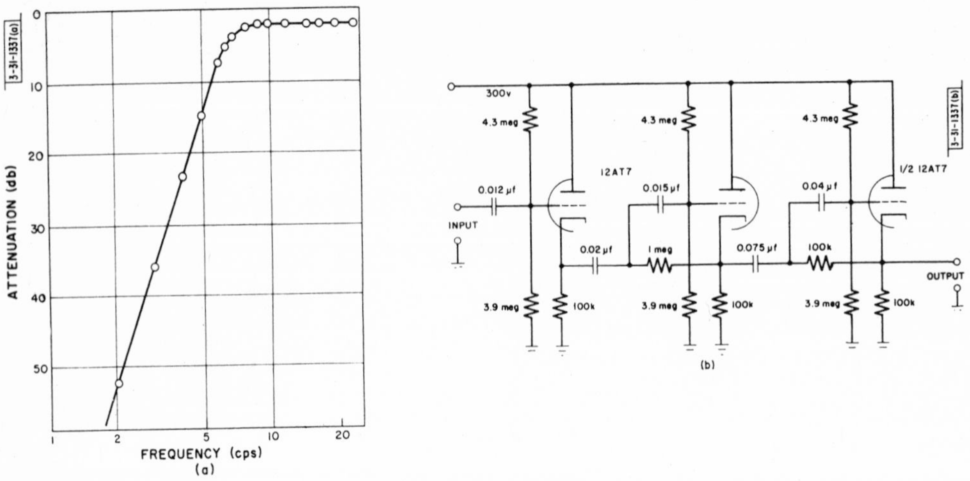

Fig. 8—Active high-pass filter.

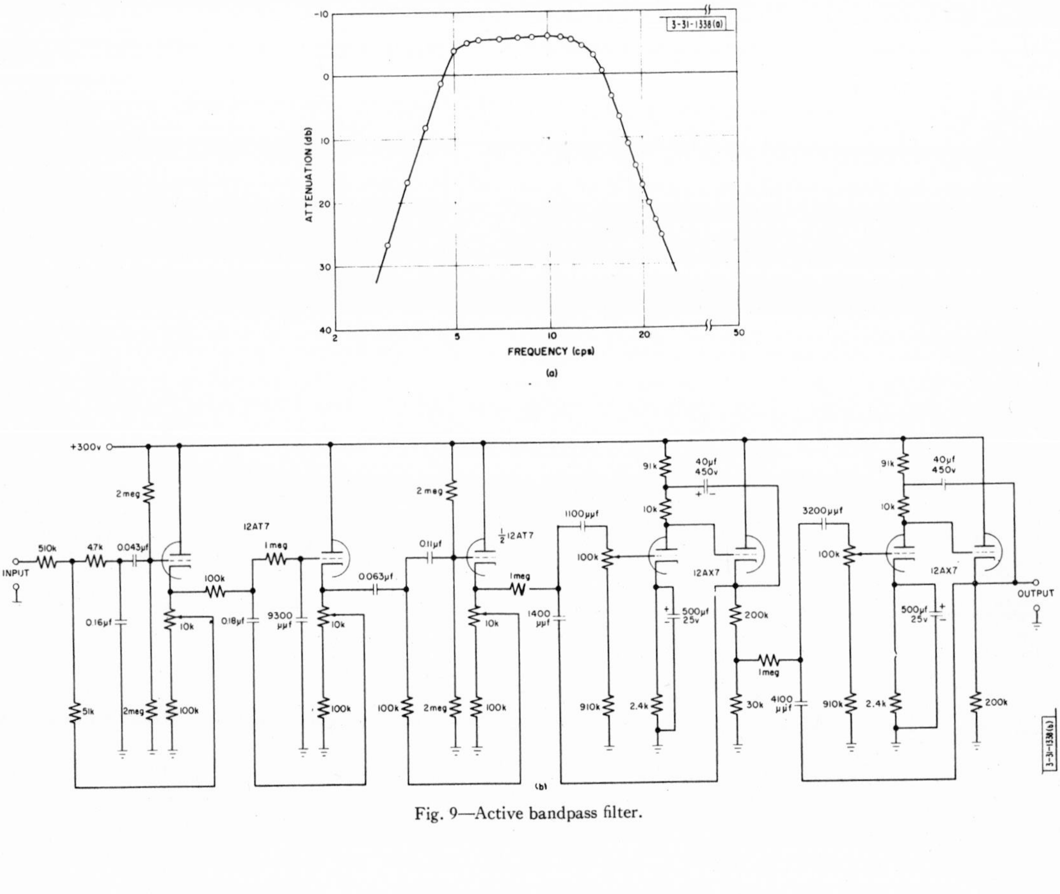

Fig. 9—Active bandpass filter.

$$\frac{1}{(s^2+ds+1)}, \quad \frac{s}{(s^2+ds+1)}, \quad \frac{s^2}{(s^2+ds+1)},$$

$$\frac{s+1}{(s^2+ds+1)} \quad \text{or} \quad \frac{s(s+1)}{(s^2+ds+1)}.$$

Because the last two functions contain zeros at $s=-1$, the establishment of which will tend to inhibit the freedom usually desired in the design for d in the denominator, we may justifiably simplify the problem by incorporating the zero at -1 in the passive network.

Reference to the catalog further indicates that networks for numerators 1 and s^2 are simple and can be achieved with positive gains less than 1. Finally, since the smaller values of d require amplifiers that more closely approximate the ideal active element, we shall associate the s^2 numerator (network No. 3) with the denominator whose $d=0.5$, so that R_1 can absorb the output impedance of the amplifier. A reasonable set of factors of $G(s)$ is then

$$G(s) = \frac{s^2}{(s^2+0.5s+1)} \times \frac{1}{(s^2+s+1)} \times \frac{s+1}{s+2}.$$

A possible circuit is shown in Fig. 6 (page 83), to which an alternative is given in Fig. 7 (page 83) where the network is based on a different set of factors of $G(s)$.

In general, there are a great number of possible circuits realizing a given $G(s)$ that can be justified on the basis of particular requirements for simplicity, output impedance, stability and so on.

Several additional points should be mentioned in regard to the cascading of second-order networks. First, a series of networks involving cathode-follower amplifiers as active elements, or a series of cathode-coupled amplifier networks, has a useful property whereby the bias given to the first amplifier can usually be carried over from stage to stage. This is important where resistive elements couple adjacent networks, since it avoids the need for coupling capacitors which, at low frequencies, might become fairly large.

In designing a circuit, it is always desirable for one to employ only capacitors whose impedance-frequency characteristics help to determine the transfer ratio; i.e., capacitors that are part of the basic networks in the catalog. Other capacitors, which might be used solely for bypassing, coupling, etc., will usually have impedances negligible at the frequencies of interest and will be appreciably larger than those in the former category. In this regard, it is often advisable, when one is designing an active RC filter to be followed by amplifier circuits, to incorporate single zeros and poles of the $G(s)$ into the amplifier interstage coupling networks, or even to add additional ones that can be cancelled by the filter network.

Finally, where possible, amplifiers in a chain of active networks should be placed so that those with the smallest dynamic range appear last.

Two examples of low-frequency filters that have been constructed and tested in the laboratory are shown in Figs. 8 and 9, on the previous page.

16

Reprinted from *IRE Trans. Circuit Theory,* **CT-7**, 62–63 (Mar. 1960)

Regenerative Modes of Active Networks*

E. S. KUH†

IN the regenerative type of circuits such as multivibrators and blocking oscillators, the maximum frequency of oscillation depends on the regenerative time.[1] When the circuit is in the regenerative operation, some natural frequencies of the circuit are in the right half of the complex frequency plane. These natural frequencies are called the regenerative modes. The regenerative time is determined by the location of the regenerative modes (usually on the positive real axis). In this paper a study is made on the permissible location in the complex plane of regenerative modes of a given active device under arbitrary passive imbedding.

We assume that an $n + 1$ terminal active device N_a can be represented by the admittance matrix of the form[2]

$$Y_a = G + pC, \tag{1}$$

where $p = \sigma + j\omega$ is the complex frequency. G is a real matrix and C is real, symmetric and positive definite. The open circuit natural frequencies are the zeros of $|Y_a|$. This active device is then imbedded in a passive but not necessarily reciprocal network N_p to form a new network N as shown in Fig. 1. The new natural frequencies

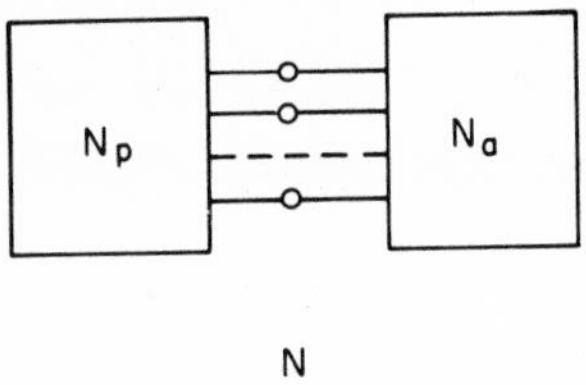

Fig. 1.

are given by the zeros of $|Y|$ where

$$Y = Y_a + Y_p. \tag{2}$$

Y_p is the admittance matrix of an $n + 1$ terminal passive, not necessarily reciprocal, network.

The regenerative modes are characterized by the natural frequencies in the right-half plane. That is, if

$$p_i = \sigma_i + j\omega_i \quad \text{with} \quad \sigma_i \quad \text{positive} \tag{3}$$

the ith mode is regenerative. Our problem is to determine 1) the largest σ_i for a given active device N_a under arbitrary passive imbedding, and 2) the network N_p to realize this regenerative mode with the largest σ_i.

Since C is positive definite, it is possible to find a transformation which simultaneously diagonalizes C and the symmetric part of G. Let N be an $n \times n$ real transformation matrix,

$$\begin{aligned} Y_a' &= N_t Y_a N = N_t(G + pC)N \\ &= -c_0[Q - pI] \end{aligned} \tag{4}$$

where c_0 is a real constant, I is the unit matrix and

$$Q = Q_d + Q_a. \tag{5}$$

Q_d is a diagonal matrix and Q_a is an antisymmetrix matrix. Physically, the operation of (4) amounts to the connection of ideal transformers to the active device to form a new circuit.[3] Hence the transformer network is a special form of a passive coupling network, N_p. However, since

$$|Y_a'| = |Y_a|\,|N|^2 \tag{6}$$

the natural frequencies of the new network are the same as the open circuit natural frequencies of the active device. If we refer to (4), we can state that the open circuit natural frequencies are the characteristic roots of the matrix Q.

We consider now the Hermitian part of Q.

$$Q_H = \tfrac{1}{2}(Q + Q^+) = Q_d \tag{7}$$

where the superscript + represents the transposed conjugate. From (4)

$$Y_{aH}' = -c_0[Q_d - \sigma I]. \tag{8}$$

Thus, the zeros of $|Y_{aH}'|$ are the characteristic roots of Q_H. If we denote the smallest and the largest characteristic roots of Q_H by m and M (they must be real), we can state that the real parts of the characteristic roots of Q are bounded by m and M.[4]

Let

$$p = \alpha + j\beta \tag{9}$$

be the characteristic roots of Q. Then

$$m \leq \alpha \leq M.$$

We therefore conclude that the real part of the open circuit natural frequencies are bounded by the smallest and the largest zeros of the determinant of the Hermitian matrix Y_{aH}' or the Hermitian part of the original active

* Manuscript received by the PGCT, September 23, 1959. This work was supported in part by the U. S. Navy under Contract N7 onr-29529.

† University of California, Berkeley, Calif.

[1] D. O. Pederson, "Regeneration analysis of junction transistor multivibrators," IRE TRANS. ON CIRCUIT THEORY, vol. CT-2, pp. 171–178; June, 1955.

[2] R. D. Thornton, "Active RC networks," IRE TRANS. ON CIRCUIT THEORY, vol. CT-4, pp. 78–89; September, 1957.

[3] V. Belevitch, "Synthése des résaux électriques passifs a n paires de bornes de matrice de répartition prédéterminée," *Annales des Télécommunications*, vol. 6, pp. 302–312; November, 1951.

[4] C. C. MacDuffee, "The Theory of Matrices," Chelsea Publishing Co., New York, N. Y.; 1956.

network, Y_{aH}. These bounds are, respectively, the smallest and the largest diagonal elements of the matrix Q.

For an arbitrary passive imbedding, the natural frequencies are given by (2). The following theorem is stated and proven.

Theorem: A given active device, N_a, under arbitrary passive imbedding can produce regenerative modes $p_i = \sigma_i + j\omega_i$. Let Y_a be the admittance matrix of N_a and Y_{aH} be its Hermitian part: then the maximum σ_i is equal to the largest zero of $| Y_{aH} |$.

First we establish the upper bound on σ_i. Since Y_p is the admittance matrix of a passive network, the Hermitian part Y_{pH} is positive semidefinite in the right-half plane.[5] From (2)

$$Y_H = Y_{aH} + Y_{pH}.$$

If the same congruent transformation of (4) is made,

$$\begin{aligned} Y_H' = N_t Y_H N &= -c_0[Q_d - \sigma I] + N_t Y_{pH} N \\ &= -c_0\left[Q_d - \frac{1}{c_0} N_t Y_{pH} N - \sigma I\right]. \end{aligned} \tag{12}$$

Thus, zeros of $| Y_H' |$ are the characteristic roots of

$$Q_d - \frac{1}{c_0} N_t Y_{pH} N. \tag{13}$$

Note that the matrix (13) is Hermitian, hence, has real characteristic roots. Since $N_t Y_{pH} N$ is positive semidefinite, it follows that the characteristic roots of (13) are not larger than the characteristic roots of Q_d. Therefore, the bounds on the natural frequencies of the over-all network cannot be moved to the right of the bounds of the natural frequencies of the open circuit case.

Next we prove that the upper bound can be realized with a passive imbedding. From (8) the zeros of $| Y_{aH}' |$ are at $\sigma = -\sigma_i$ where σ_i are the elements of the diagonal matrix Q_d. With reference to (4), it is seen that if a matrix

$$Y_p' = c_0 Q_a \tag{14}$$

is added to Y_a', then

$$Y' = -c_0(Q_d - pI). \tag{15}$$

[5] G. Raisbeck, "A definition of passive linear networks in terms of time and energy," *J. Appl. Phys.*, vol. 25, pp. 1510–1514; December, 1954.

Clearly, zeros of $| Y' |$ are then the same as zeros of $| Y_{aH}' |$. This indicates that a lossless nonreciprocal network may be used to realize the best regenerative mode. In terms of the original circuit, let

$$Y_a = G_s + G_a + pC \tag{16}$$

where G_s is real symmetric and G_a is real anti-symmetric, the passive network required is given by

$$Y_p = -G_a. \tag{17}$$

This is clear since

$$Y = Y_a + Y_p = G_s + pC \tag{18}$$

and with $p = \sigma + j\omega$

$$Y_{aH} = G_s + \sigma C. \tag{19}$$

Thus zeros of $| Y_{aH} |$, which characterize the bounds, are the natural frequencies of the new network.

For a pentode

$$Y_a = \begin{bmatrix} 0 & 0 \\ g_m & g_{m/\mu} \end{bmatrix} + p\begin{bmatrix} C_g & -C_m \\ -C_m & C_p \end{bmatrix} \tag{20}$$

the best regenerative mode is found to be

$$\sigma_1 = \frac{g_{m/2}}{C_g/\mu + C_m + \sqrt{C_p C_g + 2C_m C_g/\mu + C_g^2/\mu^2}}. \tag{21}$$

The passive network is simply a gyrator with

$$G = -g_{m/2}. \tag{22}$$

This is shown in Fig. 2.

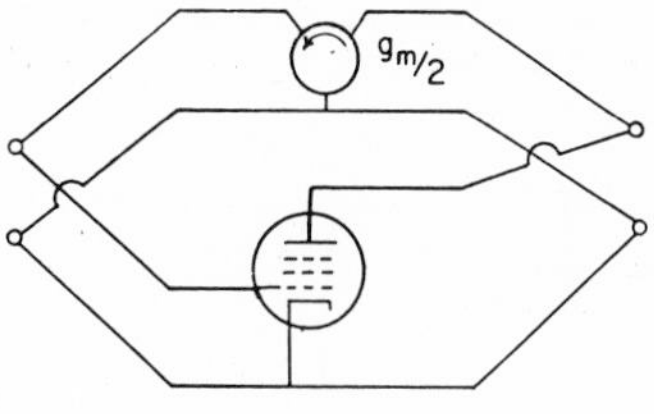

Fig. 2.

Acknowledgment

The author wishes to thank Prof. C. A. Desoer for his discussion.

V
The Scattering Matrix and Applications

Editor's Comments on Papers 17, 18, and 19

17 Belevitch: *Transmission Losses in 2n-Terminal Networks*

18 Youla, Castriota, and Carlin: *Bounded Real Scattering Matrices and the Foundations of Linear Passive Network Theory*

19 Youla: *A New Theory of Broad-Band Matching*

The use of scattering parameters to describe a transmission network with a specified load emerged after World War II, although it appears that they were used for special cases by Campbell and Foster in 1920. The Radiation Laboratory of MIT would appear to be an unlikely origin for this basic concept of circuit theory since that laboratory was involved solely in radar development. Physicists at the Radiation Laboratory who had previously studied wave-particle scatter problems applied the techniques that had been successful there to the problem of microwave waveguide junctions and the associated matching problem. Their results appeared in numerous internal reports. Fortunately, these reports were consolidated into the Radiation Laboratory series, and one volume in that series, edited by Montgomery, Dicke, and Purcell, gives an excellent summary (1948). Independently, Belevitch in Belgium studied lumped networks from a scattering point of view, although he called the scattering matrix the "efficiency matrix." His results appeared in his 1945 doctoral dissertation at Louvain University, part of which was published in 1948 (Paper 17). Belevitch treated problems of physical realizibility and the synthesis of networks containing ideal transformers. A great deal of the further progress in our understanding of the theory and applications of the scattering matrix came from engineers at the Microwave Research Laboratory at the Polytechnic Institute of Brooklyn (now the Polytechnic Institute of New York). A paper that summarizes some of their significant results was published by Youla, Castriota, and Carlin in 1959 (Paper 18).

Bode (1945) was the first to consider the problem of designing a lossless two-port network to match a source with internal resistance to a complex load consisting of a capacitor in parallel with a resistor. Fano (1950) extended Bode's result to the case of an arbitrary passive load impedance. He discussed in detail the limitation in terms of bandwidth and tolerance of mismatch and presented a method for optimal design. Youla developed an alternative approach to the Fano problem in terms of scattering parameters, offering conceptual and ease-of-derivation advantages and also being more general in the sense that it applies to the equalization of an active load (Paper 19). The broadband matching problem is an ideal vehicle for a demonstration of the advantages of the scattering matrix in the solution of transmission problems, as the reader will discover.

References

Bode, H. W. *Network Analysis and Feedback Amplifier Design,* Van Nostrand Reinhold Company, New York, 551 pp., 1945.

Campbell, G. A., and R. M. Foster. "Maximum Output Networks for Telephone Substation and Repeater Circuits," *Trans. AIEE,* **39**, 231–280 (1920), (discussion) 281–290.

Fano, R. M. "Theoretical Limitations on the Broadband Matching of Arbitrary Impedance," *J. Franklin Inst.,* **249**, 57–83, 139–154 (1950).

Montgomery, C. G., R. H. Dicke, and E. M. Purcell. *Principles of Microwave Circuits,* McGraw-Hill Book Company, 486 pp., 1948.

Reprinted from *J. Appl. Phys.*, **19**(7), 636–638 (1948)

Transmission Losses in 2*n*-Terminal Networks

VITOLD BELEVITCH
Bell Telephone Manufacturing Company, Antwerp, Belgium

(Received October 23, 1947)

The semipositive character of the dissipation function of any physical $2n$-terminal network imposes some restrictions on the realizability of prescribed transmission losses between the various pairs of terminals when they are connected to given impedances. These restrictions are most easily expressed in terms of elements of the efficiency matrix of the network, which is defined in the first part of this paper. Discussions of the application of matrix calculus to the solution of reactive and resistive, transformer-type, and general $2n$-terminal networks are presented.

A very useful tool in solving telephone transmission problems is the conventional theory of 4-terminal transducers, often called 4-pole networks. Although this theory is sufficient for the study of most circuits, some problems involve networks with a larger number of terminals, which are often referred to as $2n$-terminal, or $2n$-pole networks. This paper is a contribution to their theory.

The particular point that will be considered in this paper is the restriction imposed by physical realizability on the arbitrary choice of transmission losses between various pairs of terminals connected to a given set of terminating (generator or receiver) impedances.

THIS theory was suggested by an old paper by Campbell and Foster,[1] and a part of the following theorems is an extension of their results in modern notation using matrix calculus.

I. DEFINITION OF TRANSMISSION LOSSES

Assume a $2n$-pole network with terminal pairs numbered $(1,1')$, $(2,2')$, $\cdots$, (n,n'). Any pair (p,p') will be across a generator or a receiver impedance R_p, presumed, as is usual in transmission theory, to be a pure resistance. The network is thus associated with a given set of terminating resistances.

Suppose that a generator of voltage V_p be connected in series with resistance R_p, producing a current I_q through resistance R_q, when all other terminal pairs are across their respective resistances. We shall let S_{qp} be the square root of the ratio between the complex power $R_qI_q^2$ delivered to the receiver, and the maximum power $V_p^2/4R_p$ available from the generator. Thus

$$S_{qp}=[2(R_pR_q)^{\frac{1}{2}}I_q/V_p]. \tag{1}$$

The usual definitions of the effective loss A_{qp} in nepers, and of *phase shift* B_{qp} in radians are

$$A_{qp}+jB_{qp}=-\log_e S_{qp}. \tag{2}$$

In the following we shall use instead the expression S_{qp}, and call it the *complex efficiency factor* between terminals p and q.

[1] G. A. Campbell and R. M. Foster, "Maximum output networks for telephone substation and repeater circuits," Trans. A.I.E.E. (February, 1920).

It will also be convenient to introduce a factor S_{pp}, which will be defined as the *reflection coefficient* at the terminal pair (p,p');

$$S_{pp}=(R_p-\zeta_p/R_p+\zeta_p), \tag{3}$$

where ζ_p is the impedance measured between the terminals (p,p'), with all other terminal pairs closed on their associated resistances. By writing

$$A_{pp}+jB_{pp}=-\log_e S_{pp}, \tag{4}$$

the usual definitions of the *return loss* and *phase* are obtained. The factor S_{pp} is thus the square root of the complex ratio of reflected power to maximum power. The condition $S_{pp}=0$ expresses an impedance match at terminals (p,p').

The matrix S, composed of elements S_{pq}, will be termed the *efficiency matrix* of the $2n$-pole network associated with a given set of terminating resistances. If the main diagonal of this matrix is zero, the network is simultaneously matched at all its terminal pairs.

A simultaneous multiplication of the pth row and of the pth column by -1 is irrelevant, because it merely interchanges the terminals p and p'.

II. CALCULATION OF EFFICIENCY MATRIX

We start from the usual set of equations of a $2n$-pole network in terms of the self and transfer impedances;

$$V_p-R_pI_q=Z_{p1}I_1+Z_{p2}I_2+\cdots+Z_{pn}I_n, \tag{5}$$

where $p=1, 2, \cdots, n$. In (5), for the sake of

generality, it is supposed that generators were present in any terminal pair.

Introducing *normalized* currents, voltages, and impedances by writing

$$I_p'=(R_p)^{\frac{1}{2}}I_p,\quad V_p'=V_p/(R_p)^{\frac{1}{2}},$$
$$Z_{pq}'=Z_{pq}/(R_pR_q)^{\frac{1}{2}}, \tag{6}$$

the system of equations in (5) becomes

$$V_p'-I_p'=Z_{p1}'I_1'+Z_{p2}'I_2'+\cdots+Z_{pn}'I_n'. \tag{7}$$

Since we will always consider normalized quantities in the following equations, we may drop the primes and write (7) in matrix form:

$$V-I=ZI, \tag{8}$$

where V and I are one-column matrices, and Z is the usual normalized impedance matrix of the network.

Introducing the unit-matrix E, (8) may be written

$$V=(Z+E)I. \tag{9}$$

Solving for I,

$$I=QV, \tag{10}$$

where Q is the inverse matrix of $Z+E$;

$$Q=(Z+E)^{-1}. \tag{11}$$

A generator V_p, acting alone in the terminal pair p, produces in the terminal pair q a current $I_q=Q_{qp}V_p$. Since normalized expressions are equivalent to making the terminating resistances equal to unity, (1) is reduced to

$$S_{qp}=2Q_{qp}. \tag{12}$$

Similarly, Q_{pp} is the normalized driving-point admittance of mesh p. If we denote by ζ_p' the normalized impedance ζ_p/R_p, (3) becomes

$$S_{pp}=(1-\zeta_p')/(1+\zeta_p'). \tag{13}$$

But, by Thévenin's theorem,

$$Q_{pp}=1/(1+\zeta_p'), \tag{14}$$

and, from (13) and (14),

$$S_{pp}=2Q_{pp}-1. \tag{15}$$

Equations (12) and (15) may be combined in matrix form:

$$S=2Q-E=2(Z+E)^{-1}-E, \tag{16}$$

which may also be written

$$S=(E-Z)(E+Z)^{-1}. \tag{17}$$

This is a formal extension of (13).

In this calculation, we assumed that both matrices Z and Q exist, but this is not necessarily true. The degenerate cases will be examined later on.

It is shown in (17) that the matrix S is symmetrical, as is Z (reciprocity theorem). By replacing Z with Z^{-1} (or taking the inverse network), S is simply changed into $-S$, and this is a form of the duality principle. Inverse networks give the same transmission losses.

III. REACTIVE 2N-POLE NETWORKS

In the case of a network composed of only reactive elements, the matrix Z is purely imaginary. Taking the complex conjugate $\bar{S}$, Z is simply changed into $-Z$;

$$\bar{S}=(E+Z)(E-Z)^{-1}. \tag{18}$$

By multiplying (17) and (18),

$$S\bar{S}=E, \tag{19}$$

and *the efficiency matrix of a reactive 2n-pole network is therefore a complex orthogonal matrix.* The diagonal terms of (19), written explicitly, are

$$\exp(-2A_{p1}+\exp(-2A_{p2})+\cdots$$
$$+\exp(-2A_{pn})=1, \tag{20}$$

where $p=1, 2, \cdots, n$. This demonstrates that no power can be consumed in a reactive network; it is an extension of a familiar relation first established by Feldtkeller for reactive 4-pole networks.

IV. RESISTIVE 2N-POLE NETWORKS

In the case of a purely resistive network, the matrix Z is real. It is easily shown that Z and S may be simultaneously reduced to diagonal matrices (respectively, $\hat{Z}$ and $\hat{S}$) by means of a non-singular linear orthogonal transformation. Equation (17) is then transformed into

$$\hat{S}_{pp}=(1-\hat{Z}_{pp})/(1+\hat{Z}_{pp}). \tag{21}$$

The expression for the power consumed in a passive network is known to be of a semipositive quadratic form; when the quadratic form is transformed into a sum of squares, the coefficients are $\hat{Z}_{pp}$ and must be positive or zero. We thus have

$$-1 \leqslant \hat{S}_{pp} \leqslant 1. \tag{22}$$

The case of a degenerate network having no impedance matrix may be considered as being the limiting case of a network for which at least one of the numbers $\hat{Z}_{pp}$ becomes infinite, so that (22) must be extended:

$$-1 \leqslant \hat{S}_{pp} \leqslant 1. \tag{23}$$

In this last form the condition obviously expresses the fact that both matrices $E+S$ and $E-S$ are semipositive. This can be written explicitly in the usual way in terms of principal minors.

As an example, we will consider the case of a hexapole simultaneously matched at all its terminal pairs. The efficiency matrix is

$$S = \begin{Vmatrix} 0 & S_{12} & S_{13} \\ S_{12} & 0 & S_{23} \\ S_{13} & S_{23} & 0 \end{Vmatrix}.$$

By noting that the determinants of $E+S$ are non-negative, one obtains the two non-trivial conditions

$$1 \pm 2S_{12}S_{23}S_{13} - S_{13}^2 - S_{23}^2 - S_{12}^2 \geqslant 0. \tag{24}$$

If the hexapole has the same losses between all terminal pairs, i.e., if

$$|S_{12}| = |S_{13}| = |S_{23}| = s,$$

one of these conditions gives

$$1 - 3s^2 - 2s^3 = (1+s)^2(1-2s) \geqslant 0,$$

the other one being trivial. Thus

$$s \leqslant \tfrac{1}{2},$$

and the minimum value of the loss is 0.69 neper $=6$ decibels, or twice as great as could be expected from the division of the input energy between the two output resistances. Actually, a 3-decibel loss is obtained in a hybrid coil, but it is not symmetrical with respect to all terminal pairs.

V. IDEAL TRANSFORMERS

A network composed of only ideal transformers may be considered as being the limiting case either of a reactive or of a resistive network. As its efficiency matrix is real, (19) becomes

$$S^2 = E. \tag{25}$$

For an orthogonal symmetric matrix, the numbers $\hat{S}_{pp}$ are either $+1$ or -1. Since the sum of these numbers is equal to the sum of the main diagonal elements of S, the sum must be zero for a network simultaneously matched at all of its terminals, which is obviously impossible if the order of the matrix is odd. As a consequence, a network matching an odd number of terminal pairs and composed only of ideal transformers that match an odd number of terminal pairs does not exist.

VI. GENERAL 2*N*-POLE NETWORKS

The restrictions imposed on the efficiency matrix of a general $2n$-pole network are obtained by stating that the real part Z_{real} of the impedance matrix is semipositive. From (16) or (17),

$$Z = (E-S)(E+S)^{-1} = 2(E+S)^{-1} - E,$$

and

$$Z_{\text{real}} = \tfrac{1}{2}(\bar{Z}+Z) = (E+\bar{S})^{-1} - E + (E+S)^{-1} = (E+\bar{S})^{-1}(E-S\bar{S})(E+S)^{-1}.$$

Solving for $E-S\bar{S}$ and noting that (16) gives $E+S=2Q$, one obtains

$$E - S\bar{S} = 4\bar{Q}Z_{\text{real}}Q. \tag{26}$$

Consider now a semipositive quadratic form of matrix Z_{real} and apply to the variables the non-singular complex linear transformation defined by the matrix Q; the quadratic form becomes a Hermitian form in the new variables and has the transformed matrix (26). As the semipositive character is not changed by the transformation, the conditions of physical realizability can be obtained explicitly in terms of the efficiency factors by writing that the principal minors of $E-S\bar{S}$ must be positive or zero. The results previously obtained for reactive or resistive networks are particular cases of these conditions.

Errata

The author has provided the following list of corrections:

1. Equation (3): should read

$$S_{\mathrm{pp}} = (R_{\mathrm{p}} - \zeta_{\mathrm{p}})/(R_{\mathrm{p}} + \zeta_{\mathrm{p}})$$

2. Equation (20): the first term should read:

$$\exp(-2A_{\mathrm{p}1})$$

3. Second line following Equation (21): delete word "of."

4. Equation (22): should read

$$-1<\hat{S}_{\mathrm{pp}}\leq 1.$$

5. Second line before Section VI: delete the words "that match an odd number of terminal pairs."

18

Reprinted from *IRE Trans. Circuit Theory,* **CT-6**, 102–124 (Mar. 1959)

Bounded Real Scattering Matrices and the Foundations of Linear Passive Network Theory*

D. C. YOULA†, L. J. CASTRIOTA†, AND H. J. CARLIN†

List of Principal Symbols

Ω	= an abstract space of elements.
H or H_1	= the space of all measurable functions of a real variable t, $(-\infty < t < \infty)$.
$\|\| f \|\|$	= the norm of an element f in a Banach space Ω.
lim	= abbreviation for "limit."
sup	= abbreviation for "supremum."
inf	= abbreviation for "infimum."
Im	= abbreviation for "imaginary part of."
Re	= abbreviation for "real part of."
H_n	= the space of all n vectors whose n components are individually in H.
L_1	= the space of all functions of a real variable t, $(-\infty < t < \infty)$ absolutely integrable over $(-\infty < t < \infty)$ with respect to Lebesgue measure.
L_2	= the space of all functions of a real variable t, $(-\infty < t < \infty)$, square integrable over $(-\infty < t < \infty)$ with respect to Lebesgue measure.
L_{1n}	= the space of all n vectors whose n components are individually in L_1.
L_{2n}	= the space of all n vectors whose n components are individually in L_2.
$\mathbf{f}$, $\mathbf{g}$	= generic elements in H_n, L_{1n}, and L_{2n}.
$(\mathbf{f}, \mathbf{g})$	= the notation for the inner product of two elements $\mathbf{f}$ and $\mathbf{g}$ in L_{2n}.
	= the matrix transposition operation.
*	= the operation which conjugates and transposes a matrix.
Φ, ψ	= general operators in an abstract space Ω.
$\Omega \times \Omega$	= the cartesian product of the space Ω with itself, *i.e.*, the collection of all ordered pairs $\langle f, g \rangle$, where f and g are both in Ω.
$D(\Phi)$	= the domain of definition of the operator Φ.
$R(\Phi)$	= the set of all elements Φg, g in $D(\Phi)$.
ε	= abbreviation for "is a member of."
$G(\Phi)$	= the graph of the operator Φ, *i.e.*, the collection of all pairs $\langle f, \Phi f \rangle$, f in $D(\Phi)$.
a.e.	= abbreviation for "almost everywhere."
0	= the zero element in an abstract space Ω. Also used to denote the Roman numeral zero.
$\mathbf{0}_n$	= the zero n vector in either H_n, L_{1n}, or L_{2n}. The subscript n is dropped whenever understood.
0_n	= the zero $n \times n$ matrix.
1_n	= the identity $n \times n$ matrix.
$\cap$	= the operation of set intersection.
$\cup$	= the operation of set union.
$\Phi \otimes \psi$	= the "augmented" operator corresponding to the operators Φ and ψ.
$\mathcal{I}$	= the identity operator in an abstract space.
i	= $\sqrt{-1}$.
$\mathbf{i}$, $\mathbf{v}$, $\mathbf{e}$	= elements in either H_n, L_{1n}, or L_{2n}.
$\mathbf{I}(\omega)$, $\mathbf{V}(\omega)$, $\mathbf{E}(\omega)$	= the column-vector Fourier transforms of $\mathbf{i}(t)$, $\mathbf{v}(t)$, and $\mathbf{e}(t)$, respectively.

* Manuscript received by the PGCT, February 3, 1958; revised manuscript received, July 14, 1958. This research was supported under Contract Nos. AF-18(603)-105, AF-18(600)-1505 monitored by the AF Office of Sci. Res., ARDC.

† Microwave Res. Inst., Polytechnic Inst. of Brooklyn, Bklyn., N. Y.

z	= the complex variable, $z = \omega + i\beta$.
$I_n(\omega)$, $V_n(\omega)$, $E_n(\omega)$	= $n \times n$ matrices whose columns are column-vector Fourier transforms $\mathbf{I}(\omega)$, $\mathbf{V}(\omega)$, and $\mathbf{E}(\omega)$, respectively.
$P_n, M_n, A_n, K_n, B_n, F_n, N_n$	= $n \times n$ matrices.
T_1, T_2	= $n \times n$ orthogonal matrices.
Φ_a	= the augmented operator $\epsilon \otimes \Phi$ corresponding to the operator Φ.
Φ_s	= the operator corresponding to the ideal one-port short-circuit.
Φ_T	= the operator corresponding to the ideal $2n$-port transformer.
Φ_{02}	= the operator corresponding to the ideal two-port open circuit.
$\hat{\Phi}_a$	= the continuation of the operator Φ_a.
$Y_a(z)$	= the admittance matrix of the augmented n-port Φ_a.
$S(z)$	= the $n \times n$ scattering matrix of an n-port Φ.
$Q(z)$	= $1_n - S^*(z)S(z)$, the energy matrix.
$E(z)$	= the complex Fourier transform of a single function in either L_{1n} or L_{2n}.
Δ	= a set of points on the real axis.
$\mu(\Delta)$	= the Lebesgue measure of Δ.
$\mathbf{b}$	= a constant n vector.
α	= a complex scalar.
$\mathbf{v}_+, \mathbf{v}_-$	= the incident and reflected components of $\mathbf{v}(t)$ in the decomposition $\mathbf{v} = \mathbf{v}_+ + \mathbf{v}_-$.
$\mathbf{i}_+, \mathbf{i}_-$	= the incident and reflected components of $\mathbf{i}(t)$ in the decomposition $\mathbf{i} = \mathbf{i}_+ + \mathbf{i}_-$.
$Q_1(z)$	= $1_n - S(z)$.
$Q_2(z)$	= $1_n + S(z)$.
$0_{k,r}$	= the $k \times r$ zero matrix.
$\delta_{k,r}$	= the Kronecker delta, *i.e.*, $\delta_{k,r} = 1$, $k = r$ and $\delta_{k,r} = 0$, $k \neq r$.
$Z(z)$	= the $n \times n$ impedance matrix of an n-port Φ.
$Y(z)$	= the $n \times n$ admittance matrix of an n-port Φ.
$R_1(z)$	= $\frac{1}{2}[Z(z) + Z^*(z)]$.
$R_2(z)$	= $\frac{1}{2}[Y(z) + Y^*(z)]$.
Φ^{-1}	= the inverse operator to Φ.

I. Introduction

The principal aim of this paper is the construction of a completely rigorous theory of linear passive time-invariant n-ports from an axiomatic point of view. The notions of linearity, passivity, time-invariance and causality are defined precisely for the most general operator with domain and range in H_n, the space of all n vectors whose n components are measurable functions of a real variable t in $(-\infty < t < \infty)$. It is shown that with the exception of certain pathological cases, linearity and passivity imply causality. Moreover, the causality postulate plays absolutely no role in the subsequent development. The n-port is subject to five postulates which describe its behavior in real time. Denote the n-port by Φ and the response by $\mathbf{i}(t) = \Phi\mathbf{v}(t)$ where $\mathbf{v}(t)$ is an arbitrary element in the domain $D(\Phi)$ of Φ. Then,

P_1. Φ is linear (Definition 6, Section II).

P_2. Φ is passive (Definition 13, Section III).

P_3. Φ is time-invariant (Definition 10, Section II).

P_4. The equation

$$\mathbf{v} + \Phi\mathbf{v} = \mathbf{e}$$

is soluble for $\mathbf{v}(t)$ in $D(\Phi)$ over at least one set of $\mathbf{e}$'s dense in Hilbert space.

P_5. $\mathbf{v}(t)$ real implies that all values of $\Phi\mathbf{v}$ are real.

Postulate P_4 is indispensable, a fact already recognized by several other authors [4], [6]. The ideal one-port short-circuit reveals that P_4 can attain even when the domain of Φ contains only the zero element (of course, Φ must be multiple-valued).

Our definition of passivity is not the one usually encountered in the literature [4]–[6], *viz.*,

$$Re \int_{-\infty}^{+\infty} \mathbf{v}^*(\tau)\mathbf{i}(\tau)\, d\tau \geq 0$$

but

$$Re \int_{-\infty}^{t} \mathbf{v}^*(\tau)\mathbf{i}(\tau)\, d\tau \geq 0$$

for all $t > -\infty$, all $\mathbf{v}$ in $D(\Phi)$, and all values $\Phi\mathbf{v}$. In the authors' opinion, the latter is a more natural and accurate formulation of the very intuitive notion of passivity. Its use has enabled many arguments to be considerably simplified. Another point to be carefully noted by the reader is that our definition of linearity implies that $\mathbf{i}(t) = \mathbf{0}$ is always one of the responses to the excitation $\mathbf{v}(t) = \mathbf{0}$. This automatically excludes many networks with hidden sources that are customarily thought of as being linear, *e.g.*, a resistor in series with a battery, etc. However these are trivial failures of the definition and can be included by the simple artifice of considering $\mathbf{i} - \mathbf{i}_s$ instead of $\mathbf{i}$ to be the response of Φ to $\mathbf{v}$, where $\mathbf{i}_s$ is the short-circuit current.

Theorems 2 and 3 establish that every n-port Φ satisfying P_1 to P_4 possesses a frequency response and defines an $n \times n$ scattering matrix $S(z)$ analytic in $Im\, z > 0$ such that $Q(z) = 1_n - S^*(z)S(z)$ is the matrix of a non-negative quadratic form everywhere in $Im\, z > 0$ and almost everywhere on the real frequency ω axis.[1] Conversely any such matrix defines an n-port Φ satisfying P_1 to P_4.

A matrix $S(z)$ is said to be a bounded-real scattering matrix when

1) $S(z)$ is analytic in $Im\, z > 0$;
2) $Q(z) = 1_n - S^*(z)S(z)$ is the matrix of a non-negative quadratic form in $Im\, z > 0$;
3) $\bar{S}(z) = S(-\bar{z})$.

A physically realizable scattering matrix $S(z)$ is one which represents the scattering description of an n-port Φ subject to postulates P_1 to P_5. In Theorem 4 it is shown that $S(z)$ is physically realizable if and only if it is bounded-real. An alternative set of necessary and sufficient conditions in terms of real frequency boundary behavior are also derived:

1) $S(z)$ is analytic in $Im\, z > 0$;
2) $\sup e^{-|z|^\alpha} \mid S(z) \mid < \infty$ for $Im\, z > 0$ some $\alpha < 1$.[2]
3) $Q(\omega) = 1_n - S^*(\omega)S(\omega)$ is non-negative for almost all ω in $(-\infty < \omega < \infty)$.
4) $\bar{S}(\omega) = S(-\omega)$ almost everywhere.

Thus, any matrix $S(z)$ analytic in $Im\, z > 0$ which, *a priori*, is known to grow in the upper half-plane no faster than $e^{|z|^\alpha}$, $\alpha < 1$, is bounded real if its associated matrix $Q(z)$ is non-negative for real frequencies, *i.e.*, it is not necessary to verify the non-negativity of $Q(z)$ in $Im\, z > 0$.

The physical notion of adding positive resistance in series with the leads of an n-port in order to "smooth" or "regularize" its behavior is given a rigorous operator interpretation and called "augmentation." The insertion of unit resistors in series with the terminals of Φ is equivalent to augmenting the operator Φ with the identity operator $\mathcal{I}$; this modified n-port is denoted by $\Phi_a = \mathcal{I} \otimes \Phi$ (Fig. 2). When Φ is linear and passive, Φ_a is linear, passive and causal, a rather remarkable result. All the properties of Φ are inferred from the properties of Φ_a to which the Bochner and Wiener theorems are applicable.

A proof is also given showing that under certain conditions, $\mathbf{i} = \Phi_a \mathbf{e}$, $\mathbf{e} \in L_{1n} \cap D(\Phi_a)$, is equivalent to

$$\mathbf{i}(t) = \int_{-\infty}^{+\infty} dW_n(\tau)\mathbf{e}(t - \tau),$$

where $W_n(\tau)$ is an $n \times n$ matrix of bounded-variation if and only if $S(z)$ admits the representation

$$S(z) = \int_0^\infty e^{iz\tau}\, dK_n(\tau), \qquad Im\, z \geq 0,$$

$K_n(\tau)$ also being an $n \times n$ matrix of bounded-variation over $[0, \infty]$.

When Φ and Φ^{-1} are single-valued, the $n \times n$ matrices

$$Y(z) = (1_n - S)(1_n + S)^{-1}$$

and

$$Z(z) = (1_n + S)(1_n - S)^{-1}$$

exist respectively in $Im\, z > 0$ and are positive-real. Conversely to any positive-real matrix $A_n(z)$ corresponds two n-ports Φ_1 and Φ_2 satisfying P_1 to P_5 such that $A_n(z)$ is the admittance matrix (Y) of the first and the impedance matrix (Z) of the second. The analytic properties of $S(z)$ in $Im\, z > 0$ are also investigated in great detail.

Raisbeck [5] appears to have been one of the first to appreciate the power of the axiomatic approach in the study of physically realizable structures. In his interesting paper most of the properties of the impedance matrix $Z(z)$ are derived (nonrigorously, however). Causality and passivity are taken as distinct postulates and the n-port is, from the very outset, assumed to possess a frequency response and to be single-valued; the scattering matrix is not introduced and hence networks which do not possess immittance descriptions are not included.

Toll [7] gives a rigorous one-port analysis but also takes causality and passivity as separate axioms and makes no use of the augmented operator concept.

Wu's work [1] is undoubtedly the one most closely allied to ours. Oddly enough, after defining the notion of operator augmentation he plunges directly into the impedance formulation completely overlooking the scattering idea. Again causality and passivity are postulated separately but, apparently unaware of Bochner's L_2-theorem (Section II), he gives a proof of the existence of a frequency response (as in the previous papers, no mention is made of the very intimate connection between causality and passivity). Most of the relevant properties of $Z(z)$ are derived together with several representation theorems relating the corresponding real and imaginary parts.

Section II is designed primarily to introduce the reader to some special linear notions and theorems and is rather sketchy in character. However, the concepts of multiple-valuedness, time invariance and augmentation of operators are gone into quite deeply since they are very important and peculiar to the mode of development we have chosen. The reader is assumed to have a fair working knowledge of Lebesgue measure and integral. These tools have been invaluable in enabling the authors to fully exploit the axioms. The discussion in Section VII should convince him that the "almost everywhere" requirement is neither vacuous nor academic. In conclusion we would like to add that our subject matter is mainly deductive in nature; *i.e.*, it is possible to start with a few definitions and axioms and derive everything else from them.

[1] $S(z)$ is formed with real constant normalizing numbers. This implies that, for example, if a multiport waveguide structure is being considered, the scattering description is normalized to the characteristic impedances of short coaxial lines presumed connected to the prescribed terminal planes which constitute the accessible ports of the system.

[2] For a matrix A, the notation $\mid A \mid < \alpha$ is equivalent to saying that all the elements of A are less than α in absolute value.

II. Mathematical Preliminaries

In order to facilitate the reader's orientation and not interrupt the continuity of the proofs of the various theorems appearing in the body of the paper this section reviews some basic and special linear notions. Though the review is brief certain ideas and theorems peculiar to our later development of the theory are explained in detail.

Consider a space Ω with elements $f, g, \cdots$ which is closed with respect to a binary operation $+$ and multiplication on the left by a complex scalar, *i.e.*, if $f \in \Omega, g \in \Omega$,

$$f + g \in \Omega,$$

$$\alpha f \in \Omega, \quad (\alpha \text{ a complex scalar}).$$

Definition 1

Let f and g be any two elements of a space Ω. Then, Ω is called linear if $f + g$ and αf, (α complex), are in Ω and if

1) $f + g = g + f$,
2) $(f + g) + h = f + (g + h)$,
3) $f + g = f + h$ implies $g = h$,
4) $\alpha(f + g) = \alpha f + \alpha g$,
5) $(\alpha\beta)f = \alpha(\beta f)$, ($\alpha$ and β complex scalars),
6) $(\alpha + \beta)f = \alpha f + \beta f$,
7) $1 \cdot f = f$.

For any f in Ω the element $f - f$ is unique and is denoted by 0. Clearly, H, the space of all complex-valued measurable functions of a real parameter t, $(-\infty < t < \infty)$, is linear.

Definition 2

A linear space Ω is said to be a Banach space if to every element $f \in \Omega$ there corresponds a non-negative number $\| f \|$, called the norm of f, possessing the following properties:

1) $\| \alpha f \| = | \alpha | \cdot \| f \|$, ($\alpha$ a complex scalar),
2) $\| f + g \| \leq \| f \| + \| g \|$,
3) $\| f \| > 0$ for $f \neq 0$.

Definition 3

A linear space Ω is said to be a Hilbert space if to every pair of elements f and g belonging to it there corresponds a complex number (f, g), called the inner product of f and g, satisfying.

1) $(\alpha f, g) = \alpha(f, g)$ (α a complex scalar),
2) $(f_1 + f_2, g) = (f_1, g) + (f_2, g)$,
3) $(f, g) = \overline{(g, f)}$, where $\overline{(g, f)}$ is the complex conjugate of (g, f).
4) $(f, f) > 0$ for $f \neq 0$.

In a Banach space convergence of a sequence is defined in terms of the norm. Thus $f_n \to f$, $(n = 1, 2, \cdots)$ if

$$\lim_{n\to\infty} \| f_n - f \| = 0.$$

It is very simple to show that $f_n \to f$ and $g_n \to g$ imply

$$\lim_{n\to\infty} \| f_n \| = \| f \|,$$

$$\lim_{n\to\infty} (f_n, g_n) = (f, g).$$

In a Hilbert space $\| f \| = (f, f)^{\frac{1}{2}}$.

The only spaces of interest to us are L_1, L_2, and H: L_1 is the collection of all measurable complex-valued functions $f(t)$ of a real variable t, $(-\infty < t < \infty)$ such that

$$\int_{-\infty}^{+\infty} | f(t) | \, dt < \infty,$$

the integral being taken in the Lebesgue sense. Similarly $f(t) \in L_2$ if

$$\int_{-\infty}^{+\infty} | f(t) |^2 \, dt < \infty.$$

More accurately, our main concern is with the spaces L_{1n}, L_{2n}, H_n, n finite, where $\mathbf{f}(t) \in L_{jn}$, H_n, $(j = 1, 2)$, if

1) $$\mathbf{f}(t) = \begin{bmatrix} f_1 \\ f_2 \\ \vdots \\ f_n \end{bmatrix}.$$

and

2) $f_r(t) \in L_j, H, \quad (r = 1, 2, \cdots, n), \quad (j = 1, 2).$

The element $\mathbf{f}(t)$ whose components are identically zero in $(-\infty < t < \infty)$ is the zero of H_n and is denoted by $\mathbf{0}$. In L_{1n} the norm is defined by

$$\| \mathbf{f}(t) \|_1 = \sum_{r=1}^{n} \int_{-\infty}^{+\infty} | f_r(t) | \, dt < \infty \tag{1}$$

and in L_{2n} an inner product by

$$(\mathbf{f}, \mathbf{g}) \equiv \int_{-\infty}^{+\infty} \sum_{r=1}^{n} \bar{f}_r(t) g_r(t) \, dt < \infty. \tag{2}$$

From a previous remark, if $\mathbf{f} \in L_{2n}$,

$$\| \mathbf{f} \|_2^2 = \int_{-\infty}^{+\infty} \sum_{r=1}^{n} | f_r(t) |^2 \, dt = (\mathbf{f}, \mathbf{f}).$$

Denoting the transpose of $\mathbf{f}$ by $\mathbf{f}'$ and $\bar{\mathbf{f}}'$ by $\mathbf{f}^*$, we may write (2) as

$$(\mathbf{f}, \mathbf{g}) = \int_{-\infty}^{+\infty} \mathbf{f}^*\mathbf{g}(t) \, dt, \tag{2a}$$

in which the integrand is evaluated by ordinary matrix multiplication. For any two elements $\mathbf{f}$ and $\mathbf{g}$ belonging to L_{2n} the important Schwartz inequality

$$(\mathbf{f}, \mathbf{g}) \leq \| \mathbf{f} \|_2 \cdot \| \mathbf{g} \|_2 \tag{3}$$

is valid. Whenever there is no risk of confusion, L_{1n} and L_{2n} will be denoted simply by L_1 and L_2.

Let Ω be a linear space and $D \subset \Omega$ a subset of elements. The set D is said to be a linear manifold if $f \in D$ and $g \in D$ imply $f + g \in D$ and $\alpha f \in D$, α complex.

Definition 4

The direct product of Ω with itself, $\Omega \times \Omega$, is the set of all ordered pairs $\langle f, g\rangle$, where f and g are elements of Ω. In $\Omega \times \Omega$

1) $\alpha\langle f, g\rangle \equiv \langle \alpha f, \alpha g\rangle$, α complex,
2) $\langle f_1, g_1\rangle + \langle f_2, g_2\rangle \equiv \langle f_1 + f_2, g_1 + g_2\rangle$.

Obviously under 1) and 2), $\Omega \times \Omega$ is linear whenever Ω is linear. We are now ready to define the concept of an operator in Ω.

Definition 5

An operator in Ω is a function Φ defined over some subset D of Ω and which has one or more values Φf in Ω corresponding to each element f of D. The set D is called the domain $D(\Phi)$ of Φ. The set of all values Φf, $f \in D(\Phi)$, is called the range, $R(\Phi)$, of Φ. The set of all elements $\langle f, \Phi f\rangle$, $f \in D(\Phi)$ and $\Phi f \in R(\Phi)$, is called the graph, $G(\Phi)$, of Φ. Hence $D(\Phi) \subset \Omega$, $R(\Phi) \subset \Omega$ and $G(\Phi) \subset \Omega \times \Omega$.

To say that $\langle f, g\rangle \in G(\Phi)$ means that $f \in D(\Phi)$, $g \in R(\Phi)$, Φf exists and one of its values is g.

Definition 6

An operator Φ in Ω is called linear if $G(\Phi)$ is a linear manifold.

Since every linear set contains the zero element, $\langle 0, 0\rangle$ is in the graph of every linear operator.

An easy consequence of this definition is that if Φ is linear, then $D(\Phi)$ and $R(\Phi)$ are linear manifolds in Ω.

Definition 7

An operator Φ is called single-valued (s.v.) if there is associated exactly one value Φf with each element f in $D(\Phi)$.

Thus if Φ is s.v. and if $\langle f, g_1\rangle$ and $\langle f, g_2\rangle$ are in $G(\Phi)$, $g_1 = g_2 = \Phi f$.

A necessary and sufficient condition that a linear operator Φ be s.v. is that $\Phi(0)$ have the unique value 0. For 0 is certainly one of the values of $\Phi(0)$ so that the condition is necessary. To show that it is sufficient let $\langle f, g_1\rangle$ and $\langle f, g_2\rangle$ be in $G(\Phi)$. Then,

$$\langle f, g_1\rangle - \langle f, g_2\rangle = \langle 0, g_1 - g_2\rangle \in G(\Phi).$$

Thus $g_1 - g_2$ is one of the values of $\Phi(0)$ and by hypothesis $g_1 - g_2 = 0$ or $g_1 = g_2$.

A single-valued operator is called a transformation.

Let Φ be an operator and α a complex scalar. Then $\alpha\Phi$ is the operator which assigns to every $f \in D(\Phi)$ the values $\alpha\Phi(f)$.

If Φ is s.v. and linear and if f and g are in $D(\Phi)$, then $f + g$ and αf, α complex, are in $D(\Phi)$, $\Phi(f + g) = \Phi f + \Phi g$ and $\Phi(\alpha f) = \alpha\Phi(f)$, α complex.

The physical idea of putting a resistance in series with a network in order to "regularize" its behavior is formalized by the notion of "augmenting" an operator ψ by another operator Φ.

Definition 8

Let Φ and ψ be two operators in Ω. The operator $\Phi \otimes \psi$ is called the augmented operator corresponding to Φ and ψ and is defined in the following manner:

1) An element $f \in \Omega$ belongs to the domain of $\Phi \otimes \psi$ if there exists a unique decomposition $f = f_1 + f_2$, $f_1 \in D(\Phi)$, $f_2 \in D(\psi)$, such that at least one value of Φf_1 agrees with one value of ψf_2.
2) The set of values assigned by $\Phi \otimes \psi$ to f is precisely the intersection of the sets of values corresponding to Φf_1 and ψf_2. By an abuse of notation we can write this in the compact form

$$\Phi \otimes \psi(f) = \Phi f_1 \cap \psi f_2.$$

Clearly, $\Phi \otimes \psi = \psi \otimes \Phi$. Observe that if either Φ or ψ are s.v., $\Phi \otimes \psi$ is s.v., *i.e.*, a transformation. It does not follow however, even when Φ and ψ are linear transformations that $\Phi \otimes \psi$ is a linear transformation. As a matter of fact it doesn't even follow that it is linear. For example let Φ be a linear transformation defined on all of Ω and $\psi = -\Phi$. Then for any $g \in \Omega$, $0 = g - g$ and $\Phi(g) = \psi(-g) = -\psi(g) = \Phi(g)$ so that $0 \notin D(\Phi \otimes \psi)$. The nontrivial nature of the next lemma should now be apparent.

Lemma 1: If Φ and ψ are linear operators in Ω, a necessary and sufficient condition that $\Phi \otimes \psi$ be a linear operator is that it contain the zero element, 0, in its domain.

Proof: The necessity is trivial. Regarding sufficiency suppose $0 \in D(\Phi \otimes \psi)$. Then, by definition, there exists a unique decomposition $0 = g + h$, $g \in D(\Phi)$, $h \in D(\psi)$ such that at least one value of Φg equals some value of $\psi h = \Phi \otimes \psi(0)$. Because Φ and ψ are linear, $g = h = 0$, *i.e.*, 0 has the unique breakdown $0 = 0 + 0$. This implies that an element g cannot have two distinct decompositions $g = g_1 + g_2 = h_1 + h_2$ satisfying condition 1) of Definition 8 for then $0 = (g_1 - h_1) + (g_2 - h_2)$, $g_1 - h_1 \in D(\Phi)$, $g_2 - h_2 \in D(\psi)$ and at least one value of $\Phi(g_1 - h_1)$ equals one value of $\psi(g_2 - h_2)$. Thus, $g_1 - h_1 = g_2 - h_2 = 0$, or $g_1 = h_1$, $g_2 = h_2$. Again, by linearity, if $g = g_1 + g_2$ and $h = h_1 + h_2$ belong to $D(\Phi \otimes \psi)$, at least one value of $\Phi(g_1 + h_1)$ equals one value of $\psi(g_2 + h_2)$, and it follows that $f + g = (f_1 + g_1) + (f_2 + g_2)$ is in $D(\Phi \otimes \psi)$. Similarly $f \in D(\Phi \otimes \psi)$ implies $\alpha f \in D(\Phi \otimes \psi)$, α complex. The proof is complete.

Corollary: If in addition either Φ or ψ is s.v., $\Phi \otimes \psi$ is a linear transformation.

The next definition describes those operators in Ω that can be made single-valued by augmentation with a positive multiple of the identity operator ϵ.

Definition 9

An operator Φ in Ω is said to be "dissipatively regular" if the domain of $1/r\, g \otimes \Phi$ is nonempty for some $r > 0$.

The motivation for this definition has its roots in a very simple physical situation. Consider the ideal transformer. Certainly the currents in the primary and secondary are not uniquely determined by the terminal voltages, but yet if resistances are placed in series with the input and output leads, these currents do indeed become single-valued functions of the voltages (a more adequate discussion is given later).

A very important concept in the theory of linear transformations is that of boundedness. Let Ω be a Banach space. A set $D \subset \Omega$ is said to be dense in Ω if for any $\epsilon > 0$ and any $f \in \Omega$ there exists $ag \in D$ such that $\| f - g \| < \epsilon$; more succinctly, Ω coincides with the closure of D. The Banach space Ω is said to be complete if every Cauchy sequence in Ω converges to a limit element in Ω. Thus, if

$$\lim_{n,m\to\infty} \| f_n - f_m \| = 0,$$

there exists an element $f \in \Omega$ such that

$$\lim_{n\to\infty} f_n = f.$$

The linear transformation Φ in Ω is said to be bounded if

$$\sup_{f \in D(\Phi)} \frac{\| \Phi f \|}{\| f \|} \equiv \| \Phi \| < \infty.$$

The following standard result whose proof we omit is extremely useful.

Extension Theorem: Suppose that Ω is a complete Banach space and Φ a bounded linear transformation in Ω whose domain $D(\Phi)$ is dense in Ω. Then, Φ can be defined for every $f \in \Omega$ in such a way that its linearity and norm, $\| \Phi \|$, are preserved [2].

It is well known [2] that the spaces L_{1n} and L_{2n} are complete. Moreover the sets of vector-functions $\mathbf{f}(t)$ in L_{1n} and L_{2n} which vanish for sufficiently negative t are dense in their respective spaces.

For linear operators in H_n another notion is very important. For any set $D \subset H_n$, let D_τ, τ real, denote the set of all elements $\mathbf{v}(t + \tau)$, $\mathbf{v}(t) \in D$.

Definition 10

Suppose that Φ is an operator in H_n. Then, Φ is said to be time-invariant or to commute with the translation operator when

1) $D_\tau(\Phi) = D(\Phi)$ for any real τ;
2) If J is the set of all values $\Phi\mathbf{v}(t)$, J_τ, τ arbitrary and real is the set of all values $\Phi\mathbf{v}(t + \tau)$.

For a transformation, 1) and 2) can be written very simply as $\Phi\mathbf{v}(t) = \mathbf{i}(t)$, $\mathbf{v}(t) \in D(\Phi)$, imply $\mathbf{v}(t + \tau) \in D(\Phi)$, τ arbitrary and real, and $\Phi\mathbf{v}(t + \tau) = \mathbf{i}(t + \tau)$.

Lemma 2: Let ψ and Φ be linear time-invariant operators in Ω. Then, if $\psi \otimes \Phi$ is linear it is also time-invariant.

Proof: If $\psi \otimes \Phi$ is linear, $\mathbf{0}$ is contained in its domain. Now $\mathbf{v}(t) \in D(\psi \otimes \Phi)$ implies the existence of a unique decomposition $\mathbf{v}(t) = \mathbf{v}_1(t) + \mathbf{v}_2(t)$ such that at least one value of $\psi\mathbf{v}_1(t)$ equals some value of $\Phi\mathbf{v}_2(t)$. Thus, at least one value of $\psi\mathbf{v}_1(t + \tau)$ equals some value of $\Phi\mathbf{v}_2(t + \tau)$ from which it follows that $\mathbf{v}(t + \tau) = \mathbf{v}_1(t + \tau) + \mathbf{v}_2(t + \tau)$ is an admissible decomposition of $\mathbf{v}(t + \tau)$; it must also be unique because $\mathbf{0}$ is in the domain of $\psi \otimes \Phi$. Thus $\mathbf{v}(t + \tau) \in D(\psi \otimes \Phi)$ and the values of $\psi \otimes \Phi(\mathbf{v}(t + \tau))$ are the common values of $\psi(\mathbf{v}_1(t + \tau))$ and $\Phi(\mathbf{v}_2(t + \tau))$. Denoting this set of values by X it is easily seen that $X = J_\tau$ where J is the common set of values of $\psi(\mathbf{v}_1(t))$ and $\Phi(\mathbf{v}_2(t))$, Q.E.D.

Corollary: If Φ is a linear time-invariant operator in Ω, $\mathcal{I} \otimes \Phi$ is a linear time-invariant transformation if, and only if, it contains 0 in its domain.

As will appear in the subsequent analysis, the next four theorems by Titchmarsh, Paley-Wiener, and Bochner form the cornerstone of the entire theory of linear, passive networks. Before stating them we recall a few basic results in the theory of Fourier transforms. Let $f(t) \in L_2$. Its Fourier transform $\varphi(\omega)$ is defined by

$$\varphi(\omega) = \underset{a\to\infty}{\text{l.i.m.}} \frac{1}{\sqrt{2\pi}} \int_{-a}^{a} f(t) e^{it\omega}\, dt$$

where l.i.m. is an abbreviation for "limit in the mean." It can be shown [3] to exist, to be square-integrable and to satisfy the unitary property

$$\int_{-\infty}^{+\infty} | \varphi(\omega) |^2 \, d\omega = \int_{-\infty}^{+\infty} | f(t) |^2 \, dt.$$

More generally if $\varphi_1(\omega)$ and $\varphi_2(\omega)$ are the L_2-Fourier transforms of $f_1(t)$ and $f_2(t)$, respectively,

$$\int_{-\infty}^{+\infty} \bar{\varphi}_1(\omega)\varphi_2(\omega)\, d\omega = \int_{-\infty}^{+\infty} \bar{f}_1(t) f_2(t)\, dt \tag{4}$$

and is of course the familiar Plancherel [3] theorem. The inverse relation

$$f(t) = \underset{a\to\infty}{\text{l.i.m.}} \frac{1}{\sqrt{2\pi}} \int_{-a}^{+a} \varphi(\omega) e^{-it\omega}\, d\omega$$

also holds almost everywhere (a.e.). Using the inner product notation, (4) may be written as

$$(\varphi_1, \varphi_2) = (f_1, f_2). \tag{4a}$$

The L_1-Fourier transform of an L_1 function $f(t)$ is defined by

$$\varphi(\omega) = \frac{1}{\sqrt{2\pi}} \int_{-\infty}^{+\infty} f(t) e^{it\omega}\, dt,$$

the integral converging absolutely. The L_1 and L_2 transforms of a function are a.e. equal whenever they both exist.

If $\mathbf{f}(t) \in L_{jn}$, ($j = 1, 2$), its Fourier transform $\mathbf{F}(\omega)$ is by definition, the column-vector of Fourier transforms of its n components taken in the same order, *i.e.*, if

$$\mathbf{f}(t) = \begin{bmatrix} f_1 \\ f_2 \\ \vdots \\ f_n \end{bmatrix},$$

$$\mathbf{F}(\omega) = \begin{bmatrix} F_1 \\ F_2 \\ \cdot \\ \cdot \\ F_n \end{bmatrix}.$$

In L_{2n} Plancherel's theorem takes the form

$$\int_{-\infty}^{+\infty} \mathbf{F}_1^*(\omega)\mathbf{F}_2(\omega)\, d\omega = \int_{-\infty}^{+\infty} \mathbf{f}_1^*(t)\mathbf{f}_2(t)\, dt \tag{5}$$

or

$$(\mathbf{F}_1, \mathbf{F}_2) = (\mathbf{f}_1, \mathbf{f}_2). \tag{5a}$$

The complex Fourier transform of $f(t) \in L_2$ is

$$\varphi(z) = \frac{1}{\sqrt{2\pi}} \int_{-\infty}^{+\infty} f(t)e^{itz}\, dt, \qquad z = \omega + i\beta.$$

It is easy to see that $f(t) = 0, t < 0$, implies that $\varphi(z) \equiv \varphi(\omega + i\beta)$ is analytic for $\beta > 0$. Furthermore, for $\beta > 0$,

$$\int_{-\infty}^{+\infty} |\varphi(\omega + i\beta)|^2\, d\omega = \int_0^{\infty} |f(t)|^2 e^{-2\beta t}\, dt$$

$$\leq \int_0^{\infty} |f(t)|^2\, dt = K < \infty \tag{6}$$

where K is independent of β. The beautiful Titchmarsh theorem constitutes essentially the converse of this result.

Titchmarsh Theorem [3]: The class of square-integrable functions that vanish for negative values of their arguments is identical with the class of functions whose L_2-Fourier transforms are analytic in the upper half-plane $\beta > 0$ and satisfy (6). The

$$\lim_{\beta \to 0+} \varphi(\omega + i\beta) = \varphi(\omega)$$

exists for almost all ω and $\varphi(z)$ obeys the "dispersion" relation

$$\frac{1}{2\pi i} \int_{-\infty}^{+\infty} \frac{\varphi(x)}{x - z}\, dx = \varphi(z), \qquad Im\, z > 0 \tag{7}$$

$$= 0, \qquad Im\, z < 0.$$

An important corollary is that for any $\varphi(z)$ analytic and uniformly bounded in $\beta > 0$, $\lim_{\beta \to 0^+} \varphi(\omega + i\beta)$ exists a.e.

A sometimes more informative version of the above theorem is the Paley-Wiener theorem.

Paley-Wiener Theorem: The necessary and sufficient conditions for $a(\omega)$ to be the modulus of an L_2-Fourier transform of a square-integrable function that vanishes for negative t and is not zero a.e. are,

1) $a(\omega) \geq 0$ a.e.;

2) $\int_{-\infty}^{+\infty} a^2(\omega)\, d\omega < \infty$;

3) $$\int_{-\infty}^{+\infty} \frac{|\ln a(\omega)|}{1 + \omega^2}\, d\omega < \infty. \tag{8}$$

If, as usual, the integral of a matrix $M(t)$ depending on a parameter t is defined to be the matrix whose elements are the integrals of the corresponding elements in M etc., the Titchmarsh theorem for L_{2n} may be stated in the following way: The class of L_{2n} functions $\mathbf{f}(t)$ that vanish for negative values of their arguments is identical with the class of L_{2n} functions whose complex Fourier transforms $\mathbf{F}(\omega + i\beta)$ are analytic in $\beta > 0$ and satisfy

$$\int_{-\infty}^{+\infty} \mathbf{F}^*(\omega + i\beta)\mathbf{F}(\omega + i\beta)\, d\omega = K < \infty, \quad \beta > 0,$$

where K is independent of β. The $\lim_{\beta \to 0^+} \mathbf{F}(\omega + i\beta) = \mathbf{F}(\omega)$ a.e. and $\mathbf{F}(z)$ obeys the "dispersion" relation

$$\frac{1}{2\pi i} \int_{-\infty}^{+\infty} \frac{\mathbf{F}(x)}{x - z}\, dx = \mathbf{F}(z), \qquad Im\, z > 0, \tag{9}$$

$$= \mathbf{0}, \qquad Im\, z < 0.$$

As far as the authors are aware the existence of the next theorem by Bochner has been completely overlooked by other writers in the field. Wu [4] goes to great lengths to prove an essentially equivalent result and Raisbeck [5], Toll [6], and Van Kampen [7] assume it. It states, simply, that a bounded, linear, time-invariant transformation in L_2 whose domain is the whole space is of a very special type.

Bochner's L_2 Theorem [8]: Let Φ be a linear bounded transformation mapping all of L_2 into L_2. Then, a necessary and sufficient condition that Φ be time-invariant, *i.e.*,

$$i(t) = \Phi v(t), \qquad v(t) \in L_2,$$

imply

$$i(t + \tau) = \Phi v(t + \tau), \qquad \tau \text{ real},$$

is that there exist a measurable function $a(\omega)$ uniformly bounded for almost all ω such that

$$I(\omega) = a(\omega) V(\omega) \tag{10}$$

where $I(\omega)$ and $V(\omega)$ are the L_2-Fourier transforms of $i(t)$ and $v(t)$, respectively.

This theorem can be extended to L_{2n} with very little effort. Suppose that Φ is a linear bounded time-invariant transformation mapping all of L_{2n} into L_{2n}. By linearity,

$$\Phi\mathbf{v} \equiv \Phi \begin{bmatrix} v_1 \\ v_2 \\ \cdot \\ \cdot \\ \cdot \\ v_n \end{bmatrix} = \Phi \begin{bmatrix} v_1 \\ 0 \\ 0 \\ \cdot \\ \cdot \\ 0 \end{bmatrix} + , \cdots , + \Phi \begin{bmatrix} 0 \\ 0 \\ 0 \\ \cdot \\ \cdot \\ v_n \end{bmatrix},$$

$$= \Phi_1\mathbf{v} + \Phi_2\mathbf{v} + , \cdots , + \Phi_n\mathbf{v}$$

where

$$\Phi_r\mathbf{v} \equiv \Phi\mathbf{v}_r, \qquad (r = 1, 2, \cdots, n),$$

$\mathbf{v}_r$ being the column-vector whose components are all zero except the rth which equals v_r. Or course, Φ_r is also bounded, linear and time-invariant. Let

$$\Phi_r\mathbf{v} = \mathbf{i}_r = \begin{bmatrix} i_{1r} \\ i_{2r} \\ \vdots \\ i_{nr} \end{bmatrix}, \qquad (r = 1, 2, \cdots, n).$$

The correspondence $v_r \to i_{kr}$, $(k, r = 1, 2, \cdots, n)$, defines a linear, bounded time-invariant transformation Φ_{kr} on L_2 into L_2. Hence by the Bochner L_2 theorem, there exist n^2 measurable, essentially bounded functions $a_{kr}(\omega)$ such that

$$I_{kr}(\omega) = a_{kr}(\omega)V_r(\omega), \qquad (k, r = 1, 2, \cdots, n).$$

It is clear that if $\mathbf{i} = \Phi\mathbf{v}$,

$$I_k(\omega) = \sum_{r=1}^{n} a_{kr}(\omega)V_r(\omega), \qquad (k = 1, 2, \cdots, n).$$

$$\therefore \mathbf{I}(\omega) = A_n(\omega)\mathbf{V}(\omega)$$

where

$$A_n(\omega) \equiv \begin{bmatrix} a_{11}, & a_{12}, & \cdots, & a_{1n} \\ a_{21}, & a_{22}, & \cdots, & a_{2n} \\ \vdots & & & \\ a_{n1}, & a_{n2}, & \cdots, & a_{nn} \end{bmatrix}$$

and $\mathbf{I}(\omega)$, $\mathbf{V}(\omega)$ are the column-vector L_{2n}-Fourier transforms of $\mathbf{i}(t)$ and $\mathbf{v}(t)$, respectively.

Corollary: If Φ is a linear, bounded time-invariant transformation mapping all of L_{2n} into L_{2n} and $\mathbf{i} = \Phi\mathbf{v}$, $\mathbf{v}(t) \in L_{2n}$,

$$\mathbf{I}(\omega) = A_n(\omega)\mathbf{V}(\omega) \tag{10}$$

where $A_n(\omega)$ is a measurable $n \times n$ matrix in $(-\infty < \omega < \infty)$ for which there exists a finite, fixed, positive number α such that $|a_{kr}(\omega)| < \alpha$, $(k, r = 1, 2, \cdots, n)$, almost everywhere in $(-\infty < \omega < \infty)$. For short this is denoted by $|A_n(\omega)| < \alpha$.

Our next and last reference theorem describes the structure of linear, bounded, time-invariant transformations mapping all of L_{1n} into L_{1n}.

Bochner's L_1 Theorem [9]: Let $\mathbf{i} = \Phi\mathbf{v}$ be a transformation mapping L_{1n} into L_{1n}. Then, the necessary and sufficient condition that Φ be linear, bounded and time-invariant is that

$$\mathbf{I}(\omega) = F_n(\omega)\mathbf{V}(\omega) \tag{11}$$

where $\mathbf{I}(\omega)$ and $\mathbf{V}(\omega)$ are the L_{1n}-Fourier transforms of $\mathbf{i}(t)$ and $\mathbf{v}(t)$, respectively, and $F_n(\omega)$ is a Fourier-Stieltjes transform of an $n \times n$ matrix $B_n(\tau)$ of bounded-variation:

$$F_n(\omega) = \int_{-\infty}^{+\infty} e^{i\omega\tau}\, dB_n(\tau) \tag{12}$$

where $B_n(\tau)$ is an $n \times n$ matrix of bounded-variation over $(-\infty < \tau < \infty)$, *i.e.*,

$$\int_{-\infty}^{+\infty} |\, db_{kr}(\tau) \,| < \infty, \qquad (k, r, = 1, 2, \cdots, n). \tag{13}$$

In addition, $\mathbf{i} = \Phi\mathbf{v}$ is equivalent to

$$\mathbf{i}(t) = \int_{-\infty}^{+\infty} dB_n(\tau)\mathbf{v}(t - \tau) \tag{14}$$

for all $\mathbf{v}(t)$ belonging to L_{1n}. For short (13) is abbreviated as

$$\int_{-\infty}^{+\infty} |\, dB_n(\tau) \,| < \infty.$$

Alternately, (14) may be written more explicitly as

$$i_r(t) = \sum_{k=1}^{n} \int_{-\infty}^{+\infty} v_k(t - \tau)\, db_{rk}(\tau), \qquad (r = 1, 2, \cdots, n). \tag{12a}$$

Actually Bochner proves the above theorem for $n = 1$ only but its extension to arbitrary finite n is made by the same method used for his L_2 theorem.

III. The Linear n-Port

We now exploit the definitions, concepts, and theorems of Section I to develop a rigorus theory of linear, passive networks.

Definition 11

An n-port is an operator Φ in H_n. The correspondence $\mathbf{i}(t) = \Phi\mathbf{v}(t)$, $\mathbf{v}(t) \in D(\Phi)$ is depicted schematically in Fig. 1.

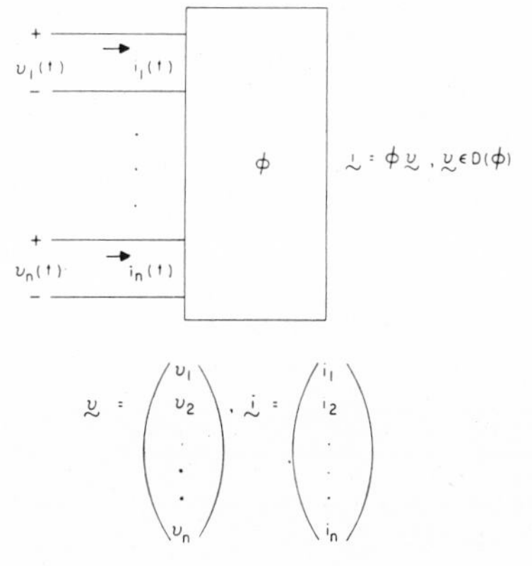

Fig. 1—The schematic of an n-port ϕ.

Definition 12

A linear n-port is a linear operator in H_n.

Many linear n-ports occurring in the applications are operators but not transformations. One interesting example is the ideal 2-port open-circuit Φ_{02}:

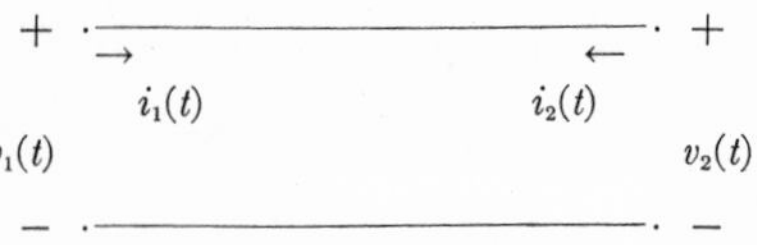

The domain of Φ_{02} is the set of all $\mathbf{v}(t) = (v_1, v_2)'$ for which $v_1(t) = v_2(t)$, $(-\infty < t < \infty)$. For any $\mathbf{v} \in D(\Phi_{02})$, $\Phi_{02}\mathbf{v}$ is the set of all $\mathbf{i} = (i_1, i_2)'$ such that $i_1(t) = -i_2(t)$, $(-\infty < t < \infty)$. Observe that $\mathbf{v}^*(t)\mathbf{i}(t) \equiv 0$. Thus Φ_{02} assigns to any $\mathbf{v}(t)$ in $D(\Phi_{02})$ an infinity of values.

Another common example is the ideal $2n$-port transformer Φ_T: $D(\Phi_T)$ is the set of all

$$\mathbf{v}(t) = (v_1, v_2, \cdots, v_n, v_{n+1}, \cdots, v_{2n})' \equiv \begin{pmatrix}\mathbf{v}_1\\ \mathbf{v}_2\end{pmatrix}$$

such that

$$\mathbf{v}_1 \equiv \begin{bmatrix} v_1 \\ v_2 \\ \vdots \\ v_n \end{bmatrix} = N_n \begin{bmatrix} v_{n+1} \\ v_{n+2} \\ \vdots \\ v_{2n} \end{bmatrix} \equiv N_n\mathbf{v}_2,$$

and the response

$$\mathbf{i}(t) = (i_1, i_2, \cdots, i_n, i_{n+1}, \cdots, i_{2n})' \equiv \begin{pmatrix}\mathbf{i}_1\\ \mathbf{i}_2\end{pmatrix}$$

is given by

$$\mathbf{i}_2 \equiv \begin{bmatrix} i_{n+1} \\ i_{n+2} \\ \vdots \\ i_{2n} \end{bmatrix} = -N_n' \begin{bmatrix} i_1 \\ i_2 \\ \vdots \\ i_n \end{bmatrix} \equiv -N_n'\mathbf{i}_1$$

where N_n is an $n \times n$ matrix of real constants. Observe that

$$\mathbf{v}_1^*\mathbf{i}_1 + \mathbf{v}_2^*\mathbf{i}_2 \equiv 0,$$

i.e., the ideal transformer is "instantaneously lossless." To any $\mathbf{v} \in D(\Phi_T)$ corresponds a whole continuum of values $\mathbf{i}$; furthermore $D(\Phi_T)$ does not contain any vector $\mathbf{v}$ with a.e. linearly independent components. It will appear shortly that the ideal transformer exhibits, to a great degree, all that is pathological in the linear n-port.

Definition 13

An n-port Φ is passive if for any $\tau > -\infty$ and any $\mathbf{v}(t) \in D(\Phi)$,

$$Re \int_{-\infty}^{\tau} \mathbf{v}^*(t)\mathbf{i}(t)\, dt \geq 0, \tag{15}$$

where $\mathbf{i} = \Phi\mathbf{v}$ is any one of the values assigned to $\mathbf{v} \in D(\Phi)$ by Φ. In particular (15) implies that

$$Re \int_{-\infty}^{+\infty} \mathbf{v}^*(t)\Phi\mathbf{v}(t)\, dt \geq 0, \tag{15a}$$

or in inner-product notation,

$$Re\, (\mathbf{v}, \Phi\mathbf{v}) \geq 0 \tag{15b}$$

whenever it exists.

The inequality (15a) is the usual passivity criterion encountered in the literature [4]-[6], [10], and suffers from two rather serious defects. First, to assume it exists restricts the domains of the admissible operators Φ. Second, it seems to admit, *a priori*, the possibility that for some instant t_0,

$$Re \int_{-\infty}^{t_0} \mathbf{v}^*\Phi\mathbf{v}(t)\, dt < 0,$$

that is to say, the possibility that in finite time the network can return more energy to the sources than it has absorbed, a situation which in the authors' opinion is at variance with the intuitive connotation of the word "passive."

As we have pointed out, a network passive in the sense of (15) is passive in the sense of (15a) only if (15a) is known to exist for all $\mathbf{v} \in D(\Phi)$. On the other hand, neither does the passivity of Φ in the sense of (15a) imply its passivity in the sense of (15). A sufficient condition for this is that $\mathbf{v}(t) \in D(\Phi)$ implies $\mathbf{v}_\tau(t) \in D(\Phi)$ for every $\tau > -\infty$ where

$$\mathbf{v}_\tau(t) = \mathbf{v}(t), \qquad t < \tau$$
$$= \mathbf{0}, \qquad t \geq \tau,$$

[Apply (15a) to $\mathbf{v}_\tau$.] We wish to avoid any such assumption because it involves a loss of generality and does not contribute anything physically significant.

The next lemma gives insight into the special nature of the linear passive n-port.

Lemma 3: If Φ is linear and passive, $1/r\, \mathcal{I} \otimes \Phi$ ($\mathcal{I}$ is the identity operator in H_n) is a linear transformation for every $r > 0$. Thus $D(1/r\, \mathcal{I} \otimes \Phi)$ contains at least $\mathbf{0} = (0, 0, \cdots, 0)'$, the zero in H_n and is therefore dissipatively regular (Definition 9).

Proof: By lemma 1 it suffices to show that $\mathbf{0}$ is in the domain of $1/r\, \mathcal{I} \otimes \Phi$. Let $\mathbf{0} = \mathbf{v}_1 + \mathbf{v}_2$, $\mathbf{v}_1 \in H_n$, $\mathbf{v}_2 \in D(\Phi)$ with $1/r\, \mathbf{v}_1$ equal to one of the values $\Phi\mathbf{v}_2$. By passivity,

$$0 \leq Re \int_{-\infty}^{\tau} \mathbf{v}_2^*\Phi\mathbf{v}_2\, dt = -\frac{1}{r}\int_{-\infty}^{\tau} \mathbf{v}_1^*\mathbf{v}_1(t)\, dt \leq 0$$

for any $\tau > -\infty$. Thus for any $\tau > -\infty$,

$$\int_{-\infty}^{\tau} \mathbf{v}_1^*\mathbf{v}_1(t)\, dt = 0$$

and this implies that the n components $v_{1k}(t)$, $(k = 1, 2, \cdots, n)$, of $\mathbf{v}_1(t)$ are almost everywhere zero in $(-\infty < t < \infty)$, *i.e.*, $\mathbf{v}_1(t) = \mathbf{0}$ a.e.; hence $\mathbf{v}_2(t) = -\mathbf{v}_1(t) = \mathbf{0}$ a.e. and $\mathbf{0}$ possesses the unique decomposition $\mathbf{0} = \mathbf{0} + \mathbf{0}$ a.e., Q.E.D.

Without any further assumption on Φ it is possible, as a trivial example shows, for $D(\mathcal{I} \otimes \Phi)$ to contain only $\mathbf{0}$. In any case the next lemma gives an explicit representation for $D(\mathcal{I} \otimes \Phi)$ when Φ is linear and passive.

Lemma 4: If Φ is linear and passive $D(\mathcal{I} \otimes \Phi)$ is the set of all elements $\mathbf{e}(t) \; \varepsilon \; H_n$ having the decomposition

$$\mathbf{e}(t) = \mathbf{v}_1(t) + \mathbf{v}_2(t), \qquad \mathbf{v}_2(t) \; \varepsilon \; D(\Phi),$$

in which $\mathbf{v}_1(t)$ is any value of $\Phi\mathbf{v}_2(t)$. Symbolically, $\mathbf{e}(t) \; \varepsilon \; D(\mathcal{I} \otimes \Phi)$ if, and only if,

$$\mathbf{e}(t) = \mathbf{v} + \Phi\mathbf{v}, \qquad \mathbf{v}(t) \; \varepsilon \; D(\Phi).$$

Proof: The decomposition $\mathbf{e} = \mathbf{v} + \Phi\mathbf{v}$, $\mathbf{v} \; \varepsilon \; D(\Phi)$, is unique since, by Lemma 4, $D(\mathcal{I} \otimes \Phi)$ contains $\mathbf{0}$. Thus $\mathbf{e} \; \varepsilon \; D(\mathcal{I} \otimes \Phi)$, Q.E.D.

Definition 14

For any n-port Φ in H_n, $\mathcal{I} \otimes \Phi$ is called the augmented n-port and denoted by Φ_a: $\Phi_a \equiv \mathcal{I} \otimes \Phi$ is a transformation and

$$\mathbf{i} = \Phi_a \mathbf{e}, \qquad \mathbf{e} \; \mathcal{I} \; D(\mathcal{I} \otimes \Phi),$$

imply that the decomposition

$$\mathbf{e} = \mathbf{i} + \mathbf{v}, \qquad \mathbf{v} \; \varepsilon \; D(\Phi)$$

is unique. The schematic of Φ_a is shown in Fig. 2 (the normalization $r = 1$ is made only for convenience).

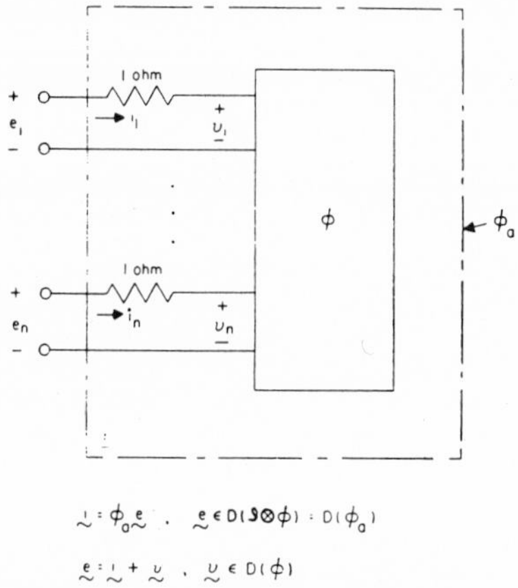

Fig. 2—The schematic of the augmented n-port ϕ_α.

The idea of using the augmented n-port Φ_a to derive properties of Φ has occurred to several other authors [4], [10], [11]. The intrinsic value of the method is revealed in Lemmas 5 and 6.

Lemma 5: Let Φ be passive. Then $\mathbf{e}(t) \; \varepsilon \; L_{2n} \cap D(\Phi_a)$ and $\mathbf{e} = \mathbf{i} + \mathbf{v}$, $\mathbf{i} = \Phi_a \mathbf{e}$, $\mathbf{v} \; \varepsilon \; D(\Phi)$, imply $\mathbf{i} \; \varepsilon \; L_{2n}$ and $\mathbf{v} \; \varepsilon \; L_{2n}$.

Proof: Suppose $\mathbf{e} = \mathbf{i} + \mathbf{v}$, $\mathbf{v} \; \varepsilon \; D(\Phi)$, $\mathbf{i} = \Phi_a \mathbf{e}$ and $(\mathbf{e}, \mathbf{e}) < \infty$. Then for any finite real τ,

$$\int_{-\infty}^{\tau} \mathbf{e}^*\mathbf{e} \, dt = \int_{-\infty}^{\tau} \mathbf{i}^*\mathbf{i} \, dt + \int_{-\infty}^{\tau} \mathbf{v}^*\mathbf{v} \, dt + 2 \, Re \int_{-\infty}^{\tau} \mathbf{v}^*\mathbf{i} \, dt.$$

Since

$$Re \int_{-\infty}^{\tau} \mathbf{v}^*\mathbf{i}(t) \, dt \geq 0,$$

the right-hand side exists, $(\mathbf{i}, \mathbf{i}) \leq (\mathbf{e}, \mathbf{e})$ and $(\mathbf{v}, \mathbf{v}) \leq (\mathbf{e}, \mathbf{e})$, Q.E.D.

From Lemma 5 we conclude that if Φ is linear and passive, Φ_a defines a linear, bounded transformation on the subset $L_{2n} \cap D(\Phi_a)$ into L_{2n} with norm, $\| \Phi_a \| \leq 1$. This observation will prove to be extremely fruitful.

Lemma 6: Suppose Φ is passive. Then $\mathbf{e}(t) \; \varepsilon \; D(\Phi_a)$ and $\mathbf{e}(t) = \mathbf{0}$ a.e. in $t < \delta$ imply $\mathbf{i}(t) = \mathbf{0} = \mathbf{v}(t)$ a.e. in $t < \delta$.

Proof: From

$$\mathbf{e} = \mathbf{i} + \mathbf{v}, \qquad \mathbf{v} \; \varepsilon \; D(\Phi),$$

it follows by hypothesis that

$$\mathbf{0} = \mathbf{i} + \mathbf{v}, \qquad \text{a.e. in} \quad t < \delta.$$

Hence by the passivity of Φ, for any $\tau < \delta$,

$$0 \leq Re \int_{-\infty}^{\tau} \mathbf{v}^*\mathbf{i} \, dt = -\int_{-\infty}^{\tau} \mathbf{v}^*\mathbf{v}(t) \, dt \leq 0.$$

$$\therefore \qquad \int_{-\infty}^{\tau} \mathbf{v}^*\mathbf{v}(t) \, dt = 0, \qquad \tau < \delta;$$

and consequently the components $v_r(t)$, $(r = 1, 2, \cdots, n)$, must be zero a.e. in $t < \delta$; *i.e.*, $\mathbf{v}(t) = \mathbf{0} = \mathbf{i}(t)$ a.e. in $t < \delta$, Q.E.D.

Lemma 6, despite its extreme simplicity, actually opens the door to the use of the Titchmarsh theorem. It is important to note that the main conclusions of the above lemmas are independent of any assumption of linearity.

Definition 15

An n-port Φ is said to be causal when for any two elements $\mathbf{v}_1(t)$ and $\mathbf{v}_2(t)$ in $D(\Phi)$ and any real τ, $\mathbf{v}_1(t) = \mathbf{v}_2(t)$ a.e. in $t \leq \tau$ implies $\mathbf{i}_1(t) = \mathbf{i}_2(t)$ a.e. in $t \leq \tau$, where $\mathbf{i}_1(t)$ is any value of $\Phi\mathbf{v}_1$ and $\mathbf{i}_2(t)$ is any value of $\Phi\mathbf{v}_2$.

Obviously if Φ is not s.v. it cannot be causal. However, the next theorem shows that a linear, passive operator can be multiple-valued only if its domain does not possess n a.e. linearly independent elements.

Theorem 1: Let Φ be a linear, passive n-port. Then, a sufficient condition for Φ to be causal (and hence s.v.) is that $D(\Phi)$ contain in its domain n elements $\mathbf{v}_k(t)$, $(k = 1, 2, \cdots, n)$, almost everywhere independent in $(-\infty < t < \infty)$, *i.e.*, the rank of the $n \times n$ matrix

$$M_n(t) = (\mathbf{v}_1, \mathbf{v}_2, \cdots, \mathbf{v}_n) \equiv \begin{bmatrix} v_{11}, & \cdots, & v_{n1} \\ v_{12}, & \cdots, & v_{n2} \\ \vdots & & \\ v_{1n}, & \cdots, & v_{nn} \end{bmatrix}$$

equals n almost everywhere in $(-\infty < t < \infty)$.

Proof: Suppose $\mathbf{v}(t) \in D(\Phi)$, $\mathbf{v}(t) = \mathbf{0}$ a.e. in $t < \delta$ and $\mathbf{i}(t)$ is any value of $\Phi\mathbf{v}$. Let $\mathbf{v}_0$ be any other member of $D(\Phi)$ and $\mathbf{i}_0(t)$ any value of $\Phi\mathbf{v}_0$. Then, for any complex α,

$$\hat{\mathbf{v}}_0(t) = \mathbf{v}_0(t) + \alpha\mathbf{v}(t)$$

coincides with $\mathbf{v}_0$ a.e. in $t < \delta$. By linearity $\hat{\mathbf{i}}_0 \equiv \mathbf{i}_0 + \alpha\mathbf{i}$ is a value of $\Phi\hat{\mathbf{v}}_0$ and by passivity,

$$Re \int_{-\infty}^{t} \hat{\mathbf{v}}_0^*(\tau)\hat{\mathbf{i}}_0(\tau)\, d\tau \geq 0, \qquad t > -\infty.$$

Thus for $t < \delta$ and all real α,

$$Re \int_{-\infty}^{t} \mathbf{v}_0^*\mathbf{i}_0(\tau)\, d\tau + \alpha\, Re \int_{-\infty}^{t} \mathbf{v}_0^*\mathbf{i}(\tau)\, d\tau \geq 0.$$

$$\therefore\ Re \int_{-\infty}^{t} \mathbf{v}_0^*\mathbf{i}(\tau)\, d\tau = 0, \qquad t < \delta.$$

$$\therefore\ Re\, \mathbf{v}_0^*(t)\mathbf{i}(t) = 0, \qquad \text{a.e. in} \quad t < \delta.$$

Substituting $i\mathbf{v}_0(t)$ for $\mathbf{v}(t)$ gives

$$Im\, \{\mathbf{v}_0^*(t)\mathbf{i}(t)\} = 0, \qquad \text{a.e. in} \quad t < \delta.$$

$$\therefore \qquad \mathbf{v}_0^*(t)\mathbf{i}(t) = 0, \qquad \text{a.e. in} \quad t < \delta. \tag{16}$$

Replacing $\mathbf{v}_0(t)$ by $\mathbf{v}_1, \mathbf{v}_2, \cdots, \mathbf{v}_n(t)$ successively we get

$$M_n^*(t)\mathbf{i}(t) = \mathbf{0} \quad \text{a.e. in} \quad t < \delta.$$

Since $M_n(t)$ is a.e. nonsingular $\mathbf{i}(t) = \mathbf{0}$ a.e. in $t < \delta$. Causality is now an immediate consequence of linearity, Q.E.D.

Corollary 1(a): Let Φ be a linear, passive n-port and suppose that for every $\mathbf{v}(t) \in D(\Phi)$, $\mathbf{0}$ is one of the values $\Phi\mathbf{v}$. Then, Φ is "instantaneously lossless;" *i.e.*, for any $\mathbf{i}(t)$ in the range $R(\Phi)$ of Φ,

$$\mathbf{v}_0^*(t)\mathbf{i}(t) = 0 \quad \text{a.e.}$$

for every $\mathbf{v}_0(t)$ in $D(\Phi)$.

Proof: If Φ is s.v. there is nothing to prove. Otherwise, to any $\mathbf{v} \in D(\Phi)$ corresponds a value $\mathbf{i} = \Phi\mathbf{v} = \mathbf{0}$. By linearity $\mathbf{i}(t) - \mathbf{0}$ is a value of $\Phi(\mathbf{v} - \mathbf{v}) = \Phi\mathbf{0}$ and by Theorem 1,

$$\mathbf{v}_0^*(t)\mathbf{i}(t) = 0 \quad \text{a.e.}$$

for all $\mathbf{v}_0(t)$ in $D(\Phi)$, Q.E.D.

Corollary 1(b): A sufficient condition for a linear, passive one-port Φ to be s.v. and causal is that $D(\Phi)$ contain at least one element $v(t) \neq 0$ a.e. in $(-\infty < t < \infty)$.

Proof: Theorem 1

Theorem 1 and Corollary 1(a) show that the ideal transformer exhibits most of the properties of those linear, passive multiple-valued n-ports for which $\mathbf{0}$ is a value of every $\Phi\mathbf{v}$, $\mathbf{v} \in D(\Phi)$ and justify a previous remark. Again, a simple example reveals that the sufficient condition of Theorem 1 is not necessary but it most certainly has the character of a "best."

The ideal short-circuit, Φ_s, is neither s.v. nor causal:

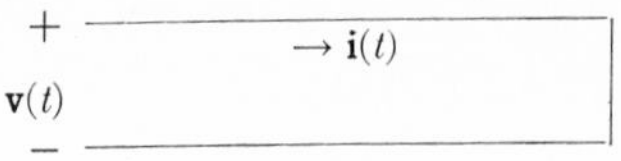

$D(\Phi_s)$ contains only the element $\mathbf{v}(t) = \mathbf{0}$; $\Phi_s\mathbf{0} =$ all $\mathbf{i}(t) \in H_1$. Clearly, the hypothesis of Corollary 1(b) is violated.

The definition of causality implies that a linear casual n-port must be single-valued but a linear, passive s.v. n-port need not be causal. The following is a counterexample. Choose $n = 2$ and let $D(\Phi)$ contain all $\mathbf{v}(t) = \alpha\mathbf{u}(t)$, α complex and arbitrary, where

$$\mathbf{u}(t) = \begin{pmatrix}0\\0\end{pmatrix}, \qquad t \leq 0,$$

$$= \begin{pmatrix}1\\1\end{pmatrix} \qquad t > 0.$$

Set

$$\Phi(\mathbf{0}_2) = \mathbf{0}_2$$

and

$$\Phi\mathbf{v} = \begin{pmatrix}\alpha\\-\alpha\end{pmatrix} = \alpha\begin{pmatrix}1\\-1\end{pmatrix}.$$

Evidently Φ is linear and single-valued. It is also passive because $\bar{v}_1 i_1 + \bar{v}_2 i_2 = 0$, $t \leq 0$ and $|\alpha|^2 - |\alpha|^2 = 0$, $t > 0$. But observe that for any $\alpha \neq 0$, $\mathbf{v}(t) = \mathbf{0}_2$, $t \leq 0$ and

$$\mathbf{i}(t) = \begin{pmatrix}\alpha\\-\alpha\end{pmatrix} \neq \mathbf{0}_2, \qquad t < 0,$$

i.e., Φ is not causal.

It is not unusual for an n-port to possess two different operational descriptions such that it is s.v. and causal with respect to one and not the other. The ideal short-circuit shown above is a good illustration.

Description 1 — Φ_s:

1) Only $\mathbf{v}(t) = \mathbf{0}$ is a member of $D(\Phi)$.
2) $\Phi_s(\mathbf{0}) =$ any $\mathbf{i}(t)$ in H_1.

Description 2 — $\tilde{\Phi}_s$:

1) Any $\mathbf{i}(t)$ in H_1 is an element of $D(\tilde{\Phi}_s)$; *i.e.*, $D(\tilde{\Phi}_s) = H_1$.
2) $\tilde{\Phi}_s\mathbf{i}(t) = \mathbf{0}$ for all $\mathbf{i}(t) \in D(\tilde{\Phi})$.

In the first description, Φ_s, the short-circuit is neither s.v. nor causal but in the second, $\tilde{\Phi}_s$, it is both.

Lemma 7: If Φ is a linear, passive and time-invariant operator, Φ_a is a linear, passive, causal and time-invariant transformation.

Proof: Lemmas 2, 3, and 6. In view of Lemmas 5 and 7, Φ_a defines a linear, bounded, time-invariant transformation on $L_{2n} \cap D(\Phi_a)$ into L_{2n} whenever Φ is linear, passive and time-invariant. When $D(\Phi_a)$ is dense in L_{2n}, the extension theorem quoted in Section II implies that Φ_a possesses a linear, bounded continuation, $\hat{\Phi}_a$, mapping all L_{2n} into L_{2n}. Thus for any $\mathbf{e} \in L_{2n}$,

$$\mathbf{i} = \hat{\Phi}_a \mathbf{e} = \underset{r\to\infty}{\text{l.i.m.}}\ \Phi_a \mathbf{e}_r = \underset{r\to\infty}{\text{l.i.m.}}\ i_r,$$

where

$$\mathbf{e}_r \in L_{2n} \cap D(\Phi_a), \qquad (r = 1, 2, \cdots, \cdots)$$

and

$$\underset{r\to\infty}{\text{l.i.m.}}\ \mathbf{e}_r = \mathbf{e}.$$

Again, if $\mathbf{v} = \mathbf{e} - \mathbf{i}$, $\mathbf{e} \in L_{2n}$, $\mathbf{i} = \hat{\Phi}_a \mathbf{e}$,

$$\mathbf{v}_r(t) = \mathbf{e}_r(t) - \mathbf{i}_r(t), \qquad (r = 1, 2, \cdots, \cdots)$$

and

$$Re \int_{-\infty}^{t} \mathbf{v}^* \mathbf{i}(\tau)\, d\tau = \lim_{r\to\infty} Re \int_{-\infty}^{t} \mathbf{v}_r^* \mathbf{i}_r(\tau)\, d\tau \geq 0$$

for all $t > -\infty$ since $\text{l.i.m.}_{r\to\infty}\ \mathbf{v}_r(t) = \mathbf{v}(t)$ and the inner-product is continuous. The proof of Lemma 6 now shows that $\mathbf{e}(t) \in L_{2n}$ and $\mathbf{e}(t) = \mathbf{0}$ a.e. in $t < \delta$ imply $\mathbf{v}(t)$ and $\mathbf{i}(t) = \hat{\Phi}_a \mathbf{e}(t) = \mathbf{0}$ a.e. in $t < \delta$: All the properties of Φ_a are transferred to $\hat{\Phi}_a$. Consequently the Bochner L_2 theorem can be applied to $\hat{\Phi}_a$ to deduce the spectral structure of Φ_a whenever $D(\Phi_a)$ is known to be dense in L_{2n}. By Lemma 4 this is the case if and cnly if

$$\mathbf{v} + \Phi \mathbf{v} = \mathbf{e}, \qquad \mathbf{v} \in D(\Phi), \tag{21}$$

is soluble for $\mathbf{v}(t)$ over a set of $\mathbf{e}$'s dense in L_{2n}. Wu [4] actually assumes that (21) is soluble for every $\mathbf{e}(t)$ in L_{2n} that vanishes for sufficiently negative t. However this may be very difficult (if not impossible) to establish in any specific problem; but in any event, the subsequent analysis indicates that it is not necessary to delineate the e's more precisely.

The assumption that $D(\Phi_a)$ is dense in L_{2n} is a physical one and appears as postulate P_4.

IV. The Postulates and the Main Theorems

The reader should now be in a position to appreciate the need for the following postulates.

P_1—The n-port Φ is linear; *i.e.*, Φ is a linear operator in H_n (Definition 6).

P_2—The n-port Φ is passive (Definition 13).

P_3—The n-port Φ is time-invariant (Definition 10).

P_4—The equation

$$\mathbf{v} + \Phi \mathbf{v} = \mathbf{e}, \qquad \mathbf{v} \in D(\Phi),$$

is soluble for $\mathbf{v}$ over a set of $\mathbf{e}$'s dense in L_{2n}.

P_5—If $\mathbf{v} \in D(\Phi)$ is real, all values of $\Phi \mathbf{v}$ are real.

Postulate P_5 may be omitted without impairing the validity of the main theorems; it is only used to impart a certain symmetry to the scattering matrix. It should be evident from the various examples in Section III that the axioms are consistent and independent.

Theorem 2: An n-port Φ satisfying postulates P_1 to P_4 possesses a frequency response in the sense that $\mathbf{e}(t) \in L_{2n} \cap D(\Phi_a)$ and $\mathbf{i} = \Phi_a \mathbf{e}$ imply

$$\mathbf{I}(\omega) = Y_a(\omega)\mathbf{E}(\omega),$$

where $\mathbf{I}(\omega)$ and $\mathbf{E}(\omega)$ are the column-vector L_{2n}-Fourier transforms of $\mathbf{i}(t)$ and $\mathbf{e}(t)$ respectively and $Y_a(\omega)$ is an essentially bounded $n \times n$ matrix over $(-\infty < \omega < \infty)$ whose elements $y_{k,r}(\omega)$, $(r, k, = 1, 2, \cdots, n)$, possess uniformly bounded analytic extensions into the upper half-plane $Im\ z > 0$.

Proof: By Lemmas 4, 5, 7 and postulate P_4, Φ_a defines a linear, bounded, time-invariant transformation on $L_{2n} \cap D(\Phi_a)$ into L_{2n} whose domain $D(\Phi_a)$ is dense in L_{2n}. The extension theorem (Section II) permits us to assert that Φ_a possesses a linear bounded continuation $\hat{\Phi}_a$ defined on all of L_{2n} such that

$$\Phi_a \mathbf{e} = \hat{\Phi}_a \mathbf{e}, \qquad \mathbf{e} \in D(\Phi_a) \cap L_{2n}.$$

From a previous discussion it is known that $\hat{\Phi}_a$ enjoys all the properties of Φ_a so that the Bochner L_2-theorem is applicable to $\hat{\Phi}_a$.

Thus

$$\mathbf{I}(\omega) = Y_a(\omega)\mathbf{E}(\omega)$$

where $Y_a(\omega)$ is an $n \times n$ matrix whose elements are essentially uniformly bounded in $(-\infty < \omega < \infty)$, *i.e.*, for some finite $\alpha > 0$, $|y_{k,r}(\omega)| < \alpha$ a.e., $(k, r = 1, 2, \cdots, n)$.

Let $\hat{\mathbf{i}}_r(t) = \hat{\Phi}_a \tilde{\mathbf{e}}_r(t)$, $(r = 1, 2, \cdots, n)$, where $\tilde{\mathbf{e}}_r(t)$, $(r = 1, 2, \cdots, n)$, is defined by

$$\begin{aligned} \tilde{e}_{rr}(t) &= \sqrt{-2\pi}\, e^{-t}, & t \geq 0, \\ &= 0, & t < 0 \end{aligned}$$

and

$$\tilde{e}_{rk}(t) \equiv 0, \qquad k \neq r.$$

The corresponding responses $\hat{\mathbf{i}}_r(t) = \hat{\Phi}_a \tilde{\mathbf{e}}_r(t)$, $(r = 1, 2, \cdots, n)$, belong to L_{2n} and vanish a.e. in $t < 0$. Denote the $n \times n$ matrices whose columns are $\tilde{\mathbf{I}}_r(\omega)$ and $\tilde{\mathbf{E}}_r(\omega)$, $(r = 1, 2, \cdots, n)$, by $\tilde{I}_n(\omega)$ and $\tilde{E}_n(\omega)$, respectively.

Then,

$$\tilde{I}_n(\omega) = Y_a(\omega)\tilde{E}_n(\omega) \tag{21}$$

where

$$\tilde{E}_n(z) = \frac{1}{z + i} 1_n, \tag{22}$$

in which 1_n is the unit $n \times n$ matrix. From the Titchmarsh theorem $\tilde{I}_n(\omega)$ has an analytic extension $\tilde{I}_n(z)$ into the upper half-plane $Im\ z > 0$ $(z = \omega + i\beta)$,

$$\lim_{\beta\to 0^+} \tilde{I}_n(z) = \tilde{I}_n(\omega) \quad \text{a.e.} \tag{23}$$

and

$$\sum_{r=1}^{n} \int_{-\infty}^{+\infty} \tilde{\mathbf{I}}_r^*(\omega + i\beta)\tilde{\mathbf{I}}_r(\omega + i\beta)\, d\omega < K, \tag{24}$$

K being a constant independent of $\beta > 0$. All this suggests that the correct analytic continuation of $Y_a(\omega)$ into $\beta > 0$ is given by

$$Y_a(z) = (z + i)\tilde{I}_n(z). \tag{25}$$

Certainly $Y_a(z)$ is analytic in $Im\, z > 0$ and takes on the correct boundary values because [using the Titchmarsh theorem and (21)]

$$\lim_{\beta \to 0+} (z + i)\tilde{I}_n(z) = (\omega + i)\tilde{I}_n(\omega) = Y_a(\omega) \quad \text{a.e.}$$

It remains to prove that for any causal $n \times n$ matrix $E_n(z)$ of complex L_{2n}-Fourier transforms,

$$I_n(z) = Y_a(z)E_n(z), \qquad Im\, z > 0, \tag{26}$$

where $I_n(z)$ is the response $n \times n$ matrix of complex L_{2n}-Fourier transforms. A causal matrix of complex L_{2n}-Fourier transforms is one whose every column $\mathbf{E}_r(z)$, $(r = 1, 2, \cdots, n)$, is a complex L_{2n}-Fourier transform of a square-integrable $\mathbf{e}_r(t)$ that vanishes for $t < 0$.

To this end note first that by the "dispersion" relation quoted in the Titchmarsh theorem,

$$\tilde{I}_n(z) = \frac{Y_a(z)}{z + i} = \frac{1}{2\pi i} \int_{-\infty}^{+\infty} \frac{Y_a(u)\, du}{(u + i)(u - z)}, \quad \beta > 0 \tag{27}$$

and

$$0_n = \frac{1}{2\pi i} \int_{-\infty}^{+\infty} \frac{Y_a(u)\, du}{(u + i)(u - \bar{z})}, \quad \beta > 0. \tag{28}$$

Subtracting (28) from (27) we get

$$\frac{Y_a(z)}{z + i} = \frac{\beta}{\pi} \int_{-\infty}^{+\infty} \frac{Y_a(u)\, du}{(u + i) \mid u - z \mid^2}, \quad \beta > 0. \tag{29}$$

$$\therefore \left| \frac{Y_a(z)}{z + i} \right| \le \frac{\beta\alpha}{\pi} \int_{-\infty}^{+\infty} \frac{du}{(u - \omega)^2 + \beta^2} = \alpha, \quad \beta > 0. \tag{30}$$

$$\therefore \mid Y_a(z) \mid \le \alpha \mid z + i \mid, \quad \beta > 0. \tag{31}$$

Since $\mid Y_a(\omega) \mid < \alpha$ a.e. in $(-\infty < \omega < \infty)$, a general form [12] of the Phragmén-Lindelöf theorem yields

$$\mid Y_a(z) \mid \le \alpha, \qquad Im\, z > 0, \tag{32}$$

i.e.,

$$\mid y_{k,r}(z) \mid \le \alpha, \qquad Im\, z > 0, \qquad (r, k = 1, 2, \cdots, n).$$

Thus the elements of $Y_a(z)$ are uniformly bounded everywhere in the upper half-plane $Im\, z > 0$ and a.e. on the real frequency ω axis. Suppose $E_n(z)$ is a causal $n \times n$ matrix. Then

$$A_n(z) \equiv Y_a(z)E_n(z)$$

is analytic in $Im\, z > 0$ and

$$\sum_{r=1}^{n} \int_{-\infty}^{+\infty} \mathbf{A}_r^*(\omega + i\beta)\mathbf{A}_r(\omega + i\beta)\, d\omega$$

$$\le n\alpha^2 \sum_{r=1}^{n} \int_{-\infty}^{+\infty} \mathbf{E}_r^*(\omega + i\beta)\mathbf{E}_r(\omega + i\beta)\, d\omega \le n\alpha^2 c, \quad \beta > 0,$$

where c is a constant independent of $\beta > 0$ (E_n is causal). Thus, by the Titchmarsh theorem,

$$A_n(z) = \frac{1}{2\pi i} \int_{-\infty}^{+\infty} \frac{Y_a(u)E_n(u)}{u - z}\, du, \qquad Im\, z > 0.$$

But also

$$I_n(z) = \frac{1}{2\pi i} \int_{-\infty}^{+\infty} \frac{I_n(u)\, du}{u - z}, \qquad Im\, z > 0,$$

$$= \frac{1}{2\pi i} \int_{-\infty}^{+\infty} \frac{Y_a(u)E_n(u)}{u - z}\, du, \qquad Im\, z > 0,$$

by (21). Hence

$$I_n(z) = Y_a(z)E_n(z), \qquad Im\, z > 0$$

and a.e. in $(-\infty < \omega < \infty)$, Q.E.D.

When P_5 attains we find easily that

$$\bar{Y}_a(z) = Y_a(-\bar{z}), \quad Im\, z > 0$$

and

$$\bar{Y}_a(\omega) = Y_a(-\omega) \quad \text{a.e.} \tag{33}$$

In particular

$$\bar{Y}_a(i\beta) = Y_a(i\beta), \quad \beta > 0,$$

i.e., $Y_a(z)$ is real for purely imaginary z in the upper half-plane. At this point it is expeditious to introduce the $n \times n$ scattering matrix $S(z)$ (normalized to one ohm):

$$S(z) \equiv 1_n - 2Y_a(z). \tag{34}$$

It follows from Theorem 2 that $S(z)$ is analytic in $Im\, z > 0$ and uniformly bounded for all z in $Im\, z > 0$ and almost all ω. However much more is true. Let $\mathbf{e} \in L_{2n} \cap D(\Phi_a)$, $\mathbf{i} = \Phi_a\mathbf{e} = \Phi\mathbf{v}$ and $\mathbf{e} = \mathbf{v} + \mathbf{i}$. Using well-established steady-state transmission line practice as a guide we define the incident and reflected voltages impinging on the n-port by

$$\mathbf{v}_+(t) = \tfrac{1}{2}(\mathbf{v} + \mathbf{i}), \tag{35a}$$

and

$$\mathbf{v}_-(t) = \tfrac{1}{2}(\mathbf{v} - \mathbf{i}), \tag{35b}$$

respectively. Of course $\mathbf{v}_+$ and $\mathbf{v}_-$ belong to L_{2n} (Lemma 5) and

$$Re \int_{-\infty}^{+\infty} \mathbf{v}^*\mathbf{i}(t)\, dt = \int_{-\infty}^{+\infty} (\mathbf{v}_+^*\mathbf{v}_+ - \mathbf{v}_-^*\mathbf{v}_-)\, dt$$

$$= (\mathbf{v}_+, \mathbf{v}_+) - (\mathbf{v}_-, \mathbf{v}_-). \tag{36}$$

Now suppose that $\mathbf{e}(t)$ is causal. Then, from Theorem 2,

$$\mathbf{I}(z) = Y_a(z)\mathbf{E}(z), \qquad Im\, z > 0$$

and a.e. in $(-\infty < \omega < \infty)$ (the symbols have their usual meanings). Since $\mathbf{v}_+ = \frac{1}{2}\mathbf{e}$,

$$\mathbf{I}(z) = 2Y_a(z)\mathbf{V}_+(z).$$

$$\therefore \quad \mathbf{V}(z) = \mathbf{E}(z) - \mathbf{I}(z) = 2(1_n - Y_a)\mathbf{V}_+(z).$$

$$\therefore \mathbf{V}_-(z) = (1_n - 2Y_a)\mathbf{V}_+(z) = S(z)\mathbf{V}_+(z) \tag{37}$$

for all z in the upper half-plane and almost all ω. By (36), P_2, and Plancherel's theorem,

$$0 \leq Re \int_{-\infty}^{+\infty} \mathbf{v}^*\mathbf{i}(t)\, dt = \int_{-\infty}^{+\infty} [\mathbf{V}_+^*\mathbf{V}_+(\omega) - \mathbf{V}_-^*S^*S\mathbf{V}_-(\omega)]\, d\omega$$

$$= \int_{-\infty}^{+\infty} \mathbf{V}_+^*(1_n - S^*S)\mathbf{V}_+(\omega)\, d\omega$$

$$= \frac{1}{4}\int_{-\infty}^{+\infty} \mathbf{E}^*(1_n - S^*S)\mathbf{E}(\omega)\, d\omega \tag{38}$$

whenever $\mathbf{e}(t) \in L_{2n} \cap D(\Phi_a)$. Actually (38) is true for all $\mathbf{e}(t) \in L_{2n}$ since $D(\Phi_a)$ is dense in L_{2n}. This observation leads to the following theorem.

Theorem 3: The $n \times n$ matrix

$$Q(z) = 1_n - S^*(z)S(z) \tag{39}$$

is the matrix of a non-negative quadratic form for all z in the strict upper half-plane and almost all ω in $(-\infty < \omega < \infty)$.

Proof: In order to lend the motivation some transparency we proceed via two lemmas.

Lemma 3(a): Let Φ satisfy P_1 to P_4 and let $\mathbf{e}(t)$ belong to L_{2n} and vanish for $t < 0$. Then, if $\mathbf{i} = \Phi_a\mathbf{e}$ and $\mathbf{v} = \mathbf{e} - \mathbf{i}$,

$$Re \int_0^{\infty} \mathbf{v}^*(t)\mathbf{i}(t)e^{-2\beta t}\, dt \geq 0 \tag{40}$$

for all $\beta \geq 0$.

Proof: From a former discussion we know that $\mathbf{v}(t)$ and $\mathbf{i}(t)$ belong to L_{2n}, vanish for $t < 0$ and that (40) is true for $\beta = 0$. If $\beta > 0$, $e^{-2\beta t} \in L_{2n}$ on $(0, \infty)$ and monotone-decreasing as $t \to \infty$. Applying the second mean-value theorem for integrals [13]

$$\int_0^t Re\, \{\mathbf{v}^*(\tau)\mathbf{i}(\tau)\}e^{-2\beta\tau}\, d\tau$$

$$= \int_0^{t_1} Re\, \{\mathbf{v}^*(\tau)\mathbf{i}(\tau)\}\, d\tau \geq 0, \qquad 0 < t_1 < t,$$

by passivity. Hence,

$$Re \int_0^{\infty} \mathbf{v}^*(\tau)\mathbf{i}(\tau)e^{-2\beta\tau}\, d\tau$$

$$= \lim_{t\to\infty} \int_0^t Re\, \{\mathbf{v}^*(\tau)\mathbf{i}(\tau)\}e^{-2\beta\tau}\, d\tau \geq 0,$$

Q.E.D.

Again, since $\mathbf{v}(t)$ and $\mathbf{i}(t)$ belong to L_{2n} and vanish for $t < 0$, the L_{2n}-Fourier transforms of $\mathbf{v}(t)\, e^{-\beta t}$ and $\mathbf{i}(t)\, e^{-\beta t}$, $\beta \geq 0$, exist and are given by $\mathbf{V}(\omega + i\beta)$ and $\mathbf{I}(\omega + i\beta)$, respectively. From (40), (37), and Plancherel's theorem,

$$\frac{1}{4}\int_{-\infty}^{+\infty} \mathbf{E}^*(z)Q(z)\mathbf{E}(z)\, d\omega \geq 0, \tag{41}$$

$$z = \omega + i\beta, \qquad \beta \geq 0,$$

for every $\mathbf{E}(z)$ that is the complex L_{2n}-Fourier transform of an $\mathbf{e}(t) \in L_{2n}$ that vanishes for $t < 0$.

Choose $\mathbf{e}(t) = e(t)\mathbf{b}$, where $e(t) \in L_2$ and vanishes for $t < 0$ and $\mathbf{b}$ is an n vector of complex constants. Denoting the complex L_2-Fourier transform of $e(t)$ by $E(z)$, (41) becomes

$$\int_{-\infty}^{+\infty} (\mathbf{b}^*Q(\omega + i\beta)\mathbf{b})\, |\, E(\omega + i\beta)\, |^2\, d\omega \geq 0 \tag{42}$$

for every n-vector $\mathbf{b}$, every $\beta \geq 0$ and every $E(\omega)$ that is the L_2-Fourier transform of an $e(t) \in L_2$ that vanishes for $t < 0$. This inspires the following lemma.

Lemma 3(b): Let $|\, k(\omega)\, | < \alpha$, α constant, a.e. in $(-\infty < \omega < \infty)$. Then $k(\omega) \geq 0$ almost everywhere is the necessary and sufficient condition that for a fixed $\beta \geq 0$,

$$\int_{-\infty}^{+\infty} k(\omega)\, |\, E(\omega + i\beta)\, |^2\, d\omega \geq 0 \tag{43}$$

for every $E(\omega)$ that is the L_2-Fourier transform of an $e(t) \in L_2$ vanishing for $t < 0$.

Proof: The sufficiency is obvious as is the fact that $k(\omega)$ must be real almost everywhere. Now define

$$e_\tau(t) = e(t), \qquad t \leq \tau,$$

$$= 0, \qquad t > \tau.$$

Clearly

$$\underset{\tau\to\infty}{\text{l.i.m.}}\, e_\tau(t) = e(t)$$

and if $E_\tau(\omega)$ denotes the transform of $e_\tau(t)$,

$$\underset{\tau\to\infty}{\text{l.i.m.}}\, E_\tau(\omega) = E(\omega).$$

The transform of $e_\tau(t)\, e^{\beta t}$ is $E_\tau(\omega - i\beta)$ and applying (43),

$$\int_{-\infty}^{+\infty} k(\omega)\, |\, E_\tau(\omega)\, |^2\, d\omega \geq 0.$$

Because $k(\omega)$ is essentially bounded

$$\underset{\tau\to\infty}{\text{l.i.m.}}\, k(\omega)E_\tau(\omega) = k(\omega)E(\omega),$$

whence, from the continuity of the inner product

$$\int_{-\infty}^{+\infty} k(\omega)\, |\, E(\omega)\, |^2\, d\omega$$

$$= \lim_{\tau\to\infty} \int_{-\infty}^{+\infty} k(\omega)\, |\, E_\tau(\omega)\, |^2\, d\omega \geq 0 \tag{44}$$

for every $E(\omega)$ meeting the requirements of the lemma.

Suppose that $k(\omega) < -\epsilon$, $\epsilon > 0$, on a set Δ of positive Lebesgue measure. Without loss of generality we can assume the measure $\mu(\Delta)$ of Δ to be finite. Define $|\, E(\omega)\, |$ as follows:

$$|\, E(\omega)\, | = 1, \qquad \omega \in \Delta$$

$$= \sqrt{\frac{\epsilon\mu(\Delta)}{4\alpha}}\,(1 + \omega^2)^{-1/2}, \qquad \omega \notin \Delta.$$

Evidently,

$$\int_{-\infty}^{+\infty} | E(\omega) |^2 \, d\omega < \infty$$

and

$$\int_{-\infty}^{+\infty} \frac{| \ln | E(\omega) ||}{1 + \omega^2} \, d\omega < \infty .$$

By the Paley-Wiener theorem, $| E(\omega) |$ is the modulus of a square-integrable $e(t)$ vanishing for $t < 0$ and must therefore satisfy (44). But,

$$\int_{-\infty}^{+\infty} k(\omega) \, | E(\omega) |^2 \, d\omega = \int_{\omega \varepsilon \Delta} + \int_{\omega \not\varepsilon \Delta}$$

$$\leq -\epsilon\mu(\Delta) + \frac{\epsilon\mu(\Delta)}{4\alpha} \int_{-\infty}^{+\infty} \frac{| k(\omega) |}{1 + \omega^2} \, d\omega$$

$$\leq -\epsilon\mu(\Delta) + \frac{\epsilon\mu(\Delta)\pi}{4} < 0,$$

a contradiction. Thus $k(\omega) \geq 0$ a.e., Q.E.D.

Identifying the factor $\mathbf{b}^*Q(\omega + i\beta)\mathbf{b}$ in (42) with the $k(\omega)$ of Lemma 3(b) we get

$$\mathbf{b}^*Q(\omega + i\beta)\mathbf{b} \geq 0, \qquad \beta \geq 0, \tag{45}$$

for almost all ω. However, because $Q(z)$ is continuous in $Im\ z > 0$, (45) is true for all ω when $\beta > 0$. From the analyticity and uniform boundedness of $S(z)$ in $Im\ z > 0$ it follows that

$$\lim_{\beta \to 0^+} Q(\omega + i\beta) = Q(\omega)$$

exists for almost all ω. Thus,

$$\lim_{\beta \to 0^+} \mathbf{b}^*Q(\omega + i\beta)\mathbf{b} = \mathbf{b}^*Q(\omega)\mathbf{b} \geq 0$$

almost everywhere, the exceptional set being independent of $\mathbf{b}$. In other words we have finally shown that $Q(z)$ is the matrix of a non-negative quadratic form everywhere in $Im\ z > 0$ and almost everywhere in $(-\infty < \omega < \infty)$, Q.E.D.

Henceforth, in order to avoid needless repetition, a statement that is true everywhere in the upper half-plane and almost everywhere on the real axis is said to be true throughout.

If $S(z)$ is an arbitrary $n \times n$ matrix, $Q(z) = 1_n - S^*(z)S(z)$ is called its associated matrix: $Q(z)$ is said to be non-negative in a region of the z plane when $\mathbf{b}^*Q(z)\mathbf{b} \geq 0$ for all n vectors $\mathbf{b}$ of complex constants and all z in the region.

Definition 16

An $n \times n$ matrix $S(z)$ is said to be bounded-real when

1) $S(z)$ is analytic in $Im\ z > 0$;
2) $Q(z) = 1_n - S^*(z)S(z)$ is non-negative throughout;
3) $\bar{S}(z) = S(-\bar{z})$ throughout.

The content of Theorems 2 and 3 can be summarized by saying that any n-port Φ satisfying postulates P_1 to P_5 possesses a bounded-real scattering description. As a matter of fact this description is unique up to sets of measure zero on the real ω axis [this is easily seen by referring to (25)]. Furthermore the reader should have no difficulty in convincing himself that if the bounded-real matrix $S(z)$ represents the scattering descriptions of two n-ports Φ_1 and Φ_2 satisfying P_1 to P_5, then $\hat{\Phi}_{a1} = \hat{\Phi}_{a2}$. Physically this means that for any $\mathbf{e}$ in $L_{2n} \cap D(\Phi_{a1})$, $\Phi_{a1}\mathbf{e} = \text{l.i.m.}_{r\to\infty}\ \Phi_{a2}\mathbf{e}_r$ where $\mathbf{e}_r(t)\ \varepsilon\ L_{2n} \cap D(\Phi_{a2})$, $(r = 1, 2, \cdots, \cdots)$ and $\text{l.i.m.}_{r\to\infty}\ \mathbf{e}_r(t) = \mathbf{e}(t)$. In short, all the values $\Phi_{a1}\mathbf{e}$, $\mathbf{e}\ \varepsilon\ L_{2n} \cap D(\Phi_{a1})$, are known when all the values $\Phi_{a2}\mathbf{e}$, $\mathbf{e}\ \varepsilon\ L_{2n} \cap D(\Phi_{a2})$, are known, and conversely.

Definition 17

An $n \times n$ matrix $S(z)$ is said to be a physically-realizable scattering matrix when it represents the scattering description of an n-port Φ satisfying postulates P_1 to P_5.

Theorem 4: The necessary and sufficient conditions for an $n \times n$ matrix $S(z)$ to be a physically realizable scattering matrix are

1) $S(z)$ is analytic in $Im\ z > 0$;
2) $\sup_{Im\, z > 0} e^{-|z|^\alpha} \, | S(z) | < \infty$ for some $\alpha < 1$;
3) $Q(\omega) = 1_n - S^*(\omega)S(\omega)$ is non-negative for almost ω in $(-\infty < \omega < \infty)$, where $S(\omega) \equiv \lim_{\beta\to 0^+} S(\omega + i\beta)$ [that this limit exists a.e. is a consequence of 1) and 2)].
4) $\bar{S}(\omega) = S(-\omega)$ for almost all ω.

Proof: Necessity has already been demonstrated in Theorems 2 and 3 and in (33). As regards sufficiency, note first that 3) implies that $| S(\omega) | \leq 1$ a.e. in $(-\infty < \omega < \infty)$. Hence by 1), 2), and the Phragmén-Lindelöf theorem already referred to in the proof of Theorem 2, $| S(z) | \leq 1$ throughout.

By the Titchmarsh theorem $\mathbf{I}(z) = Y_a(z)\mathbf{E}(z)$, $(2Y_a(z) \equiv 1_n - S(z))$ is a causal complex L_{2n}-Fourier transform whenever $\mathbf{E}(z)$ is a causal complex L_{2n}-Fourier transform. This suggests that we define Φ_a in the following manner:

a) $D(\Phi_a) = L_{2n}$;
b) For $\mathbf{e}(t)\ \varepsilon\ L_{2n}$, $\mathbf{i}(t) = \Phi_a\mathbf{e}$ is that element of L_{2n} whose L_{2n}-Fourier transform is $Y_a(\omega)\mathbf{E}(\omega)$, $\mathbf{E}(\omega)$ being the L_{2n}-Fourier transform of $\mathbf{e}(t)$. Thus,

$$\mathbf{i}(t) = \Phi_a\mathbf{e} = \text{l.i.m.}_{T\to\infty} \frac{1}{\sqrt{2\pi}} \int_{-T}^{+T} e^{-i\omega t} \, Y_a(\omega)\mathbf{E}(\omega) \, d\omega,$$

where $2Y_a(\omega) \equiv 1_n - S(\omega)$.

Next we define Φ:

c) $D(\Phi)$ = the set of all $\mathbf{v}(t)$ in L_{2n} such that

$$\mathbf{v}(t) = \mathbf{e}(t) - \Phi_a\mathbf{e}(t), \tag{46}$$

$$\mathbf{e}(t)\ \varepsilon\ L_{2n};$$

d) For $\mathbf{v} \varepsilon D(\Phi)$, the values $\Phi\mathbf{v}$ are all the values $\Phi_a\mathbf{e}$ where $\mathbf{e}$ ranges over all L_{2n} solutions of (46).

Obviously Φ_a is a linear, bounded, time-invariant transformation mapping all of L_{2n} into L_{2n} and Φ is a linear time-invariant operator in L_{2n}. The necessary and sufficient condition that Φ be s.v. is that (46) admit for $\mathbf{v}(t) = \mathbf{0}$ the unique solution $\mathbf{e}(t) = \mathbf{0}$. Suppose

$$\mathbf{0} = \mathbf{e}(t) - \Phi_a\mathbf{e}(t), \qquad \mathbf{e}(t) \varepsilon L_{2n}.$$

Then, by b)

$$[1_n + S(\omega)]\mathbf{E}(\omega) = \mathbf{0} \quad \text{a.e.} \tag{47}$$

This system of equations admits an L_{2n} solution $\mathbf{E}(\omega) \neq \mathbf{0}$ a.e. if, and only if, the determinant of the $n \times n$ matrix $1_n + S(\omega)$ equals zero on an ω-set of positive Lebesgue measure. Since $S(z)$ is analytic and bounded in $Im\ z > 0$ a theorem of F. Riez shows that this is possible if, and only if, the determinant of $1_n + S(z)$ is identically zero in the closed upper half-plane, $Im\ z \geq 0$. Thus Φ as defined in c) and d) is s.v. if, and only if, the determinant of $1_n + S(z)$ is unequal to zero for at least one z in $Im\ z > 0$ (later it will be shown that this implies that it is unequal to zero for all z, $Im\ z > 0$).

Observe that it is not yet certain that Φ_a is the augmented operator corresponding to Φ. That it is, follows rigorously from Lemma 3 once the passivity of Φ has been established.

If $\mathbf{e}(t)$ is real, $\bar{\mathbf{E}}(\omega) = \mathbf{E}(-\omega)$ a.e. and from 4),

$$\bar{\mathbf{I}}(\omega) = \bar{Y}_a(\omega)\bar{\mathbf{E}}(\omega) = Y_a(-\omega)\mathbf{E}(-\omega)$$
$$= \mathbf{I}(-\omega) \quad \text{a.e.},$$

so that $\mathbf{i}(t)$ is also real. The remaining step is to prove that Φ is passive, *i.e.*,

$$Re \int_{-\infty}^{t} \mathbf{v}^*(\tau)\Phi\mathbf{v}(\tau)\, d\tau \geq 0 \tag{48}$$

for all $t > -\infty$, all $\mathbf{v} \varepsilon D(\Phi)$ and all values $\Phi\mathbf{v}$. Let $\mathbf{e}(t)$ be any L_{2n} solution of (46) corresponding to $\mathbf{v}(t) \varepsilon D(\Phi)$ and set

$$\mathbf{v} = \mathbf{v}_+ + \mathbf{v}_-,$$
$$\mathbf{i} = \mathbf{v}_+ - \mathbf{v}_- = \Phi_a\mathbf{e} = \text{a value of } \Phi\mathbf{v}.$$

Then, the left-hand side of (48) may be written as

$$\int_{-\infty}^{t} [\mathbf{v}_+^*\mathbf{v}_+(\tau) - \mathbf{v}_-^*\tau[\mathbf{v}_-(\tau)]][d\tau \geq 0. \tag{49}$$

Clearly $\mathbf{V}_-(\omega) = S(\omega)\ \mathbf{V}_+(\omega)$. Let

$$\hat{\mathbf{v}}_+(\tau) = \mathbf{v}_+(\tau), \qquad \tau \leq t,$$
$$= \mathbf{0}, \qquad \tau > t.$$

Hence if $\hat{\mathbf{v}}_-(\tau)$ is defined to be that element whose L_{2n}-Fourier transform $\hat{\mathbf{V}}_-(\omega)$ is given by

$$\hat{\mathbf{V}}_-(\omega) = S(\omega)\hat{\mathbf{V}}_+(\omega),$$

an application of the Titchmarsh theorem yields

$$\mathbf{v}_-(\tau) = \hat{\mathbf{v}}_-(\tau), \quad \text{a.e. in} \quad \tau < t.$$

By Plancherel's theorem and 3)

$$\int_{-\infty}^{+\infty} [\hat{\mathbf{v}}_+^*\hat{\mathbf{v}}_+(\tau) - \hat{\mathbf{v}}_-^*\hat{\mathbf{v}}_-(\tau)]\, d\tau$$
$$= \int_{-\infty}^{+\infty} \hat{\mathbf{V}}_+^*(\omega)Q(\omega)\hat{\mathbf{V}}_+(\omega)\, d\omega \geq 0.$$

$$\therefore \int_{-\infty}^{t} [\mathbf{v}_+^*\mathbf{v}_+(\tau) - \mathbf{v}_-^*\mathbf{v}_-(\tau)]\, d\tau - \int_{t}^{\infty} \hat{\mathbf{v}}_-^*\hat{\mathbf{v}}_-(\tau)\, d\tau \geq 0.$$

$$\therefore \int_{-\infty}^{t} [\mathbf{v}_+^*\mathbf{v}_+(\tau) - \mathbf{v}_-^*\mathbf{v}_-(\tau)]\, d\tau \geq 0, \qquad t > -\infty$$

and is precisely the statement embodied in (48), Q.E.D.

An appeal to Theorem 3 now permits us to conclude that $Q(z)$ is non-negative throughout and that $\bar{S}(z) = S(-\bar{z})$ throughout. These remarks are subsumed in the following corollary.

Corollary 4(a): A necessary and sufficient condition that an $n \times n$ matrix $S(z)$ be physically realizable is that it be bounded-real.

The importance of Theorem 4 stems from the fact that the non-negativeness of $Q(z)$ need be ascertained only on the real ω axis. Oono [11] has succeeded in showing that if $S(z)$ is rational and bounded-real it possesses a network representation in terms of a finite number of lumped resistors, inductors, capacitors, ideal transformers and ideal gyrators [15]. Unfortunately Theorem 4 offers no clues concerning the building blocks that can occur in nature and it appears, from the paucity of significant results in this direction, that for nonrational $S(z)$ the problem belongs to an entirely new order of ideas.

In network analysis it is customary to express the response of the augmented n-port Φ_a in terms of a convolution integral of the type

$$\mathbf{i}(t) = \int_{-\infty}^{+\infty} W_n(\tau)\mathbf{e}(t - \tau)\, d\tau \tag{50}$$

in which

$$\mathbf{e}(t) \varepsilon L_{1n} \cap D(\Phi_a),$$
$$\mathbf{i}(t) = \Phi_a\mathbf{e}(t)$$

and $W_n(\tau)$ is known as the $n \times n$ matrix weighting-function of the system. It may not always be possible to find such a representation since it imposes severe limitations on the nature of the operator Φ_a. Theorem 5 completely answers the question, but first we lay down another definition.

Definition 18

A linear passive n-port Φ is said to possess a "Faltung-representation" if for every $\mathbf{e}(t) \varepsilon L_{1n} \cap D(\Phi_a)$, $\mathbf{i} = \Phi_a\mathbf{e}(t)$ is equivalent to

$$\mathbf{i}(t) = \int_{-\infty}^{+\infty} dW_n(\tau)\mathbf{e}(t-\tau), \tag{51}$$

the $n \times n$ matrix $W_n(\tau)$ being of bounded variation over $(-\infty < \tau < \infty)$, *i.e.*,

$$\int_{-\infty}^{+\infty} | dW_n(\tau) | < \infty .$$

Theorem 5: Let Φ be an n-port satisfying postulates P_1 to P_4 and which is such that $L_{1n} \cap L_{2n} \cap D(\Phi_a)$ contains n elements $\mathbf{e}_1(t), \cdots, \mathbf{e}_n(t)$, almost everywhere independent in $(-\infty < t < \infty)$, *i.e.*,

$$\det (\mathbf{e}_1 \mid \mathbf{e}_2 \mid \cdots \mid \mathbf{e}_n) \neq 0, \quad \text{a.e.}$$

Then, (51) is valid for every $\mathbf{e}(t)$ in $L_{1n} \cap L_{2n} \cap D(\Phi_a)$ if, and only if,

$$S(z) = \int_0^{\infty} e^{iz\tau}\, dK_n(\tau), \qquad Im\, z \geq 0, \tag{52}$$

where $K_n(\tau)$ is an $n \times n$ matrix of bounded-variation over $(-\infty < \tau < \infty)$.

Proof: Necessity: Clearly, the L_{1n} and L_{2n}-Fourier transforms of the $\mathbf{e}_r(t)$, $(r = 1, 2, \cdots, n)$, are a.e. equal. Denote them by $\mathbf{E}_r(\omega)$, $(r = 1, 2, \cdots, n)$. When (51) attains, the corresponding responses $\mathbf{i}_r(t) = \Phi_a\mathbf{e}_r(t)$, $(r = 1, 2, \cdots, n)$, are in L_{1n}. Since Φ is passive they are also in L_{2n}. Denote their Fourier transforms by $\mathbf{I}_r(\omega)$, $(r = 1, 2, \cdots, n)$ and let

$$I_n(\omega) \equiv (\mathbf{i}_1 \mid \mathbf{i}_2 \mid \cdots \mid \mathbf{i}_n(\omega)),$$

$$E_n(\omega) \equiv (\mathbf{E}_1 \mid \mathbf{E}_2 \mid \cdots \mid \mathbf{E}_n(\omega)).$$

Transforming both sides of (51) (using the Fubini theorem) yields

$$I_n(\omega) = \tilde{Y}_a(\omega)E_n(\omega), \tag{53}$$

where

$$\tilde{Y}_a(\omega) = \int_{-\infty}^{+\infty} e^{i\omega\tau}\, dW_n(\tau). \tag{54}$$

Since

$$\det (\mathbf{e}_1 \mid \mathbf{e}_2 \mid \cdots \mid \mathbf{e}_n) \neq 0 \quad \text{a.e.},$$

$$\det E_n(\omega) \neq 0, \quad \text{a.e.}$$

By Theorem 2 we know that

$$I_n(\omega) = Y_a(\omega)E_n(\omega) \quad \text{a.e.}$$

Thus,

$$Y_a(\omega) = \tilde{Y}_a(\omega) \quad \text{a.e.}$$

Consequently,

$$S(\omega) = 1_n - 2Y_a(\omega) = 1_n - 2\tilde{Y}_a(\omega)$$

$$= \int_{-\infty}^{+\infty} e^{i\omega\tau}\, d\tilde{K}_n(\tau) \quad \text{a.e.}, \tag{55}$$

where

$$\tilde{K}_n(\tau) = V_n(\tau) - 2W_n(\tau) \tag{56}$$

and the $n \times n$ matrix $V_n(\tau)$ is defined by

$$V_n(\tau) = 1_n, \qquad \tau > 0,$$
$$= 0_n, \qquad \tau \leq 0.$$

Obviously $\tilde{K}_n(\tau)$ is of bounded-variation in $(-\infty < \omega < \infty)$. Again by Theorem 2 we know that $S(z)$ is analytic in $Im\, z > 0$. Thus the matrix functions

$$A_n(z) \equiv S(z) - \int_0^{\infty} e^{iz\tau}\, d\tilde{K}_n(\tau)$$

and

$$P_n(z) \equiv \int_{-\infty}^{0} e^{iz\tau}\, d\tilde{K}_n(\tau)$$

are analytic and uniformly bounded in $Im\ z > 0$ and $Im\ z < 0$, respectively. Furthermore [see (55)]

$$A_n(\omega) \equiv \lim_{\beta \to 0^+} A_n(\omega + i\beta) = P_n(\omega) \quad \text{a.e.}$$

Because a Cauchy theorem holds for both $A_n(z)$ and $P_n(z)$ the matrix function $N_n(z)$ defined by

$$N(z) = A_n(z), \qquad Im\, z \geq 0,$$
$$= P_n(z), \qquad Im\, z \leq 0,$$

is analytic and uniformly bounded in the entire z plane. By the Liouville theorem $N_n(z) = C_n$, a constant $n \times n$ matrix. In particular,

$$S(z) = C_n + \int_0^{\infty} e^{iz\tau}\, d\tilde{K}_n(\tau), \qquad Im\, z \geq 0.$$

$$\therefore\ S(z) = \int_0^{\infty} e^{iz\tau}\, dK_n(\tau), \qquad Im\, z \geq 0,$$

with

$$K_n(\tau) \equiv C_nV(\tau) + \tilde{K}_n(\tau), \qquad \text{Q.E.D.}$$

Sufficiency: Let $S(z)$ be representable as in (52) and let $\mathbf{e}(t)\ \varepsilon\ L_{1n} \cap L_{2n} \cap D(\Phi_a)$. Then,

$$\mathbf{I}(\omega) = Y_a(\omega)\mathbf{E}(\omega) \tag{57}$$

where

$$2Y_a(\omega) = 1_n - S(\omega).$$

$$\therefore\ Y_a(\omega) = \int_{-\infty}^{+\infty} e^{i\omega\tau}\, dW_n(\tau)$$

in which

$$2W_n(\tau) = V_n(\tau) - K_n(\tau), \qquad \tau \geq 0,$$
$$= -K_n(0), \qquad \tau < 0,$$

is of bounded-variation over $(-\infty < \omega < \infty)$. From the Fourier-Stieltjes convolution theorem, (57) is equivalent to

$$\mathbf{i}(t) = \int_{-\infty}^{+\infty} dW_n(\tau)\mathbf{e}(t - \tau),$$

Q.E.D.

Observe that if $\bar{S}(\omega) = S(-\omega)$, the $K(\tau)$ of (52) must be real.

Theorem 5(a): Let Φ be a linear, passive, time-invariant n-port such that $D(\Phi_a)$ is dense in L_{1n} (in L_{1n} norm). Then a necessary and sufficient condition for Φ to possess a "Faltung-representation" is that Φ_a be bounded when considered as a transformation from L_{1n} into L_{1n}.

Proof: The Bochner L_1 theorem. If in addition Φ satisfies postulate P_4 and if it is known *a priori* that $L_{1n} \cap L_{2n} \cap D(\Phi_a)$ contains n a.e. linearly independent elements, the necessary and sufficient condition of Theorem 5(a) can be stated alternatively in terms of $S(z)$ (Theorem 5).

The above theorem suggests a very obvious question: What is the physical significance of assuming that Φ_a maps L_{1n} into L_{1n} in a bounded manner? Our Theorem 5 does not provide the answer because it deals with $S(z)$, a derived notion. To date, the authors have been unsuccessful in their attempts to find a postulate comparable to P_2 in either simplicity or naturalness and the problem is open. The utility of (52) for the microwave approximation problem has been exploited to great advantage by one of the authors in a doctoral dissertation [18].

Definition 19

Let Φ be a linear, passive n-port, $\mathbf{i} = \Phi_a\mathbf{e} = \Phi\mathbf{v}$ and $\mathbf{e} = \mathbf{v} + \mathbf{i}$. Then, Φ is said to be lossless if

$$Re \int_{-\infty}^{+\infty} \mathbf{v}^*\mathbf{i}(\tau)\, d\tau = 0$$

for every $\mathbf{e}(t) \in L_{2n} \cap D(\Phi_a)$.

In words, Φ is lossless if the energy supplied to the augmented n-port Φ_a by a square-integrable $\mathbf{e}(t)$ in $L_{2n} \cap D(\Phi_a)$ is completely dissipated as heat in the one-ohm resistors in series with the ports. The usefulness of this definition is illustrated by the next theorem.

Theorem 6: An n-port Φ satisfying postulates P_1 to P_4 is lossless if, and only if,

$$Q(\omega) = 1_n - S^*(\omega)S(\omega) = 0_n \quad \text{a.e.} \tag{58}$$

Proof: From (38) we see that an n-port Φ satisfying P_1 to P_4 is lossless if, and only if,

$$\int_{-\infty}^{+\infty} \mathbf{E}^*(\omega)Q(\omega)\mathbf{E}(\omega)\, d\omega = 0 \tag{59}$$

for every $\mathbf{E}(\omega)$ that is the L_{2n}-Fourier transform of an $\mathbf{e}(t) \in L_{2n}$. The theorem now follows by the same technique used in Theorem 3, Q.E.D.

V. Analytic Properties of the Scattering Matrix

In Theorems 2 and 3 it was shown that an n-port Φ obeying P_1 to P_4 admits a scattering description $S(z)$ analytic in $Im\ z > 0$ such that the associated matrix $Q(z) = 1_n - S^*(z)S(z)$ is non-negative throughout, *i.e.*, for all n-vectors $\mathbf{b}$,

$$\mathbf{b}^*Q(z)\mathbf{b} \geq 0 \tag{60}$$

for all z, $Im\ z > 0$ and almost all ω.

Theorem 7: The ranks of the $n \times n$ matrices

$$Q(z) = 1_n - S^*(z)S(z),$$

$$Q_1(z) = 1_n - S(z)$$

and

$$Q_2(z) = 1_n + S(z)$$

are invariant in the strict upper half-plane Im $z > 0$ and can decrease only on the real ω axis. Moreover the ranks of Q_1 and Q_2 can diminish at most on an ω set of measure zero.

Proof: Note first that (60) may be written as

$$\sum_{r=1}^{n} | s_r(z, \mathbf{b}) |^2 \leq \mathbf{b}^*\mathbf{b}, \tag{61}$$

where

$$s_r(z, \mathbf{b}) = \sum_{k=1}^{n} s_{rk}(z)b_k, \qquad (r = 1, 2, \cdots, n),$$

$(b_1, b_2, \cdots, b_n)' = \mathbf{b}$ and $s_{rk}(z)$ is the element in the rth row and kth column of the $n \times n$ matrix $S(z)$. Clearly, the functions $s_r(z, \mathbf{b})$, $(r = 1, 2, \cdots, n)$, are, for a fixed constant $\mathbf{b}$, analytic in $Im\ z > 0$. Now suppose that for some $z = z_0$, $Im\ z_0 > 0$, $Q(z)$ is semi-positive definite. Then, there exists a $\mathbf{b}_0 \neq \mathbf{0}$ such that

$$\sum_{r=1}^{n} | s_r(z_0, \mathbf{b}_0) |^2 = \mathbf{b}_0^*\mathbf{b}_0,$$

and by a familiar variant of the Maximum-Modulus theorem [17], this implies that each $s_r(z, \mathbf{b}_0)$, $(r = 1, 2, \cdots, n)$, is constant in $Im\ z > 0$. Thus,

$$s_r(z, \mathbf{b}_0) = k_r, \qquad (r = 1, 2, \cdots, n),$$

throughout and

$$\sum_{r=1}^{n} | k_r |^2 = \mathbf{b}_0^*\mathbf{b}_0.$$

$$\therefore \quad S(z)\mathbf{b}_0 = \mathbf{k}, \quad \text{throughout}, \tag{62}$$

$$\mathbf{b}_0^*\mathbf{b}_0 = \mathbf{k}^*\mathbf{k},$$

$$\mathbf{k} = (k_1, k_2, \cdots, k_n)'.$$

Again, let the rank of $Q(z_0)$, $Im\ z_0 > 0$, be r_0. Then, the number of linearly independent vectors $\mathbf{b}$ for which $\mathbf{b}^*\, Q(z_0)\, \mathbf{b} = 0$ is exactly $n - r_0$. For, $Q(z_0)$ being hermitian is reducible to diagonal form by a unitary matrix U. Hence the linear transformation $\mathbf{b} = U\mathbf{a}$ converts $\mathbf{b}^*\, Q(z_0)\, \mathbf{b} = 0$ into

$$\sum_{m=1}^{r_0} \lambda_m \mid a_m \mid^2 = 0, \tag{63}$$

where $\lambda_m > 0$, $(m = 1, 2, \cdots, r_0)$. Therefore, $a_m = 0$, $(m = 1, 2, \cdots, r_0)$ and the number of linearly independent vectors $\mathbf{a}$ satisfying (63) is precisely $n - r_0$. Since U is nonsingular, the number of linearly independent vectors $\mathbf{b}$ is also $n - r_0$. Denote them by $\mathbf{b}_m$, $(m = 1, 2, \cdots, n - r_0)$. By (62),

$$S(z)\mathbf{b}_m = \mathbf{k}_m, \quad \text{throughout}$$

and

$$\mathbf{b}_m^*\mathbf{b}_m = \mathbf{k}_m^*\mathbf{k}_m, \qquad (m = 1, 2, \cdots, n - r_0).$$

Thus $\mathbf{b}_m^*\, Q(z)\, \mathbf{b}_m = 0$, $m = (1, 2, \cdots, n - r_0)$, for all z in $Im\ z > 0$ so that the rank of $Q(z)$ is certainly not greater than r_0 and *a fortiori*, not less. This completes the proof for $Q(z)$.

Let the rank of $Q_1(z_0)$, $Im\ z_0 > 0$ be r_0. Then there exist $n - r_0$ linearly independent vectors $\mathbf{b}_m$, $(m = 1, 2, \cdots, n - r_0)$, such that

$$S(z_0)\mathbf{b}_m = \mathbf{b}_m, \qquad (m = 1, 2, \cdots, n - r_0). \tag{64}$$

$$\therefore\ \mathbf{b}_m^* Q(z_0)\mathbf{b}_m = 0, \qquad (m = 1, 2, \cdots, n - r_0).$$

From (62),

$$S(z)\mathbf{b}_m = \mathbf{b}_m, \quad \text{throughout,}$$

or

$$Q_1(z)\mathbf{b}_m = \mathbf{0}, \qquad (m = 1, 2, \cdots, n - r_0),$$

throughout, whence rank $Q_1(z) = r \leq r_0$.

Reversing the argument, $r_0 \leq r$. Thus rank $Q_1(z) = r_0$ everywhere in the strict upper half-plane. Now suppose that rank $Q_1(\omega) < r_0$ on an ω-set Δ of positive measure, *i.e.*, rank $Q_1(\omega) \leq r_0 - 1$ for $\omega \in \Delta$. This means that all minors of order r_0 or greater vanish for $\omega \in \Delta$ and by a theorem of Riez these minors must vanish identically in $Im\ z > 0$. But then rank $Q_1(z) < r_0$, $Im\ z > 0$, a contradiction, Q.E.D. The proof for $Q_2(z)$ is similar and the theorem is established.

As any number of simple examples reveal, the rank of $Q(z)$ can decrease everywhere on the entire ω axis.

Corollary 7(a): The determinants $D_1(z)$ and $D_2(z)$ of the matrices $Q_1(z)$ and $Q_2(z)$, respectively, are either nonsingular or singular throughout.

Corollary 7(b): The ranks of $Q_1(\omega)$ and $Q_2(\omega)$ are constant a.e. in $(-\infty < \omega < \infty)$.

Corollary 7(c): If $S(z)$ is bounded-real, $S'(z)$ is bounded-real.

Proof: S^*S and $\bar{S}S'$ have the same eigenvalues, Q.E.D.

Theorem 8: Let $S(z)$ be bounded-real and r the rank of its associated matrix $Q(z)$ in $Im\ z > 0$. Then, there exist two constant orthogonal $n \times n$ matrices T_1 and T_2 such that

$$T_2'S(z)T_1 = \left[\begin{array}{c|c} 1_{n-r} & 0_{n-r,r} \\ \hline 0_{r,n-r} & S_r(z) \end{array}\right] \tag{65}$$

and

$$T_1'Q(z)T_1 = \left[\begin{array}{c|c} 0_{n-r,n-r} & 0_{n-r,r} \\ \hline 0_{r,n-r} & 1_r - S_r^*S_r(z) \end{array}\right], \tag{66}$$

where $S_r(z)$ is an $r \times r$ bounded-real scattering matrix whose associated matrix $Q_r(z) = 1_r - S_r^*(z)\ S_r(z)$ is nonsingular in $Im\ z > 0$; $0_{k,m}$ denotes the $k \times m$ zero matrix.

Proof: By hypothesis rank $Q(z) = r$ in $Im\ z > 0$. Thus there exist $n - r$ constant vectors $\mathbf{b}_1, \mathbf{b}_2, \cdots, \mathbf{b}_{n-r}$ and $n - r$ constant vectors $\mathbf{k}_1, \mathbf{k}_2, \cdots, \mathbf{k}_{n-r}$ such that

$$S(z)\mathbf{b}_m = \mathbf{k}_m, \tag{67}$$

$$\mathbf{b}_m^*\mathbf{b}_m = \mathbf{k}_m^*\mathbf{k}_m, \qquad (m = 1, 2, \cdots, n - r),$$

throughout. The vectors $\mathbf{b}_m$, $(m = 1, 2, \cdots, n - r)$, can be assumed orthonormal and taken to be the linearly independent solutions of

$$Q(z_0)\mathbf{b} = \mathbf{0} \tag{68}$$

at any point z_0, $Im\ z_0 > 0$. Since $Q(z_0)$ is real for $z_0 = i\beta$, $\beta > 0$, these solutions can also be assumed real without loss of generality. From (67) and (68),

$$\mathbf{k}_m^*\mathbf{k}_l = \mathbf{b}_m^*S^*(z_0)S(z_0)\mathbf{b}_l = \mathbf{b}_m^*\mathbf{b}_l = \delta_{m,l},$$
$$(m, l, = 1, 2, \cdots, n - r)$$

and the k's are also orthonormal. Again by (67) and (68),

$$\mathbf{b}_m = S^*(z_0)S(z_0)\mathbf{b}_m$$
$$= S^*(z_0)\mathbf{k}_m = S'(-\bar{z}_0)\mathbf{k}_m, \qquad (m = 1, 2, \cdots, n - r)$$

which, since $S'(z)$ is also bounded-real [Corollary 7(c)], implies that

$$\mathbf{b}_m = S'(z)\mathbf{k}_m, \qquad (m = 1, 2, \cdots, n - r),$$

throughout. Denote the (non-unique) $n \times n$ orthogonal matrices whose first $n - r$ columns are $\mathbf{b}_1, \mathbf{b}_2, \cdots, \mathbf{b}_{n-r}$ and $\mathbf{k}_1, \mathbf{k}_2, \cdots, \mathbf{k}_{n-r}$ by T_1 and T_2, respectively. Then,

$$\{T_2'S(z)T_1\}_{m,l} = \mathbf{k}_m'S(z)\mathbf{b}_l$$
$$= \mathbf{k}_m'\mathbf{k}_l = \delta_{m,l},$$
$$(m = 1, 2, \cdots, n), \quad (l = 1, 2, \cdots, n - r),$$

where $\delta_{m,l}$ is the familiar Kronecker delta. Again,

$$\{T_2'S(z)T_1\}_{m,l} = \{T_1'S'(z)T_2\}_{l,m}$$
$$= \mathbf{b}_l'S'(z)\mathbf{k}_m$$
$$= \mathbf{b}_l'\mathbf{b}_m = \delta_{l,m},$$
$$(m = 1, 2, \cdots, n - r), \quad (l = 1, 2, \cdots, n).$$

Consequently $T_2'S(z)T_1$ has the required form (65). By direct manipulation with (65) we get (66). Since the rank of $Q(z)$ is r, $Im\ z > 0$, $Q_r(z)$ is nonsingular in $Im\ z > 0$ and the positive-definiteness of $Q_r(z)$ in $Im\ z > 0$ follows from the non-negativeness of $Q(z)$. Hence $S_r(z)$ is a bounded-real $r \times r$ scattering matrix, Q.E.D.

Corollary 8(a): Let the ranks of $Q_1(z) = 1_n - S(z)$ and $Q_2(z) = 1_n + S(z)$ be r_1 and r_2, respectively. Then, there exist two constant orthogonal matrices T_1 and T_2 such that

$$T_1'S(z)T_1 = \left[\begin{array}{c|c} 1_{n-r_1} & 0_{n-r_1,r_1} \\ \hline 0_{r_1,n-r_1} & S_{r_1}(z) \end{array}\right] \quad (70)$$

and

$$T_2'S(z)T_2 = \left[\begin{array}{c|c} -1_{n-r_2} & 0_{n-r_2,r_2} \\ \hline 0_{r_2,n-r_2} & S_{r_2}(z) \end{array}\right], \quad (71)$$

where $S_{r_1}(z)$ and $S_{r_2}(z)$ are bounded-real scattering matrices of orders $r_1 \times r_1$ and $r_2 \times r_2$. Moreover the matrices $1_{r_1} - S_{r_1}(z)$ and $1_{r_2} - S_{r_2}(z)$ are nonsingular throughout.

Proof: Denote the solutions (assumed orthonormalized) of

$$Q_1(z)\mathbf{b} = \mathbf{0}, \qquad Im\, z > 0,$$

and

$$Q_2(z)\mathbf{h} = \mathbf{0}, \qquad Im\, z > 0,$$

by $\mathbf{b}_1, \mathbf{b}_2, \cdots, \mathbf{b}_{n-r_1}$ and $\mathbf{h}_1, \mathbf{h}_2, \cdots, \mathbf{h}_{n-r_2}$, respectively. Reviewing the proof of the above theorem we see that T_1 is that orthogonal matrix whose first $n - r_1$ columns are the **b**'s and T_2 is the one whose first $n - r_2$ columns are the **h**'s.

Definition 20

An n-port Φ satisfying P_1 to P_4 is said to be reciprocal if its scattering matrix $S(z)$ is symmetric, *i.e.*, $S(z) = S'(z)$.

Theorem 9: Let $S(z)$ be a symmetric bounded-real scattering matrix and r, r_1, r_2 the ranks of $Q(z)$, $Q_1(z)$, and $Q_2(z)$, respectively, in $Im\, z > 0$. Then a set of $n - r$ linearly independent solutions of $Q(z)\,\mathbf{b} = \mathbf{0}$ can be formed by choosing any $n - r_1$ linearly independent solutions of $Q_1(z)\,\mathbf{b} = \mathbf{0}$ and any $n - r_2$ linearly independent solutions of $Q_2(z)\,\mathbf{b} = \mathbf{0}$. Hence $n - r = (n - r_1) + (n - r_2)$.

Proof: Clearly, any solution **b** of $Q_1(z)\,\mathbf{b} = \mathbf{0}$ is orthogonal to any solution **h** of $Q_2(z)\,\mathbf{h} = \mathbf{0}$. For, if

$$\mathbf{b} = S(z)\mathbf{b},$$

$$\mathbf{h}^*\mathbf{b} = \mathbf{h}^*S(z)\mathbf{b} = \mathbf{h}^*S'(-\bar{z})\mathbf{b} = \mathbf{h}^*S^*(z)\mathbf{b}$$

$$= -\mathbf{h}^*\mathbf{b}$$

since

$$\mathbf{h} = -S(z)\mathbf{h}.$$

$$\therefore \mathbf{h}^*\mathbf{b} = 0.$$

Again any solution of either $Q_1(z)\,\mathbf{b} = \mathbf{0}$ or $Q_2(z)\,\mathbf{b} = \mathbf{0}$, $Im\, z > 0$, is a solution of $Q(z)\,\mathbf{b} = \mathbf{0}$. For example, if

$$\mathbf{b} = -S(z)\mathbf{b},$$

$$S^*S\mathbf{b} = -S^*\mathbf{b} = -S(-\bar{z})\mathbf{b} = \mathbf{b}.$$

$$\therefore Q(z)\mathbf{b} = \mathbf{0}$$

and similarly for Q_1: Q_1 and Q_2 contribute $(n - r_1) + (n - r_2)$ solutions of $Q(z)\,\mathbf{b} = \mathbf{0}$ whence $n - r \geq (n - r_1) + (n - r_2)$. From $S' = S$ and the identity

$$Q(z) = Q_1^*Q_2(z) + S^*(z) - S(z),$$

we see that $Q(z)\,\mathbf{b} = \mathbf{0}$ implies $Q_1^*\,Q_2(z)\,\mathbf{b} = \mathbf{0}$. Thus $n - r \leq n -$ rank $Q_1^*\,Q_2$. But by the law of nullity for matrices, $n \geq r_1 + r_2 -$ rank $Q_1^*\,Q_2$ or rank $Q_1\,Q_2 \geq r_1 + r_2 - n$ which when substituted in the above inequality gives $n - r \leq (n - r_1) + (n - r_2)$. Thus $n - r = (n - r_1) + (n - r_2)$, Q.E.D.

The ideal gyrator

$$S = \begin{bmatrix} 0 & 1 \\ -1 & 0 \end{bmatrix}$$

shows that Theorem 9 may fail for nonreciprocal networks.

VI. Impedance and Admittance Matrices

Let Φ be an n-port satisfying P_1 to P_4. We have seen that $\hat{\Phi}_a$ defines a bounded, linear transformation on all of L_{2n} into L_{2n} such that

$$\mathbf{i}(t) = \hat{\Phi}_a\mathbf{e}, \qquad \mathbf{e} \,\varepsilon\, L_{2n},$$

and

$$\mathbf{v}(t) = \mathbf{e}(t) - \hat{\Phi}_a\mathbf{e}$$

imply

$$\mathbf{e}^*\mathbf{e}(t) = \mathbf{v}^*\mathbf{v}(t) + \mathbf{i}^*\mathbf{i}(t) + 2\,Re \int_{-\infty}^{t} \mathbf{v}^*(\tau)\mathbf{i}(\tau)\, d\tau,$$

where

$$Re \int_{-\infty}^{t} \mathbf{v}^*(\tau)\mathbf{i}(\tau)\, d\tau \geq 0$$

for all $t > -\infty$. Clearly,

$$\mathbf{I}(\omega) = Y_a(\omega)\mathbf{E}(\omega) = \tfrac{1}{2}Q_1(\omega)\mathbf{E}(\omega) \quad (72)$$

and

$$\mathbf{V}(\omega) = (1_n - Y_a)\mathbf{E}(\omega) = \tfrac{1}{2}Q_2(\omega)\mathbf{E}(\omega), \quad (73)$$

where

$$Q_1(\omega) = 1_n - S(\omega),$$

$$Q_2(\omega) = 1_n + S(\omega).$$

First, if Q_1^{-1} exists a.e.,

$$\mathbf{E}(\omega) = 2Q_1^{-1}(\omega)\mathbf{I}(\omega)$$

and

$$\mathbf{V}(\omega) = Q_2(\omega)Q_1^{-1}(\omega)\mathbf{I}(\omega). \quad (74)$$

Second, if $Q_2^{-1}(\omega)$ exists, a.e.,

$$\mathbf{E}(\omega) = 2Q_2^{-1}(\omega)\mathbf{V}(\omega)$$

and

$$\mathbf{I}(\omega) = Q_1(\omega)Q_2^{-1}(\omega)\mathbf{V}(\omega) \tag{75}$$

The quantities

$$Z(z) = Q_2(z)Q_1^{-1}(z) \tag{76}$$

and

$$Y(z) \equiv Q_1(z)Q_2^{-1}(z), \tag{77}$$

are known, respectively, as the impedance and admittance matrices associated with the n-port Φ (provided, of course, that they exist).

The ideal transformer possesses neither an impedance nor an admittance matrix. Thus, these descriptions of Φ are less general than the one in terms of $S(z)$ (which always exist), the scattering matrix.

$Y(\omega)$ permits the calculation of the response $\mathbf{i} = \Phi\mathbf{v}$ for all $\mathbf{v}$'s of the form

$$\mathbf{v} = (\mathcal{I} - \Phi_a)\mathbf{e}, \qquad \mathbf{e} \,\varepsilon\, L_{2n} \cap D(\Phi_a), \tag{77a}$$

and $Z(\omega)$ enables the excitation $\mathbf{v}(t)$ to be calculated for all responses $\mathbf{i}(t)$ of the form

$$\mathbf{i}(t) = \Phi_a\mathbf{e}, \qquad \mathbf{e} \,\varepsilon\, L_{2n} \cap D(\Phi_a), \tag{78}$$

i.e., $Z(\omega)$ represents Φ^{-1} for a suitably restricted class of elements $\mathbf{v}(t) \,\varepsilon\, L_{2n} \cap D(\Phi)$. Denote the set of $\mathbf{v}$'s generated in (77a) and the set of $\mathbf{i}$'s generated in (78) by $\tilde{D}(\Phi)$ and $\tilde{D}(\Phi^{-1})$, respectively. Are the sets $\tilde{D}(\Phi)$ and $\tilde{D}(\Phi^{-1})$ dense in L_{2n}? A complete answer is available, but first consider Theorem 10.

Theorem 10: The impedance and admittance matrices

$$Z(z) = (1_n + S)(1_n - S)^{-1} \tag{79}$$

and

$$Y(z) = (1_n - S)(1_n + S)^{-1} \tag{80}$$

either exists throughout or nowhere.

Proof: Theorem 7, Corollary 7(a) Obviously, if both $Y(z)$ and $Z(z)$ exist, $Y(z)\,Z(z) = 1_n$, throughout.

Definition 21

An $n \times n$ matrix $A_n(z)$ is said to be positive-real if

1) $A_n(z)$ is analytic in $Im\, z > 0$;
2) $\frac{1}{2}[A_n(z) + A_n^*(z)] \equiv R_n(z)$ is the matrix of a non-negative quadratic form in $Im\, z > 0$, *i.e.*,

$$\mathbf{b}^*R_n(z)\mathbf{b} \equiv Re\,\{\mathbf{b}^*A_n(z)\mathbf{b}\} \geq 0, \qquad Im\, z > 0,$$

for all constant n-vectors $\mathbf{b}$.
3) $\bar{A}_n(z) = A_n(-\bar{z})$, $Im\, z > 0$, *i.e.*, $A_n(i\beta)$ is real, $\beta > 0$.

Theorem 11: The impedance and admittance matrices of an n-port Φ satisfying P_1 to P_5 are positive-real whenever they exist.

Proof: Suppose $Z(z)$ exists. Properties 1) and 3) are obvious from (79). Since $Q_1(z) = 1_n - S(z)$ is nonsingular throughout, the linear system

$$Q_1(z)\,\mathbf{b} = \mathbf{c}$$

has a unique solution $\mathbf{b}$ for every n-vector $\mathbf{c}$ (throughout). By (79)

$$Q_1^*Z(z)Q_1(z) = Q(z) + (S - S^*).$$

$$\therefore Re\,\{\mathbf{b}^*Q_1^*Z(z)Q_1(z)\mathbf{b}\} = \mathbf{b}^*Q(z)\mathbf{b} \geq 0, \qquad Im\, z > 0.$$

Thus,

$$Re\,\{\mathbf{c}^*Z(z)\mathbf{c}\} \geq 0,$$

or

$$\mathbf{c}^*\left[\frac{Z(z) + Z^*(z)}{2}\right]\mathbf{c} \equiv \mathbf{c}^*R(z)\mathbf{c} \geq 0$$

in $Im\, z > 0$ for every constant n-vector $\mathbf{c}$, Q.E.D. The proof for $Y(z)$ is similar. Note that it now follows that $R(z)$ is actually nonnegative throughout since

$$\lim_{\beta\to 0^+} Re\,\{\mathbf{c}^*Z(\omega + i\beta)\mathbf{c}\} = \mathbf{b}^*Q(\omega)\mathbf{b} \geq 0$$

almost everywhere.

Corollary 11(a): The ranks of

$$R_1(z) \equiv \frac{Z(z) + Z^*(z)}{2}$$

and

$$R_2(z) \equiv \frac{Y(z) + Y^*(z)}{2}$$

are invariant in the strict upper half-plane, $Im\, z > 0$.

Theorem 12: $\tilde{D}(\Phi)$ and $\tilde{D}(\Phi^{-1})$ are dense in L_{2n} if, and only if, $Y(z)$ and $Z(z)$ exist respectively.

Proof: Recall that $L_{2n} \cap D(\Phi_a)$ is a linear manifold dense in L_{2n}, which in turn implies that $\tilde{D}(\Phi)$ and $\tilde{D}(\Phi^{-1})$ are linear. Necessity is obvious [see (72) and (73)]. As regards sufficiency, suppose that $Y(z)$ exists and that the set of all $\mathbf{v}$'s of the form

$$\mathbf{v}(t) = (\mathcal{I} - \Phi_a)\mathbf{e}, \qquad \mathbf{e} \,\varepsilon\, L_{2n} \cap D(\Phi_a),$$

is not dense in L_{2n}. Then, $[\tilde{D}(\Phi)]$, the closure of $\tilde{D}(\Phi)$, does not coincide with L_{2n} and there exists a square-integrable $\mathbf{e}_0(t) \neq \mathbf{0}$ orthogonal to all elements in $[\tilde{D}(\Phi)]$. Evidently, because of the boundedness of Φ_a, $[\tilde{D}(\Phi)]$ is the set of all elements of the form

$$\mathbf{v} = (\mathcal{I} - \hat{\Phi}_a)\mathbf{e}, \qquad \mathbf{e} \,\varepsilon\, L_{2n}.$$

Thus, $(\mathbf{e}_0, (\mathcal{I} - \hat{\Phi}_a)\mathbf{e}) = 0$ for all $\mathbf{e} \,\varepsilon\, L_{2n}$. In particular, for $\mathbf{e} = \mathbf{e}_0$,

$$Re\,(\mathbf{e}_0, \mathbf{v}_0) = 0,$$

where $\mathbf{v}_0 = \mathbf{e}_0 - \hat{\Phi}_a\mathbf{e}_0$. But

$$Re\,(\mathbf{e}_0, \mathbf{v}_0) = (\mathbf{v}_0, \mathbf{v}_0) + Re\,(\mathbf{v}_0, \mathbf{i}_0),$$

$$\mathbf{i}_0(\tau) = \mathbf{e}_0(\tau) - \mathbf{v}_0(\tau) = \hat{\Phi}_a\mathbf{e}_0.$$

By passivity, $Re\ (\mathbf{v}_0, \mathbf{i}_0) \geq 0$ so that $Re\ (\mathbf{e}_0, \mathbf{v}_0) = 0$ implies $\mathbf{v}_0 = \mathbf{0}$, *i.e.*,

$$\mathbf{0} = (1_n - Y_a)\mathbf{E}_0(\omega).$$

Since, by hypothesis, $(1_n - Y_a)^{-1}$ exists a.e., $\mathbf{E}_0(\omega) = \mathbf{0}$, a contradiction. The considerations for $Z(z)$ are the same, Q.E.D.

Corollary 12(a): Let Φ be an n-port satisfying P_1 to P_4. Then, the existence of $Y(z)$ is the necessary and sufficient condition for Φ to be s.v. and causal.

Proof: Sufficiency—Since $\mathbf{0}\ \varepsilon\ \tilde{D}(\Phi)$ and $\mathbf{I}(\omega) = Y(\omega)\mathbf{V}(\omega)$, $\Phi\mathbf{0}$ has the unique value $\mathbf{0}$, *i.e.*, Φ is s.v. Now $D(\Phi)$ includes $\tilde{D}(\Phi)$ which is dense in L_{2n} and therefore surely contains n elements almost everywhere independent in $(-\infty < t < \infty)$. The rest follows from Theorem 1. Necessity is obvious.

For an n-port Φ, with domain $D(\Phi)$ and range $R(\Phi)$ the generalized inverse Φ^{-1} is defined in the following manner: for any $\mathbf{i}_0(t)\ \varepsilon\ R(\Phi)$, $\Phi^{-1}\mathbf{i}_0$ is the set of all $\mathbf{v}$'s belonging to $D(\Phi)$ such that $\Phi\mathbf{v} = \mathbf{i}_0$. It is quite possible for Φ^{-1} to be a transformation even when Φ is not (the ideal one-port short-circuit is a good example). Obviously Φ^{-1} is linear if Φ is linear and moreover, in this latter case, Φ^{-1} is s.v. if, and only if, $\Phi^{-1}\mathbf{0}$ has the unique value $\mathbf{0}$.

Corollary 12(b): Let Φ be an n-port satisfying P_1 to P_4. A necessary and sufficient condition for Φ^{-1} to be single-valued is that $Z(z)$ exist.

Proof: Clearly $\mathbf{i} = \mathbf{0}$ belongs to $\tilde{D}(\Phi^{-1})$ and since $Z(\omega)\ \mathbf{I}(\omega) = \mathbf{V}(\omega)$, $\Phi^{-1}\ \mathbf{0}$ has the unique value $\mathbf{0}$, Q.E.D.

Corollary 12(c): Let Φ be an n-port satisfying P_1 to P_4. Then, if $Y(z)$ exists, $D(\Phi)$ is dense in L_{2n}; if $Z(z)$ exists, $D(\Phi^{-1})$ is dense in L_{2n}.

Definition 22

An $n \times n$ matrix $A_n(z)$ is said to be a physically realizable immittance matrix if it represents either the impedance description, or admittance description of an n-port Φ satisfying P_1 to P_5.

Theorem 13: An $n \times n$ matrix $A_n(z)$ is a physically realizable immittance matrix if, and only if, it is positive-real.

Proof: Necessity is a consequence of Theorem 11. Sufficiency—Note that $(1_n + A_n(z))^{-1}$ exists for all z in the strict upper-half plane. For, suppose $(A_n(z_0) + 1_n)$ is singular, $Im\ z_0 > 0$. Then the system

$$(1_n + A_n(z_0))\mathbf{b} = \mathbf{0}$$

possesses a nontrivial solution $\mathbf{b}$ and

$$Re\ \{\mathbf{b}^* A_n(z)\mathbf{b}\} = -\mathbf{b}^*\mathbf{b} < 0,$$

a contradiction with the positive-real character of $A_n(z)$. Hence the two matrices

$$S_1(z) = -S(z) = (1_n + A_n(z))^{-1}(1_n - A_n(z)) \tag{81a}$$

and

$$S(z) = (1_n + A_n(z))^{-1}(A_n(z) - 1_n), \tag{81b}$$

exist for all z in $Im\ z > 0$. Clearly $S(z)$ is analytic in $Im\ z > 0$ and $\bar{S}(z) = S(-\bar{z})$. From (81),

$$A_n(z)(1_n - S(z)) = 1_n + S(z).$$

$$\therefore (1_n - S^*)A_n(z)(1_n - S) = 1_n - S^*S + S - S^*.$$

Setting $(1_n - S(z))\ \mathbf{b} = \mathbf{c}$,

$$0 \leq Re\ \{\mathbf{c}^* A_n(z)\mathbf{c}\} = \mathbf{b}^*(1_n - S^*S(z))\mathbf{b}$$

and so the matrix $Q(z) = 1_n - S^*\ S(z)$ is non-negative everywhere in $Im\ z > 0$ (and thus throughout). Hence $\pm S(z)$ are bounded-real scattering matrices. By Theorem 4, they represent the scattering descriptions of two n-ports Φ and Φ_1 satisfying P_1 to P_5. Since

$$(1_n - S)^{-1} = (1_n + S_1)^{-1} = \tfrac{1}{2}(1_n + A_n), \quad Im\ z > 0,$$

exist,

$$A_n(z) = (1_n + S)(1_n - S)^{-1} \tag{82a}$$

$$= (1_n - S_1)(1_n + S_1)^{-1}, \tag{82b}$$

i.e., $A_n(z)$ represents the impedance description of Φ and the admittance description of Φ_1, Q.E.D.

In a general way, Theorem 13 asserts that every positive-real matrix has two network representations which are duals of each other.

VII. Closing Remarks

Though $S(z)$ cannot admit poles on the real ω axis, it can possess essential singularities and branch points. For example $s(z) = e^{-i/z}$ is bounded-real and has an essential singularity at $z = 0$. As a matter of fact it is not difficult to construct bounded-real scattering matrices with essential singularities at every rational point of the ω axis so that continuation into the lower half-plane is impossible [19].

For matrices $S(z)$ representing lossless n-ports, *i.e.*, matrices satisfying

$$S^*(\omega)S(\omega) = 1_n \quad \text{a.e.},$$

it can be shown that their continuity from above on the ω axis actually implies that they are meromorphic and verify the relation

$$S(z)S'(-z) = 1_n$$

in the entire z plane [19].

In principle it should be possible to develop an axiomatic approach to the theory of active n-ports. The real difficulty is that of discovering the correct weakening of the passivity postulate P_2. It is tempting to define an active n-port as one that is nonpassive, *i.e.*, one for which

$$Re \int_{-\infty}^{t} \mathbf{v}^*\mathbf{i}(\tau)\ d\tau < 0$$

for at least one t in $(-\infty < t < \infty)$. However this criterion appears to be much too general to allow any positive nontrivial conclusions. Since time-invariance is the all-important characteristic of a network possessing a frequency response, it is very probable that a judicious use of known results in the theory of semigroups of operators commuting with translations will lead to a physically acceptable resolution of the problem.

Bibliography

[1] J. Von Neumann, "Functional Operators," in "Annals of Mathematics Studies," Princeton Univ. Press, Princeton, N. J., vol. 2; 1950.
[2] A. C. Zaanen, "Linear Analysis," Interscience Publishers, New York, N. Y.; 1953.
[3] E. C. Titchmarsh, "Theory of Fourier Integrals," Clarendon Press, Oxford, Eng., 2nd ed.; 1948.
[4] T. T. Wu, "Causality and the Radiation Condition," Cruft Lab., Harvard Univ., Cambridge, Mass., Tech. Rept. No. 211; November, 1954.
[5] G. Raisbeck, "A definition of passive linear networks in terms of time and energy," *J. Appl. Phys.*, vol. 25, pp. 1510–1514; December, 1954.
[6] J. S. Toll, "Causality and Dispersion Relation, I. Logical Foundations," Physics Dept., Univ. of Maryland, Tech. Rept. No. 27; January, 1956.
[7] N. G. Van Kampen, "*S*-matrix and causality condition, I. Maxwell field," *Phys. Rev.*, vol. 89; March, 1953.
[8] S. Bochner and K. Chandrasekharan, "Fourier Transforms," in "Annals of Mathematics Studies," Princeton Univ. Press, Princeton, N. J.; 1949.
[9] E. Hille, "Functional Analysis and Semi-Groups," American Mathematical Society; 1948.
[10] H. J. Carlin, "The scattering matrix in network theory," IRE Trans. on Circuit Theory, vol. CT-3, pp. 88–96; June, 1956.
[11] Y. Oono, "Application of scattering matrices to synthesis of *n*-ports," IRE Trans. on Circuit Theory, vol. CT-3, pp. 111–120; June, 1956.
[12] D. C. Youla, L. J. Castriota, and H. J. Carlin, "Scattering Matrices and the Foundations of Linear Network Theory," Microwave Res. Inst., Polytechnic Inst. Bklyn., Bklyn., N. Y., Res. Rept. R-594-57, PIB-522; 1957.
[13] E. C. Titchmarsh, "The Theory of Functions," Oxford Univ. Press, New York, N. Y.; 1952.
[14] G. Valiron, "Functions Analytique," Presses Universitares De France; 1954.
[15] H. J. Carlin, "Synthesis of nonreciprocal networks," *Proc. Symp. Modern Network Synthesis*; Microwave Res. Inst., Polytechnic Inst. Bklyn., Bklyn., N. Y.; April, 1955.
[16] D. V. Widder, "The LaPlace Transform," Princeton Univ. Press, Princeton, N. J.; 1946.
[17] A. Zygmund, "Trigonometrical Series," Mathematics Monograph, Z Subwencji Funduszu Kultury Narodowej, Warsaw, Poland; 1935.
[18] L. J. Castriota, "Network Realizability Theory and Its Application to the Design of Distributed Parameter Matching Networks," Doctoral dissertation, Polytechnic Inst. Bklyn., Bklyn., N. Y.; June, 1958.
[19] D. C. Youla, "Representation Theory of Linear Passive Networks," Polytechnic Inst. Bklyn., Bklyn., N. Y., R-655-58, PIB-583; 1958.

19

Reprinted from *IEEE Trans. Circuit Theory*, **CT-11**, 30–50 (Mar. 1964)

A New Theory of Broad-band Matching*

D. C. YOULA†, SENIOR MEMBER, IEEE

Summary—**This paper solves the problem of designing an optimum, passive, reactive two-port equalizer to match out a dissipative frequency sensitive load to a resistive generator. The method is simple and elementary and completely avoids any recourse to the Darlington equivalent of the load. Moreover the technique is also applicable to the design of active equalizers.**

I. Introduction

THE CLASSIC PROBLEM of designing an optimum, lumped, reciprocal reactive equalizer N to match an arbitrary passive, lumped load to a resistive generator (Fig. 1) was initiated by Bode[1] and definitively solved, in its full generality, by Fano.[2]

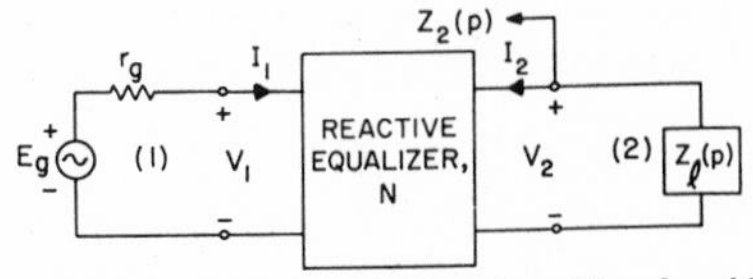

Fig. 1—Schematic of the Bode-Fano broad-band problem.

The key idea underlying Fano's solution is that of replacing the load $z_l(p)$ by its Darlington equivalent. Needless to say, this procedure has been extended and elaborated upon by many authors.[3–6] Nevertheless, Fano's idea, although very ingenious and elegant, has two main drawbacks, both from a practical and theoretical point of view:

1) A single function $z_l(p)$ is replaced by three functions—those necessary for the delineation of its Darlington equivalent. Most of the complications that arise are due to the necessity of translating certain properties of $z_l(p)$ into structural properties of its associated Darlington.
2) The technique is not applicable to the design of general active equalizers since Darlington's theorem is valid only for passive impedances. With the advent of the tunnel diode and varactor this consideration is no longer academic.

The object of this paper is to develop a new and elementary broad-band matching theory based on the principle of complex normalization which completely circumvents objection 1) and to illustrate every phase of the theory with fully worked nontrivial examples. The restrictions on transducer power gain are expressed in both nonintegral and integral form, the latter being rigorously derived in an Appendix. The application of the method to the design of active equalizers is reserved for a future publication.

II. Real-Frequency Complex Normalization[7]

Suppose a linear, time-invariant $2n$-terminal network N (Fig. 2) is excited at its n ports by n generators E_1, E_2, $\cdots$, E_n, with prescribed internal impedances $z_1(j\omega)$, $z_2(j\omega)$, $\cdots$, $z_n(j\omega)$, respectively. How does one decide upon an appropriate electrical description for N when interest centers mainly on the distribution of average ac power throughout the structure? More specifically, if the z's are not real and positive but are complex over the frequency band under consideration, is it still possible to describe N in terms of a "normalized" $n \times n$ scattering matrix $S = (s_{kl})$ which is unitary when N is lossless and such that under matched terminations (Fig. 3), the transducer power gain $G_{kl}(\omega^2)$ from port k to port l is measured by[11]

$$G_{kl}(\omega^2) = |s_{lk}(j\omega)|^2? \tag{1}$$

Interestingly enough, the answer turns out to be in the affirmative if the z's have positive-real parts over the frequency range of interest. Before embarking on the proof we introduce some matrix notation.

* Received July 1, 1963. The work reported in this memorandum was sponsored by the Rome Air Development Center, Air Force Systems Command under Contract No. AF-30(602)-2213.

† Department of Electrophysics, Networks and Waveguide Group, Polytechnic Institute of Brooklyn, N. Y.

[1] H. Bode, "Network Analysis and Feedback Amplifier Design," D. Van Nostrand Company, Inc., New York, N. Y.; 1947.

[2] R. M. Fano, "Theoretical limitations on the broadband matching of arbitrary impedances," *J. Franklin Inst.*, vol. 249, pp. 57–83, January, 1960; and pp. 139–155, February, 1960.

[3] R. LaRosa and H. J. Carlin, "A Class of Broadband Dissipative Matching Networks Designed on an Insertion—Loss Basis," Microwave Res. Inst., Polytechnic Inst. of Brooklyn, N. Y., Rept. R-264-52, PIB-203, Contract No. Nobsr-43360 Index NE-100-42; 1952.

[4] D. C. Fielder, "Numerical determination of cascaded LC network elements from return loss coefficients," IRE Trans. on Circuit Theory, vol. CT-5, pp. 356–359; December, 1958.

[5] B. K. Kinariwala, "Realization of Broadband Matching Networks for Arbitrary Impedances," Office of Naval Research, Rept. No. 59, Contract NRonr-29529; 1957.

[6] L. Smilen, "Broadband Equalization of Two-Ports," Res. Rept. R-667-58, PIB-595; 1958.

[7] This idea has already been exploited by P. Penfield in a recent and interesting paper.[8] He has also extended the notion in several unpublished internal memoranda (private communication). However, the author, unaware that he had been anticipated, published his own views of the matter in "On scattering matrices normalized to complex port numbers," Proc. IRE, (*Correspondence*) vol. 49, p. 1221; July, 1961. Further enlargements and discussion appear in Youla.[9,10]

[8] P. J. Penfield, "Noise in Negative-Resistance Amplifiers," IRE Trans. on Circuit Theory, vol. CT-7, pp. 166–171; June, 1960.

[9] D. C. Youla, "Solution to the Problem of Complex Normalization," Polytechnic Inst. of Brooklyn, N. Y., Memo 48, PIBMRI-891-61; 1961.

[10] D. C. Youla, "Solution to the Problem of Complex Normalization-Concluded," Polytechnic Inst. of Brooklyn, N. Y., Memo 66, PIBMRI-1033-62; 1962.

[11] H. J. Carlin, "An Introduction to the Use of the Scattering Matrix in Network Theory." Polytechnic Institute of Brooklyn, N. Y., Rept. R-366-54, PIB-300; June, 1954.

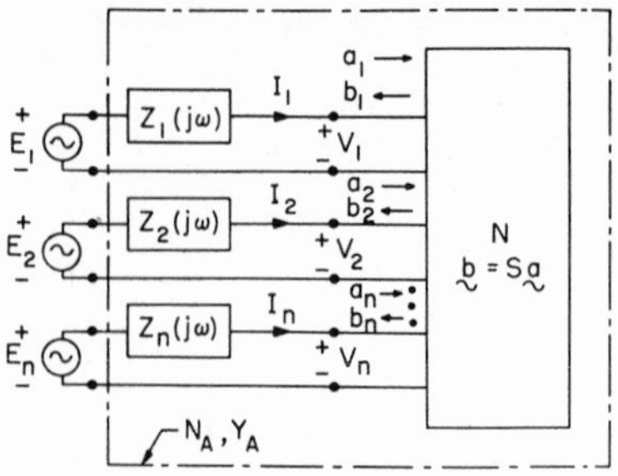

Fig. 2—Schematic of a linear, time-invariant $2n$-terminal network excited at its n ports by n generators $E_1, E_2, \cdots, E_n$ with respective internal impedance $z_1, z_2, \cdots, z_n$.

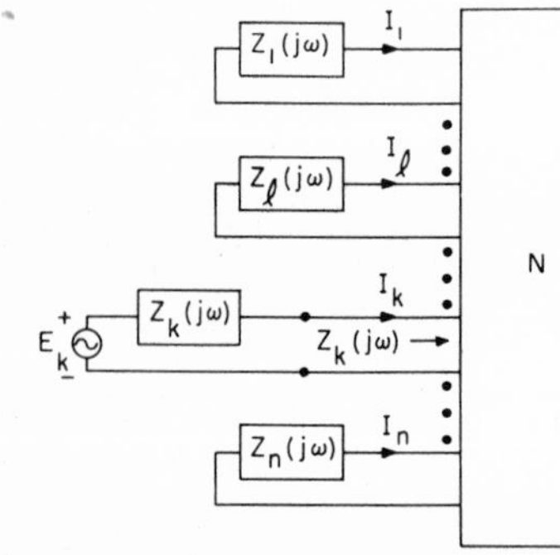

Fig. 3—Schematic illustrating the meaning of the transducer power gain $G_{kl}(\omega^2)$ from port k to port l under matched terminations.

Let A be an arbitrary matrix. Then A' and $\bar{A}' = A^*$ denote the transpose and the complex conjugate transpose (also called the adjoint) of A, respectively. Column vectors are written $\mathbf{a}$, $\mathbf{b}$, $\mathbf{x}$, etc., and in the alternative form $\mathbf{x} = (x_1, x_2, \cdots, x_n)'$ whenever it is desirable to exhibit the components explicitly. The matrices 1_n and 0_n are, in the same order, the $n \times n$ identity matrix and the $n \times n$ zero matrix. For an hermetian matrix $A = A^*$, $A \geq 0_n$ means that A is the matrix of a non-negative quadratic form.

Refer to Fig. 2 and assume that the internal impedances $z_1(j\omega), z_2(j\omega), \cdots, z_n(j\omega)$ have positive-real parts over the frequency band W; *i.e.*,

$$\text{Real}\, z_k(j\omega) = r_k(\omega) > 0, \quad \omega \,\varepsilon\, W, \quad (k = 1, 2, \cdots, n). \quad (2)$$

The normalized incident and reflected wave amplitudes $\mathbf{a}_k$ and $\mathbf{b}_k$ impinging on the kth port of N are defined as linear combinations of the associated port voltages and currents according to the scheme,

$$2\sqrt{r_k}a_k = V_k + z_k I_k \quad (3)$$

$$2\sqrt{r_k}b_k = V_k - \bar{z}_k I_k, \quad (k = 1, 2, \cdots, n). \quad (4)$$

All square roots have been chosen positive. The $n \times n$ scattering matrix $S(j\omega)$ of N, normalized with respect to the n impedances $z_1(j\omega), z_2(j\omega), \cdots, z_n(j\omega)$, is defined by means of the linear matrix equation,

$$\mathbf{b} = S\mathbf{a}, \quad (5)$$

where $\mathbf{a} = (a_1, a_2, \cdots, a_n)'$ and $\mathbf{b} = (b_1, b_2, \cdots, b_n)'$. A more explicit expression for $S(j\omega)$ appears later.

Let P_k equal the average power entering N through port k, *i.e.*, $P_k = \text{Real}\,(\bar{V}_k I_k)$. Using (3) and (4), an easy calculation yields

$$P_k = |a_k|^2 - |b_k|^2, \quad (k = 1, 2, \cdots, n). \quad (6)$$

Thus, P_{AV}, the total average power absorbed by N, is given by

$$P_{AV} = \sum_{k=1}^{n} P_k = \sum_{k=1}^{n} |a_k|^2 - \sum_{k=1}^{n} |b_k|^2 = \mathbf{a}^*\mathbf{a} - \mathbf{b}^*\mathbf{b}, \quad (7)$$

which, with the aid of (5), transforms into

$$P_{AV} = \mathbf{a}^*(1_n - S^*S)\mathbf{a}. \quad (8)$$

Consequently, if N is lossless over W, $P_{AV}(\omega^2) = 0$, $\omega \,\varepsilon\, W$, for any $\mathbf{a}$, and this in turn, implies that

$$1_n - S^*(j\omega)S(j\omega) = 0_n, \quad (9)$$

i.e., $S(j\omega)$ is unitary for $\omega \,\varepsilon\, W$. Again, if N is dissipative, $P_{AV}(\omega^2) \geq 0$ for all $\mathbf{a}$, and so

$$1_n - S^*(j\omega)S(j\omega) \geq 0_n, \quad (10)$$

$\omega \,\varepsilon\, W$. Thus the normalization procedure subsumed in (3) and (4) has succeeded in preserving two of the most important properties possessed by a scattering matrix normalized to real-positive port numbers.[11]

When the r_k's are not all positive over the band W, (9) and (10) go into

$$\sigma - S^*(j\omega)\sigma S(j\omega) = 0_n \quad (9a)$$

and

$$\sigma - S^*(j\omega)\sigma S(j\omega) \geq 0_n, \quad (10a)$$

respectively, where σ is an $n \times n$ diagonal matrix whose diagonal elements are either plus or minus one. If $r_k > 0$, $(\sigma)_{kk} = +1$ but if $r_k < 0$, $(\sigma)_{kk} = -1$. It is not difficult to show that (9a) implies the equality of complementary minors in S.

Let all ports of N except the kth port be closed on their respective normalization impedances and suppose that port k is driven as shown in Fig. 3. Then,

$$V_l = -z_l I_l, \quad l \neq k. \quad (11)$$

According to (3), (11) is equivalent to

$$a_l = 0, \quad l \neq k. \quad (12)$$

Hence, from (5),

$$b_l = s_{lk}(j\omega)a_k = s_{lk}(j\omega)\frac{E_k}{2\sqrt{r_k(j\omega)}}, \quad (l = 1, 2, \cdots, n), \quad (13)$$

since, as is obvious from Fig. 3, $E_k = V_k + z_k I_k$. Eq. (11) coupled with (4) yields

$$b_l = -\sqrt{r_l(j\omega)}I_l, \qquad l \neq k, \tag{14}$$

and, therefore, from (13),

$$|s_{lk}(j\omega)|^2 = \frac{r_l(j\omega)\,|I_l|^2}{\dfrac{|E_k|^2}{4r_k(j\omega)}} = G_{kl}(\omega^2), \qquad l \neq k, \tag{15}$$

the transducer power gain from port k to port l. To determine $s_{kk}(j\omega)$, let $Z_k(j\omega)$ represent the impedance seen looking into port k under matched terminations (Fig. 3). Then $V_k = Z_k I_k$ and the division of (4) by (3) yields, with the help of (13),

$$\frac{b_k}{a_k} = s_{kk}(j\omega) = \frac{V_k - \bar{z}_k I_k}{V_k + z_k I_k} = \frac{Z_k - \bar{z}_k}{Z_k + z_k}. \tag{16}$$

In other words, s_{kk} is the input reflection coefficient at port k with all other ports matched. It is important to note that

$$s_{kk} \neq \frac{Z_k - z_k}{Z_k + z_k}, \tag{17}$$

which is the result that is obtained by normalizing in the usual (but generally incorrect) manner.

Also observe that if port l is terminated in $-\bar{z}_l(j\omega)$ instead of $z_l(j\omega)$, $V_l = \bar{z}_l I_l$. Thus, by (3), $b_l = 0$. In short, closing a port on its respective normalization impedance obliterates the corresponding incident wave whereas termination in the negative complex conjugate of the normalization impedance obliterates the reflected wave. In the first case the port is said to be "matched" and in the second, to be "paraconjugate matched."

It is possible to express S in terms of $Y_A(j\omega)$, the admittance matrix of N_A, the "augmented" n port associated with N. From Fig. 2,

$$\mathbf{I} = Y_A\mathbf{E} = Y_A(\mathbf{V} + Z\mathbf{I}). \tag{18}$$

$$\begin{aligned} \therefore\quad \mathbf{V} - \bar{Z}\mathbf{I} &= (\mathbf{V} + Z\mathbf{I}) - (Z + \bar{Z})\mathbf{I} \\ &= \mathbf{V} + Z\mathbf{I} - 2R\mathbf{I} \\ &= (1_n - 2RY_A)(\mathbf{V} + Z\mathbf{I}). \end{aligned}$$

Hence,

$$2R^{1/2}\mathbf{b} = 2(1_n - 2RY_A)R^{1/2}\mathbf{a}$$

$$\therefore\quad S = 1_n - 2R^{1/2}Y_A R^{1/2}, \tag{19}$$

where

$$Z = \text{diag}\,[z_1, z_2, \cdots, z_n]$$

$$R = \text{diag}\,[r_1, r_2, \cdots, r_n] = \text{Real } Z$$

and

$$R^{\pm 1/2} = \text{diag}\,[r_1^{\pm 1/2}, r_2^{\pm 1/2}, \cdots, r_n^{\pm 1/2}].$$

Under our assumptions, Y_A certainly exists if N is passive. Now if N is reciprocal, Y_A is symmetric ($Y_A = Y_A'$) and therefore S is also symmetric since $R^{1/2}$ is diagonal.

Similarly, if N possesses an impedance matrix Z_N, $\mathbf{V} = Z_N\mathbf{I}$ and

$$2R^{1/2}\mathbf{a} = (Z_N + Z)\mathbf{I},$$

$$2R^{1/2}\mathbf{b} = (Z_N - \bar{Z})\mathbf{I}.$$

$$\therefore\quad 2R^{1/2}S\mathbf{a} = R^{1/2}SR^{-1/2}(Z_N + Z)\mathbf{I} = (Z_N - \bar{Z})\mathbf{I}.$$

$$\therefore S = R^{-1/2}(Z_N - \bar{Z})(Z_N + Z)^{-1}R^{1/2}. \tag{20}$$

Again note that the conventional formula[11]

$$S = R^{-1/2}(Z_N - Z)(Z_N + Z)^{-1}R^{1/2} \tag{21}$$

is incorrect if Z is complex.

III. A New Theory of Broad-Band Matching

Refer to Fig. 1. Let

$$S(j\omega) = \begin{bmatrix} s_{11}(j\omega) & s_{12}(j\omega) \\ s_{21}(j\omega) & s_{22}(j\omega) \end{bmatrix} \tag{22}$$

denote the scattering matrix of the reactive two-port equalizer N normalized to $z_g(j\omega)$ on the left and $z_l(j\omega)$ on the right. Of course in our particular case z_g and z_l are non-Foster and passive and z_g is a constant positive resistor r_g. Since N is lossless, $S(j\omega)$ is unitary according to (9). Thus,

$$S^*(j\omega)S(j\omega) = 1_2$$

or, in expanded form,

$$\bar{s}_{11}(j\omega)s_{11}(j\omega) + \bar{s}_{21}(j\omega)s_{21}(j\omega) = 1 \tag{23}$$

$$\bar{s}_{11}(j\omega)s_{12}(j\omega) + \bar{s}_{21}(j\omega)s_{22}(j\omega) = 0 \tag{24}$$

$$\bar{s}_{12}(j\omega)s_{12}(j\omega) + \bar{s}_{22}(j\omega)s_{22}(j\omega) = 1. \tag{25}$$

Therefore,

$$s_{12}(j\omega) = -\bar{s}_{21}(j\omega)\frac{s_{22}(j\omega)}{\bar{s}_{11}(j\omega)}. \tag{26}$$

Substituting (26) in (25) yields

$$|s_{11}(j\omega)| = |s_{22}(j\omega)|, \tag{27}$$

a result that is valid irrespective of whether or not N is reciprocal. Together with (23) and (25), (27) gives

$$|s_{12}(j\omega)| = |s_{21}(j\omega)|. \tag{28}$$

Eq. (15) tells us that the transducer power gain from port 1 to port 2 is

$$\begin{aligned} G(\omega^2) &= |s_{21}(j\omega)|^2 \\ &= |s_{12}(j\omega)|^2 \\ &= 1 - |s_{22}(j\omega)|^2 \qquad (29) \\ &= 1 - \left|\frac{Z_2(j\omega) - \bar{z}_l(j\omega)}{Z_2(j\omega) + z_l(j\omega)}\right|^2. \qquad (30) \end{aligned}$$

The last two equations are a consequence of (25) and (16). Since N is a passive, lossless, lumped two-port and $z_g(p)$ and $z_l(p)$ are two rational non-Foster positive-real functions, $Z_2(p)$ is also rational and positive-real. Clearly, the power gain is completely determined by the real-frequency magnitude of the rational function

$$s_{22}(p) = \frac{Z_2(p) - z_l(-p)}{Z_2(p) + z_l(p)}. \tag{31}$$

Because of the positive-real character of $Z_2(p)$ and $z_l(p)$,

$$|s_{22}(j\omega)| \leq 1.$$

Now the poles of $s_{22}(p)$ in $Re\ p > 0$ are *precisely those* of $z_l(-p)$. Denote these poles by $\nu_1, \nu_2, \cdots, \nu_m$ and set

$$b(p) = \prod_{r=1}^{m} \frac{p - \nu_r}{p + \bar{\nu}_r}; \tag{32}$$

$b(p)$ is a regular all pass; *i.e.*, $b(p)$ is real for real p, analytic in $Re\ p \geq 0$ and $b(-p)b(p) = 1$. Moreover

$$s(p) = b(p)s_{22}(p) \tag{33}$$

is analytic in $Re\ p \geq 0$. In other words, $s(p)$ is a bounded-real scattering coefficient. Since $|b(j\omega)| = 1$, $|s(j\omega)| = |s_{22}(j\omega)|$ and

$$G(\omega^2) = 1 - |s(j\omega)|^2. \tag{34}$$

Consequently to delimit the class of transducer power gains compatible with a prescribed $z_l(p)$ it suffices to study the rational, bounded-real scattering coefficient

$$s(p) = b(p) \cdot \frac{Z_2(p) - z_l(-p)}{Z_2(p) + z_l(p)}. \tag{35}$$

Eqs. (34) and (35), simple as they may be, form the cornerstone of the new theory of classical broad-band matching.

Let

$$r_l(p) = \frac{z_l(p) + z_l(-p)}{2} \tag{36}$$

represent the "even" part of the load impedance. Then evidently,

$$b(p) - s(p) = \frac{2r_l(p)b(p)}{Z_2(p) + z_l(p)}. \tag{37}$$

The notion of an inherent restriction emerges immediately from (37). Set

$$\lambda(p) = \frac{r_l(p)}{z_l(p)}. \tag{38}$$

Then obviously, every zero of $\lambda(p)$ in $Re\ p \geq 0$ must also be a zero of $b(p) - s(p)$! Stated differently, regardless of the choice of reactive equalizer N, there exist points p_0 in $Re\ p \geq 0$, dictated solely by the choice of load $z_l(p)$, such that

$$b(p_0) = s(p_0). \tag{39}$$

Recall that $b(p)$ is completely specified via (32) by the poles of $z_l(-p)$ in $Re\ p > 0$.

Definition 1: A zero p_0 of $\lambda(p)$ in $Re\ p \geq 0$ of multiplicity k is said to be a zero of transmission of the load of order k.

The restrictions are formulated most compactly in terms of the coefficients in the power series expansions of the following quantities:

$$s(p) = \sum_{r=0}^{\infty} S_r(p - p_0)^r \tag{40}$$

$$b(p) = \sum_{r=0}^{\infty} B_r(p - p_0)^r \tag{41}$$

$$f(p) = 2r_l(p)b(p) = \sum_{r=0}^{\infty} F_r(p - p_0)^r. \tag{42}$$

Also,

$$z_l(j\omega) \equiv r_l(\omega) + jx_l(\omega), \tag{43}$$

$$x_l'(\omega) \equiv \frac{dx_l}{d\omega}$$

and the residue of $z_l(p)$ at a pole $p = j\omega_0$ is

$$a_{-1} = \lim_{p \to j\omega_0} (p - j\omega_0)z_l(p) \tag{44}$$

$$= \lim_{p \to \infty} \frac{z_l(p)}{p} \tag{45}$$

depending on whether ω_0 is finite or infinite, respectively.

For convenience the zeros of transmission are divided into four mutually exclusive classes. Class I contains all those in the strict right half plane, $Re\ p > 0$. Class II contains all those on the real-frequency axis which are *simultaneously* zeros of $z_l(p)$. Class III contains all those on the real-frequency axis for which $0 < |z_l(j\omega_0)| < \infty$ and Class IV all those for which $|z_l(j\omega_0)| = \infty$.

The restrictions are now written down practically by inspection.

Class I: Let $p_0 = \sigma_0 + j\omega_0$, $\sigma_0 > 0$, be a zero of transmission of order k. Then,

$$B_r = S_r, \qquad (r = 0, 1, \cdots, k-1). \tag{46}$$

Class II: Let $p_0 = j\omega_0$ be a zero of transmission of order k. Then,

$$B_r = S_r, \qquad (r = 0, 1, \cdots, k-1) \tag{47}$$

and

$$\frac{B_k - S_k}{F_{k+1}} \geq 0. \tag{48}$$

Class III: Let $p_0 = j\omega_0$ be a zero of transmission of order k. Then,

$$B_r = S_r, \qquad (r = 0, 1, \cdots, k-2), \tag{49}$$

and

$$\frac{B_{k-1} - S_{k-1}}{F_k} \geq 0. \tag{50}$$

Class IV: Let $p_0 = j\omega_0$ be a zero of transmission of order k. Then,

$$B_r = S_r, \qquad (r = 0, 1, \cdots, k-1) \tag{51}$$

and

$$\frac{F_{k-1}}{B_k - S_k} \geq a_{-1}(\omega_0). \tag{52}$$

Observe that a zero of transmission of order k belonging to either Class I, II or IV is a zero of $b - s$ of *at least order* k, but one belonging to Class III is a zero of $b - s$ of *at least order* $k - 1$.

Eq. (46) merely states that at a right half-plane zero p_0 of $r_l(p)$ of order k, the difference $b - s$ must also have a zero of order k. Equivalently, the first k coefficients in the power series expansions of $b(p)$ and $s(p)$ about $p = p_0$ must coincide. Upon examination of (37), these facts become fairly obvious.

The Class II restrictions are a bit trickier to derive. At a zero $p_0 = j\omega_0$ of this kind, $z_l(j\omega_0) = 0$. Hence, as a zero of $f(p) = 2r_l(p)b(p)$, p_0 must be of multiplicity $k + 1$. If $Z_2(j\omega_0) = 0$, the entire denominator of (37) vanishes and $p_0 = j\omega_0$ is a zero of $b - s$ of order k. If $Z_2(j\omega_0) \neq 0$, p_0 is a zero of $b - s$ of order $k + 1$ and

$$B_k = S_k. \tag{53}$$

If $Z_2(j\omega_0) = 0$, $p_0 = j\omega_0$ is a zero of $b - s$ of order k and (47) follows. To find the relation tying together B_k and S_k we write

$$z_l(p) = x_l'(\omega_0)(p - j\omega_0) + 0(|p - j\omega_0|^2)$$

and

$$Z_2(j\omega_0) = X_2'(\omega_0)(p - j\omega_0) + 0(|p - j\omega_0|^2)$$

where (remember that Z_2 and z_l are positive-real)

$$x_l'(\omega_0) > 0,$$

$$X_2'(\omega_0) > 0.$$

From the right-hand side of (37) we see that the coefficient of $(p - j\omega_0)^k$ is

$$\frac{F_{k+1}}{X_2'(\omega_0) + x_l'(\omega_0)}.$$

Comparing this to the left-hand side of (37) yields the equality

$$B_k - S_k = \frac{F_{k+1}}{X_2'(\omega_0) + x_l'(\omega_0)}.$$

Since $F_{k+1} \neq 0$,

$$\frac{B_k - S_k}{F_{k+1}} = \frac{1}{X_2'(\omega_0) + x_l'(\omega_0)} > 0$$

and therefore,

$$\frac{B_k - S_k}{F_{k+1}} > 0.$$

Thus the two cases $Z_2(j\omega_0) \neq 0$ and $Z_2(j\omega_0) = 0$ are subsumed in the single statement

$$\frac{B_k - S_k}{F_{k+1}} \geq 0,$$

the equality sign being attained if and only if $Z_2(j\omega_0) \neq 0$. This establishes (48). The restrictions for Class III and IV zeros are derived in an entirely similar manner and we omit the details.

If $|z_l(j\omega_0)| \neq \infty$, the equalizer is said to be *nondegenerate* if and only if $Z_2(j\omega_0) + z_l(j\omega_0) \neq 0$. If $|z_l(j\omega_0)| = \infty$, the equalizer is said to be nondegenerate if and only if $|Z_2(j\omega_0)| \neq \infty$. *Thus we see that a real-frequency zero of transmission leads to a nondegenerate equalizer if and only if the equality sign is chosen in formulas* (48), (50) *and* (52).

The classical Fano broad-band problem may be stated in the following manner. A 1-ohm generator is to be equalized to a prescribed lumped, passive, non-Foster load $z_l(p)$ by means of a lumped, reciprocal, reactive equalizer, to achieve a preassigned transducer power gain $G(\omega^2)$. Design the equalizer N. The procedure is outlined in five steps.

Step 1: First, it is necessary to verify that $G(\omega^2)$ is rational in ω^2 and satisfies

$$0 \leq G(\omega^2) \leq 1 \tag{54}$$

along the entire real-frequency axis.

Step 2: Since $|s(j\omega)|^2 = 1 - G(\omega^2)$,

$$s(-p)s(p) = 1 - G(-p^2). \tag{55}$$

Let $s_0(p)$ be that factorization of (55) which is analytic and devoid of zeros is $Re\ p > 0$. That is to say, $s_0(p)$ is minimum-phase and uniquely determined by $G(-p^2)$ up to within a plus or minus sign. The most general bounded-real scattering coefficient satisfying (55) is given by

$$s(p) = \pm\eta(p)s_0(p) \tag{56}$$

where $\eta(p)$ is an arbitrary, real, rational regular all pass of the form

$$\eta(p) = \prod_{r=1}^{\nu} \frac{p - \mu_r}{p + \bar{\mu}_r},\ Re\ \mu_r > 0, \quad (r = 1, 2, \cdots, \nu). \tag{57}$$

Step 3: From $z_l(p)$ we calculate $\lambda(p)$, $b(p)$ and $f(p)$. From $\lambda(p)$ we find the zeros of transmission and their multiplicities and we divide them into their respective classes. About each zero of transmission we form the power series expansions of $b(p)$ and $f(p)$. Then, *if possible*, we choose an $\eta(p)$ so that the $s(p)$ defined in (56) meets the restrictions imposed by (46)–(52).

Step 4: Having successfully carried out Step 3, the equalizer back-end impedance $Z_2(p)$ is determined from (38),

$$Z_2(p) = \frac{2r_l(p)b(p)}{b(p) - s(p)} - z_l(p). \tag{58}$$

In Theorem 1 it is proven that (58) always defines a rational positive-real function.

Step 5: The equalizer N is now obtained by synthesizing $Z_2(p)$, preferably by the Darlington technique, as the input impedance of a lossless 2-port N terminated in the positive resistor r_g. This completes the design.

The five basic steps outlined above will be amply illustrated later in a variety of nontrivial problems. But for the present the immediate task is to establish the fundamental assertion of Step 4—the positive-reality of $Z_2(p)$.

Theorem 1

Let $z_l(p)$ be a prescribed rational, non-Foster positive-real function and $s(p)$ a real rational function of p. Then the formula

$$Z_2(p) = \frac{2r_l(p)b(p)}{b(p) - s(p)} - z_l(p) \tag{59}$$

defines a positive-real function $Z_2(p)$ *if and only if* $s(p)$ *is a bounded-real scattering coefficient satisfying the constraints* (46)–(52).

Proof

Only sufficiency requires verification. The even part of $Z_2(p)$ is given by

$$\begin{aligned} r_2(p) &= \frac{Z_2(p) + Z_2(-p)}{2} \\ &= r_l(p)\left[\frac{b(p)}{b(p) - s(p)} + \frac{b(-p)}{b(-p) - s(-p)} - 1\right] \\ &= \frac{r_l(p)[1 - s(-p)s(p)]}{[b(p) - s(p)][b(-p) - s(-p)]}, \end{aligned} \tag{60}$$

in which use has been made of the identity $b(p)b(-p) = 1$. Thus,

$$Re\, Z_2(j\omega) = r_2(j\omega) = \frac{r_l(j\omega)[1 - |s(j\omega)|^2]}{|b(j\omega) - s(j\omega)|^2} \geq 0, \tag{61}$$

since $r_l(j\omega) \geq 0$ and $|s(j\omega)| \leq 1$. Set

$$z(p) = \frac{2r_l(p)b(p)}{b(p) - s(p)}. \tag{62}$$

Evidently, $Re\, Z_2(j\omega) \geq 0$ implies that

$$Re\, z(j\omega) \geq Re\, z_l(j\omega) \geq 0. \tag{63}$$

Our immediate aim is to prove that $z(p)$ is positive-real and this is equivalent to showing that

$$\begin{aligned} y(p) = z^{-1}(p) &= \frac{b(p) - s(p)}{2r_l(p)b(p)} \\ &= \frac{b(p) - s(p)}{f(p)} \end{aligned} \tag{64}$$

is positive-real. Since $Re\, y(j\omega) \geq 0$, we must prove, according to a classic theorem,[12] that $y(p)$ is analytic in $Re\, p > 0$ and that its real-frequency poles are simple with positive residues.

Because of its non-Foster positive-real character, $z_l(p)$ is neither zero nor infinite in $Re\, p > 0$. Hence the Class I zeros of $\lambda(p)$ are precisely the zeros of $f(p)$, multiplicity included. At such a zero, (46) guarantees that $b(p) - s(p)$ *vanishes to at least the same order*. Thus $y(p)$ is analytic at all the zeros of $f(p)$ in $Re\, p > 0$ and consequently analytic in the *entire* open right-half p plane.

On the real-frequency axis, the only possible poles of $y(p)$ are again the zeros of $f(p)$. Consider a zero $p_0 = j\omega_0$ of $f(p)$ which is also a zero of $z_l(p)$. This is a Class II zero and its multiplicity as a zero of $\lambda(p)$, k say, is one less than its multiplicity as a zero of $f(p)$. Hence, as a zero of $f(p)$, $p_0 = j\omega_0$ is of order $k + 1$. According to (47), $p_0 = j\omega_0$ is a zero of $b(p) - s(p)$ of *at least order* k; and therefore such a zero is a pole of $y(p)$ of *at most order one*. Obviously, the associated residue is

$$\frac{B_k - S_k}{F_{k+1}} \geq 0,$$

the inequality being a consequence of the left-hand side of (48).

Consider now a real-frequency zero of $f(p)$ at which $0 < |z_l(j\omega_0)| < \infty$. This zero is in Class III and its order k is the same for $\lambda(p)$ and $f(p)$. Invoking (49), $p_0 = j\omega_0$ is a zero of $b(p) - s(p)$ of at least order $k - 1$ and therefore is a pole of $y(p)$ of at most order one. Its residue is

$$\frac{B_{k-1} - S_{k-1}}{F_k} \geq 0,$$

by (50).

Finally, if $p_0 = j\omega_0$ is a zero of $f(p)$ which is simultaneously a pole of $z_l(p)$, its order as a zero of $f(p)$ is $k - 1$, where k is its order as a zero of $\lambda(p)$. According to (51) its order as a zero of $b(p) - s(p)$ is at least k and $y(p)$ is therefore actually zero at this point. Gathering everything together, $y(p)$ and $z(p) = y^{-1}(p)$ are positive-real.

It is now evident that to prove $Z_2(p) = z(p) - z_l(p)$ positive-real it is merely necessary to ascertain that its residues at the real-frequency poles of $z_l(p)$ are nonnegative. Such a pole is a Class IV zero of $\lambda(p)$ and we have just shown that it is also a zero of $y(p)$ and hence a pole of $z(p)$. The residue of $z(p)$ at this pole is

$$\frac{F_{k-1}}{B_k - S_k}.$$

[12] O. Brune, "Synthesis of a finite two-terminal network whose driving-point impedance is a prescribed function of frequency," *J. Math. Phys.*, vol. 10, pp. 191–236; 1931.

The corresponding residue of $Z_2(p)$ is therefore

$$\frac{F_{k-1}}{B_k - S_k} - a_{-1}(\omega_0) \geq 0$$

by (52). This terminates the proof, Q.E.D.

IV. Illustrative Examples

Before working out some concrete examples we wish to point out that the realization of the equalizer N is obtained by synthesizing the back-end impedance $Z_2(p)$ as a lossless two-port terminated in the generator resistance r_g. One practical way to do this is to use a method presented by the author.[13]

Now

$$Z_2(p) = \frac{2r_l b}{b - s} - z_l(p) = \frac{b(p)z_l(-p) + s(p)z_l(p)}{b(p) - s(p)} \quad (65)$$

and is constructed from a knowledge of $z_l(p)$ and $s(p)$. To synthesize $Z_2(p)$ as a cascade of lossless two-ports terminated in a resistor it is necessary to know the zeros of its even part $r_2(p)$. From (60)

$$r_2(p) = \frac{r_l(p)[1 - s(-p)s(p)]}{[b(p) - s(p)][b(-p) - s(-p)]}. \quad (66)$$

Suppose that we are broad-banding a load $z_l(p)$ which is minimum reactance and whose even part $r_l(p)$ has no zeros on the *entire* real-frequency axis, infinity included. Such an impedance is realized by terminating a finite cascade of type C and D sections in a positive resistor.[13] Obviously the zeros of $\lambda(p) = r_l/z_l$ are all in Class I and the reader should have no difficulty in convincing himself that in this case the zeros of $r_2(p)$ in $Re\ p \geq 0$ are exactly those of

$$\beta(p) = \frac{1 - s(-p)s(p)}{s(-p)} \quad (67)$$

and possess the same multiplicity.

Example 1: It is desired to equalize the parallel RC load

$$z_l(p) = \frac{R}{1 + pRC} \quad (68)$$

to a resistive generator to achieve the nth-order Butterworth transducer power gain

$$G(\omega^2) = \frac{K}{1 + \left(\frac{\omega}{\omega_c}\right)^{2n}}, \quad 0 \leq K \leq 1; \quad (69)$$

ω_c is the 3 db radian bandwidth. Determine the gain-bandwidth restrictions.

Solution: For this load we have

$$r_l(p) = \frac{R}{1 - p^2R^2C^2} \quad (70)$$

$$b(p) = \frac{p - 1/RC}{p + 1/RC} \quad (71)$$

$$f(p) = 2r_l b = -\frac{1}{RC^2}\cdot\frac{1}{(p + 1/RC)^2} \quad (72)$$

and

$$\lambda(p) = \frac{r_l}{z_l} = \frac{1}{1 - pRC}. \quad (73)$$

The only zero of $\lambda(p)$ is at $p = \infty$ and is of order one. Since $z_l(p)$ has a coincident zero, $p = \infty$ is in Class II. Hence, from (47) and (48),

$$B_0 - S_0 = 0 \quad (74)$$

$$\frac{B_1 - S_1}{F_2} \geq 0. \quad (75)$$

The expansions of $b(p)$ and $f(p) = 2r_l b$ about infinity are, respectively,

$$b(p) = 1 - \frac{2}{RCp} + 0(p^{-2}) \quad (76)$$

and

$$f(p) = -\frac{2}{RC^2p^2} + 0(p^{-3}), \quad (77)$$

whence

$$B_0 = 1, \quad B_1 = -\frac{2}{RC} \quad (78)$$

$$F_0 = F_1 = 0, \quad F_2 = -\frac{2}{RC^2}. \quad (79)$$

Clearly,

$$G(-p^2) = \frac{K}{1 + (-1)^n\left(\frac{p}{\omega_c}\right)^{2n}} \quad (80)$$

and

$$s(-p)s(p) = 1 - G(-p^2) = \frac{1 - K + (-1)^n\left(\frac{p}{\omega_c}\right)^{2n}}{1 + (-1)^n\left(\frac{p}{\omega_c}\right)^{2n}}. \quad (81)$$

The polynomial $Q(p) = 1 + (-1)^n p^{2n}$ may be factored in terms of the roots of -1[14]

$$Q(p) = \Delta_n(-p)\Delta_n(p)$$

where

$$\Delta_n(p) = a_0 + a_1p + \cdots + a_np^n \quad (82)$$

is the nth order Hurwitz Butterworth polynomial. It is well known[14] that $a_0 = a_n = 1$, coefficients equidistant from the end are equal, and

$$a_{n-1} = \frac{1}{\sin\frac{\pi}{2n}}. \quad (83)$$

[13] D. C. Youla, "A new theory of cascade synthesis," IRE Trans. on Circuit Theory, vol. CT-8, pp. 244–260; September, 1961.

[14] L. Weinberg, "Network design by use of modern synthesis techniques and tables," *Proc. Nat'l Electronics Conf.*, vol. 12; October, 1956.

Clearly, (81) may be written as

$$s(-p)s(p) = (1 - K)\frac{Q(x)}{Q(y)} \tag{84}$$

where

$$x = \frac{(1-K)^{-1/2n}}{\omega_c} p \tag{85}$$

$$y = p/\omega_c. \tag{86}$$

The minimum-phase solution $s_0(p)$ of (84) is

$$s_0(p) = (1 - K)^{1/2}\frac{\Delta_n(x)}{\Delta_n(y)} \tag{87}$$

and any other solution $s(p)$ analytic in $Re\, p \geq 0$ is given by

$$s(p) = \pm\eta(p)s_0(p), \tag{88}$$

$$\eta(p) = \prod_{r=1}^{\nu}\frac{p-\mu_r}{p+\bar{\mu}_r}, \quad Re\,\mu_r > 0, \quad (r = 1, 2, \cdots, m),$$

being an arbitrary, real regular all pass.

To find S_0 and S_1 we must expand $s(p)$ about infinity. The expansion of $s_0(p)$ about $p = \infty$ is

$$s_0(p) = 1 + a_{n-1}\left(\frac{1}{x} - \frac{1}{y}\right) + 0\left(\frac{1}{xy}\right) \tag{89}$$

$$= 1 - \frac{\omega_c[1-(1-K)^{1/2n}]}{\sin\frac{\pi}{2n}}\cdot\frac{1}{p} + 0(p^{-2}). \tag{90}$$

The expansion of $\eta(p)$ about infinity is

$$\eta(p) = 1 - \frac{2}{p}\sum_{r=1}^{\nu}\mu_r + 0(p^{-2}). \tag{91}$$

Thus the expansion of $s(p)$ about infinity is given by

$$s(p) = \pm\left[1 - \frac{1}{p}\left[2\sum_{r=1}^{\nu}\mu_r + \frac{\omega_c[1-(1-K)^{1/2n}]}{\sin\frac{\pi}{2n}}\right] + 0(p^{-2})\right]. \tag{92}$$

Since $B_0 = 1$, we must choose the plus sign to secure agreement with (73). Hence,

$$S_0 = 1$$

$$-S_1 = 2\sum_{r=1}^{\nu}\mu_r + \frac{\omega_c[1-(1-K)^{1/2n}]}{\sin\frac{\pi}{2n}} \tag{93}$$

and substituting (93) in (75) yields the inequality

$$1 - (1-K)^{1/2n} \leq \frac{2\sin\frac{\pi}{2n}}{\omega_c}\left(\frac{1}{RC} - \sum_{r=1}^{\nu}\mu_r\right). \tag{94}$$

In order that (94) possess a solution K in the range $0 \leq K \leq 1$ it is necessary that the zeros μ_r not be chosen arbitrarily but in accordance with the obvious requirement

$$\frac{1}{RC} \geq \sum_{r=1}^{\nu}\mu_r. \tag{94a}$$

But even so, an adequate discussion of (94) demands that two cases be distinguished.

Case 1:

$$2\sin\frac{\pi}{2n}/\omega_c RC > 1.$$

By choosing the zeros μ_r appropriately we can guarantee that

$$\frac{2\sin\frac{\pi}{2n}}{\omega_c}\left(\frac{1}{RC} - \sum_{r=1}^{\nu}\mu_r\right) = 1, \tag{95}$$

so that we can simultaneously achieve a dc gain of unity ($K = 1$) with a *nondegenerate equalizer*. On the other hand if

$$2\sin\frac{\pi}{2n}/\omega_c RC > 1$$

and we do not pick the right half-plane zeros μ_r of $s(p)$ in such a way as to satisfy (95) but choose instead to work with the inequality

$$\frac{2\sin\frac{\pi}{2n}}{\omega_c}\left(\frac{1}{RC} - \sum_{r=1}^{\nu}\mu_r\right) > 1, \tag{96}$$

then we can still obtain unity dc gain with a *degenerate* equalizer.

Case 2:

$$2\sin\frac{\pi}{2n}/\omega_c RC \leq 1.$$

By manipulating (94) we now get the *bona fide* gain-bandwidth restriction

$$K \leq 1 - \left[1 - \frac{2\sin\frac{\pi}{2n}}{\omega_c}\left(\frac{1}{RC} - \sum_{r=1}^{\nu}\mu_r\right)\right]^{2n}. \tag{97}$$

It is clear that choosing $\mu_r \neq 0$, $(r = 1, 2, \cdots, m)$, will lead to a reduction in dc gain K for a preassigned bandwidth ω_c. Setting all $\mu_r = 0$, the gain-bandwidth restriction (94) simplifies to

$$K \leq 1 - \left\{\frac{2\sin\frac{\pi}{2n}}{\omega_c RC}\right\}^{2n} = K_{\max}. \tag{98}$$

In the limit as $n \to \infty$, $K_{\max} \to K_\infty$, where

$$K_\infty = 1 - e^{-1/RCf_c}, \qquad f_c = \omega_c/2\pi. \tag{99}$$

The limiting form of the transducer gain is

$$G(\omega^2) = 1 - e^{-1/RCf_c}, \qquad 0 \leq \omega < \omega_c, \tag{100}$$

$$= 0, \qquad \omega > \omega_c. \tag{101}$$

In this example it is *never* necessary to insert right half-plane zeros in $s(p)$.

In the special case $n = 1$, and with $s(p)$ chosen minimum phase,

$$s(p) = s_0(p) = \frac{p + \omega_c\sqrt{1-K}}{p + \omega_c}.$$

By direct substitution in (65),

$$Z_2(p) = \frac{R\omega_c(1 - \sqrt{1-K})}{[2 - RC\omega_c(1 - \sqrt{1-K})]p + \omega_c(1 + \sqrt{1-K})}. \tag{102}$$

Evidently, $Z_2(p)$ is positive-real if and only if

$$1 - \sqrt{1-K} \leq \frac{2}{\omega_c RC},$$

which is precisely the inequality obtained by setting $n = 1$ in (94). The corresponding matching network is shown in Fig. 4. If $\omega_c \leq 2/RC$, we may always set $K = 1$ and the resulting parameters are

$$C_N = \frac{2}{\omega_c R} - C \geq 0 \tag{103}$$

$$t = \sqrt{R}. \tag{104}$$

If $\omega_c \geq 2/RC$, the maximum gain is

$$K_{\max} = \frac{4(\omega_c RC - 1)}{\omega_c^2 R^2 C^2}, \tag{105}$$

$$C_N = 0 \tag{106}$$

and

$$t = \left(\frac{R}{\omega_c RC - 1}\right)^{1/2}. \tag{107}$$

This equalizer is nondegenerate. One last point deserves mention. In the cascade synthesis of $Z_2(p)$, the final terminating resistance is $Z_2(0)$. From (65),

$$\frac{Z_2(0)}{z_l(0)} = \frac{b(0) + s(0)}{b(0) - s(0)}.$$

However,

$$z_l(0) = R$$

$$b(0) = -1$$

and from (81)

$$s(0) = (1 - K)^{1/2}.$$

Hence,

$$Z_2(0) = R\frac{1 - (1-K)^{1/2}}{1 + (1-K)^{1/2}}. \tag{108}$$

In order to transform it into 1 ohm (the internal resistance of the generator), it is necessary to use an ideal transformer of ratio

$$t = \sqrt{R}\left(\frac{1 - \sqrt{1-K}}{1 + \sqrt{1-K}}\right)^{1/2}, \tag{109}$$

a result in agreement with the value quoted in Fig. 4 for the special case $n = 1$.

The next example deals with a load possessing a single zero of transmission on the positive σ – axis ($p = +j\omega$).

Example 2: It is desired to equalize the Darlington type-C load (shown in Fig. 5) to a resistive generator to achieve the nth-order Butterworth transducer power gain of (69). Determine the gain-bandwidth restrictions and discuss the design.

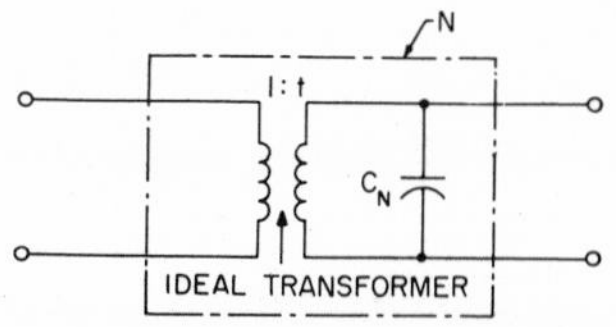

Fig. 4—Schematic of the equalizer corresponding to a first-order Butterworth gain in Example 1.

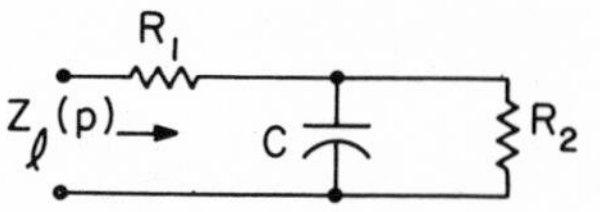

Fig. 5—Schematic of a load with a simple zero of transmission on the real σ axis.

Solution: Refer to Fig. 5.

Proceeding as before,

$$z_l(p) = \frac{(R_1R_2C)p + (R_1 + R_2)}{R_2Cp + 1} \tag{110}$$

$$b(p) = \frac{p - 1/R_2C}{p + 1/R_2C}, \tag{111}$$

$$r_l(p) = \frac{(R_1 + R_2) - R_1R_2^2C^2p^2}{1 - p^2R_2^2C^2}, \tag{112}$$

$$\tfrac{1}{2}f(p) = \frac{R_1R_2^2C^2p^2 - (R_1 + R_2)}{(1 + pR_2C)^2} \tag{113}$$

and

$$\lambda(p) = \frac{1}{pR_2C - 1}\cdot\frac{R_2Cp^2 - a}{p + a},$$

where

$$a = \frac{R_1 + R_2}{R_1R_2C}. \tag{114}$$

The only zero of $\lambda(p)$ in $Re\ p \geq 0$ is

$$p_0 = \sigma_0 = \frac{1}{R_2C}\left(1 + \frac{R_2}{R_1}\right)^{1/2} \tag{115}$$

and is real and *simple* ($k = 1$). This zero is in Class I. According to (46), the only restriction is

$$b(\sigma_0) = s(\sigma_0) \tag{116}$$

or

$$B_0 = S_0. \tag{117}$$

From (111),

$$B_0 = b(\sigma_0) = \frac{(1 + R_2/R_1)^{1/2} - 1}{(1 + R_2/R_1)^{1/2} + 1}. \tag{118}$$

Setting

$$x_0 = \frac{(1-K)^{1/2n}}{\omega_c}\cdot\sigma_0 \tag{118a}$$

and

$$y_0 = \sigma_0/\omega_c, \tag{118b}$$

we get, by using (87) and (88),

$$S_0 = \pm\eta(\sigma_0)(1 - K)^{1/2} \frac{\Delta_n(x_0)}{\Delta_n(y_0)}. \tag{119}$$

The constraint (117) takes the form

$$B_0 = b(\sigma_0) = \frac{(1 + R_2/R_1)^{1/2} - 1}{(1 + R_2/R_1)^{1/2} + 1} = \pm\eta(\sigma_0)(1 - K)^{1/2} \frac{\Delta_n(x_0)}{\Delta_n(y_0)} \tag{120}$$

or more succinctly,

$$b(\sigma_0) = \pm\eta(\sigma_0)s_0(\sigma_0), \tag{121}$$

$s_0(p)$ again being given by (87). Now since both $b(\sigma_0)$ and $s_0(\sigma_0)$ are positive, the choice of sign in (121) depends on the sign of $\eta(\sigma_0)$. For the sake of definiteness it is assumed that $\eta(\sigma_0) > 0$ and (121) becomes

$$b(\sigma_0) = \eta(\sigma_0)(1 - K)^{1/2} \frac{\Delta_n(x_0)}{\Delta_n(y_0)}. \tag{122}$$

Let

$$K = 1 - \mu^{2n}. \tag{123}$$

Then, using (118a), (118b) it is possible to rewrite (122) in the form

$$b(\sigma_0) = \eta(\sigma_0)\mu^n \frac{\Delta_n(y_0/\mu)}{\Delta_n(y_0)}. \tag{124}$$

But for any τ, $\Delta_n(\tau) = \tau^n \Delta_n(1/\tau)$. Thus

$$b(\sigma_0) = \eta(\sigma_0)\cdot\frac{\Delta_n(\mu/y_0)}{\Delta_n(1/y_0)} \tag{125}$$

or

$$\frac{b(\sigma_0)\Delta_n\left(\frac{\omega_c}{\sigma_0}\right)}{\eta(\sigma_0)} = \Delta_n\left(\frac{\mu\omega_c}{\sigma_0}\right). \tag{126}$$

This last expression contains all the information conveniently packaged.

As is obvious from (82), $\Delta_n(\tau)$ increases monotonically for non-negative τ and $\Delta_n(0) = 1$ for all n. Since[14] the coefficients $a_0, a_1, \cdots, a_n$ are known numerically up to $n = 10$, it is an easy matter to plot $\Delta_n(\tau)$ as a function of τ. A typical curve is shown qualitatively in Fig. 6. The design procedure may now be outlined very quickly. Suppose as is usual, that $z_l(p)$, ω_c and n are prescribed in advance. Then,

1) if

$$b(\sigma_0)\Delta_n\left(\frac{\omega_c}{\sigma_0}\right) \leq 1, \tag{127}$$

choose

$$\eta(\sigma_0) = b(\sigma_0)\Delta_n\left(\frac{\omega_c}{\sigma_0}\right). \tag{128}$$

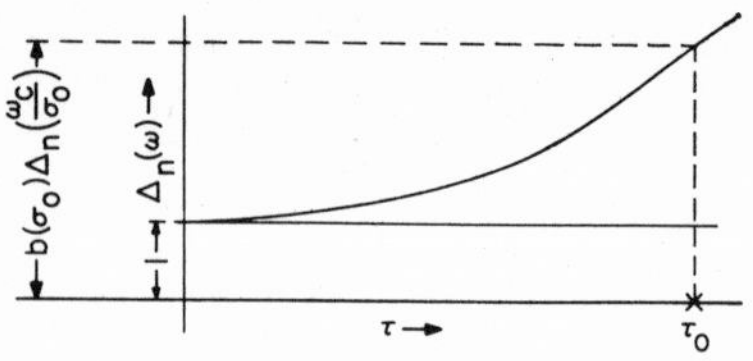

Fig. 6—Qualitative graph of a Butterworth polynomial.

From (126) and the properties of $\Delta_n(\tau)$ we get

$$\Delta_n\left(\frac{\mu\omega_c}{\sigma_0}\right) = 1;$$

i.e., $\mu = 0$ and $K = 1$. Again as in Example 1, unity dc gain is feasible provided the 3 db bandwidth ω_c is sufficiently small. But *unlike* Example 1, this unity gain *is not obtainable without the insertion of right half-plane zeros in* $s(p)$.

Fortunately, it is never necessary to insert anything but a single positive σ-axis zero. For if we set

$$\eta(p) = \frac{p - \xi_0}{p + \xi_0} \tag{129}$$

where ξ_0 is to be determined from (128), we obtain

$$\xi_0 = \sigma_0 \cdot \frac{1 - b(\sigma_0)\Delta_n\left(\frac{\omega_c}{\sigma_0}\right)}{1 + b(\sigma_0)\Delta_n\left(\frac{\omega_c}{\sigma_0}\right)}. \tag{130}$$

2) If

$$b(\sigma_0)\Delta_n\left(\frac{\omega_c}{\sigma_0}\right) > 1, \tag{131}$$

the insertion of right half-plane zeros in $s(p)$ actually decreases the gain. This is easily understood by referring to (126). Clearly $\eta(\sigma_0) < 1$ increases the left-hand side over its value for $\eta(\sigma_0) = 1$ and consequently μ increases. Accordingly to (123) this increase in μ decreases K. Thus when (131) prevails the best we can do is to have $\eta(\sigma_0) = 1$, *i.e.*, we choose $s(p) = s_0(p)$, a minimum-phase coefficient. With $b(\sigma_0)\ \Delta_n(\omega_0/\sigma_0)$ as ordinate, the value τ_0 is read off Fig. 6. From τ_0 we get μ:

$$\mu = \frac{\sigma_0\tau_0}{\omega_c}. \tag{132}$$

From μ we get K,

$$K = 1 - \left(\frac{\sigma_0\tau_0}{\omega_c}\right)^{2n}. \tag{133}$$

Moreover, since $b(\sigma_0) \leq 1$,

$$b(\sigma_0)\Delta_n\left(\frac{1}{y_0}\right) \leq \Delta_n\left(\frac{1}{y_0}\right),$$

and the equality

$$\Delta_n\left(\frac{\mu}{y_0}\right) = b(\sigma_0)\Delta_n\left(\frac{1}{y_0}\right)$$

implies

$$\Delta_n\left(\frac{\mu}{y_0}\right) \leq \Delta_n\left(\frac{1}{y_0}\right).$$

Invoking the monotonic character of $\Delta_n(\tau)$, $\mu/y_0 \leq 1/y_0$ or $\mu \leq 1$. That is, $0 \leq K \leq 1$. In other words the solution is *always* physical.

3) Substituting $s(p) = \eta(p)s_0(p)$ in (58) yields $Z_2(p)$ which is synthesized as a cascade of lossless 2 ports terminated in r_g. This concludes the design.

To illustrate the numerical details, set $R_2 = R_1 = 10^6\Omega$, $C = 1\mu uf$, $n = 2$ and $\omega_c = 2\pi \times 10^6$ rad/sec. Then, from (115) and (118),

$$\sigma_0 = \sqrt{2} \times 10^6$$

$$b_0 = \frac{\sqrt{2}-1}{\sqrt{2}+1} = 0.170$$

and

$$\frac{\omega_c}{\sigma_0} = \frac{2\pi \times 10^6}{\sqrt{2} \times 10^6} = 4.454.$$

Now $a_0 = a_2 = 1$ and $a_1 = \sqrt{2}$. Thus,

$$\Delta_2(\tau) = 1 + \sqrt{2}\tau + \tau^2.$$

Therefore,

$$\Delta_2\left(\frac{\omega_c}{\sigma_0}\right) = 27.13,$$

$$b(\sigma_0)\Delta_2\left(\frac{\omega_c}{\sigma_0}\right) = 0.170 \times 27.13 = 4.61 > 1.$$

Hence $\eta(\sigma_0) = 1$ (no right half-plane zeros in s) and τ_0 must be determined from the equation

$$\tau_0^2 + \sqrt{2}\tau_0 - 3.61 = 0.$$

Solving this quadratic we get $\tau_0 = 1.323$ and $\mu = \tau_0\sigma_0/\omega_c = 0.29$. Finally, $K = 1 - \mu^4 = 1 - (0.3)^4 = 0.9919$ and is practically unity. To find $s(p)$ we go back to (87) and (125),

$$s(p) = s_0(p) = \frac{\Delta_2\left(\frac{\mu\omega_c}{p}\right)}{\Delta_2\left(\frac{\omega_c}{p}\right)} = \frac{p^2 + \sqrt{2}\mu\omega_c p + (\mu\omega_c)^2}{p^2 + \sqrt{2}p\omega_c + \omega_c^2}$$

$$= \frac{p^2 + (2.95) \times 10^6 p + (4.41) \times 10^{12}}{p^2 + (8.85) \times 10^6 p + (39.44) \times 10^{12}}.$$

Suppose instead that $\omega_c = 2\pi \times 10^5$ rad/sec, everything else being the same. Now we have

$$\frac{\omega_c}{\sigma_0} = \frac{2\pi \times 10^5}{\sqrt{2} \times 10^6} = 0.4454,$$

$$\Delta_2\left(\frac{\omega_c}{\sigma_0}\right) = 1.833,$$

$$b(\sigma_0)\Delta_2\left(\frac{\omega_c}{\sigma_0}\right) = 0.208 < 1.$$

Thus $K = 1$ is achievable provided we use an all pass $\eta(p)$. From (130)

$$\eta(p) = \frac{p - 0.927 \times 10^6}{p + 0.927 \times 10^6}.$$

Correspondingly,

$$s(p) = \eta(p)s_0(p)$$

$$= \frac{p^2(p - 0.927 \times 10^6)}{(p + 0.927 \times 10^6)(p^2 + 8.85 \times 10^6 p + 39.44 \times 10^{10})}.$$

Loads with zeros of transmission restricted to lie exclusively in the strict right-half p-plane have another very interesting feature which is the content of Theorem 2.

Theorem 2

Suppose the load $z_l(p)$ has *only one* pair of complex zeros or one σ-axis zero of transmission in $Re\ p \geq 0$. Then it is possible to match it to a resistive generator with a *lumped* reactive equalizer to achieve a transducer power gain $G(\omega^2)$ that is *constant* over the *entire frequency spectrum*! That is,

$$G(\omega^2) = K^2, \qquad 0 \leq \omega^2,$$

K a real constant.

Proof

By hypothesis, the equation $\lambda(p) = 0$ has either one right half-plane pair of solutions, p_0 and $\bar{p}_0$, $Re\ p > 0$, $Im\ p_0 \neq 0$, or the single solution σ_0, $\sigma_0 > 0$. Since $\lambda(p) = r_l(p)/z_l(p)$, the above equation is equivalent to $r_l(p) = 0$. The complex pair of zeros corresponds to a Darlington type D load $z_l(p)$ whereas the single zero corresponds to a Darlington type C. We assume that $z_l(p)$ is type D, so that $p_0 = \sigma_0 + j\omega_0$, $\omega_0 \neq 0$, is complex. Clearly p_0 and $\bar{p}_0$ are in Class I and (46) yields

$$B_0 \equiv b(p_0) = s(p_0) \equiv S_0. \tag{134}$$

Because of the reality of $b(p)$ and $s(p)$, the restriction at the conjugate point $\bar{p}_0$ is automatically satisfied and it may be assumed, without loss of generality, that $\omega_0 > 0$. Now recall that $G(p^2) = 1 - s(-p)s(p)$. Hence $G(\omega^2) = K^2$, $0 \leq \omega^2 < \infty$ implies that

$$s(p) = \pm(1 - K^2)^{1/2}d(p) \equiv \pm cd(p), \tag{135}$$

where $d(p)$ is a regular all pass and

$$c = +(1 - K^2)^{1/2}. \tag{136}$$

Maximum K^2 implies minimum c so that the problem reduces to finding the minimum value of c subject to the limitations (134) and (135). We choose the plus sign in (135), incorporating the possible minus sign in $d(p)$.

The first observation is that when $s(p)$ has the form (135), the function

$$Z(p) \equiv \frac{c + s(p)}{c - s(p)} = \frac{1 + d(p)}{1 - d(p)} \tag{137}$$

is *Foster*. Moreover, according to (134),

$$Z(p_0) = \frac{c + b(p_0)}{c - b(p_0)}. \tag{138}$$

Set $Z(p_0) = R_0 + jX_0$. Invoking a classic result[15]

$$\frac{R_0}{\sigma_0} \geq \left|\frac{X_0}{\omega_0}\right| \tag{139}$$

with equality if and only if $Z(p) = pL$ or $1/pC$, L and C non-negative. Using (138) and setting $b(p_0) = \mu_0 + j\nu_0$,

$$R_0 = \frac{c^2 - |b(p_0)|^2}{(c - \mu_0)^2 + \nu_0^2} \tag{140}$$

and

$$X_0 = \frac{2c\nu_0}{(c - \mu_0)^2 + \nu_0^2}. \tag{141}$$

The substitution of (140) and (141) in (139) gives the inequality,

$$c^2 - 2\left|\frac{\sigma_0\nu_0}{\omega_0}\right| c - |b(p_0)|^2 \geq 0. \tag{142}$$

Obviously the smallest positive value of c consistent with (142) occurs when the equality sign is attained,

$$c_{\min} = \left|\frac{\sigma_0\nu_0}{\omega_0}\right| + \sqrt{\left(\frac{\sigma_0\nu_0}{\omega_0}\right)^2 + |b(p_0)|^2} \tag{143}$$

and

$$K_{\max}^2 = 1 - c_{\min}^2. \tag{144}$$

To find the $d(p)$ associated with $c_{\min}$, $d_{\min}(p)$, say, we must make use of the fact that equality is attainable in (142) if and only if $Z(p)$ represents either a pure inductor or a pure capacitor. If we choose $Z(p) = pL$,

$$L = \left|\frac{\sigma_0 X_0}{\omega_0}\right| = \frac{\sigma_0}{\omega_0}\cdot\frac{2c_{\min}|\nu_0|}{(c_{\min} - \mu_0)^2 + \nu_0^2} \tag{145}$$

and

$$s(p) = c_{\min}d_{\min}(p) = c_{\min}\frac{pL - 1}{pL + 1}. \tag{146}$$

If, on the other hand, we choose $Z(p) = C^{-1}/p$,

$$C^{-1} = \left|\frac{p_0^2 X_0}{\omega_0}\right| = \frac{|p_0|^2}{\omega_0}\cdot\frac{2c_{\min}|\nu_0|}{(c_{\min} - \mu_0)^2 + \nu_0^2} \tag{147}$$

and

$$s(p) = c_{\min}d_{\min}(p) = c_{\min}\cdot\frac{1 - pC}{1 + pC}. \tag{148}$$

[15] Youla, *op. cit.*, see p. 255.

Two last points are worth mentioning. First, does (143) always guarantee that $c_{\min} \leq 1$? The answer is yes. For $c_{\min} \leq 1$ is equivalent to

$$\left(\frac{\sigma_0\nu_0}{\omega_0}\right)^2 + |b(p_0)|^2 \leq 1 - 2\left|\frac{\sigma_0\nu_0}{\omega_0}\right| + \left(\frac{\sigma_0\nu_0}{\omega_0}\right)^2,$$

or

$$|b(p_0)|^2 \leq 1 - 2\left|\frac{\sigma_0\nu_0}{\omega_0}\right|. \tag{149}$$

But, $b(p)$ is known *a priori* to be a physical all pass, whence

$$z(p) \equiv \frac{1 + b(p)}{1 - b(p)}$$

is positive-real. Consequently the inequality (142) is applicable with $c = 1$ and this yields precisely (149). Secondly, (67) shows that the zeros of $R_2(p)$, the even part of the back-end impedance $Z_2(p)$, are *exactly* those of $d(p)$, since

$$\frac{1 - s(-p)s(p)}{s(-p)} = \frac{1 - c^2}{c}\,d(p).$$

Thus, the *equalizer N which matches $z_l(p)$ to a resistive generator to achieve maximum flat transducer power gain over the entire frequency range is always a (possibly degenerate) single Darlington type C section.*

In the above derivation it has been assumed that $\omega_0 > 0$. If we permit $\omega_0 \to 0$, then in the limit, $p_0 = \sigma_0$ and the load $z_l(p)$ is one possessing a *double-order* zero of transmission at σ_0. The corresponding formula are obtained by a simple limiting process. For example, consider (143) for $c_{\min}$. Obviously (as a Taylor expansion shows),

$$\operatorname*{limit}_{p_0 \to \sigma_0}\left(\frac{\nu_0}{\omega_0}\right) = b'(\sigma_0).$$

Hence, in the limit (143) goes into

$$c_{\min} = \sigma_0\,|b'(\sigma_0)| + \sqrt{[\sigma_0 b'(\sigma_0)]^2 + b^2(\sigma_0)}, \tag{143a}$$

etc.

If the load is type C, *i.e.*, $z_l(p)$ has a single zero of transmission σ_0 on the positive σ axis, it is clear from (134) that $c_{\min} = |b(\sigma_0)|$ and $s(p) = \pm c_{\min}$. In other words $s(p)$ is a constant. Hence, the even part of $Z_2(p)$ is totally devoid of zeros and *the equalizer is simply an ideal transformer*. More generally, even for type D loads, the equalizer turns out to be an ideal transformer whenever $b(p_0)$ is real. In this case, again $c_{\min} = |b(p_0)|$ and $s(p) = \pm b(p_0)$. Example 4 illustrates this possibility.

Example 3: Consider the load

$$z_l(p) = \frac{25p^2 + 4p + 1}{p^2 + 8p + 1}. \tag{150}$$

Design the equalizer which achieves the largest *flat* power gain over the *entire* frequency spectrum when working between a 1-ohm generator and $z_l(p)$.

Solution: From (150) we find that

$$r_l(p) = \frac{(5p^2 - 4p + 1)(5p^2 + 4p + 1)}{(p^2 - 8p + 1)(p^2 + 8p + 1)} \tag{151}$$

and

$$b(p) = \frac{p^2 - 8p + 1}{p^2 + 8p + 1}. \tag{152}$$

Thus $\lambda(p) = 2r_l b/z_l$ has a pair of complex zeros at

$$p_0 = \frac{2}{5} \pm \frac{j}{5}, \tag{153}$$

and $z_l(p)$ falls within the scope of Theorem 2. Using (153) and (152),

$$b(p_0) = -\frac{9}{17} - j\frac{2}{17} = \mu_0 + j\nu_0.$$

$$\therefore \quad \mu_0 = -\frac{9}{17}, \quad \nu_0 = -\frac{2}{17}.$$

From (143),

$$c_{\min} = \frac{4 + \sqrt{101}}{17}$$

and

$$K_{\max}^2 = 1 - c_{\min}^2 = 0.318.$$

Using (147), (148),

$$C^{-1} = \frac{2}{9 + \sqrt{101}},$$

$$d(p) = \frac{2 - (9 + \sqrt{101})p}{2 + (9 + \sqrt{101})p}. \tag{154}$$

Thus,

$$s(p) = \frac{4 + \sqrt{101}}{17} \cdot \frac{2 - (9 + \sqrt{101})p}{2 + (9 + \sqrt{101})p}. \tag{155}$$

The irrational terms have been retained in order to facilitate the synthesis of $Z_2(p)$. From (58),

$$Z_2(p) = \frac{(40 + 10\sqrt{101})p + 21 + \sqrt{101}}{(29 + 3\sqrt{101})p + 13 - \sqrt{101}} \tag{156}$$

and $R_2(p)$ possesses a first order zero at

$$p_0 = \sigma_0 = \frac{2}{9 + \sqrt{101}} = 0.105.$$

Thus,

$$Z_2(\sigma_0) = 4.99,$$

$$Z_2'(\sigma_0) = -16.90.$$

The indices of $Z_2(p)$ are[13]

$$I_1 = 2.103, \qquad I_3 = 0.0310$$

$$I_2 = 1.625, \qquad I_4 = I_1^{-1}$$

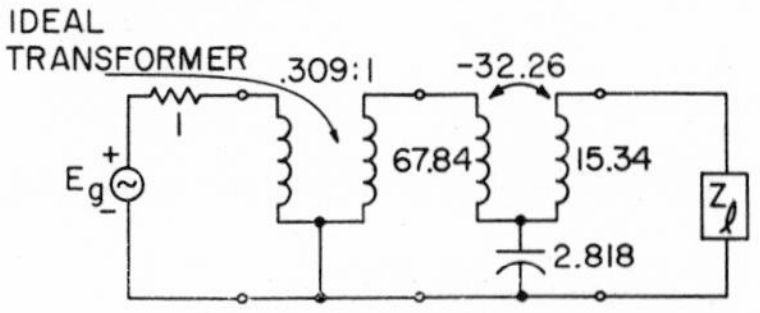

Fig. 7—Schematic of a Darlington type D load, $z_l(p)$, equalized to a truly flat transducer power gain.

and the elements of the type C equalizer section are

$$L_P = 15.34, \qquad M = -32.26$$

$$L_S = 67.84, \qquad C = 2.818.$$

The generator, equalizer, and load are shown in Fig. 7. The ideal transformer performs the usual function of changing the resistance level at the generator end.

Example 4: Equalize the load

$$z_l(p) = \frac{p^2 + 4p + 10}{p^2 + 9p + 10} \tag{147}$$

to a truly flat transducer power gain $G(\omega^2) = K^2, 0 \leq \omega^2$. Find $K_{\max}^2$ and the equalizer that achieves it.

Solution:

$$r_l(p) = \frac{(p^2 - 6p + 10)(p^2 + 6p + 10)}{(p^2 - 9p + 10)(p^2 + 9p + 10)} \tag{158}$$

and

$$b(p) = \frac{p^2 - 9p + 10}{p^2 + 9p + 10}. \tag{159}$$

Clearly, $\lambda(p) = 2r_l b/z_l$ has a pair of complex zeros at $p_0 = 3 \pm j$. The restriction is

$$B_0 \equiv b(p_0) = s(p_0) \equiv S_0.$$

By direct calculation, $b(p_0) = -1/5$ and is *real*. Hence, it suffices to choose $s(p) = -b(p_0) = 1/5$ and therefore

$$c_{\min} = |b(p_0)| = 1/5,$$

$$K_{\max}^2 = 1 - \frac{1}{25} = \frac{24}{25} = 0.96.$$

Since $Z_2(p)$ is a constant, it equals its value for $p = 0$. From (157), (158) and (35),

$$1/5 = \frac{Z_2 - 1}{Z_2 + 1}.$$

$$\therefore \quad Z_2(p) = \tfrac{3}{2}.$$

If the internal resistance of the generator is r_g, the matching network is an ideal transformer of turns ratio

$$n = \pm\sqrt{\frac{2r_g}{3}}. \tag{160}$$

The equalizer is shown in Fig. 8.

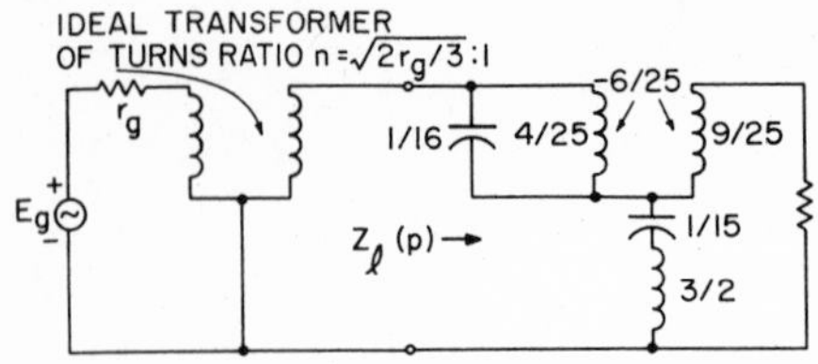

Fig. 8—Schematic of a Darlington type D load equalized to a truly flat transducer power gain by means of an ideal transformer.

Theorem 2 deals *solely* with loads devoid of zeros of transmission on $p = j\omega$ and possessing either a *single* real or one *pair* of complex zeros in $Re\ p > 0$. The natural question arises as to whether or not it is possible to equalize such a load to a flat gain (over the entire frequency spectrum), if it has an *arbitrary* number of zeros of transmission in $Re\ p > 0$. It can be shown that the answer is always in the affirmative but a proper understanding of what is involved requires an appreciation of the problem of *interpolation with positive-real functions*. This topic is important enough to merit a paper of its own and does not fall within the scope of the present work.

V. The Restrictions in Integral Form

The broad-band procedure described in Section III is, in essence, a coefficient matching technique and in general offers the simplest and most direct analytic approach to the design of an equalizer. However, it is possible to translate the restrictions (46)–(52) into an equivalent set of integral equations and integral inequalities which emphasize the real-frequency values of the transducer power gain $G(\omega^2)$. Their main contribution stems from the fact that they sometimes permit the designer to draw very rapid conclusions concerning the relative merits and potentialities of the various power shapes under consideration. Their derivation rests on a simple and basic representation theorem established in Appendix A. To enunciate the result in an intelligible manner we must first introduce some auxiliary definitions and notation.

With the understanding that $s(p)$, $b(p)$, $\eta(p)$ and $r_l(p)$ retain the meanings assigned to them in (35), (32), (56) and (57), *i.e.*,

$$s(p) = b(p)\frac{Z_2(p) - z_l(-p)}{Z_2(p) + z_l(p)}, \tag{161}$$

$$r_l(p) = \frac{z_l(p) + z_l(-p)}{2}, \tag{161a}$$

$$b(p) = \prod_{l=1}^{m} \frac{p - \nu_l}{p + \bar{\nu}_l} \tag{162}$$

and

$$\eta(p) = \prod_{l=1}^{\nu} \frac{p - \mu_l}{p + \bar{\mu}_l}, \tag{163}$$

set, for any complex number $p_0 = \sigma_0 + j\omega_0$,

$$\ln s(p) = \sum_{r=0}^{\infty} s_r(p_0)\zeta^r, \tag{161b}$$

$$\ln b(p) = \sum_{r=0}^{\infty} b_r(p_0)\zeta^r, \tag{162a}$$

$$\ln \eta(p) = \sum_{r=0}^{\infty} \eta_r(p_0)\zeta^r, \tag{163a}$$

$$\frac{1}{\pi}\cdot\frac{p}{p^2 + \omega^2} = \sum_{r=0}^{\infty} f_r(p_0, \omega)\zeta^r \tag{164}$$

$$r_l(p) = \sum_{r=0}^{\infty} R_r(p_0)\zeta^r, \tag{164a}$$

where $\zeta = p - p_0$, $|p_0| \neq \infty$, $= p^{-1}$, $|p_0| = \infty$. The $s_r(p_0)$, $b_r(p_0)$, $\eta_r(p_0)$, $f_r(p_0, \omega)$ and $R_r(p_0)$ are of course nothing more than the coefficients in the power series expansions of $\ln s(p)$, $\ln b(p)$, $\ln \eta(p)$, $1/\pi\cdot p/(p^2 + \omega^2)$ and $r_l(p)$, respectively, about the point p_0. The coefficients $b_r(p_0)$, $\eta_r(p_0)$ and $f_r(p_0, \omega)$ may be calculated explicitly and for the sake of convenience and ease of reference are presented below in tabular form.

$\boxed{\ln b(p)}$:

$$b_0(p_0) = \ln \prod_{l=1}^{m} \frac{p_0 - \nu_l}{p_0 + \bar{\nu}_l}, \tag{165}$$

$$b_r(p_0) = \frac{(-1)^{r+1}}{r} \sum_{l=1}^{m} \frac{(p_0 + \nu_l)^r - (p_0 - \nu_l)^r}{(p_0^2 - \nu_l^2)^r}, \tag{166}$$

$(r = 1, 2, \cdots, \cdots)$.

$$Re\ b_0(p_0) = \sum_{l=1}^{m} \ln \left|\frac{p_0 - \nu_l}{p_0 + \bar{\nu}_l}\right| \leq 0, \quad Re\ p_0 \geq 0, \tag{167}$$

with equality if and only if $p_0 = j\omega_0$, ω_0 real. In particular,

$$b_0(0) = j\pi, \quad m \text{ odd} \tag{168}$$

$$= 0, \quad m \text{ even} \tag{169}$$

and

$$b_0(\infty) = 0. \tag{170}$$

For $p_0 = j\omega_0$, ω_0 real,

$$b_0(j\omega_0) = \ln \prod_{l=1}^{m} \frac{j\omega_0 - \nu_l}{j\omega_0 + \bar{\nu}_l} \equiv j\theta_0(\omega_0), \tag{171}$$

$$b_r(j\omega_0) = -\frac{2j}{r} \sum_{l=1}^{m} \frac{Im\ \Gamma_l^r}{|\Gamma_l|^{2r}}, \quad 0 \neq r \text{ even}, \tag{172}$$

and

$$b_r(j\omega_0) = -\frac{2}{r} \sum_{l=1}^{m} \frac{Re\ \Gamma_l^r}{|\Gamma_l|^{2r}}, \quad r \text{ odd}, \tag{173}$$

where $0 \leq |\omega_0| < \infty$ and

$$\Gamma_l = j\omega_0 + \nu_l, \quad (l = 1, 2, \cdots, m). \tag{174}$$

In particular,

$$b_r(0) = 0, \qquad 0 \neq r \text{ even}, \tag{175}$$

$$= -\frac{2}{r}\sum_{l=1}^{m} \nu_l^{-r}, \qquad r \text{ odd}. \tag{176}$$

Observe that $b_1(j\omega_0) < 0$ and all $b_r(0)$ are real, $r \geq 1$.

When $|\omega_0| = \infty$,

$$b_r(\infty) = 0, \qquad r \text{ even} \tag{177}$$

$$= -\frac{2}{r}\sum_{l=1}^{m} \nu_l^{r}, \qquad r \text{ odd}. \tag{178}$$

Thus all $b_r(\infty)$ are real and $b_1(\infty) < 0$.

The coefficients $\eta_r(p_0)$, $(r = 0, 1, \cdots, \cdots)$, in the expansion of $\ln \eta(p)$ are obtained from the above formulas by substituting μ_l for ν_l and ν for m.

$\boxed{\frac{1}{\pi}\cdot\frac{p}{p^2+\omega^2}}$:

$$f_r(p_0, \omega) = \frac{(-1)^r}{2\pi}\cdot\frac{(p_0 + j\omega)^{r+1} + (p_0 - j\omega)^{r+1}}{(p_0^2 + \omega^2)^{r+1}}, \tag{179}$$

$(r = 0, 1, \cdots, \cdots)$. Note that

$$Re\, f_0(p_0, \omega) = \frac{\sigma_0}{\pi}\cdot\frac{|p_0|^2 + \omega^2}{|p_0^2 + \omega^2|^2} \geq 0, \qquad \sigma_0 \geq 0. \tag{180}$$

For $p_0 = j\omega_0$, ω_0 real,

$$f_r(j\omega_0, \omega) = -\frac{j^{r+1}}{2\pi}\frac{(\omega_0 + \omega)^{r+1} + (\omega_0 - \omega)^{r+1}}{(\omega_0^2 - \omega^2)^{r+1}}, \tag{181}$$

$(r = 0, 1, \cdots, \cdots)$, provided $0 \leq |\omega_0| < \infty$. Hence the f's are either all purely imaginary or purely real.

When $\omega_0 = 0$,

$$f_r(0, \omega) = -\frac{j^{r+1}}{2\pi}\left[\frac{1 + (-1)^{r+1}}{\omega^{r+1}}\right] \tag{182}$$

and therefore $f_r(0, \omega) = 0$, r even. To find the appropriate form for $f_r(j\omega_0, \omega)$ when $|\omega_0| = \infty$ it is merely necessary to expand the function

$$\frac{1}{\pi}\frac{p}{p^2 + \omega^2}$$

in a Laurent series about $p = \infty$. This yields

$$f_r(\infty, \omega) = 0, \qquad r \text{ even}, \tag{183}$$

$$= \frac{1}{\pi}(-1)^{\frac{r-1}{2}}\omega^{r-1}, \qquad r \text{ odd}. \tag{184}$$

We list the important deductions:

1) $f_r(j\omega_0, \omega)$, $0 < |\omega_0| < \infty$ has a pole of order $r + 1$ at $\omega = \pm\omega_0$.
2) $f_r(0, \omega) = 0$, r even, but has a pole of order $r + 1$ at $\omega = 0$, r odd.
3) $f_r(\infty, \omega) = 0$, r even but has a pole of order $r - 1$ at $\omega = \infty$, r odd.

The next theorem restates Theorem 1 in a form exhibiting the real-frequency behavior of the transducer power gain $G(\omega^2)$. The reader must always bear in mind that all logarithms are principal-valued; *i.e.*, for any complex number a,

$$-\pi < Im\,(\ln a) \leq \pi.$$

The statement $\ln a = \ln b$, modulo $2\pi j$, means that for some integer q, $\ln a = \ln b + 2q\pi j$ in the ordinary sense.

Theorem 3

Let $z_l(p)$ be a rational non-Foster positive-real function, $r_l(p)$ its even part and $b(p)$ the real, regular all pass formed with the right half-plane poles of $z_l(-p)$. Then, there exists a lumped, reactive, reciprocal 2-port N which equalizes $z_l(p)$ to a resistive generator to achieve the preassigned transducer power gain $G(\omega^2)$, if and only if

1) $G(\omega^2)$ is real and rational and $0 \leq G(\omega^2) \leq 1$, $\omega^2 \geq 0$.
2) There exists a real rational all-pass $\eta(p)$ of the form (163) and an integer ϵ equal to 0 or 1 such that:

a) At every Class I zero p_0 of order k (refer to Definition 1), we have modulo $2\pi j$,

$$b_0(p_0) = \epsilon\pi j + \eta_0(p_0) + \int_0^\infty f_0(p_0, \omega) \ln\,[1 - G(\omega^2)]\, d\omega \tag{185}$$

and

$$b_r(p_0) = \eta_r(p_0) + \int_0^\infty f_r(p_0, \omega) \ln\,[1 - G(\omega^2)]\, d\omega \tag{186}$$

$$(r = 1, 2, \cdots, k - 1).$$

If $G(\omega^2)$ is not identically constant over the entire frequency spectrum, ϵ must be set equal to zero. In other words, one need worry about ϵ only when dealing with *truly* flat gains. The integer ϵ makes its appearance partly because of our insistence in dealing solely with principal-valued logarithms. (See also Appendix).

b) At every Class II zero, $p_0 = j\omega_0$, of order k,

$$b_r(p_0) = \eta_r(p_0) + \int_0^\infty f_r(j\omega_0, \omega) \ln\,[1 - G(\omega^2)]\, d\omega, \tag{187}$$

$$(r = 0, 1, \cdots, k - 1),$$

and

$$\frac{b_k(j\omega_0) - s_k(j\omega_0)}{R_{k+1}(j\omega_0)} \geq 0, \tag{188}$$

with equality if and only if the equalizer is nondegenerate. If $|\omega_0| = 0$ or ∞, (188) may be replaced by

$$\frac{b_k(j\omega_0) - \eta_k(j\omega_0) - \int_0^\infty f_k(j\omega_0, \omega) \ln\,[1 - G(\omega^2)]\, d\omega}{R_{k+1}(j\omega_0)} \geq 0. \tag{189}$$

c) At every Class III zero, $p_0 = j\omega_0$, of order k,

$$b_r(j\omega_0) = \eta_r(j\omega_0) + \int_0^\infty f_r(j\omega_0, \omega) \ln [1 - G(\omega^2)] \, d\omega, \quad (190)$$

$$(r = 0, 1, \cdots, k-2),$$

and

$$\frac{b_{k-1}(j\omega_0) - s_{k-1}(j\omega_0)}{R_k(j\omega_0)} \geq 0, \quad (191)$$

with equality if and only if the equalizer is nondegenerate. If $|\omega_0| = 0$ or ∞, (191) may be replaced by

$$\frac{b_{k-1}(j\omega_0) - \eta_{k-1}(j\omega_0) - \int_0^\infty f_{k-1}(j\omega_0, \omega) \ln [1 - G(\omega^2)] \, d\omega}{R_k(j\omega_0)} \geq 0. \quad (192)$$

d) At every Class IV zero, $p_0 = j\omega_0$, of order k,

$$b_r(j\omega_0) = \eta_r(j\omega_0) + \int_0^\infty f_r(j\omega_0, \omega) \ln [1 - G(\omega^2)] \, d\omega, \quad (193)$$

$$(r = 0, 1, \cdots, k-1),$$

and

$$\frac{2R_{k-1}(j\omega_0)}{b_k(j\omega_0) - s_k(j\omega_0)} \geq a_{-1}(\omega_0), \quad (194)$$

with equality if and only if the equalizer is nondegenerate. Recall that $a_{-1}(\omega_0)$ is the residue of $z_l(p)$ at the pole $p_0 = j\omega_0$. If $|\omega_0| = 0$ or ∞, (194) may be replaced by

$$\frac{2R_{k-1}(j\omega_0)}{b_k(j\omega_0) - \eta_k(j\omega_0) - \int_0^\infty f_k(j\omega_0, \omega) \ln [1 - G(\omega^2)] \, d\omega} \geq a_{-1}(\omega_0). \quad (195)$$

The proof of the above theorem is relegated to Appendix A.

Observe that a Class I zero always contributes a set of integral equalities whereas a zero belonging to any of the three other classes always yields an additional *inequality*. *In general the inequality is not expressible in integral form unless the zero of transmission occurs either at dc* ($\omega_0 = 0$) *or* ∞ ($\omega_0 = \infty$). This hybrid behavior is, at first glance, really quite puzzling and subtle and is due to the fact that the power series expansion of $\ln |s(j\omega)|$ about either $\omega_0 = 0$ or $\omega_0 = \infty$ *contains only even powers* of the variable (see Appendix A for a full discussion). Another noteworthy point is that the integer ϵ does not appear in the restrictions for Class II, III and IV zeros because if a real-frequency zero of transmission exists $G(\omega^2)$ *cannot be truly flat!*

The chart in Fig. 9 has been compiled to summarize all the information in a compact and readily accessible manner. By referring to the chart the designer can immediately pick out the formulas pertinent to his load and specifications.

Example 5: For the same Darlington type C load considered in Example 2, determine the optimum value possible for a power shape that is flat over $0 \leq \omega < \omega_c$ (Fig. 10).

Solution: The only zero of transmission σ_0 is simple and positive. Thus there is a unique Class I restriction. In nonintegral form, $s(\sigma_0) = b(\sigma_0)$. To obtain its integral equivalent we refer to the chart and read off, for $k = 1$,

$$b_0(\sigma_0) \equiv \ln b(\sigma_0) \equiv \epsilon\pi j + \eta_0(\sigma_0) + \int_0^\infty f_0(\sigma_0, \omega) \ln [1 - G(\omega^2)] \, d\omega,$$

where (refer to Example 2)

$$b_0(\sigma_0) = \ln \frac{(1 + R_2/R_1)^{1/2} - 1}{(1 + R_2/R_1)^{1/2} + 1} = \ln b(\sigma_0).$$

Taking real parts of both sides of the above equation yields (use the chart again),

$$b_0(\sigma_0) = Re\, \eta_0(\sigma_0) + \frac{\sigma_0}{\pi} \int_0^\infty \frac{\ln [1 - G(\omega^2)] \, d\omega}{\omega^2 + \sigma_0^2}. \quad (196)$$

Now $\eta(p)$ is a regular all pass and therefore $Re\, \eta_0(\sigma_0) \leq 0$. Since the object is to make $G(\omega^2)$ as close to unity as possible, the best choice is $Re\, \eta_0(\sigma_0) = 0$, *i.e.*, $\eta(p) = 1$. In other words $s(p)$ is minimum phase (this corresponds to case 2 in Example 2).

Rewriting (196),

$$\frac{\sigma_0}{\pi} \int_0^{\omega_c} \frac{\ln [1 - G(\omega^2)] \, d\omega}{\omega^2 + \sigma_0^2} \geq b_0(\sigma_0) - \frac{\sigma_0}{\pi} \int_{\omega_c}^\infty \frac{\ln [1 - G(\omega^2)] \, d\omega}{\omega^2 + \sigma_0^2}, \quad (197)$$

with equality if and only if $\eta(p) \equiv 1$. Since $G(\omega^2) \equiv G_m$, a constant, over $0 \leq \omega^2 < \omega_c^2$, the above integrates out to read

$$\frac{1}{\pi} \arctan \frac{\omega_c}{\sigma_0} \cdot \ln (1 - G_m) \geq b_0(\sigma_0), \quad (198)$$

with equality if and only if $G(\omega^2) \equiv 1$, $\omega^2 > \omega_c^2$. In any case, note that the contribution of the second term on the right side of (197) is *always non-negative*. Simplifying (198) and choosing the equality sign, gives

$$G_m = 1 - \exp \left[\frac{\pi \ln b(\sigma_0)}{\arctan \dfrac{\omega_c}{\sigma_0}} \right], \quad (199)$$

the fundamental gain-bandwidth limitation for a power shape that is flat over the pass band. In the limit, as $\omega_c \to \infty$, $G_m \to G_\infty$:

$$G_\infty = 1 - |b(\sigma_0)|^2, \quad (200)$$

a result we already know to be correct. Eq. (199) represents the limiting value of gain for a Butterworth shape with an infinite number of poles. A direct derivation is considerably more difficult and the reader is invited to verify this for himself.

A zero of p_o of $\lambda(p)$ in Re $p \geq 0$, of order k, is a zero of transmission of $z_\ell(p)$ of order k.

Class I zero: Re $p_o > 0$	Class II zero: $p_o = j\omega_o$, $z_\ell(j\omega_o) = 0$	Class III zero: $p_o = j\omega_o$, $0 < \lvert z_\ell(j\omega_o)\rvert < \infty$	Class IV zero: $p_o = j\omega_o$, $\lvert z_\ell(j\omega_o)\rvert = \infty$.

$b_r(p_o)$

$$b_o(p_o) = \ln \prod_{\ell=1}^{m} \frac{p_o - \nu_\ell}{p_o + \bar{\nu}_\ell}$$

$$b_r(p_o) = \frac{(-1)^{r+1}}{r} \sum_{\ell=1}^{m} \frac{(p_o+\nu_\ell)^r - (p_o-\nu_\ell)^r}{(p_o^2 - \nu_\ell^2)^r}, \quad r = 1, 2, \ldots$$

$$\text{Re}\, b_o(p_o) = \sum_{\ell=1}^{m} \ln \left| \frac{p-\nu_\ell}{p+\bar{\nu}_\ell} \right| \leq 0, \ \text{Re}\, p_o \geq 0^*$$

$b_o(0) = j\pi$, m odd; $b_o(\infty) = 0$

$= 0$, m even

$$b_o(j\omega_o) = \ln \prod_{\ell=1}^{m} \frac{j\omega_o - \nu_\ell}{j\omega_o + \bar{\nu}_\ell} \equiv j\theta_o(\omega_o).$$

$$b_r(j\omega_o) = \frac{2}{jr} \sum_{\ell=1}^{m} \frac{\text{Im}\, \Gamma_\ell^r}{\lvert\Gamma_\ell\rvert^{2r}}, \quad 0 \neq r \text{ even},$$

$$= -\frac{2}{r} \sum_{\ell=1}^{m} \frac{\text{Re}\, \Gamma_\ell^r}{\lvert\Gamma_\ell\rvert^{2r}}, \quad r \text{ odd},$$

where $0 \leq \lvert\omega_o\rvert < \infty$ and

$\Gamma_\ell = j\omega_o + \nu_\ell$, $(\ell = 1, 2, \ldots, m)$.

In particular,

$$b_r(0) = -\frac{2}{r} \sum_{\ell=1}^{m} \nu_\ell^{-r}, \quad r \text{ odd}$$

$= 0$, r even.

Observe that $b_1(j\omega_o) < 0$ and $b_r(0)$ are real.

When $\lvert\omega_o\rvert = 0$

$$b_r(\infty) = -\frac{2}{r} \sum_{\ell=1}^{m} \nu_\ell^{r}, \quad r \text{ odd},$$

$= 0$, r even.

Thus all $b_r(\infty)$ are real and $b_1(\infty) < 0$.

$f_r(p_o, \omega)$

$$f_r(p_o, \omega) = \frac{(-1)^r}{2\pi} \cdot \frac{(p_o+j\omega)^{r+1} + (p_o-j\omega)^{r+1}}{(p_o^2+\omega^2)^{r+1}}.$$

$$\text{Re}\, f_o(p_o, \omega) = \frac{\sigma_o}{\pi} \cdot \frac{\lvert p_o\rvert^2 + \omega^2}{\lvert p_o^2+\omega^2\rvert^2} \geq 0, \quad \sigma_o \geq 0.$$

$$f_r(j\omega_o, \omega) = -\frac{j^{r+1}}{2\pi} \cdot \frac{(\omega_o+\omega)^{r+1} + (\omega_o-\omega)^{r+1}}{(\omega_o^2-\omega^2)^{r+1}}$$

$$0 \leq \lvert\omega_o\rvert < \infty$$

For $\omega_o = 0$,

$$f_r(0, \omega) = -\frac{j^{r+1}}{2\pi} \left[\frac{1+(-1)^{r+1}}{\omega^{r+1}} \right]$$

Hence, $f_r(0, \omega) = 0$, r even.

For $\lvert\omega_o\rvert = \infty$,

$f_r(\infty, \omega) = 0$, r even,

$$= \frac{1}{\pi} (-1)^{\frac{r-1}{2}} \omega^{r-1}, \quad r \text{ odd}.$$

$\eta_r(p_o)$

These formulas are obtained from the first column by replacing ν_ℓ with μ_ℓ and m with ν. For example,

$$\eta_o(p_o) = \ln \prod_{\ell=1}^{\nu} \frac{p_o - \mu_\ell}{p_o + \bar{\mu}_\ell},$$

$$\eta_r(p_o) = \frac{(-1)^{r+1}}{r} \sum_{\ell=1}^{\nu} \frac{(p_o+\mu_\ell)^r - (p_o-\mu_\ell)^r}{(p_o^2-\mu_\ell^2)^r},$$

etc.

The Restrictions In Integral Form

Class I: $p_o = \sigma_o + j\omega_o$, $\sigma_o > 0$, is of order k.

$$b_r(p_o) = \epsilon\pi j + \eta_r(p_o) + \int_0^\infty f_r(p_o, \omega) \ln\left[1 - G(\omega^2)\right] d\omega, \quad (r = 0, 1, \ldots, k-1).$$

$\epsilon = 0$, $r > 0$. For $r = 0$, $\epsilon = 1$ if $s(p) = -c$, $c > 0$ and zero otherwise.

Class II: $p_o = j\omega_o$ is of order k.

$$b_r(p_o) = \eta_r(p_o) + \int_0^\infty f_r(j\omega_o, \omega) \ln\left[1 - G(\omega^2)\right] d\omega, \quad (r = 0, 1, \ldots, k-1), \text{ and}$$

$$\frac{b_k(j\omega_o) - s_k(j\omega_o)}{R_{k+1}(j\omega_o)} \geq 0,$$

with equality if and only if N is non-degenerate. If $\lvert\omega_o\rvert = 0$ or ∞, this inequality may be replaced by

$$\frac{b_k(j\omega_o) - \eta_k(j\omega_o) - \int_0^\infty f_k(j\omega_o, \omega) \ln\left[1 - G(\omega^2)\right] d\omega}{R_{k+1}(j\omega_o)} \geq 0.$$

Class III: $p_o = j\omega_o$ is of order k.

$$b_r(j\omega_o) = \eta_r(j\omega_o) + \int_0^\infty f_r(j\omega_o, \omega) \ln\left[1 - G(\omega^2)\right] d\omega, \quad (r = 0, 1, \ldots, k-2),$$

and

$$\frac{b_{k-1}(j\omega_o) - s_{k-1}(j\omega_o)}{R_k(j\omega_o)} \geq 0.$$

If $\lvert\omega_o\rvert = 0$ or ∞, this inequality may be replaced by

$$\frac{b_{k-1}(j\omega_o) - \eta_{k-1}(j\omega_o) - \int_0^\infty f_{k-1}(j\omega_o, \omega) \ln\left[1 - G(\omega^2)\right] d\omega}{R_k(j\omega_o)} \geq 0$$

Class IV: $p_o = j\omega_o$ is of order k.

$$b_r(j\omega_o) = \eta_r(j\omega_o) + \int_0^\infty f_r(j\omega_o, \omega) \ln\left[1 - G(\omega^2)\right] d\omega, \quad (r = 0, 1, \ldots, k-1),$$

and

$$\frac{2R_{k-1}(j\omega_o)}{b_k(j\omega_o) - s_k(j\omega_o)} \geq a_{-1}(\omega_o).$$

If $\lvert\omega_o\rvert = 0$ or ∞, this inequality may be replaced by

$$\frac{2R_{k-1}(j\omega_o)}{b_k(j\omega_o) - \eta_k(j\omega_o) - \int_0^\infty f_k(j\omega_o, \omega) \ln\left[1 - G(\omega^2)\right] d\omega} \geq a_{-1}(\omega_o).$$

Symbols and Notation

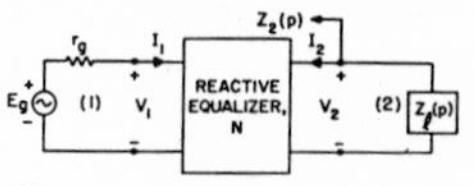

$$r_\ell(p) = \frac{z_\ell(p) + z_\ell(-p)}{2}; \quad b(p) = \prod_{\ell=1}^{m} \frac{p - \nu_\ell}{p + \bar{\nu}_\ell}; \quad f(p) = 2r_\ell(p)\, b(p)$$

$\nu_1, \nu_2, \ldots, \nu_m$ are the poles of $z_\ell(-p)$ in Re $p > 0$; $Z_2(p)$ is the back end equalizer impedance seen with E_g shorted.

$$s(p) = b(p) \cdot \frac{Z_2(p) - z_\ell(-p)}{Z_2(p) + z_\ell(p)}; \quad 1 - \lvert s(j\omega)\rvert^2 = G(\omega^2), \text{ the}$$

transducer power gain: $0 \leq G(\omega^2) \leq 1$, $\omega^2 \geq 0$.

$$b - s = \frac{2r_\ell b}{Z_2 + z_\ell}; \quad Z_2 = \frac{2r_\ell b}{b-s} - z_\ell; \quad \eta(p) = \prod_{\ell=1}^{\nu} \frac{p - \mu_\ell}{p + \bar{\mu}_\ell}, \quad \text{Re}\, \mu_\ell > 0.$$

$$\frac{Z_2(p) + Z_2(-p)}{2} = \frac{r_\ell(p)\left[1 - s(-p)\,s(p)\right]}{\left[b(p) - s(p)\right]\left[b(-p) - s(-p)\right]} \equiv r_2(p).$$

$z_\ell(j\omega) = r_\ell(\omega) + jx_\ell(\omega)$; $x'_\ell(\omega) = \dfrac{dx_\ell}{d\omega}$;

$\lambda(p) = \dfrac{r_\ell(p)}{z_\ell(p)}$; $p_o = \sigma_o + j\omega_o$; $\xi = p - p_o$.

$$a_{-1}(\omega_o) = \lim_{p \to j\omega_o} (p - j\omega_o)\, z_\ell(p),$$

$$= \lim \frac{z_\ell(p)}{p}, \quad \lvert\omega_o\rvert = \infty.$$

$$s(p) = \sum_{r=0}^{\infty} S_r (p-p_o)^r; \quad b(p) = \sum_{r=0}^{\infty} B_r (p-p_o)^r; \quad f(p) = \sum_{r=0}^{\infty} F_r (p-p_o)^r.$$

$$\ln s(p) = \sum_{r=0}^{\infty} s_r(p_o)\, \xi^r; \quad \ln b(p) = \sum_{r=0}^{\infty} b_r(p_o)\, \xi^r; \quad \ln \eta(p) = \sum_{r=0}^{\infty} \eta_r(p_o)\, \xi^r.$$

$$\frac{1}{\pi} \frac{p}{p^2+\omega^2} = \sum_{r=0}^{\infty} f_r(p_o, \omega)\, \xi^r; \quad r_\ell(p) = \sum_{r=0}^{\infty} R_r(p_o)\, \xi^r.$$

$$\beta(p) = \frac{1 - s(-p)\,s(p)}{s(-p)}.$$

$s(p) = \eta(p)\, s_o(p)$; $s_o(p)$ is minimum-phase; $\eta(p)$ is a regular all-pass.

$$\ln s(p) = \epsilon\pi j + \ln \eta(p) + \frac{2p}{\pi} \int_0^\infty \frac{\ln \lvert s(j\omega)\rvert\, d\omega}{p^2+\omega^2}, \quad \text{Re}\, p > 0,$$

modulo $2\pi j$: $\epsilon = 0$ except when $s(p) = -c\eta(p)$, $c > 0$, in which case, $\epsilon = 1$.

Fig. 9—Chart of Broad-band formulas.

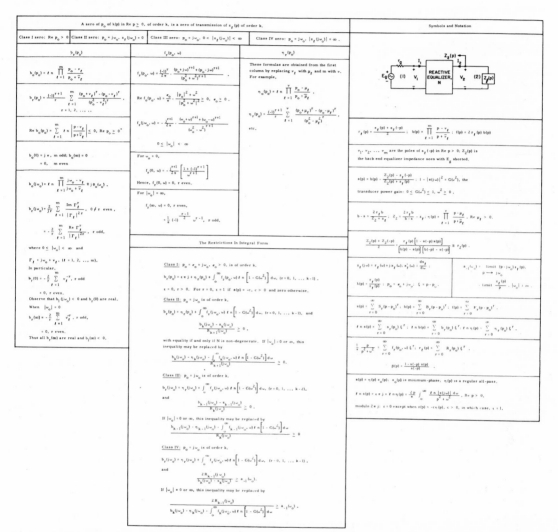

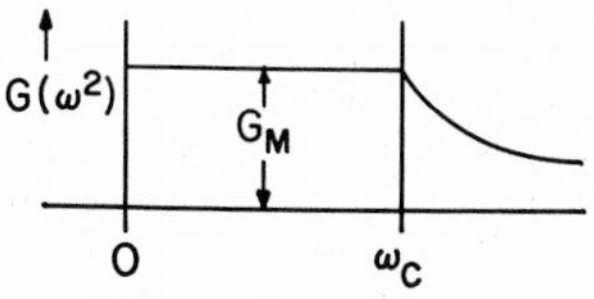

Fig. 10.

VI. Conclusions

A direct method has been presented for broad-banding an arbitrary, lumped, passive load $z_l(p)$ energized through a resistive generator. This technique eliminates the intermediate step of finding the Darlington equivalent of the impedance $z_l(p)$ and leads to simple realization procedure for the reactive equalizer. All restrictions have been exhibited in nonintegral and integral form.

Appendix A

The object of this Appendix is to supply the details necessary to establish Theorem 3.

Consider (37). Clearly,

$$\ln b(p) - \ln s(p) = -\ln\left[1 - \frac{2r_l(p)}{Z_2(p) + z_l(p)}\right]. \quad (201)$$

Let p_0 denote a zero of transmission of order k. Then, in its immediate neighborhood,

$$\frac{2r_l(p)}{Z_2(p) + z_l(p)} = 0(|p - p_0|^k), \qquad |p_0| \neq \infty$$

$$= 0(|p|^{-k}), \qquad |p_0| = \infty,$$

regardless of the choice of equalizer, provided p_0 is in either Class I, II or IV. If p_0 is in Class III, k must be replaced by $k - 1$.

Thus, in the neighborhood of p_0,

$$\ln b(p) - \ln s(p) = \frac{2r_l}{Z_2 + z_l} + \frac{1}{2}\left(\frac{2r_l}{Z_2 + z_l}\right)^2 + \cdots. \quad (202)$$

Of course we have employed the expansion

$$-\ln(1 - x) = x + \frac{x^2}{2} + \frac{x^3}{3} + \cdots, \qquad |x| < 1.$$

From (202) we read off immediately that

$$b_r(p_0) = s_r(p_0), \qquad (r = 0, 1, \cdots k - 1), \quad (203)$$

for Class I, II and IV zeros and

$$b_r(p_0) = s_r(p_0), \qquad (r = 0, 1, \cdots, k - 2), \quad (204)$$

for Class III zeros.

Conversely, suppose (203) and (204) are valid. Then,

$$\ln b(p) - \ln s(p) = 0(\zeta^k) \quad (205)$$

and

$$\ln b(p) - \ln s(p) = 0(\zeta^{k-1}), \quad (206)$$

respectively. Raising both sides of (205) and (206) to the exponential power, yields, in turn, and without ambiguity,

$$b(p) - s(p) = 0(\zeta^k) \quad (207)$$

and

$$b(p) - s(p) = 0(\zeta^{k-1}). \quad (208)$$

From (207) and (208) we recover (46), (47), (51) and (49). To summarize, (203) is *equivalent* to (46), (47) and (51) and (204) is *equivalent* to (49).

We must now find a way to express the inequalities (48), (50) and (52) in terms of logarithmic coefficients. First, let $p_0 = j\omega_0$ belong to Class II. According to (202),

$$b_k(j\omega_0) - s_k(j\omega_0) = \frac{2R_{k+1}(j\omega_0)}{X_2'(j\omega_0) + x_l'(j\omega_0)}.$$

$$\therefore \quad \frac{b_k(j\omega_0) - s_k(j\omega_0)}{R_{k+1}(j\omega_0)} = \frac{2}{X_2'(j\omega_0) + x_l'(j\omega_0)} > 0.$$

Therefore we must always have

$$\frac{b_k - s_k}{R_{k+1}} \geq 0, \quad (209)$$

with equality if and only if $Z_2(j\omega_0) \neq 0$; *i.e.*, the equalizer is nondegenerate. For a Class III zero we derive in an entirely similar manner

$$\frac{b_{k-1} - s_{k-1}}{R_k} \geq 0. \quad (210)$$

For a Class IV zero, $|z_l(j\omega_0)| = \infty$. Denote the residue of $Z_2(p)$ at this pole by $A_{-1}(\omega_0)$. Then, from (202),

$$b_k(j\omega_0) - s_k(j\omega_0) = \frac{2R_{k-1}(j\omega_0)}{A_{-1}(\omega_0) + a_{-1}(\omega_0)}.$$

$$\therefore \quad \frac{2R_{k-1}(j\omega_0)}{b_k(j\omega_0) - s_k(j\omega_0)} = A_{-1}(\omega_0) + a_{-1}(\omega_0) \geq a_{-1}(\omega_0).$$

In short, at a Class IV zero,

$$\frac{2R_{k-1}}{b_k - s_k} \geq a_{-1}(\omega_0), \quad (211)$$

with equality if and only if the equalizer is nondegenerate. For convenience all the logarithmic restrictions are collected together:

Class I:

$$p_0 = \sigma_0 + j\omega_0, \qquad \sigma_0 > 0.$$

$$b_r(p_0) = s_r(p_0), \qquad (r = 0, 1, \cdots, k - 1). \quad (212)$$

Class II:

$$p_0 = j\omega_0, \qquad \omega_0 \text{ real}.$$

$$b_r(j\omega_0) = s_r(j\omega_0), \qquad (r = 0, 1, \cdots, k - 1), \quad (213)$$

and

$$\frac{b_k(j\omega_0) - s_k(j\omega_0)}{R_{k+1}(j\omega_0)} \geq 0. \quad (214)$$

Class III:

$$p_0 = j\omega_0, \qquad \omega_0 \text{ real}.$$

$$b_r(j\omega_0) = s_r(j\omega_0), \qquad (r = 0, 1, \cdots, k - 2), \quad (215)$$

and

$$\frac{b_{k-1}(j\omega_0) - s_{k-1}(j\omega_0)}{R_k(j\omega_0)} \geq 0. \quad (216)$$

Class IV:

$$p_0 = j\omega_0, \qquad \omega_0 \text{ real}.$$

$$b_r(j\omega_0) = s_r(j\omega_0), \qquad (r = 0, 1, \cdots, k-1), \tag{217}$$

and

$$\frac{2R_{k-1}(j\omega_0)}{b_k(j\omega_0) - s_k(j\omega_0)} \geq a_{-1}(\omega_0). \tag{218}$$

In order to prove that (212)–(218) are equivalent to (46)–(52) it only remains to show that the former equations imply (48), (50) and (52). We carry this through only for a Class II zero. Thus suppose $p_0 = j\omega_0$ is a Class II zero of order k and that (213) and (214) are valid. Then the power series expansion of $\ln b(p) - \ln s(p)$ with $p_0 = j\omega_0$ as center begins with a term that is at least $0(\zeta^k)$. *Define* the function $\hat{Z}_2(p)$ by means of the equation

$$\ln b(p) - \ln s(p) = \frac{2r_l(p)}{\hat{Z}_2(p) + z_l(p)}. \tag{219}$$

Naturally, $\hat{Z}_2(p)$ is not necessarily positive-real. But in any case, it follows from (214) that if $\hat{Z}_2(p)$ does possess a zero at $p = j\omega_0$, *i.e.*, if in the neighborhood of $p = j\omega_0$,

$$Z_2(p) = M(p - j\omega_0) + 0(|p - j\omega_0|^2)$$

then,

$$M + x_l'(j\omega_0) \geq 0.$$

Taking antilogarithms of both sides of (219) yields

$$s = b\left[1 - \frac{2r_l}{\hat{Z}_2 + z_l} + \frac{1}{2!}\left(\frac{2r_l}{\hat{Z}_2 + z_l}\right)^2 - \cdots\right],$$

or, rearranging,

$$b - s = \frac{2r_l b}{\hat{Z}_2 + z_l} + \text{higher order terms}.$$

Thus,

$$\frac{B_k(j\omega_0) - S_k(j\omega_0)}{F_{k-1}(j\omega_0)} = \frac{1}{M + x_l'(j\omega_0)} \geq 0,$$

Q.E.D. In view of Theorem 1, enough material is on hand for a lemma.

Lemma 1

Let $z_l(p)$ be a rational non-Foster positive-real function. Then (58) defines a rational positive-real function $Z_2(p)$ if and only if 1) $s(p)$ is a rational bounded-real scattering coefficient and 2) the logarithmic restrictions (212)–(218) are satisfied at every zero of transmission.

To translate (212)–(218) into integral restrictions we need the following representation theorem.

Representation Theorem

Let $s(p)$ be a rational, real function of p that is *analytic and uniformly bounded* in $Re\ p \geq 0$ and let $\eta(p)$ denote the regular (normalized) all-pass constructed from its strict right half-plane zeros. Then, for any p in $Re\ p > 0$,

$$\ln s(p) = \epsilon\pi j + \ln \eta(p) + \frac{2p}{\pi}\int_0^\infty \frac{\ln |s(j\omega)|\, d\omega}{p^2 + \omega^2}, \tag{220}$$

modulo $2\pi j$, where ϵ is an integer equal to 0 or 1. This integer is always 0 unless $s(p) = -c\eta(p)$, where $c > 0$, in which case $\epsilon = +1$.[16]

Proof

Suppose first that $s(p)$ is devoid of zeros in $Re\ p > 0$; *i.e.*, $\eta(p) = 1$ and $s(p) \equiv s_0(p)$, a minimum-phase function. Then *any* determination of $\ln s_0(p)$ is analytic in $Re\ p > 0$ and has at most logarithmic singularities on $p = j\omega$. Let us agree to choose the principal value of the logarithm. Then by Cauchy's theorem,

$$-\ln s_0(p) = \frac{1}{2\pi j}\oint \frac{\ln s_0(z)}{z - p}\, dz, \tag{221}$$

the closed contour being the one shown in Fig. 11. Since $1/(z + p)$ is analytic within this contour,

$$0 = \frac{1}{2\pi j}\oint \frac{\ln s_0(z)}{z + p}\, dz. \tag{222}$$

Subtracting (222) from (221) yields

$$\ln s_0(p) = -\frac{p}{\pi j}\oint \frac{\ln s_0(z)}{z^2 - p^2}\, dz, \tag{223}$$

$Re\ p > 0$. By hypothesis, $\ln s_0(z)$ becomes infinite like $\ln R$ for large R. Hence it is easy to see that the contribution due to the integration over the semicircle tends to zero as $R \to \infty$. Therefore, going to the limit in the usual way,

$$\ln s_0(p) = \frac{p}{\pi}\int_{-\infty}^{+\infty} \frac{\ln s_0(j\omega)}{p^2 + \omega^2}\, d\omega, \tag{224}$$

$Re\ p > 0$. Writing

$$s_0(j\omega) = |s_0(j\omega)|\, e^{j\theta_0(j\omega)},$$

$$\ln s_0(j\omega) = \ln |s_0(j\omega)| + j\theta_0(j\omega). \tag{225}$$

Because of our decision to work with the principal value of $\ln s_0(p)$,

$$-\pi < \theta_0(j\omega) \leq \pi. \tag{226}$$

Since $s_0(p)$ is real for real p, $|s_0(j\omega)|$ is an *even* function of ω and with *one exception*, $\theta_0(j\omega)$ is an *odd* function. Ingnoring this exception for the moment, we can write

$$\ln s_0(p) = \frac{2p}{\pi}\int_0^\infty \frac{\ln |s_0(j\omega)|\, d\omega}{p^2 + \omega^2}, \qquad Re\ p > 0. \tag{227}$$

Note that the validity of (227) hinges on the assumption that $\theta_0(j\omega)$ is odd. To appreciate the above remark suppose $s_0(p) \equiv -1$. Then to be consistent with (226) we must choose $\theta_0(j\omega) = +\pi$ *for all* ω, an even function! The difficulty does not disappear by changing branches. For example if it is agreed that

$$-3\pi < \theta_0(j\omega) \leq 3\pi,$$

[16] This theorem is valid under much more general conditions. See R. Boas, "Entire Functions," Academic Press, Inc.; New York, N. Y.; 1954.

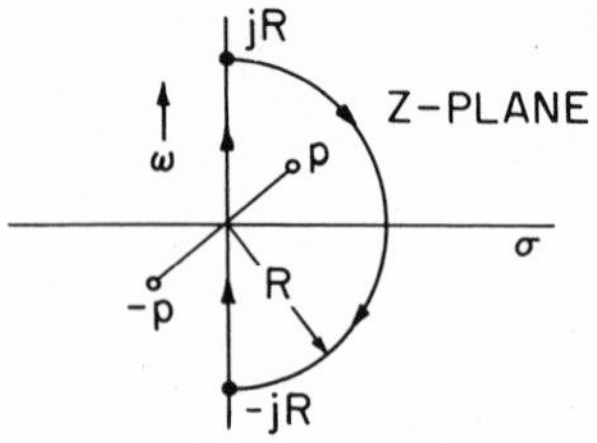

Fig. 11.

then in the previous illustration, $\theta_0(j\omega) \equiv 3\pi$, again an even function. Because Cauchy's theorem is only applicable to a determinate branch of $\ln s_0(p)$ we cannot choose different branches for different values of ω and it becomes necessary to sort out in advance the class of functions $s_0(p)$ which do not admit continuous phase functions $\theta_0(j\omega)$, $-\pi < \theta_0(j\omega) \leq \pi$, which are odd functions of ω.

Decompose $s_0(p)$ as shown below:

$$s_0(p) = \frac{s_0(p) + s_0(-p)}{2} + \frac{s_0(p) - s_0(-p)}{2}.$$

The first factor is even in p and the second is odd. Clearly,

$$\theta_0(j\omega) = \arctan\left\{\frac{Im\ s_0(j\omega)}{Re\ s_0(j\omega)}\right\}$$

and

$$\theta_0(-j\omega) = \arctan\left\{\frac{-Im\ s_0(j\omega)}{Re\ s_0(j\omega)}\right\}.$$

Thus, if $Im\ s_0(j\omega) \neq 0$, $\theta_0(j\omega) = -\theta_0(-j\omega)$. As we follow the excursion of $\theta_0(j\omega)$ it is seen that we only encounter continuity difficulties if for some ω_1, $Im\ s_0(j\omega_1) = 0$ and $Re\ s_0(j\omega_1) < 0$, because in this case both $\theta_0(j\omega_1)$ and $\theta_0(-j\omega_1)$ must be set equal to $+\pi$. In other words $\theta_0(j\omega)$ suffers a jump discontinuity at $\omega = -\omega_1$. If the set of points at which this occurs is of finite measure, the contribution of $\theta_0(j\omega)$ to the integral in (227) is no longer zero. However, if $Im\ s_0(j\omega) = 0$ on a set of positive measure, $s_0(p) \equiv s_0(-p)$. Thus $s_0(p)$ must, by Liouville's theorem, be equal to a constant. Since $Re\ s_0(j\omega) < 0$, this constant must be negative; *i.e.*, $s_0(p) = -c$, $c > 0$. In conclusion (227) is valid for all minimum-phase $s_0(p)$ except those of the type $s_0(p) = -c$, c a positive constant. Since $\ln(-c) = \pi j + \ln|c|$, it is easy to modify (227) so that it is universally correct,

$$\ln s_0(p) = \epsilon\pi j + \frac{2p}{\pi}\int_0^\infty \frac{\ln|s_0(j\omega)|}{p^2 + \omega^2}\,d\omega, \tag{228}$$

$Re\ p > 0$, where $\epsilon = 0$ except when $s_0(p) = -c$, $c > 0$, in which case $\epsilon = 1$.

To complete the proof of (220) we must remove the minimum-phase restriction on $s(p)$. Let $\eta(p)$ be the regular all pass (normalized to $+1$ at $p = \infty$) formed with the strict right half-plane zeros of $s(p)$. Then $s(p) = \eta(p)s_0(p)$, where $s_0(p)$ is minimum phase and

$$\ln s(p) = \ln \eta(p) + \ln s_0(p).$$

Using (228) and noticing that $|s(j\omega)| = |s_0(j\omega)|$, we immediately obtain (220), Q.E.D.

Proof of Theorem 3

This is accomplished by using (220) and Lemma 1. Let p_0 be a Class I zero. Then $Re\ p_0 > 0$ and equating the respective coefficients in the power series expansions of both sides of (220) we obtain immediately

$$s_0(p_0) = \epsilon\pi j + \eta_0(p_0) + \int_0^\infty f_0(p_0, \omega)\ln[1 - G(\omega^2)]\,d\omega \tag{229}$$

and

$$s_r(p_0) = \eta_r(p_0) + \int_0^\infty f_r(p_0, \omega)\ln[1 - G(\omega^2)]\,d\omega, \tag{230}$$

$(r = 1, 2, \cdots, \cdots)$. Invoking Lemma 1, we get (185) and (186).

A Class II zero, $p_0 = j\omega_0$ of order k is a zero of $r_l(p)$ of order $k + 1$ since $z_l(j\omega_0) = 0$. From (213),

$$\ln s(j\omega) = \ln b(j\omega) + 0(|\omega - \omega_0|^k). \tag{231}$$

Taking real parts of both sides of (228),

$$\ln|s(j\omega)| = \ln|b(j\omega)| + 0(|\omega - \omega_0|^k).$$

Since $|b(j\omega)| = 1$, $\ln|b(j\omega)| = 0$ and

$$\ln|s(j\omega)| = 0(|\omega - \omega_0|^k). \tag{232}$$

Thus $\ln|s(j\omega)|$ must posses a zero of at least order k at $\omega = \omega_0$. It is possible to strengthen this statement if $\omega_0 = 0$ or ∞, because in these two latter cases. the right-hand side of (232) contains only even powers. Hence, if k is an even integer, (232) stands but if k is odd,

$$\ln|s(j\omega)| = 0(\omega^{k+1}), \qquad \omega_0 = 0, \tag{233}$$

$$= 0(\omega^{-k-1}), \qquad |\omega_0| = \infty. \tag{234}$$

Similar considerations apply to Class III and IV zeros. The results are summarized below. In all cases, $p_0 = j\omega_0$ is a zero of transmission of order k.

Class II:

$$\ln|s(j\omega)| = 0(|\omega - \omega_0|^k), \qquad |\omega_0| \neq 0, \quad \infty, \tag{235}$$

$$= 0(\omega^{k+1}), \qquad \omega_0 = 0, \quad k \text{ odd},$$

$$= 0(\omega^{-k-1}), \qquad |\omega_0| = \infty, \quad k \text{ odd}. \tag{236}$$

Class III:

$$\ln|s(j\omega)| = 0(|\omega - \omega_0|^{k-1}), \qquad |\omega_0| \neq 0, \quad \infty, \tag{237}$$

$$= 0(\omega^k), \qquad \omega_0 = 0, \quad k \text{ odd}, \tag{238}$$

$$= 0(\omega^{-k}), \qquad |\omega_0| = \infty, \quad k \text{ odd}. \tag{239}$$

Class IV: Same as Class II.

Suppose that in (229) and (230) we let p_0 tend to a point $j\omega_0$ on the real-frequency axis. Formally, the resulting formulas are

$$s_0(j\omega_0) = \epsilon\pi j + \eta_0(j\omega_0) + 2\int_0^\infty f_0(j\omega_0, \omega) \ln |s(j\omega)|\, d\omega \quad (240)$$

$$s_r(j\omega_0) = \eta_r(j\omega_0) + 2\int_0^\infty f_r(j\omega_0, \omega) \ln |s(j\omega)|\, d\omega.$$

Are they valid? In the first place, $f_r(j\omega_0, \omega)$ has a pole of order $r + 1$ at $\omega = \omega_0$ if $0 < |\omega_0| < \infty$. In order that the integral in (235) be convergent, $\ln |s(j\omega)|$ must have a zero of at least the same order. Let $p_0 = j\omega_0$ be a Class II zero of transmission of order k. Then if $|\omega_0| \neq 0, \infty$, $\ln |s(j\omega)|$ has a zero of order k at $\omega = \omega_0$. Consequently, the integral certainly converges for $r = 0, 1, \cdots, k - 1$ and it is not hard to show that (235) is indeed valid for the same range of r. If $\omega_0 = 0$, and k is *odd*, $f_k(0, \omega)$ again has a pole of order $k + 1$ at $\omega = \omega_0$. But for k odd and $\omega_0 = 0$, $\ln |s(j\omega)|$ has a zero of order $k + 1$ at $\omega = \omega_0$ and the integral again converges. Lastly, if $|\omega_0| = \infty$, $f_k(\infty, \omega)$ has a pole of order $k - 1$ at $\omega = \infty$ for k odd. According to (231), $\ln |s(j\omega)|$ has a zero of order $k + 1$ at $\omega = \infty$ and therefore the integrand in (232) behaves at least like ω^{-2} for $|\omega| \to \infty$. The integral is again convergent. In short, if $p_0 = j\omega_0$ is a Class II zero of order k, (235) is valid for $r = 0, 1, \cdots, k - 1$, if $|\omega_0| \neq 0, \infty$ and for $r = 0, 1, \cdots, k$, otherwise. Class III and IV zeros are treated in the same manner. Combining these remarks with Lemma 1 yields the assertions of Theorem 3, Q. E. D.

Acknowledgment

The author wishes to take this opportunity to thank his colleague Dr. L. Smilen of the Polytechnic Institute of Brooklyn, N. Y., for acting as a willing and enthusiastic captive of one and for subjecting many of the ideas in this paper to close and careful scrutiny.

VI
Some Foundation Concepts

Editor's Comments on Papers 20 Through 23

20 Tellegen: *The Gyrator, a New Electric Network Element*

21 Tellegen: *A General Network Theorem, with Applications*

22 Raisbeck: *A Definition of Passive Linear Networks in Terms of Time and Energy*

23 Bashkow: *The A Matrix, New Network Description*

Here are grouped together four papers that do not directly relate to each other but have in common the fact that they are significant ideas that have become foundation concepts in circuit theory. The first of these by Tellegen (Paper 20), introduces the gyrator, which marked the beginning of research on nonreciprocal circuits, especially by Carlin (1955) and his colleagues. The gyrator was realized early at microwave frequencies by the use of ferrites by Hogan (1952) and approximated at lower frequencies by exploiting the Hall effect. Other new elements have since been added to the field of circuit theory—the nullator of Carlin, for example. For this pioneer achievement of Tellegen and his contributions to circuit theory in general, Tellegen was awarded the 1973 Edison Medal of the IEEE.

Another of Tellegen's many contributions is a result that has been popularized—even in sophomore-level textbooks—as Tellegen's theorem. The theorem sounds direct enough. It states that the sum of the products of branch voltages in one network and the corresponding branch currents in the same network or another topologically equivalent network sum to zero. Its many applications have been summarized in book form by Penfield, Spence, and Duinker (1970). Important applications include its use in introducing the adjoint network approach to computing network sensitivities and in computer-aided design, by Director and Rohrer (1969; see also Kuh and Rohrer 1965).

In establishing the foundations of circuit theory, considerable effort has been devoted to the characterization of an n-port network in terms of such quantities as passive and active, causality and noncausality, and stability and instability. An axiomatic time–energy approach to the characterization of n-port networks was first proposed by Raisbeck in 1954 (Paper 22), building on the approach of Brune (Paper 2). Further progress on the Raisbeck approach was given by Youla, Castriota, and Carlin in 1959 (Paper 18), who translate the criteria into the frequency domain (Paper 18; see also Carlin and Giordano, 1946).

Early in the 1960s there was a general growth of interest in the method of formulation of circuit equations which has come to be known as the state-variable approach. The first to apply state variables to circuit analysis was Bashkow in 1957 (Paper 23), followed by Bryant (1962) and others. Since that time this approach has found extensive use in both analysis and synthesis, as exemplified by the work of Dervisoglu (1964) and Pottle (1969). State variables have been used extensively in other fields of electrical engineering, such as control systems, starting with the pioneer paper of Kalman and Bertram (1960) and others.

References

Bryant, P. R. "The Explicit Form of Bashkow's *A* Matrix," *IRE Trans. Circuit Theory,* **CT-9**, 303–306 (1962).

Carlin, H. J. "Synthesis of Nonreciprocal Networks," *Proc. Symp. Modern Network Synthesis* (Polytechnic Institute of Brooklyn), **5**, 11–44 (1955).

Carlin, H. J., and A. B. Giordano. *Network Theory: An Introduction to Reciprocal and Nonreciprocal Circuits,* Prentice-Hall, Inc., Englewood Cliffs, N.J., 477 pp., 1964.

Dervisoglu, Ahmet. "Bashkow's *A* matrix for Active *RLC* Networks," *IEEE Trans. Circuit Theory,* **CT-11**, 404–406 (1964).

Director, S. W., and R. A. Rohrer. "The Generalized Adjoint Network and Network Sensitivities," *IEEE Trans. Circuit Theory,* **CT-16**, 318–323 (1969). See also S. W. Director, *Computer-Aided Design: Circuit Stimulation and Optimization,* Dowden, Hutchinson & Ross., Inc., Stroudsburg, Pa., 1974.

Hogan, C. L. "The Ferromagnetic Faraday Effect at Microwave Frequencies and Its Applications—The Microwave Gyrator," *Bell System Tech. J.,* **31**, 1–31 (1952).

Kalman, R. E., and J. E. Bertram. "Control System Analysis and Design via the 'Second Method' of Lyapunov: I. Continuous-Time Systems, II. Discrete-Time Systems," *Trans. ASME,* **D82**, 371–400 (1960).

Kuh, E. S., and R. A. Rohrer. "The State Variable Approach to Network Analysis," *Proc. IEEE,* **53**, 672–686 (1965).

Penfield, Paul, Jr., Robert Spence, and Simon Duinker. *Tellegen's Theorem and Electrical Networks,* MIT Press, Cambridge, Mass., 143 pp., 1970.

Pottle, Christopher. "A 'Textbook' Computerized State-Space Network Analysis Algorithm," *IEEE Trans. Circuit Theory,* **CT-16**, 566–568 (1969).

20

Reprinted from *Philips Res. Rept.*, **3**, 81–101 (Apr. 1948)

THE GYRATOR, A NEW ELECTRIC NETWORK ELEMENT

by B. D. H. TELLEGEN

538.55:621.392

Summary

Besides the capacitor, the resistor, the inductor, and the ideal transformer a fifth, linear, constant, passive network element is conceivable which violates the reciprocity relation and which is defined by (10). We have denoted it by the name of "ideal gyrator". By its introduction the system of network elements is completed and network synthesis is much simplified. The gyrator can be realized by means of a medium consisting of particles carrying both permanent electric and permanent magnetic dipoles or by means of a gyromagnetic effect of a ferromagnetic medium.

Résumé

A côté de la capacité, de la résistance, de l'inductance et du transformateur idéal on peut concevoir un cinquième élément de circuit, linéaire, constant et passif, qui transgresse la relation de réciprocité et qui est défini par (10). Nous l'avons appelé „gyrateur idéal". En l'introduisant, le système d'éléments de circuit est complété et la synthèse de circuits s'en trouve très simplifiée. Le gyrateur peut être réalisé à l'aide d'un milieu composé de particles portant à la fois des dipôles électriques permanents et des dipôles magnétiques permanents ou au moyen d'un effet gyromagnétique d'un milieu ferromagnétique.

1. Introduction

The physical investigation of electric phenomena led to the creation of several devices which afterwards were used in engineering for the construction of electric networks.

From electrostatics resulted the *capacitor*, whose properties are described by

$$i = C \frac{\mathrm{d}v}{\mathrm{d}t}, \qquad (1)$$

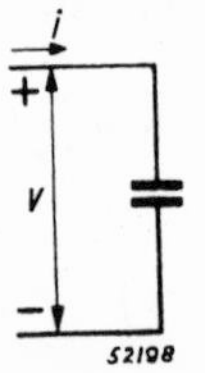

Fig. 1. Capacitor.

where i is the current and v the voltage. The coefficient C is called the *capacitance* and is always positive.

From the investigation of currents in conductors resulted the *resistor*, described by

$$v = Ri . \qquad (2)$$

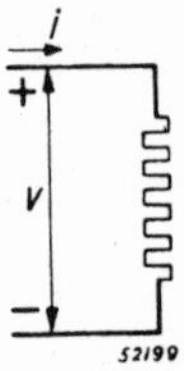

Fig. 2. Resistor.

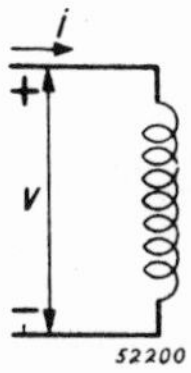

Fig. 3. Coil.

The coefficient R is called the *resistance* and is always positive.

From electromagnetism resulted the *inductor* or *coil*, described by

$$v = L \frac{di}{dt}. \tag{3}$$

The coefficient L is called the *self-inductance* and is always positive.

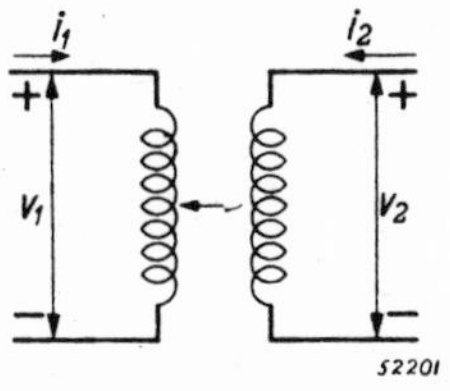

Fig. 4. Two coupled coils.

Two *coupled coils* are described by a set of two equations

$$\left.\begin{aligned} v_1 &= L_1 \frac{di_1}{dt} + M \frac{di_2}{dt}; \\ v_2 &= M \frac{di_1}{dt} + L_2 \frac{di_2}{dt}. \end{aligned}\right\} \tag{4}$$

The coefficient M is called the *mutual inductance* and is restricted by $M^2 \leqq L_1L_2$.

Further development of physics has not resulted in the creation of other similar devices and coefficients. Engineering has taken these devices into use as network elements, and we shall begin by recalling broadly what has been done with them.

1·1. The network elements

As to the *network elements* themselves, many types of capacitor, resistor and inductor are constructed for various purposes, with different values, for different currents and voltages, fixed and variable. It was soon discovered that elements whose properties are given by (1), (2), and (3) must be considered as *ideal elements*, which in practice can only be approximated; every capacitor, for instance, has some losses and some inductance, and similar remarks hold for resistors and inductors.

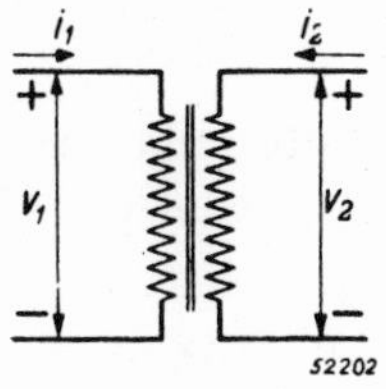

Fig. 5.
Ideal transformer.

From the system of two coupled coils a more radical development started. By making the coupling coefficient M^2/L_1L_2 as nearly as possible equal to 1, the transformer was created, whose properties in the ideal state are described by

$$\left.\begin{aligned} i_1 &= -ui_2, \\ v_2 &= uv_1. \end{aligned}\right\} \tag{5}$$

The coefficient u is called the *transformation ratio*.

For general considerations on networks we can better regard this *ideal transformer* as being the fourth network element rather than the general system of two coupled coils.

As to energy relations: an ideal capacitor and an ideal inductor can only store energy; an ideal resistor can only dissipate energy; an ideal transformer, however, can neither store nor dissipate energy but can only transfer energy, for, by (5), the absorbed power is always zero:

$$i_1 v_1 + i_2 v_2 = 0\,. \tag{6}$$

1·2. **The networks**

By connecting the four types of network element together, *networks* can be constructed which give rise to several problems. In the first place, when given voltage or current sources are connected to a given network one may ask for the voltages and currents of the various branches of the network. These problems are denoted by the name of network *analysis* and may be considered generally solved.

For engineering purposes networks are often not given, but it may be asked to construct networks with specified properties. The problems arising therefrom are denoted by the name of network *synthesis*, and we shall go into them in some detail.

To solve these problems the networks are classified in several ways. In the first place the networks are classified by the number of terminals, which we shall suppose always to be combined in pairs such that the current entering the network by one terminal of a pair is always equal to the current leaving the network by the other terminal of the same pair. Thus we distinguish between two-poles, four-poles, six-poles, ..., $2n$-poles when a network has 1, 2, 3, ..., n terminal pairs, respectively. In the second place the networks are classified as resistanceless networks and as networks with resistance. In the third place the networks are classified, according to the order of the differential equation to which they give rise, as networks of zeroth order, first order, second order, and so on.

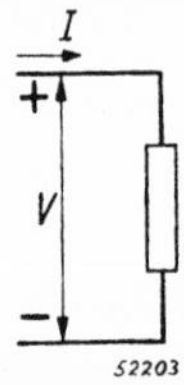

Fig. 6. Two-pole.

A two-pole is characterized by a relation between the voltage and the current of the terminals, which may be written in complex form, for instance, as

$$V = ZI\,. \tag{7}$$

The two-pole parameter Z is the impedance of the two-pole and is a function of frequency.

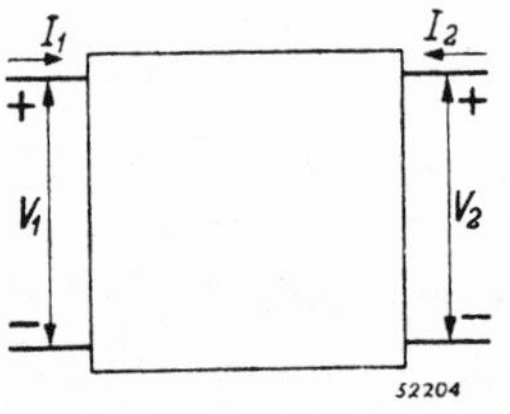

Fig. 7. Four-pole.

A four-pole is characterized by two relations between the voltages and the currents of the terminals, which may be written in complex form, for instance, as

$$\left.\begin{aligned} V_1 &= Z_{11} I_1 + Z_{12} I_2\,, \\ V_2 &= Z_{21} I_1 + Z_{22} I_2\,. \end{aligned}\right\} \tag{8}$$

The four-pole parameters Z_{11}, Z_{12}, Z_{21}, Z_{22} are functions of frequency.

A $2n$-pole is thus characterized by n similar relations between the voltages and the currents of the terminals. The corresponding parameters of two $2n$-poles with different networks may be equal functions of frequency. Such $2n$-poles are said to be *equivalent*.

The problem of network synthesis may be stated as the problem of finding the necessary and sufficient conditions for a system of functions of frequency in order that these may represent the parameters of a $2n$-pole composed of the above-mentioned four types of network element, and further as the problem of indicating for each such set of parameters at least one way to construct a corresponding network. However, when it is required to construct a $2n$-pole with specified properties, as a rule its parameters cannot be considered as given but must first be found. Now, in general, the more complicated we are prepared to make the network, the better we shall be able to satisfy given requirements. As the classification of networks by their order is essentially a classification by their complexity, the problem of network synthesis may also be stated as the problem of finding the parameters of the most general $2n$-poles of a certain order that are realizable with the help of the four types of network element, and of indicating for each such set of parameters at least one way to construct a corresponding network [1]).

The synthesis problem is solved in both senses for resistanceless two-poles, for resistanceless four-poles, and for two-poles with resistance. For four-poles with resistance it is only solved in the first sense.

The synthesis of resistanceless two-poles was accomplished by Foster [2]). The result, for the zeroth to the fourth order, is given by fig. 8. For every order there are two types of two-pole. Fig. 8 contains all those two-pole networks of these orders composed of L's and C's for which the sum of the numbers of L's and C's is equal to the order of the two-pole. For the third order both types of two-pole can be realized by two such equivalent networks, for the fourth order by four.

The synthesis of resistanceless four-poles of a certain order was accomplished by the author [1]). Apart from order zero there are four types of four-pole of odd order and five types of even order.

The synthesis of two-poles with resistance was accomplished by Brune [3]). For the 0th, 1st, 2nd, 3rd, and 4th orders there are, respectively, 1, 2, 5, 12, and 29 types of two-pole.

The synthesis of four-poles with resistance was studied by Gewertz [4]), who succeeded in finding necessary and sufficient conditions for the four-pole parameters. He did not, however, tackle the problem of finding all four-pole parameters of a certain order; so we might try to solve this problem and to construct the simplest corresponding networks. This problem is not an academic one, as four-poles are most extensively used in engineering. However, we have seen above that the synthesis both of resistanceless four-poles and of two-poles with resistance is more complicated than that of resistanceless two-poles, and we must therefore expect the synthesis of four-poles with resistance to be still more complicated.

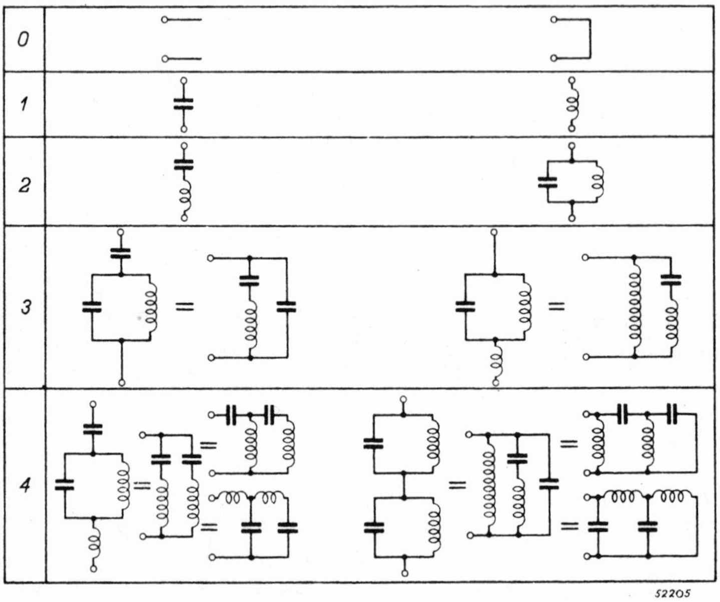

Fig. 8. Synthesis of resistanceless two-poles from the zeroth to the fourth order.

2. Statement of the problem — the ideal gyrator

For these reasons, before starting to try solving this problem, let us stop for a moment and look backwards. What are we doing as a matter of fact? Our four network elements have their origin in physics, and engineering has taken them for granted. It is worth while asking whether the system of four network elements is complete. The physicist has no reason to raise this question. He studies the phenomena of nature as they present themselves to him. For the technician, who wants to create useful systems, this question is of the utmost importance, and so we ask: *"Besides the four known network elements are other, similar elements conceivable?"*

At first this question seems rather vague: what is similar? — what is conceivable? However, the question appears to be a very definite one if we pay attention to the methods and results of network synthesis.

These are based upon the finding, first, of a number of general properties of $2n$-poles composed of the four known network elements, and then trying to construct any $2n$-pole possessing these properties by means of these elements. These properties are:

(a) the relation between the voltages and the currents of the terminals is formed by a system of ordinary *linear* differential equations, with

(b) *constant* coefficients;

(c) the $2n$-pole is *passive*, i.e., it can deliver no energy;

(d) the *reciprocity* relation.

This last property is expressed by the equality of those coefficients of the four-pole equations that lie symmetrically with respect to the principal diagonal if these equations express both voltages in both currents, or vice versa — the voltages and currents being taken positive according to fig. 7. Thus in (8) $Z_{21} = Z_{12}$, and also (4) shows an example of this form of the reciprocity relation. If, however, the current of one pair of terminals and the voltage of the other pair are expressed in the voltage of the first pair and the current of the second pair, the reciprocity relation is expressed by the opposite equality of the corresponding coefficients; for from (8) it follows that

$$\left.\begin{aligned} I_1 &= (1/Z_{11})\ V_1 - (Z_{12}/Z_{11})\ I_2\,, \\ V_2 &= (Z_{21}/Z_{11})\ V_1 + (Z_{22} - Z_{21}Z_{12}/Z_{11})\ I_2\,. \end{aligned}\right\} \tag{9}$$

Equations (5) exhibit an example of this form of the reciprocity relation.

The above-mentioned investigations on network synthesis show that any two-pole and four-pole possessing these four properties can be realized by a network composed of the four elements. It seems very unlikely that this will be otherwise for $2n$-poles with $n > 2$. Therefore, if we restrict ourselves to $2n$-poles possessing the four mentioned properties the answer to our principal question is in the negative: no other similar network elements are conceivable. If we want to extend the possibilities we must drop one or more of the four properties.

If we drop the first property, the linearity, the principle of superposition will no longer hold and the systems will become much more complicated. Consequently, we want to keep this property.

If we drop the second property, the coefficients may become functions of the time, for instance periodic functions. Frequency conversion may then arise and this also complicates the system considerably, so that we want to keep the second property too.

If we drop the third property, the system must contain some source of energy. Amplifying valves, for instance, whose properties when dealing with small alternating voltages and currents are described by linear equations with constant coefficients, need D.C. sources of energy and thus constitute elements more complicated than those passive elements considered so far. Consequently, we want to keep also the third property.

As to the fourth property, the reciprocity, however, compared with the former three properties this is of much less importance. A $2n$-pole possessing the first three properties but lacking the fourth may very well be termed similar to the $2n$-poles composed of the four normal network elements. Therefore it seems worth while to investigate what it will lead to if we maintain the first three properties but drop the fourth.

We shall need some new type of network element to realize these $2n$-poles, in particular an element violating the reciprocity relation. This requirement has no significance for a two-pole element, such as C, R and L, so we must look for a new four-pole element. The simplest types of four-pole are the resistanceless ones of order zero. These are types for which $i_1v_1 + i_2v_2 = 0$, as this expresses that energy can neither be dissipated nor stored in the four-pole. The ideal transformer, whose equations are given by (5):

$$\left.\begin{aligned} i_1 &= -u\, i_2 , \\ v_2 &= u\, v_1 , \end{aligned}\right\} \qquad (5)$$

is an example of such a four-pole satisfying the reciprocity relation. Another such four-pole, but violating the reciprocity relation, is described by

$$\left.\begin{aligned} v_1 &= -s\, i_2 , \\ v_2 &= s\, i_1 . \end{aligned}\right\} \qquad (10)$$

In fact, from (10) it follows that $i_1v_1 + i_2v_2 = 0$ whereas the coefficients in (10) are not equal, as the reciprocity relation requires, but oppositely equal.

For reasons given below we shall denote such a four-pole by the name of *ideal gyrator*. We shall consider the ideal gyrator as a fifth network element.

3. Properties of the ideal gyrator

The ideal gyrator has the property that it "gyrates" a current into a voltage, and vice versa. The coefficient s, which has the dimension of a resistance, we shall call the *gyration resistance*, whilst $1/s$ we shall call the *gyration conductance*. In circuit diagrams we shall represent the ideal gyrator by the symbol of fig. 9.

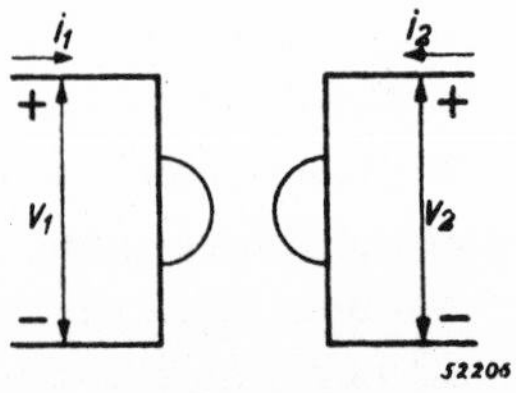

Fig. 9. Proposed symbol for the ideal gyrator.

The following properties of the ideal gyrator are easily derived from (10).

If we leave the secondary terminals open, $i_2 = 0$, the primary terminals are short-circuited, $v_1 = 0$, and vice versa. If we connect the secondary terminals by an inductance L, between the primary terminals we find a capacitance $C = L/s^2$. Conversely, if we connect the secondary terminals by a capacitance C, between the primary terminals we find an inductance $L = s^2C$. Generally, if we connect the secondary terminals by an impedance Z we find between the primary terminals an impedance s^2/Z. An impedance Z in series with or in parallel to the secondary terminals is equivalent to an impedance s^2/Z in parallel to respectively in series with the primary terminals, and vice versa (fig. 10).

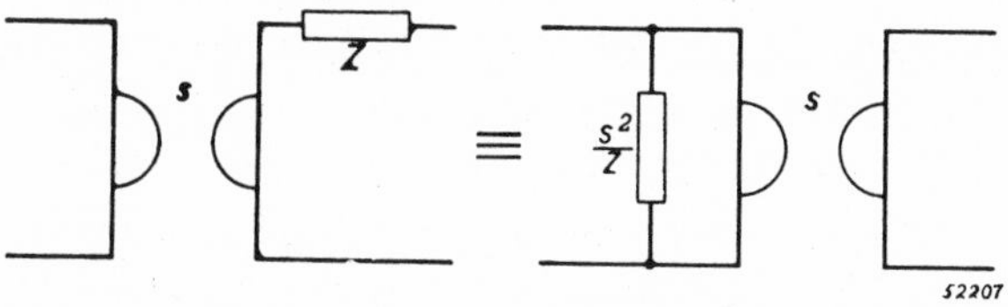

Fig. 10. An impedance in series with one pair of terminals of an ideal gyrator is equivalent to another impedance in parallel to the other pair of terminals.

Two ideal gyrators in cascade constitute an ideal transformer; an ideal gyrator and an ideal transformer in cascade constitute another ideal gyrator.

4. Networks with ideal gyrators

If ideal gyrators were available we could investigate anew all network problems arising in engineering, and since the extension of the system of four network elements to five is a relatively large one, we may expect considerably improved solutions to be possible for most network problems. As an example we may mention the system of two equal, critically coupled, tuned circuits such as are commonly used in the intermediate-frequency stages of superheterodyne radio receivers. If the circuits are coupled in an appropriate way by a gyrator and a resistance, the amplification per stage, compared with that obtained with inductively or capacitively coupled circuits under similar conditions, can be made larger by a factor $1 + \sqrt{2}$. Further details will be given in a subsequent paper.

Before investigating how an ideal gyrator might be realized or approximated, we shall first devote attention to the theory of networks that may contain ideal gyrators. As the reciprocity relation is of only subordinate importance in the methods of network analysis, this part of network theory is not much influenced by the introduction of the gyrator. Network synthesis, however, is influenced by it to a great extent and proves to be much simplified.

The synthesis of resistanceless two-poles does not change by the introduction of the gyrator. We may add that, by connecting them to an ideal gyrator, the two types of two-pole of a certain order are transformed one into the other, as mentioned above for the two-poles of the first order, the L and the C.

The synthesis of resistanceless four-poles is much simplified by the introduction of the gyrator. For every order there are two types, which can be transformed one into the other by connecting an ideal gyrator to any one of their terminal pairs.

As to the synthesis of two-poles with resistance, the addition of the gyrator does not create new possibilities, for the properties of a two-pole do not depend upon the reciprocity relation. However, the number of networks necessary for the realization of the most general two-pole of a certain order can be reduced to one by the use of ideal gyrators. We can construct this general type of two-pole of a certain order by taking any one of the two types of resistanceless four-pole of the same order and connecting any one of its two pairs of terminals by a resistance, whereby the four-pole changes into a two-pole.

The same simplicity is expected to hold for $2n$-poles. For every order of a resistanceless $2n$-pole there are probably two types, which can be transformed one into the other by connecting a gyrator to any one of their terminal pairs. For every order of a $2(n-1)$-pole with resistance there is probably one type, which can be constructed by taking any one of the two types of resistanceless $2n$-pole of the same order and connecting any one of its n pairs of terminals by a resistance.

These results of network synthesis will be fully dealt with in subsequent papers. They show how much network synthesis is simplified by the introduction of the ideal gyrator, and demonstrate that it is only by adding the ideal gyrator to the four known network elements that a complete set of elements arises.

5. Related problems in mechanics and electromechanics

Before turning to the problem of realizing the gyrator we shall give a short survey of related problems in mechanics and electromechanics.

Systems whose properties are described by a set of ordinary linear differential equations with constant coefficients were first studied in mechanics under the theory of small vibrations. A full account of these studies is given by Thomson and Tait in their "Treatise on natural philosophy". In these equations special terms may occur, called by those authors "gyroscopic" or "gyrostatic" terms, "because they occur when fly-wheels each given in a state of rapid rotation form part of the system by being mounted on frictionless bearings connected throug framework with other parts of the system; and because they occur when the motion considered is motion of the given system relatively to a rigid body revolving with a constrainedly constant angular velocity round a fixed axis" [5]).

As an example let us consider a system described by

$$\left.\begin{aligned} a_1 \frac{d^2\psi_1}{dt^2} + c_1\psi_1 - b\frac{d\psi_2}{dt} &= \Psi_1, \\ a_2 \frac{d^2\psi_2}{dt^2} + c_2\psi_2 + b\frac{d\psi_1}{dt} &= \Psi_2. \end{aligned}\right\} \tag{11}$$

In these equations ψ_1 and ψ_2 denote two general coordinates of the system, Ψ_1 and Ψ_2 the corresponding general forces; a_1 and a_2 are the general masses, c_1 and c_2 the general stiffnesses. The terms $-b\, d\psi_2/dt$ and $b\, d\psi_1/dt$, whose coefficients are oppositely equal, are called gyrostatic terms.

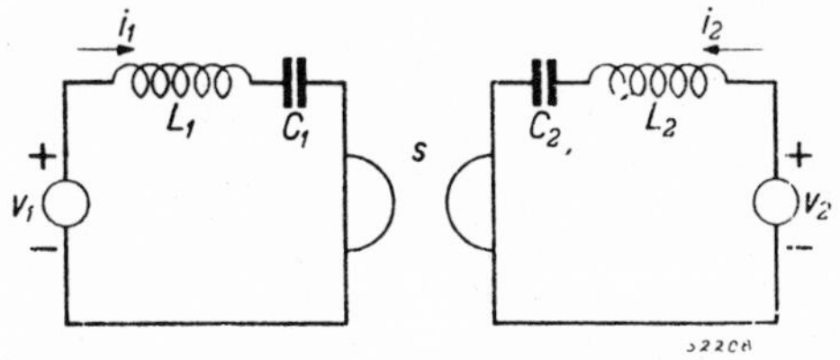

Fig. 11. Two circuits coupled by an ideal gyrator.

Let us compare with such a mechanical system the electrical system of fig. 11 in which two tuned circuits, L_1-C_1 and L_2-C_2, incorporating e.m.f.'s v_1 and v_2, respectively, are coupled by an ideal gyrator with gyration resistance s. The equations of this system are

$$\left.\begin{aligned} L_1 \frac{di_1}{dt} + \frac{1}{C_1}\int i_1 dt - s\, i_2 &= v_1, \\ L_2 \frac{di_2}{dt} + \frac{1}{C_2}\int i_2 dt + s\, i_1 &= v_2. \end{aligned}\right\} \tag{12}$$

If we put $i_1 = dQ_1/dt$ and $i_2 = dQ_2/dt$, we come to

$$\left.\begin{aligned} L_1 \frac{d^2Q_1}{dt^2} + \frac{1}{C_1} Q_1 - s\frac{dQ_2}{dt} &= v_1, \\ L_2 \frac{d^2Q_2}{dt^2} + \frac{1}{C_2} Q_2 + s\frac{dQ_1}{dt} &= v_2, \end{aligned}\right\} \tag{13}$$

which is of quite the same form as (11). The gyrostatic terms of (11) correspond to the terms of (13) arising from the gyrator, and it is because of this correspondence that we have chosen the name of gyrator for the new network element. Recently Bloch [6]) devised an ideal gyroscopic coupler which is the exact mechanical equivalent of our gyrator.

Another field where gyrostatic terms arise is in the theory of electromechanical transducers. Poincaré, in his theory of the telephone receiver [7] [8]), deduced equations that in complex notation are of the form

$$\left.\begin{aligned} V &= Z_e\, I - A\, W\,, \\ F &= A\, I + Z_m\, W\,. \end{aligned}\right\} \qquad (14)$$

V, F, I, W are, respectively, the (complex) voltage, force, current, velocity; Z_e is the (complex) electrical and Z_m the (complex) mechanical impedance; $-AW$ and AI are the coupling terms, in which A is real.

We can derive from (14) the equations of a corresponding mechanical system by replacing the voltage by a force and the current by a velocity. The coupling terms then become a pair of gyrostatic terms, and so in equation (14) they are also often called gyrostatic terms.

Equations of the form (14) occur in the theory of magnetic or moving-coil transducers. In the theory of electrostatic or piezo-electric transducers the coefficients of the coupling terms have equal signs [8]). Thus, if a magnetic and an electrostatic transducer are connected in cascade, so that by the former electrical oscillations are transduced into mechanical oscillations and by the latter the mechanical oscillations are again transduced into electrical oscillations, the resultant four-pole will violate the reciprocity relation of ordinary networks. This was explicitly stated by Jefferson [9]) and by Mc. Millan [10]).

So we could try to approximate the ideal gyrator by some special electromechanical apparatus. However, if we want to develop a gyrator that can also be used in high-frequency networks we must remember that high-frequency oscillations in electromechanical systems are only possible by the use of resonance (such as in piezo-quartz oscillators), and so we could at most arrive at a gyrator for a very narrow frequency range, contrary to the ideal gyrator, which has an unlimited frequency range. Our object must therefore be to approximate the ideal gyrator by other than electromechanical means.

6. The origin of the reciprocity relation

As the gyrator violates the reciprocity relation of ordinary networks we begin by recalling the origin of this relation, in the expectation that this will show us how to dispense with it *).

Let us first consider two insulated conductors 1 and 2 with potentials v_1 and v_2 and charges Q_1 and Q_2. The charges are linear functions of the potentials of the form

$$\left.\begin{aligned} Q_1 &= C_{11}\, v_1 + C_{12}\, v_2\,, \\ Q_2 &= C_{21}\, v_1 + C_{22}\, v_2\,. \end{aligned}\right\} \qquad (15)$$

*) As the ideal gyrator has no dissipation we restrict our investigation to systems without dissipation. The reciprocity relation of systems with dissipation has been studied by Onsager [11]).

To change the charges by amounts dQ_1 and dQ_2 there must be supplied an energy

$$v_1\,dQ_1 + v_2\,dQ_2 = (C_{11}\,v_1 + C_{21}\,v_2)\,dv_1 + (C_{12}\,v_1 + C_{22}\,v_2)\,dv_2\,. \tag{16}$$

As this must be a total differential dU,

$$\left.\begin{aligned} C_{11}\,v_1 + C_{21}\,v_2 &= \partial U/\partial v_1\,, \\ C_{12}\,v_1 + C_{22}\,v_2 &= \partial U/\partial v_2\,, \end{aligned}\right\} \tag{17}$$

from which we arrive at

$$C_{21} = \frac{\partial^2 U}{\partial v_1 \partial v_2} = \frac{\partial^2 U}{\partial v_2 \partial v_1} = C_{12}\,, \tag{18}$$

the reciprocity relation of electrostatics.

We may generalize this result (Ehrenfest [12]).

There are many more pairs of quantities in physics, such as Q and v, whose product is an energy; e.g., magnetic flux Φ and current i, mechanical displacement s and force K, turning angle φ and moment M, entropy S and temperature T. If a pair of such quantities characterizing the state of a system are slightly changed, the energy supplied to the system is equal to one of these quantities multiplied by the increment of the other quantity:

$$v\mathrm{d}Q,\ i\mathrm{d}\Phi,\ K\mathrm{d}s,\ M\mathrm{d}\varphi,\ T\mathrm{d}S.$$

The quantities are thereby divided into two classes: v, i, K, M, T belong to one class of quantities and Q, Φ, s, φ, S belong to the other class.

Suppose now, we have a system whose state is characterized by two pairs of such quantities, denoted by x_1, y_1 and x_2, y_2, and let a linear relationship exist between them such as

$$\left.\begin{aligned} y_1 &= a_{11}\,x_1 + a_{12}\,x_2\,, \\ y_2 &= a_{21}\,x_1 + a_{22}\,x_2\,. \end{aligned}\right\} \tag{19}$$

Then, if x_1 and x_2 belong to one class of quantities and therefore y_1 and y_2 to the other class, by a reasoning similar to that in the electrostatic case we may deduce $a_{12} = a_{21}$. On the other hand, if x_1 and x_2 belong to different classes of quantities we arrive at $a_{12} = -a_{21}$.

7. The violation of the reciprocity relation

The network equations, such as (8), are usually written with voltages and currents as variables. Now, these do not form a pair of quantities such as those mentioned above, for their product is not an energy but a power. We can change from the former to the latter pairs of quantities by differentiation with respect to the time. From the electrostatic equations (15) we thus come to

$$\left.\begin{aligned} i_1 &= C_{11}\frac{dv_1}{dt} + C_{12}\frac{dv_2}{dt}, \\ i_2 &= C_{21}\frac{dv_1}{dt} + C_{22}\frac{dv_2}{dt}. \end{aligned}\right\} \tag{20}$$

By comparing this with what has been written in section 2 we see that $C_{12} = C_{21}$ corresponds to the reciprocity relation of networks.

If we start from the equations of two coupled coils

$$\left.\begin{aligned} \Phi_1 &= L_{11}\, i_1 + L_{12}\, i_2, \\ \Phi_2 &= L_{21}\, i_1 + L_{22}\, i_2, \end{aligned}\right\} \tag{21}$$

where the i's are the currents and the Φ's are the fluxes through the coils, we arrive at $L_{12} = L_{21}$. By differentiation we come to

$$\left.\begin{aligned} v_1 &= L_{11}\frac{di_1}{dt} + L_{12}\frac{di_2}{dt}, \\ v_2 &= L_{21}\frac{di_1}{dt} + L_{22}\frac{di_2}{dt}. \end{aligned}\right\} \tag{22}$$

Here, too, the relation $L_{12} = L_{21}$ corresponds to the reciprocity relation of networks.

To come to a violation of the reciprocity relation of networks we must start from a system characterized by a pair of quantities Q_1, v_1 and a pair of quantities Φ_2, i_2. The equations will then take the form

$$\left.\begin{aligned} Q_1 &= C\, v_1 + A\, i_2, \\ \Phi_2 &= A\, v_1 + L\, i_2, \end{aligned}\right\} \tag{23}$$

where the two coefficients A will be equal. By differentiation we now arrive at

$$\left.\begin{aligned} i_1 &= C\frac{dv_1}{dt} + A\frac{di_2}{dt}, \\ v_2 &= A\frac{dv_1}{dt} + L\frac{di_2}{dt}, \end{aligned}\right\} \tag{24}$$

which, according to section 2, violates the reciprocity relation of networks if $A \neq 0$.

The equations (24) bear much resemblance to the equations (4). The energy of the system is equal to

$$U = \int i_1 v_1\, dt + \int i_2 v_2\, dt = \tfrac{1}{2} C\, v_1^2 + A\, v_1 i_2 + \tfrac{1}{2} L\, i_2^2, \tag{25}$$

if we take this energy as zero when $v_1 = 0$ and $i_2 = 0$. As this energy can never become negative in a passive system, C and L must be positive and A is restricted by $A^2 \leqq CL$.

In the same way as we can approximate the ideal transformer from the system described by (4), we can approximate the ideal gyrator from a system described by (24) by making the "coupling coefficient" A^2/CL as nearly as possible equal to 1.

8. The realization of the gyrator

Our object, therefore, is to devise a system described by (23) with a coupling coefficient as nearly as possible equal to 1. To investigate how this can be done we first show how such a system could be constructed if a medium were available characterized by relations between the field vectors of the type

$$\left.\begin{aligned} \mathbf{D} &= \varepsilon\,\mathbf{E} + \gamma\,\mathbf{H}\,, \\ \mathbf{B} &= \gamma\,\mathbf{E} + \mu\,\mathbf{H}\,, \end{aligned}\right\} \qquad (26)$$

and with $\gamma^2/\varepsilon\mu$ nearly equal to 1.

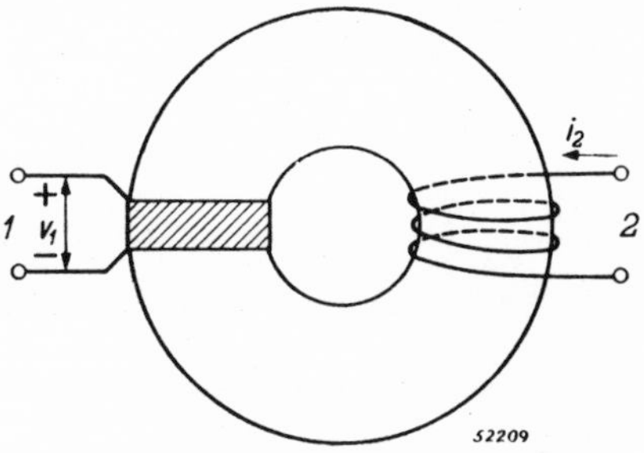

Fig. 12. A construction of the gyrator.

Let us consider the system of fig.12. This consists of two flat parallel electrodes the space between which is filled with a medium described by (26). Besides, there is a yoke of magnetic material, having a very large permeability, on which a coil is wound. The electrodes constitute the terminals 1 of the system and the coil terminals constitute the terminals 2.

Let the area of the electrodes and of the cross-section of the yoke be S, then the charge Q_1 on the electrodes will be *)

$$Q_1 = S\,D = \varepsilon S\,E + \gamma S\,H \qquad (27)$$

and the flux Φ_2 through the coil will be

$$\Phi_2 = nS\,B = \gamma nS\,E + \mu nS\,H\,, \qquad (28)$$

where n is the number of turns of the coil.
The field vectors in the medium will all be perpendicular to the electrodes.

Let the distance between the electrodes be l, then the voltage v_1 between the electrodes will be

$$v_1 = lE \qquad (29)$$

and the current i_2 through the coil will be

$$i_2 = H\,l/n\,. \qquad (30)$$

*) We use the rationalized Giorgi system of units.

Putting (29) and (30) in (27) and (28) we get

$$\left.\begin{aligned} Q_1 &= \frac{\varepsilon S}{l} \cdot v_1 + \frac{\gamma n S}{l} \cdot i_2 , \\ \Phi_2 &= \frac{\gamma n S}{l} \cdot v_1 + \frac{\mu n^2 S}{l} \cdot i_2 . \end{aligned}\right\} \tag{31}$$

This corresponds to (23) and shows that $A^2/CL = \gamma^2/\varepsilon\mu$. Therefore, if $\gamma^2/\varepsilon\mu$ is nearly equal to 1, the same holds for A^2/CL.

8·1. The medium

To investigate media as described by (26) we first introduce the electric polarization **P** and the magnetic polarization **J** by putting

$$\mathbf{D} = \varepsilon_0 \mathbf{E} + \mathbf{P} \quad \text{and} \quad \mathbf{B} = \mu_0 \mathbf{H} + \mathbf{J}$$

and further putting $\varepsilon - \varepsilon_0 = \varkappa$ and $\mu - \mu_0 = \chi$.
By substituting this in (26) we get

$$\left.\begin{aligned} \mathbf{P} &= \varkappa \mathbf{E} + \gamma \mathbf{H} , \\ \mathbf{J} &= \gamma \mathbf{E} + \chi \mathbf{H} . \end{aligned}\right\} \tag{32}$$

Now, what does this represent? The coefficients $\varkappa$ and χ are ordinary electric and magnetic susceptibilities, the coefficient γ is something new. The term $\gamma\mathbf{E}$ expresses that when the medium is exposed to an electric field it will become magnetically polarized; the term $\gamma\mathbf{H}$ expresses that when the medium is exposed to a magnetic field it will become electrically polarized.

How can we get a medium with such properties? When a material medium is placed in a field, polarization of the medium may result from two different causes. Firstly, the elements of the medium, e.g. the molecules or the atoms, can acquire a dipole moment by the action of the field, i.e., they are polarizable; and, secondly, the elements of the medium can be permanently polarized and are oriented by the action of the field.

It is difficult to imagine how magnetic and electric polarizations due to the first cause can be coupled to each other in such a way that a medium with a coefficient γ results. When we consider polarization due to the second cause, however, we see immediately that our purpose can be attained if the elements of the medium bear both permanent electric and permanent magnetic dipoles, and if in all elements these dipoles are parallel or are anti-parallel. If we place such a medium in an electric field the electric dipoles will be oriented and thus the magnetic dipoles will be oriented at the same time, and the same will take place if the medium is placed in a magnetic field. The magnetic and the electric polarizations will then be proportional to each other. This means that **P/J** will be independent of **E** and **H**, from which we deduce by (23) that $\varkappa/\gamma = \gamma/\chi$ or $\gamma^2/\varkappa\chi = 1$. If, besides bearing permanent dipoles, the elements of the medium have some polarizability, $\varkappa$ and χ will in general become larger, so that $\gamma^2/\varkappa\chi \leqq 1$.

As mentioned above, $\gamma^2/\varepsilon\mu$ should be nearly equal to 1. Now

$$\frac{\gamma^2}{\varepsilon\mu} = \frac{\gamma^2}{(\varepsilon_0 + \varkappa)(\mu_0 + \chi)} \leqq \frac{\gamma^2}{\varkappa\chi} \leqq 1.$$

So we see that to reach this end we must make $\varkappa \gg \varepsilon_0$ and $\chi \gg \mu_0$, whilst these large susceptibilities must exclusively result from the orientation of permanent dipoles.

Media with large permanent, orientable, electric dipoles exist in abundance, e.g. water, which maintains its great electric susceptibility up to very high frequencies.

Media with permanent, orientable, magnetic dipoles are the paramagnetic media, whose magnetic susceptibility, however, is small compared with μ_0 and thus does not comply with the requirements just mentioned. In ferromagnetic media the susceptibility is due to the orientation of the electron spins, but it is hard to imagine how these could be directly coupled to a permanent electric dipole. We can, however, imagine a medium consisting of small ferromagnetic particles with permanent moments, small permanent magnets, suspended in some appropriate liquid.

We made some preliminary experiments in this direction by grinding some magnet steel and sifting out the smallest particles, which under a microscope were found to have dimensions of about one micron. From this a stable suspension was made which was put into a test-tube carrying a coil, and the self-inductance of the coil was measured. Then the suspension was magnetized by pouring it between the poles of a permanent magnet, after which it was again put into the test-tube. The self-inductance was increased by a factor 1·3. However, it is rather difficult to achieve a suspension that is stable during a long time, and one calls for very small magnetized particles.

9. Another realization of the gyrator

As experiments in this direction seem to be rather difficult it was considered worth while first to investigate whether there are conceivable fundamentally different possibilities of realizing the gyrator. Since the ideal transformer shows some analogy to the ideal gyrator we may include this in these investigations.

From the point of view of network theory the ideal transformer and the ideal gyrator are network elements defined by the equations (5) and (10), irrespective of any method of realizing them physically. To investigate how this could be done we first turn our attention to the terminals. Any terminal pair consists of two terminals situated closely together, so that even at very high frequencies we can speak of the voltage between the terminals. From such a terminal pair there start two wires of well-conducting material connecting the terminals with the element proper. Within the element these wires may be separated or connected. If the wires are separated we can deform the ends of them into two parallel electrodes. If we supply a voltage to the terminals, an electric charge Q will flow to the electrodes, and thus a current $i = \mathrm{d}Q/\mathrm{d}t$ will arise. If the wires are connected we can deform them into a coil. If we supply a current to the terminals, a magnetic flux Φ will flow through the coil, and thus a voltage $v = \mathrm{d}\Phi/\mathrm{d}t$ will arise. Therefore, we shall refer to the first type of terminal pair as the electric type and to the second as the magnetic type. According to the types of terminal pair of a four-pole element we shall refer to it as a double-electric, a double-magnetic, or an electromagnetic type. We could try to realize the transformer and the gyrator by means of each of these three types.

9·1. The general medium

Between the electrodes of an electric pair of terminals and within the coil of a magnetic pair of terminals we may now introduce some material medium. The type of four-pole that then results will depend to a large extent upon the properties of the medium. Therefore we ask what are the most general, linear relations by which the electric and magnetic properties of a medium can be phenomenologically described. These relations will consist of relations between the electric and magnetic polarizations **P** and **J** and the electric and magnetic field strengths **E** and **H** through which these polarizations may arise.

An example of relations of this kind is presented by the equations (32), but these do not constitute the most general form these relations can take. In the first place the medium may be anisotropic. For such a medium the equations must take a form wherein the three components of **P** and the three components of **J** along the directions of some rectangular system of coordinates are expressed in the three components of **E** and the three components of **H** along the same directions. In the second place we can suppose all components to vary sinusoidally with time with the same angular frequency ω. Then the components can be represented by complex quantities whereby the amplitude and the phase of each component can be expressed. We thus arrive at a system of six linear, homogeneous equations between twelve complex quantities and with complex coefficients.

As the ideal transformer and the ideal gyrator have no dissipation, we shall confine our conderations to media without dissipation. For such a medium some properties of the above-mentioned coefficients are deduced in the appendix. The equations for a medium without dissipation may thus be written in full as:

$$
\begin{aligned}
P_x &= \varkappa_{xx}E_x + (\varkappa_{xy}+j\lambda_{xy})E_y + (\varkappa_{zx}+j\lambda_{zx})E_z + (\gamma_{xx}+j\delta_{xx})H_x + (\gamma_{xy}+j\delta_{xy})H_y + (\gamma_{xz}+j\delta_{xz})H_z,\\
P_y &= (\varkappa_{xy}-j\lambda_{xy})E_x + \varkappa_{yy}E_y + (\varkappa_{yz}+j\lambda_{yz})E_z + (\gamma_{yx}+j\delta_{yx})H_x + (\gamma_{yy}+j\delta_{yy})H_y + (\gamma_{yz}+j\delta_{yz})H_z,\\
P_z &= (\varkappa_{zx}-j\lambda_{zx})E_x + (\varkappa_{yz}-j\lambda_{yz})E_y + \varkappa_{zz}E_z + (\gamma_{zx}+j\delta_{zx})H_x + (\gamma_{zy}+j\delta_{zy})H_y + (\gamma_{zz}+j\delta_{zz})H_z,\\
J_x &= (\gamma_{xx}-j\delta_{xx})E_x + (\gamma_{yx}-j\delta_{yx})E_y + (\gamma_{zx}-j\delta_{zx})E_z + \chi_{xx}H_x + (\chi_{xy}+j\zeta_{xy})H_y + (\chi_{zx}+j\zeta_{zx})H_z,\\
J_y &= (\gamma_{xy}-j\delta_{xy})E_x + (\gamma_{yy}-j\delta_{yy})E_y + (\gamma_{zy}-j\delta_{zy})E_z + (\chi_{xy}-j\zeta_{xy})H_x + \chi_{yy}H_y + (\chi_{yz}+j\zeta_{yz})H_z,\\
J_z &= (\gamma_{xz}-j\delta_{xz})E_x + (\gamma_{yz}-j\delta_{yz})E_y + (\gamma_{zz}-j\delta_{zz})E_z + (\chi_{zx}-j\zeta_{zx})H_x + (\chi_{yz}-j\zeta_{yz})H_y + \chi_{zz}H_z.
\end{aligned} \tag{33}
$$

The coefficients in the principal diagonal are real; the coefficients lying symmetrically with respect to this diagonal are conjugate complex.

The coefficients consist of six groups of quantities, denoted by $\varkappa$, λ, γ, δ, χ, ζ, representing six different properties of the medium. If the terminal pairs of a four-pole are coupled to each other by means of one of these properties of the medium constituting the four-pole, this four-pole will be of a certain type.

In the first place this four-pole will or will not satisfy the reciprocity relation. To investigate this we differentiate the equations (33) with respect to t. The left-hand sides will then become $\mathrm{d}P_x/\mathrm{d}t$, etc. and $\mathrm{d}J_x/\mathrm{d}t$, etc., and the right-hand sides we may multiply by $j\omega$. Now $\mathrm{d}P/\mathrm{d}t$ is a part of a current, $\mathrm{d}J/\mathrm{d}t$ is a part of a voltage, E is a part of a voltage, and H is a part of a current. So, bearing in mind what has been said in section 2 about the way the reciprocity relation is expressed by equality or opposite equality of certain four-pole coefficients, we see that those four-poles of which the terminal pairs are coupled to each other by means of the property of the medium represented by $\varkappa$, δ, or χ, respectively by λ, γ, or ζ, will, respectively will not, satisfy the reciprocity relation of networks.

Furthermore, P and E are related to electric pairs of terminals and J and H to magnetic pairs. Therefore, if a transformer or a gyrator could be realized by coupling two terminal pairs by means of one of the above-mentioned six properties of a medium, coupling by

$\varkappa$ could lead only to a double-electric transformer,
λ to a double-electric gyrator,
γ to an electromagnetic gyrator,
δ to an electromagnetic transformer,
χ to a double-magnetic transformer,
ζ to a double-magnetic gyrator.

Thus we see that both for the transformer and for the gyrator there are three fundamentally different ways in which we could try to realize them.

Let us examine the properties of the medium more closely. The $\varkappa$'s represent the electric susceptibility, which in the general, anisotropic case has the form of a symmetric tensor, characterized by six components. Likewise, the χ's represent the magnetic susceptibility. The γ's are the generalizations of the coefficient γ discussed in section 8·1, where we pointed out the experimental difficulties to be overcome in realizing a medium with γ-properties. The λ's and ζ's show two other ways of realizing the gyrator.

As previously mentioned, the λ's can only lead to a double-electric gyrator. Double-electric elements depend on the properties of the electric field. Now the electric field, because of curl $\mathbf{E} = 0$, has limited possibilities: a voltage step-up is impossible in a dielectric. As a consequence the only ideal transformer realizable by means of a dielectric is a transformer with a one-to-one ratio, which hardly deserves the name of transformer. As similar limitations may be expected when trying to realize an ideal gyrator by means of a dielectric with a λ-property, we shall not go into the questions what this λ-property represents and how a medium with such a property could be achieved.

Thus we are left with the problem of investigating the ζ-property of a medium and the realization of a gyrator by means of it. To investigate the ζ-property we suppose the medium to be characterized by χ's and ζ's only, the $\varkappa$'s, λ's, γ's, and δ's being zero. The equations (33) then reduce to three, expressing the components of $\mathbf{J}$ in the components of $\mathbf{H}$. Instead of studying these equations it is simpler to study the inverse equations by which the components of $\mathbf{H}$ are expressed in the components of $\mathbf{J}$. If in these equations we replace j by $(1/\omega)\mathrm{d}/\mathrm{d}t$ we arrive at equations of the form

$$\left.\begin{aligned}
H_x &= \xi_{xx} J_x + (\xi_{xy} - \eta_z\,\mathrm{d}/\mathrm{d}t)\,J_y + (\xi_{zx} + \eta_y\,\mathrm{d}/\mathrm{d}t)\,J_z\,,\\
H_y &= (\xi_{xy} + \eta_z\,\mathrm{d}/\mathrm{d}t)\,J_x + \xi_{yy} J_y + (\xi_{yz} - \eta_x\,\mathrm{d}/\mathrm{d}t)\,J_z\,,\\
H_z &= (\xi_{zx} - \eta_y\,\mathrm{d}/\mathrm{d}t)\,J_x + (\xi_{yz} + \eta_x\,\mathrm{d}/\mathrm{d}t)\,J_y + \xi_{zz} J_z\,.
\end{aligned}\right\} \tag{34}$$

The ξ's constitute an inverse susceptibility. It is the terms with η in which we are particularly interested. The equations show that these terms give a contribution to the field that may be written in vector notation as

$$\mathbf{H}' = \vec{\eta} \times \frac{\mathrm{d}\mathbf{J}}{\mathrm{d}t}, \tag{35}$$

where $\times$ denotes the vector product and the components of $\vec{\eta}$ are η_x, η_y, η_z.

This shows that the η's describe some transverse effect in the medium: a rate of change of the magnetic polarization of the medium, $d\mathbf{J}/dt$, has to give rise to a component of the magnetic field, $\mathbf{H}'$, at right angles to it. Such an effect can be expected in a ferromagnetic medium when this is magnetized to saturation in a certain direction, e.g. by placing it in a sufficiently strong constant magnetic field. The spins of the electrons contributing to the ferromagnetism will then all be parallel to one another. When the magnetization changes in a direction perpendicular to the direction of saturation the spins will turn. As the electrons carry not only a magnetic moment but also an angular momentum, there is a tendency for the spins to deviate in a transverse direction, perpendicular to the direction of the saturation and the direction in which we want them to turn. This tendency is equivalent to the action of a magnetic-field component in that direction and can thus be balanced by a component in the opposite direction. This effect is related to the various known types of gyromagnetic effect. In a subsequent paper we shall show how this effect enables us to realize the gyrator.

10. Appendix

The mean dissipated energy in a medium per unit of time and per unit of volume is equal to the real part of

$$\left(\mathbf{E}^* \cdot \frac{d\mathbf{D}}{dt} + \mathbf{H}^* \cdot \frac{d\mathbf{B}}{dt}\right),$$

where the asterisk denotes the conjugate complex value and the dot the scalar product. For a medium without dissipation this energy must be zero, so we come to

$$\mathrm{Re}\,(\mathbf{E}^* \cdot j\omega\,\mathbf{D} + \mathbf{H}^* \cdot j\omega\mathbf{B}) = 0.$$

Substituting $\mathbf{D} = \varepsilon_0\mathbf{E} + \mathbf{P}$ and $\mathbf{B} = \mu_0\mathbf{H} + \mathbf{J}$, we get

$$\mathrm{Im}\,(\mathbf{E}^* \cdot \mathbf{P} + \mathbf{H}^* \cdot \mathbf{J}) = 0. \tag{36}$$

Let us suppose first that $\mathbf{H} = 0$ and that $\mathbf{E}$ has only a component E_x. Then (36) becomes

$$\mathrm{Im}\,(E_x^*\, P_x) = 0. \tag{37}$$

In this case the equation for P_x reduces to an equation of the form

$$P_x = (\varkappa_{xx} + j\lambda_{xx})\, E_x\,. \tag{38}$$

Substituting this in (37) we arrive at $\lambda_{xx} = 0$. Thus the coefficient of E_x in the equation for P_x is real. By similar reasoning we may show that also the coefficient of E_y in the equation for P_y, etc., and the coefficient of H_x in the equation for J_x, etc., are real.

Let us now suppose that **H** $= 0$ and that **E** has only components E_x and E_y. Then (36) becomes

$$\mathrm{Im}\,(E_x{}^*\,P_x + E_y{}^*\,P_y) = 0\,. \tag{39}$$

In this case the equations for P_x and P_y reduce to equations of the form

$$\left.\begin{aligned} P_x &= \varkappa_{xx}\,E_x + (\varkappa_{xy} + j\lambda_{xy})\,E_y\,, \\ P_y &= (\varkappa_{yx} + j\lambda_{yx})\,E_x + x_{yy}\,E_y\,, \end{aligned}\right\} \tag{40}$$

the coefficient of E_x in the first equation and the coefficient of E_y in the second equation being real on account of what has just been proved. Substituting this in (39) we get

$$\mathrm{Im}\,\{(\varkappa_{xy} + j\lambda_{xy})\,E_x{}^*\,E_y + (\varkappa_{yx} + j\lambda_{yx})\,E_x\,E_y{}^*\} = 0\,.$$

As $\mathrm{Re}(E_xE_y{}^*) = \mathrm{Re}(E_x{}^*E_y)$ and $\mathrm{Im}(E_xE_y{}^*) = -\,\mathrm{Im}(E_x{}^*E_y)$, we get

$$(\varkappa_{xy} - \varkappa_{yx})\,\mathrm{Im}(E_x{}^*E_y) + j(\lambda_{xy} + \lambda_{yx})\,\mathrm{Re}(E_x{}^*E_y) = 0.$$

As this must be true for every value of $E_x{}^*E_y$, we finally get

$$\varkappa_{yx} = \varkappa_{xy} \text{ and } \lambda_{yx} = -\,\lambda_{xy}\,.$$

Thus the coefficient of E_y in the equation for P_x and the coefficient of E_x in the equation for P_y are conjugate complex. By similar reasoning we may show that many other analogous pairs of coefficients are conjugate complex.

Eindhoven, September 1947

REFERENCES

1) B. D. H. Tellegen, Network synthesis, especially the synthesis of resistanceless four-terminal networks, Philips Res. Rep. **1**, 169-184, 1946.

2) R. M. Foster, A reactance theorem, Bell Syst. tech. J. **3**, 259-267, 1924.

3) O. Brune, Synthesis of a finite two-terminal network whose driving-point impedance is a prescribed function of frequency, J. Math. Phys. **10**, 191-236, 1931.

4) C. Gewertz, Synthesis of a finite four-terminal network from its prescribed driving-point functions and transfer function, J. Math. Phys. **12**, 1-257, 1933.

5) W. Thomson and P. G. Tait, Treatise on natural philosophy, Part I, section 345VI.

6) A. Bloch, A new approach to the dynamics of systems with gyroscopic coupling terms, Phil. Mag. **35**, 315-334, 1944.

7) H. Poincaré, Etude du récepteur téléphonique, Eclair. élect. **50**, 221-234, 1907.

8) Ph. le Corbeiller, Origine des termes gyroscopiques dans les équations des appareils électromécaniques, Ann. P.T.T. **18**, 1-22, 1929.

9) H. Jefferson, Gyroscopic coupling terms, Phil. Mag. **36**, 223-224, 1945.

10) E. M. Mc. Millan, Violation of the reciprocity theorem in linear passive electro-mechanical systems, J. Acous. Soc. Am. **18**, 344-347, 1946.

11) See, for example, H. B. G. Casimir, On Onsager's principle of microscopic reversibility, Philips Res. Rep. **1**, 185-196, 1946.

12) P. Ehrenfest, Das Prinzip von Le Chatelier-Braun und die Reziprozitätssätze der Thermodynamik, Z. phys. Chem. **77**, 227-244, 1911.

Errata

The author has provided the following list of corrections:

1. The last sentence of the next-to-last paragraph of Section 4 should read: "For every order of a $2n$-pole with resistance there is one type, which can be constructed by taking any one of the two types of resistanceless $4n$-poles of the same order and connecting any n of its $2n$ pairs of terminals by a resistance."

2. The last sentence of Section 9 • 1 mentions a subsequent paper. That paper has never been written.

21

Reprinted from *Philips Res. Rept.*, **7**, 259–269 (Aug. 1952)

A GENERAL NETWORK THEOREM, WITH APPLICATIONS

by B. D. H. TELLEGEN **621.392.1**

Summary

It is proved that in a network configuration, for branch currents i satisfying the node equations and branch voltages v satisfying the mesh equations, $\sum iv$ summed over all branches is zero. By this theorem it is possible to prove the energy theorem and the reciprocity relation of networks, and to show that if given, arbitrarily varying voltages are applied on a $2n$-pole at rest the difference between the electric and the magnetic energy will at any instant depend only on the admittance matrix of the $2n$-pole and not on the particular network used for realizing it.

Résumé

Il a été démontré que dans une configuration de réseau pour courants de branche i répondant aux équations nodales et pour tensions de branche v répondant aux équations de maille, $\sum iv$ sommé sur tous les branches est égal à zéro. Ce théorème permet de prouver le théorème de l'énergie et la relation de réciprocité des réseaux et de démontrer que si des tensions données, variant arbitrairement, sont appliquées à un $2n$-pôle au repos, la différence entre les énergies électrique et magnétique dépendra à tout instant seulement de la matrice d'admittance du $2n$-pôle et non pas du réseau particulier, utilisé pour sa réalisation.

Zusammenfassung

Es wird gezeigt, daß in einer Netzwerkkonfiguration bezüglich Zweigströme i, die den Knotengleichungen genügen, und Zweigspannungen v, die den Maschengleichungen genügen, $\sum iv$, summiert über alle Zweige, gleich Null ist. Dieses Theorem ermöglicht es, das Energietheorem und die Reziprozitätsbeziehung von Netzwerken nachzuweisen, sowie zu zeigen, daß bei Anwendung gegebener, willkürlich veränderbarer Spannungen auf einen $2n$-Pol in Ruhe, die Differenz zwischen der elektrischen und der magnetischen Energie in jedem Zeitpunkt ausschließlich von der Admittanzmatrix des $2n$-Poles abhängt und nicht von dem besonderen Netzwerk, das zu ihrer Verwirklichung benutzt wird.

1. Introduction

Many properties of electrical networks can be derived from their impedance functions alone, without a detailed knowledge of their circuit diagrams. Certain properties, however, cannot be established in this way; for example, in order to prove the energy theorem and the reciprocity relation of networks and to investigate the distribution of the supplied energy in electric, magnetic, and dissipated energy, the inner structure of the network has to be taken into account. In this paper a general theorem is given which can be used in such cases. Considerations leading to the theorem are inherently present in various network investigations; the theorem itself was, so far as the author is aware, never explicitly stated.

2. The general theorem

We prove the following

Theorem:

In a network configuration, imagine branch currents i such that for every node $\Sigma i = 0$, imagine branch voltages v such that for every mesh $\Sigma v = 0$, and for every branch let the positive direction of the current be from the + to the — denoting the positive polarity of the voltage (fig. 1). Then $\Sigma iv = 0$, where the summation is over all branches.

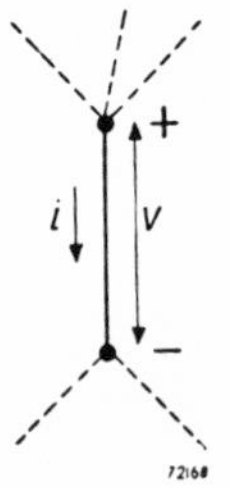

Fig. 1. Positive senses of current and voltage of a branch.

Since for every mesh $\Sigma v = 0$, there exist node potentials V such that the voltage on each branch is equal to the difference between the potentials of its end points. We denote the potentials of the nodes k and l by V_k and V_l, and the voltage on the branch connecting k and l by v_{kl}, where $v_{kl} = V_k - V_l$. We denote the current in the branch flowing from k to l by i_{kl}. Then

$$i_{kl}v_{kl} = i_{kl}(V_k - V_l) = i_{kl}V_k + i_{lk}V_l\,.$$

After applying this to all terms of Σiv, we collect the terms containing V_k, $\Sigma i_{kl}V_k = V_k\Sigma i_{kl}$, where the summation is over all nodes that are connected to k by a branch. Since the i_{kl} now considered are the currents flowing away from k, their sum is zero and thus the sum of the considered terms of Σiv is zero. The same holds for the sum of the terms containing any other node potential, so that $\Sigma iv = 0$.

The theorem is the network equivalent of the well-known theorem that the volume integral of the scalar product of a solenoidal vector (comparable with i) and an irrotational vector (comparable with v) is zero. The theorem holds for all types of network, linear and nonlinear, constant and variable, passive and active. The i's and v's in the theorem may be complex quantities satisfying the node equations and the mesh equations, respectively. The theorem can be applied to networks provided with terminal pairs by considering the terminal pairs as constituting branches of the network. If we take the positive senses of the current and the voltage of a terminal pair as indicated in fig. 2, we can write the theorem as $\Sigma i_t v_t = \Sigma i_b v_b$, where the first summation is over the terminal pairs and the second over the internal branches.

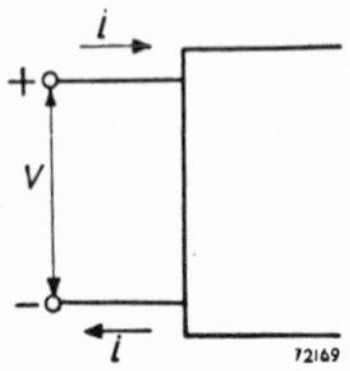

Fig. 2. Positive senses of current and voltage of a terminal pair.

3. The energy theorem of networks

If i and v denote branch currents and voltages simultaneously present in a network, $\Sigma iv = 0$ means that at any instant the total power consumption is zero. This constitutes the energy theorem of networks. For networks with terminal pairs it implies that the power absorbed by the branches is equal to the power delivered to the network through the terminal pairs.

If I and V are complex quantities representing sinusoidal branch currents and voltages of the same frequency in a constant, linear network we may write

$$\Sigma I_t V_t^* = \Sigma I_b V_b^* , \tag{1}$$

where the asterisk denotes the conjugate complex quantity, and the first summation is again over the terminal pairs and the second over the internal branches. The left-hand side of (1) equals the active power $+j$ reactive power delivered to the network. The right-hand side of (1) may be written as

$$\sum_k R_k I_k I_k^* + j\omega \sum_p C_p V_p V_p^* - j\omega \sum_{m,n} L_{mn} I_m I_n^* = W_{\mathrm{av}} + 2j\omega(U_{\mathrm{av}} - T_{\mathrm{av}}),$$

where the R_k are the resistances, the C_p are the capacitances, and the L_{mn} are the self and mutual inductances of the network; W_{av}, U_{av}, and T_{av} are the average dissipated power, the average electric energy, and the average magnetic energy, respectively. Thus we find

$$\text{active power} = W_{\mathrm{av}}, \quad \text{reactive power} = 2\omega\,(U_{\mathrm{av}} - T_{\mathrm{av}}) .$$

These results are related to results obtained by Bode [1]).

4. The reciprocity relation

Let us consider two states of a constant, linear four-pole, the first state being denoted by unprimed quantities, the second by primed quantities We can then write

$$I_1 V_1' + I_2 V_2' = \Sigma I_b V_b' ,$$
$$I_1' V_1 + I_2' V_2 = \Sigma I_b' V_b ,$$

the subscripts 1 and 2 denoting the terminal pairs, the subscript b the internal branches. Considering a branch with impedance Z, for that branch we have

$$I_b V_b' = I_b Z I_b' = I_b' Z I_b = I_b' V_b .$$

Considering a set of n coupled coils for which

$$V_{bi} = \sum_{k=1}^{n} Z_{ik} I_{bk} , \qquad (i = 1, \ldots, n ; Z_{ik} = Z_{ki}) ,$$

we have

$$\sum_{i=1}^{n} I_{bi} V_{bi}' = \sum_{i=1}^{n} I_{bi} \sum_{k=1}^{n} Z_{ik} I_{bk}' = \sum_{k=1}^{n} I_{bk}' \sum_{i=1}^{n} Z_{ki} I_{bi} = \sum_{k=1}^{n} I_{bk}' V_{bk} .$$

Considering an ideal transformer for which

$$I_{b1} = - T I_{b2} ,$$
$$V_{b2} = T V_{b1} ,$$

we have

$$I_{b1} V_{b1}' + I_{b2} V_{b2}' = - T I_{b2} V_{b1}' + I_{b2} T V_{b1}' = 0 ,$$

and also

$$I_{b1}' V_{b1} + I_{b2}' V_{b2} = 0 .$$

Thus we see that for a four-pole composed of the elements under consideration

$$\Sigma I_b V_b' = \Sigma I_b' V_b ,$$

and hence

$$I_1 V_1' + I_2 V_2' = I_1' V_1 + I_2' V_2 . \tag{2}$$

This is the reciprocity relation of four-poles in its general form. The proof shows that for this relation to hold, the four-pole need not be passive. It may contain negative resistances.

There are four important special cases of (2).
With $I_2 = 0$ and $I_1' = 0$, we get $V_2/I_1 = V_1'/I_2'$.
With $V_2 = 0$ and $V_1' = 0$, we get $I_2/V_1 = I_1'/V_2'$.
With $I_2 = 0$ and $V_1' = 0$, we get $V_2/V_1 = -I_1'/I_2'$.
With $V_2 = 0$ and $I_1' = 0$, we get $I_2/I_1 = -V_1'/V_2'$.

5. The distribution of energy

Let us consider a linear, constant, passive $2n$-pole the admittance matrix of which is given as a function of the frequency parameter λ, and let it contain no energy: no voltages on the capacitors, no currents through the coils. At $t = 0$ we apply arbitrarily varying voltages to the $2n$-pole, which causes a certain power to be consumed by it. This power is partly dissipated in the resistors, partly stored as electric energy in the capacitors and as magnetic energy in the coils. If we know the $2n$-pole network we can at any instant calculate this distribution of energy. We know that a given admittance matrix can be realized by different networks. We now ask whether anything can be said about the distribution of energy that depends only on the admittance matrix and not on the particular network used for realizing it *).

To answer the question, imagine a $2n$-pole network composed of resistances, capacitances, inductances, and ideal transformers **). At $t = 0$ constant voltages $v_1, \ldots, v_n$ are applied to the $2n$-pole. We suppose the admittance matrix to have no pole at $\lambda = \infty$ (as would be the case if the network contained capacitances in parallel to its terminal pairs). Then at $t = 0$ the $2n$-pole currents $i_1, \ldots, i_n$ remain finite.

As a consequence, immediately after applying the voltages the $2n$-pole will contain no energy, i.e. the currents through the inductances and the voltages on the capacitances will still be zero. We further suppose the admittance matrix to have no pole at $\lambda = 0$ (as would be the case if the network contained inductances in parallel to its terminal pairs). Then at $t = \infty$ the $2n$-pole currents remain finite. Admittance matrices with poles at $\lambda = \infty$ or $\lambda = 0$ can be considered as limiting cases of those without such poles, and therefore these will not be investigated separately.

Let us begin by considering a $2n$-pole the admittance matrix of which has only two poles, λ_a and λ_b. The poles may be real or conjugate complex. We suppose them to be unequal (equal poles can be considered as a limiting case of unequal ones). The poles determine the free oscillations of the $2n$-pole with short-circuited terminal pairs. For positive t every $2n$-pole current is the sum of a term with $e^{\lambda_a t}$, a term with $e^{\lambda_b t}$, and a constant term. As the $2n$-pole voltages are constant, we can write the power delivered to the $2n$-poles as

$$i_1 v_1 + \ldots + i_n v_n = G_a e^{\lambda_a t} + G_b e^{\lambda_b t} + G_0. \tag{3}$$

*) It was this question, for a two-pole composed of resistors and capacitors only, that induced me to undertake the investigations reported in this paper. This question was put to me by Dr A. J. Staverman and Dr F. Schwarzl of "Kunststoffeninstituut T.N.O." Delft, Netherlands, who were led to it by their study of visco-elastic matter, the electric equivalent of which is a system composed of resistors and capacitors.

**) Networks containing gyrators [2]) have to be excluded from the following considerations. Since a gyrator of which one terminal pair is connected to a capacitance behaves as an inductance, in such networks the distinction between electric and magnetic energy loses its interest.

If the admittance matrix and $v_1, \ldots, v_n$ are known, G_a, G_b, G_0 can be calculated in several ways.
The current and the voltage of a resistance R_k we write as

$$i_{Rk} = A_{ak}e^{\lambda_a t} + A_{bk}e^{\lambda_b t} + A_{0k}, \quad v_{Rk} = R_k(A_{ak}e^{\lambda_a t} + A_{bk}e^{\lambda_b t} + A_{0k}).$$

The current and the voltage of an inductance L_m we write as

$$i_{Lm} = B_{am}(e^{\lambda_a t}-1) + B_{bm}(e^{\lambda_b t}-1), \quad v_{Lm} = L_m(B_{am}\lambda_a e^{\lambda_a t} + B_{bm}\lambda_b e^{\lambda_b t}),$$

since $i_{Lm} = 0$ at $t = 0$.
The voltage and the current of a capacitance C_p we write as

$$v_{Cp} = D_{ap}(e^{\lambda_a t}-1) + D_{bp}(e^{\lambda_b t}-1), \quad i_{Cp} = C_p(D_{ap}\lambda_a e^{\lambda_a t} + D_{bp}\lambda_b e^{\lambda_b t}),$$

since $v_{Cp} = 0$ at $t = 0$.

We now apply our general theorem and take the currents at $t = t'$ and the voltages at $t = t''$. In section 4 it has been seen that for the ideal transformers $\Sigma iv = 0$, and so we can disregard them here. Thus we can write

$$\begin{aligned} G_a e^{\lambda_a t'} + G_b e^{\lambda_b t'} + G_0 = \\ = \sum_k R_k(A_{ak}e^{\lambda_a t'} + A_{bk}e^{\lambda_b t'} + A_{0k})(A_{ak}e^{\lambda_a t''} + A_{bk}e^{\lambda_b t''} + A_{0k}) + \\ + \sum_m L_m\{B_{am}(e^{\lambda_a t'}-1) + B_{bm}(e^{\lambda_b t'}-1)\}(B_{am}\lambda_a e^{\lambda_a t''} + B_{bm}\lambda_b e^{\lambda_b t''}) + \\ + \sum_p C_p(D_{ap}\lambda_a e^{\lambda_a t'} + D_{bp}\lambda_b e^{\lambda_b t'})\{D_{ap}(e^{\lambda_a t''}-1) + D_{bp}(e^{\lambda_b t''}-1)\}. \end{aligned}$$

Since this must hold for all positive values of t' and t'', we find

$$\begin{aligned} G_a &= \sum_k R_k A_{ak} A_{0k} - \lambda_a \sum_p C_p D_{ap}^2 - \lambda_a \sum_p C_p D_{ap} D_{bp}, \\ G_b &= \sum_k R_k A_{bk} A_{0k} - \lambda_b \sum_p C_p D_{bp}^2 - \lambda_b \sum_p C_p D_{ap} D_{bp}, \\ 0 &= \sum_k R_k A_{ak} A_{0k} - \lambda_a \sum_m L_m B_{am}^2 - \lambda_a \sum_m L_m B_{am} B_{bm}, \\ 0 &= \sum_k R_k A_{bk} A_{0k} - \lambda_b \sum_m L_m B_{bm}^2 - \lambda_b \sum_m L_m B_{am} B_{bm}, \\ 0 &= \sum_k R_k A_{ak} A_{bk} + \lambda_a \sum_m L_m B_{am} B_{bm} + \lambda_b \sum_p C_p D_{ap} D_{bp}, \\ 0 &= \sum_k R_k A_{ak} A_{bk} + \lambda_b \sum_m L_m B_{am} B_{bm} + \lambda_a \sum_p C_p D_{ap} D_{bp}, \end{aligned}$$

and three other relations which we do not need. From the last two equations it follows that

$$\sum_m L_m B_{am} B_{bm} = \sum_p C_p D_{ap} D_{bp}. \qquad (4)$$

From the first four equations we then deduce

$$\lambda_a(\sum_m L_m B_{am}^2 - \sum_p C_p D_{ap}^2) = G_a, \quad \lambda_b(\sum_m L_m B_{bm}^2 - \sum_p C_p D_{bp}^2) = G_b. \qquad (5)$$

The magnetic energy T is given by

$$\begin{aligned} T = \sum_m \tfrac{1}{2} L_m i_{Lm}^2 = \tfrac{1}{2}\sum_m L_m B_{am}^2(e^{\lambda_a t}-1)^2 + \tfrac{1}{2}\sum_m L_m B_{bm}^2(e^{\lambda_b t}-1)^2 + \\ + \sum_m L_m B_{am} B_{bm}(e^{\lambda_a t}-1)(e^{\lambda_b t}-1). \end{aligned}$$

The electric energy U is given by

$$U = \sum_p \tfrac{1}{2} C_p v_{Cp}^2 = \tfrac{1}{2} \sum_p C_p D_{ap}^2 (e^{\lambda_a t} - 1)^2 + \tfrac{1}{2} \sum_p C_p D_{bp}^2 (e^{\lambda_b t} - 1)^2 + \\ + \sum_p C_p D_{ap} D_{bp} (e^{\lambda_a t} - 1)(e^{\lambda_b t} - 1) .$$

By subtraction, using (4) and (5), we arrive at

$$T - U = \frac{G_a}{2\lambda_a} (e^{\lambda_a t} - 1)^2 + \frac{G_b}{2\lambda_b} (e^{\lambda_b t} - 1)^2. \tag{6}$$

The quantities G_a, G_b, λ_a, λ_b occurring in the right-hand side of (6) depend, as seen above, only on $v_1, \ldots, v_n$ and the admittance matrix, but not on the particular network used to realize the latter, and so this also holds for $T - U$.

This result, derived for a $2n$-pole at rest on which at $t = 0$ constant voltages are applied and of which the admittance matrix has only two poles, remains valid in more general cases. To investigate $2n$-poles the admittance matrices of which have more than two poles, let us consider the expressions derived for the energy more closely. For an admittance matrix with only one pole, λ_a or λ_b, T only contains terms with $\sum_m L_m B_{am}^2$ or $\sum_m L_m B_{bm}^2$ respectively. If λ_a and λ_b are both poles, T also contains terms with $\sum_m L_m B_{am} B_{bm}$. Similar remarks apply to U. Because of (4), in $T - U$ these latter terms cancel, and thus (6) consists of the sum of a term arising from λ_a and a term arising from λ_b *). Thus, if the admittance matrix has more than two poles, the corresponding expression for $T - U$ will be equal to the sum of the terms arising from each pole separately, and thus (6) can immediately be extended to any number of poles. It can also be extended to variable voltages, as will be shown in section 7. Thus we have the

Theorem:

If on a linear, constant, passive, gyratorless 2n-pole at rest given, arbitrarily varying voltages are applied, the difference between the magnetic and the electric energy will at any instant depend only on the admittance matrix of the 2n-pole and not on the particular network used for realizing it.

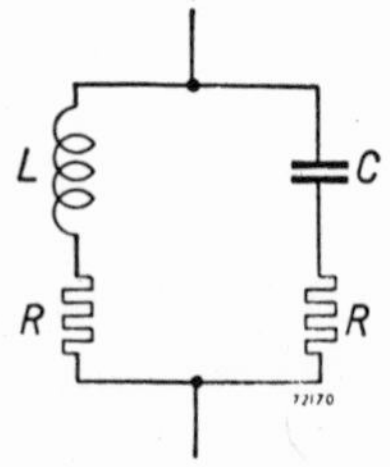

Fig. 3. Two-pole network.

*) This property is related to the conjugate property derived by Heaviside [3]).

Since for the energy A delivered to the $2n$-pole we can write $A = W+T+U$, where W is the dissipated energy, the above theorem not only holds for $T-U$ but also for $W+2T$ and for $W+2U$. If one of the energies is zero, as in resistanceless, inductanceless, or capacitanceless $2n$-poles, the theorem holds for any of the two remaining energies.

That in the general case W, T, and U are not separately determinable from the admittance matrix alone, is shown by the two-pole network of fig. 3, where $L/C = R^2$. As is well known, the impedance of the two-pole equals R. Thus, when starting from rest, T and U of the two-pole network are not determinable from the impedance function; only $T-U$ is, and here it is zero.

6. A theorem of Heaviside

From (6) we can draw an interesting conclusion. If we disregard admittance matrices that have poles at imaginary values of λ, the real parts of λ_a and λ_b will both be negative. Therefore for large t we get

$$T - U = \frac{G_a}{2\lambda_a} + \frac{G_b}{2\lambda_b}. \tag{7}$$

The energy A delivered to the $2n$-pole is determined according to (3) by

$$\frac{\mathrm{d}A}{\mathrm{d}t} = G_a\, e^{\lambda_a t} + G_b e^{\lambda_b t} + G_0 .$$

Integrating with respect to t between the limits 0 and t, we find

$$A = \frac{G_a}{\lambda_a}(e^{\lambda_a t} - 1) + \frac{G_b}{\lambda_b}(e^{\lambda_b t} - 1) + G_0 t .$$

For large t we get

$$A = G_0 t - \frac{G_a}{\lambda_a} - \frac{G_b}{\lambda_b}. \tag{8}$$

The quantity G_0 is the power delivered to the $2n$-pole at large t, which power is dissipated. Therefore $G_0 t$ is the amount of energy that would have been dissipated up to the time t if from $t = 0$ the dissipation rate would have had the value G_0, which energy we denote by W'. From (7) and (8) it then follows that

$$A - W' = 2(U - T) . \tag{9}$$

This relation remains valid if the admittance matrix has more than two poles. It expresses a theorem due to Heaviside and proved by Lorentz [2]), which may be stated as follows:

Theorem:

If on a linear, constant, passive, gyratorless 2n-pole at rest suddenly constant voltages are applied, then, when the final state has been reached, the total energy delivered to the 2n-pole exceeds the energy representing the loss by dissipation at the final rate, supposed to start at once, by twice the excess of the electric over the magnetic energy.

7. Variable voltages

To investigate the distribution of energy when variable voltages are applied to a $2n$-pole at rest, we first consider a four-pole the admittance matrix of which has two poles, on whose first terminal pair a constant voltage v_1 is applied at $t = t_1$, and on whose second terminal pair a constant voltage v_2 is applied at $t = t_2$. For $t > t_1$ and $t > t_2$ the currents and the voltages are the superposition of those due to v_1 and those due to v_2. The currents to the terminal pairs can be written as

$$\begin{aligned} i_1 = {} & \{G_{a11}\, e^{\lambda a(t-t_1)} + G_{b11}\, e^{\lambda b(t-t_1)} + G_{011}\}\, v_1 + \\ & + \{G_{a12}\, e^{\lambda a(t-t_2)} + G_{b12}\, e^{\lambda b(t-t_2)} + G_{012}\}\, v_2\,, \\ i_2 = {} & \{G_{a12}\, e^{\lambda a(t-t_1)} + G_{b12}\, e^{\lambda b(t-t_1)} + G_{012}\}\, v_1 + \\ & + \{G_{a22}\, e^{\lambda a(t-t_2)} + G_{b22}\, e^{\lambda b(t-t_2)} + G_{022}\}\, v_2 \end{aligned}$$

If the admittance matrix is known, the G's can be calculated. They do not depend on v_1, v_2, t_1, t_2. The power delivered to the four-pole amounts to

$$\begin{aligned} i_1 v_1 + i_2 v_2 = {} & G_{a11}\, v_1^2\, e^{\lambda a(t-t_1)} + G_{a12}\, v_1 v_2\, \{e^{\lambda a(t-t_1)} + e^{\lambda a(t-t_2)}\} + G_{a22}\, v_2^2\, e^{\lambda a(t-t_2)} + \\ & + G_{b11}\, v_1^2\, e^{\lambda b(t-t_1)} + G_{b12}\, v_1 v_2\, \{e^{\lambda b(t-t_1)} + e^{\lambda b(t-t_2)}\} + G_{b22}\, v_2^2\, e^{\lambda a(t-t_2)} + \\ & + G_{011}\, v_1^2 + 2\, G_{012}\, v_1 v_2 + G_{022}\, v_2^2\,. \end{aligned}$$

For the current through the resistance R_k we take the same expression as in section 5.

The current through the inductance L_m we now write as

$$\begin{aligned} i_{Lm} = {} & B_{a1m} v_1\, \{e^{\lambda a(t-t_1)} - 1\} + B_{a2m}\, v_2\, \{e^{\lambda a(t-t_2)} - 1\}, \\ & + B_{b1m} v_1\, \{e^{\lambda b(t-t_1)} - 1\} + B_{b2m}\, v_2\, \{e^{\lambda b(t-t_2)} - 1\}, \end{aligned} \tag{10}$$

since $i_{Lm} = 0$ at $t = t_1$ if $v_2 = 0$, and at $t = t_2$ if $v_1 = 0$.
The voltage on the capacitance C_p we write accordingly as

$$\begin{aligned} v_{Cp} = {} & D_{a1p}\, v_1\, \{e^{\lambda a(t-t_1)} - 1\} + D_{a2p}\, v_2\, \{e^{\lambda a(t-t_2)} - 1\} + \\ & + D_{b1p}\, v_1\, \{e^{\lambda b(t-t_1)} - 1\} + D_{b2p}\, v_2\, \{e^{\lambda b(t-t_2)} - 1\}. \end{aligned}$$

As in section 5, we apply our general theorem with the currents at $t = t'$ and the voltages at $t = t''$. Instead of (4) we now arrive at

$$\sum_m L_m\,(B_{a1m}\,v_1 e^{-\lambda_a t_1}+B_{a2m}\,v_2 e^{-\lambda_a t_2})(B_{b1m}v_1 e^{-\lambda_b t_1}+B_{b2m}v_2 e^{-\lambda_b t_2}) =$$
$$= \sum_p C_p\,(D_{a1p}\,v_1 e^{-\lambda_a t_1}+D_{a2p}\,v_2 e^{-\lambda_a t_2})\,(D_{b1p}v_1 e^{-\lambda_b t_1}+D_{b2p}v_2 e^{-\lambda_b t_2})\,.$$

Since this must hold for all values of t_1 and t_2, we find

$$\left.\begin{aligned}\sum_m L_m B_{a1m}B_{b1m} = \sum_p C_p D_{a1p}D_{b1p},\ \sum_m L_m B_{a2m}B_{b2m} = \sum_p C_p D_{a2p}D_{b2p}\,,\\ \sum_m L_m B_{a1m}B_{b2m} = \sum_p C_p D_{a1p}D_{b2p},\ \sum_m L_m B_{a2m}B_{b1m} = \sum_p C_p D_{a2p}D_{b1p}\,.\end{aligned}\right\}\quad (11)$$

By means of these relations, instead of (5) we now arrive at

$$\lambda_a \Big\{ \sum_m L_m(B_{a1m}v_1 e^{-\lambda_a t_1}+B_{a2m}v_2 e^{-\lambda_a t_2})\,(B_{a1m}v_1+B_{a2m}v_2) -$$
$$- \sum_p C_p(D_{a1p}v_1 e^{-\lambda_a t_1}+D_{a2p}v_2 e^{-\lambda_a t_2})\,(D_{a1p}v_1+D_{a2p}v_2)\Big\} =$$
$$= G_{a11}\,v_1^2 e^{-\lambda_a t_1} + G_{a12}v_1v_2(e^{-\lambda_a t_1}+e^{-\lambda_a t_2}) + G_{a22}v_2^2 e^{-\lambda_a t_2},$$

and a similar equation derived therefrom by changing the index a into b. Since this must hold for all values of v_1 and v_2, we find

$$\left.\begin{aligned}\lambda_a(\sum_m L_m B_{a1m}^2 - \sum_p C_p D_{a1p}^2) = G_{a11}\,,\\ \lambda_a(\sum_m L_m B_{a1m}B_{a2m} - \sum_p C_p D_{a1p}D_{a2p}) = G_{a12}\,,\\ \lambda_a(\sum_m L_m B_{a2m}^2 - \sum_p C_p D_{a2p}^2) = G_{a22}\,,\end{aligned}\right\}\quad (12)$$

and three similar relations by changing the index a into b. The magnetic energy and the electric energy are given by

$$T = \tfrac{1}{2}\sum_m L_m[B_{a1m}v_1\{e^{\lambda_a(t-t_1)}-1\} + B_{a2m}v_2\{e^{\lambda_a(t-t_2)}-1\} +$$
$$+ B_{b1m}v_1\{e^{\lambda_b(t-t_1)}-1\} + B_{b2m}v_2\{e^{\lambda_b(t-t_2)}-1\}]^2,$$
$$U = \tfrac{1}{2}\sum_p C_p[D_{a1p}v_1\{e^{\lambda_a(t-t_1)}-1\} + D_{a2p}v_2\{e^{\lambda_a(t-t_2)}-1\} +$$
$$+ D_{b1p}v_1\{e^{\lambda_b(t-t_1)}-1\} + D_{b2p}v_2\{e^{\lambda_b(t-t_2)}-1\}]^2.$$

By subtraction, using (11) and (12), we arrive at

$$T - U = \frac{1}{2\lambda_a}\,[G_{a11}v_1^2\{e^{\lambda_a(t-t_1)}-1\}^2 + 2G_{a12}v_1v_2\{e^{\lambda_a(t-t_1)}-1)\}\,\{e^{\lambda_a(t-t_2)}-1\} +$$
$$+ G_{a22}\,v_2^2\{e^{\lambda_a(t-t_2)}-1\}^2] +$$
$$+ \frac{1}{2\lambda_b}\,[G_{b11}\,v_1^2\{e^{\lambda_b(t-t_1)}-1\}^2 + 2G_{b12}v_1v_2\{e^{\lambda_b(t-t_1)}-1\}\,\{e^{\lambda_b(t-t_2)}-1\} +$$
$$+ G_{b22}\,v_2^2\{e^{\lambda_b(t-t_2)}-1\}^2]\,.$$

As is well known, the application of a variable voltage can be conceived as the successive application of incremental constant voltages. Therefore, if a constant voltage v, applied at $t = 0$ in some point A of a network at rest, gives rise, for $t > 0$ in some point B of the network, to a voltage or current $vf(t)$, then a variable voltage v, applied at $t = 0$ in the point

A of the network at rest, will give rise, for $t > 0$ in the point B, to a voltage or current $\int_0^t v'(\tau) f(t-\tau) d\tau$. If variable voltages v_1 and v_2 are applied at $t = 0$ to the four-pole at rest, we can thus for $t = 0$ write instead of (10)

$$i_{Lm} = B_{a1m} \int_0^t v_1'(\tau) \{e^{\lambda_a(t-\tau)} - 1\} d\tau + B_{a2m} \int_0^t v_2'(\tau) \{e^{\lambda_a(t-\tau)} - 1\} d\tau +$$
$$+ B_{b1m} \int_0^t v_1'(\tau) \{e^{\lambda_b(t-\tau)} - 1\} d\tau + B_{b2m} \int_0^t v_2'(\tau) \{e^{\lambda_b(t-\tau)} - 1\} d\tau .$$

The B's are the same as in (10), but the time functions have changed. Similarly, in v_{Cp}, T, U, and T—U the coefficients remain the same, and only the time functions are altered. This proves our statement in section 5 that also in the case of variable voltages T—U does not depend on the particular network used for realizing a given admittance matrix. This result can be extended to networks with any number of terminal pairs and with admittance matrices having any number of poles.

Eindhoven, February 1952

REFERENCES

[1]) H. W. Bode, Network analysis and feedback amplifier design, 1945, p. 130.
[2]) B. D. H. Tellegen, Philips Res. Rep. **3**, 81, 1948.
[3]) O. Heaviside, Electrical Papers, Vol. II, p. 202.
[4]) O. Heaviside, Electrical Papers, vol. II, p. 412;
H. A. Lorentz, Proc. nat. Acad. Sciences **8**, 333, 1922 = Collected Papers, Vol. III, p. 331. See also Balth. van der Pol, Physica, 's-Grav. **4**, 585, 1937.

Errata

The author has provided the following list of corrections:

1. Equation (1): should read

$$\sum V_t I_t^* = \sum V_b I_b^*$$

2. The first sentence of the third paragraph of Section 5: "As a consequence . . . zero" should be used as the last sentence of the second paragraph. The third paragraph then starts with "We further . . . ."

3. Line preceding Equation (3): change "2n-poles" to "2n-pole."

4. Equation (10), first line: the comma at the end of the first line should be replaced by a plus sign.

5. Page 11, third line: change "t = 0" to "t > 0."

Reprinted from *J. Appl. Phys.*, **25**(12), 1510–1514 (1954)

A Definition of Passive Linear Networks in Terms of Time and Energy

G. Raisbeck
Bell Telephone Laboratories, Inc., Murray Hill, New Jersey
(Received March 3, 1954)

A definition of passive linear network is made:

(a) The network is linear.
(b) If currents of any wave form are fed to the terminals of the network, the total energy delivered to the network is not negative.
(c) No voltages appear between any pair of terminals before a current is fed to the network.

When this definition is applied to two terminal networks, i.e., impedances, a necessary and sufficient condition that a two-terminal network be linear passive is that its impedance function be a positive real function.

An analysis of multiterminal networks yields as a necessary and sufficient condition from the foregoing hypotheses that a certain Hermitian quadratic form be positive definite. In the case of three-terminal networks, it reduces to

$$4R_{11}R_{22}-(R_{12}+R_{21})^2-(X_{12}-X_{21})^2\geq 0,$$

where $R_{ij}+jX_{ij}=Z_{ij}$ are the terms of the matrix of impedances of the network. The relation of this formula to similar but not identical formulas of Gewertz and Llewellyn, and other consequences of the condition, are discussed.

In a search for a suitable definition of a passive network it seems philosophically desirable to let the definition rest on broad properties involving energy, time, and the like, rather than specific properties such as the analytic or positive character of impedance functions. A good deal is known, however, about passive two-terminal linear networks. Whatever reasonable definition of passivity is assumed, the impedance function is a positive real function.[1] It follows from this fact that if a current is passed through the impedance, the total energy delivered to the impedance is positive, and that no voltage appears across the terminals before a current is impressed. Now these conditions are of the type desirable for a good definition of *linear passive network*. We shall accordingly generalize them for n-terminal networks.

This paper approaches in some respects the work of McMillan[2] from which it differs in one main respect. McMillan deals with realizable networks, and consequently restriction to rational impedance functions and reciprocity (i.e., symmetric impedance matrices) is inevitable. The present paper assumes neither rationality nor reciprocity. The definition of passive linear network which is proposed is apparently much broader than any definition involving realizability, in that the hypotheses are less restrictive; yet it is notable that if a network satisfying the present definition of passive linear network also has a rational symmetric impedance matrix, it satisfies the hypotheses of McMillan for a realizable network. Thus, the restrictions on a passive linear network required to make it realizable are only the obvious conditions that its impedance matrix be rational and symmetrical. Because of this, the chief interest of the paper must lie in its results about nonreciprocal devices and in the fact that the results are cast in a form identical for rational and nonrational impedances.

It will be assumed throughout that the network in question has an impedance matrix. In the case of two-terminal devices, the impedance may be identically zero or the admittance may be identically zero. In either case, if the impedance is not defined, the admittance is, and vice versa. However, a general network may have both kinds of bad behavior at once: it may be that neither the impedance matrix nor the admittance matrix exists. This will happen, for example, if it has one pair of open-circuited and one pair of short-circuited terminals. The classic way to deal with such problems is to suppose infinitesimal external admittances connected among the terminals.

McMillan has shown another way out of the dilemma for networks obeying reciprocity. It seems likely that his technique would also be applicable to nonreciprocal devices. The question is avoided because the chief interest of the present paper is believed to lie in application to cases where there is no question of the existence of the impedance matrix.

Before proceeding to n-terminal networks, we shall first prove a converse of our original statement about positive real impedance functions, to wit:

If a two-terminal network has the following properties: (a) it is linear, (b) if a current of any wave form is impressed on the impedance, the total energy delivered to the impedance is not negative, (c) no voltage appears across the terminals before a current is passed through them; then its impedance function is a positive real function.

A function $Z(\lambda)$, $\lambda=\sigma+j\omega$, is called positive real if[1]

[1] Guillemin, *A Summary of Modern Methods of Network Analysis, Advances in Electronics* (Academic Press, New York, 1951), Vol. 3, pp. 261–303.

[2] McMillan, Bell System Tech. J. **31**, Nos. 2 and 3, pp. 217–279, 541–600, March and May, 1952.

(a′) $Z(\lambda)$ is real for real λ; (b′) $Z(\lambda)$ is analytic in the region $\sigma>0$; (c′) $Re[Z(\sigma+j\omega)]\geq 0$ if $\sigma\geq 0$.

A proof will be outlined, but no attempt will be made at rigor. It is assumed that integrals converge and that limits may be interchanged throughout. These conditions can be assured by more or less stringent assumptions about continuity, boundedness, or absolute integrability, depending on the degree of mathematical sophistication one is willing to assume.

The idea of the proof is as follows.

An expression for the energy delivered to the network will be found (15). An expression for the real part of the impedance in the right-hand half-plane in terms of the impedance on the axis of real frequencies will be found (18). A particular excitation (19) will be found such that the two expressions above are equal. Since the energy delivered to the network is positive, then it follows that the real part of the impedance in the right-hand half-plane is also positive.

Because the circuit is linear we can describe the relation between the voltage $v(t)$ and the current $i(t)$ by

$$v(t)=\int_{-\infty}^{\infty} i(\tau)K(t-\tau)d\tau. \tag{1}$$

Furthermore, because the voltage is zero until the current begins, it follows simply that

$$K(t)=0 \quad \text{for} \quad t<0. \tag{2}$$

In fact, $K(t)$ is no more than the impulse response of the network. The impedance function is simply

$$Z(j\omega)=\int_{-\infty}^{\infty} K(t)e^{-j\omega t}dt. \tag{3}$$

Because of the fact that $K(t)$ vanishes for negative arguments,

$$Z(\sigma+j\omega)=\int_{0}^{\infty} K(t)e^{-\sigma t}e^{-j\omega t}dt \tag{4}$$

is uniformly convergent for $\sigma\geq\delta>0$, and hence is an analytic function of $\sigma+j\omega$ in the right half-plane. Furthermore, inasmuch as $K(t)$ is the pulse response of the impedance, it is a real function, and consequently $Z(\sigma+j\omega)$ is real when $\omega=0$. It remains only to prove that

$$Re[Z(\sigma+j\omega)]\geq 0 \quad \text{for} \quad \sigma\geq 0. \tag{5}$$

Now consider the energy delivered to the impedance

$$E=\int_{\infty}^{\infty} v(t)i(t)dt. \tag{6}$$

Note that by hypotheses

$$E\geq 0. \tag{7}$$

Form the function

$$e(u)=\int_{-\infty}^{\infty} i(t-u)v(t)dt \tag{8}$$

$$=\int_{-\infty}^{\infty} i(t-u)dt\int_{-\infty}^{\infty} i(\tau)K(t-\tau)d\tau \tag{9}$$

$$=\int_{-\infty}^{\infty}\int_{-\infty}^{\infty} i(t-u)i(\tau)K(t-\tau)dtd\tau. \tag{10}$$

Note that

$$E=e(0). \tag{11}$$

Now take the Fourier transforms of both sides. The right side is the convolution of three functions

$$i(-u)*i(u)*K(u), \tag{12}$$

and its Fourier transform is the product of the transforms of the three separate functions, i.e.,

$$E(\omega)=I^*(\omega)\cdot I(\omega)\cdot Z(j\omega)$$
$$=|I(\omega)|^2\cdot Z(j\omega). \tag{13}$$

Inverting again, we get

$$e(u)=\frac{1}{2\pi}\int_{-\infty}^{\infty} Z(j\omega)|I(\omega)|^2e^{j\omega u}d\omega. \tag{14}$$

Setting $u=0$, we get

$$E=e(0)=\frac{1}{2\pi}\int_{-\infty}^{\infty} Z(j\omega)|I(\omega)|^2d\omega. \tag{15}$$

Now, recalling Poisson's integral for the case of a half-plane,

$$Z(\sigma+j\omega)=\frac{1}{\pi}\int_{-\infty}^{\infty} Z(j\eta)\frac{\sigma}{\sigma^2+(\omega-\eta)^2}d\eta, \quad \sigma>0 \tag{16}$$

for arbitrary ω and arbitrary $\sigma>0$. Inasmuch as Z is typically real,[3] (see Eq. (4))

$$Z(\sigma+j\omega)=Z^*(\sigma-j\omega), \tag{17}$$

we can form both functions, add, and divide by 2 to get

$$Re[Z(\sigma+j\omega)]=\frac{1}{2\pi}\int_{-\infty}^{\infty} Z(j\eta)\times\left[\frac{\sigma}{\sigma^2+(\omega+\eta)^2}+\frac{\sigma}{\sigma^2+(\omega-\eta)^2}\right]d\eta. \tag{18}$$

Now let

$$\begin{aligned} i_0(t)&=C_1e^{-\sigma t}\cos\omega(t-C_2), & t>0\\ &=0 & t<0. \end{aligned} \tag{19}$$

[3] A function is called *typically real* if it assumes conjugate values for conjugate values of its argument.

For suitable real values of C_1 and C_2,

$$|I(\eta)|^2=\frac{1}{2\pi}\left[\frac{\sigma}{\sigma^2+(\omega+\eta)^2}+\frac{\sigma}{\sigma^2+(\omega-\eta)^2}\right]. \qquad (20)$$

Hence, from (7), (15), (18), and (20),

$$Re|Z(\sigma+j\omega)|=E\geq 0 \quad \text{for} \quad \sigma>0. \qquad (21)$$

The same inequality for $\sigma=0$ follows from continuity arguments. This completes the proof.

Note that the impedance function need not be assumed rational. The rationality of $Z(\lambda)$ arises only when the question of realization with lumped elements is brought up.

Let us now imagine a network having $n+1$ terminals. Suppose it satisfies the following hypotheses.

(a″) It is linear.

(b″) If currents of any wave form are fed to the terminals of the network, the total energy delivered to the network is not negative.

(c″) No voltages appear between any pair of terminals before a current is fed to the network.

We shall call a network satisfying (a″), (b″), and (c″) a *linear passive network*.

It will be proved by a method exactly analagous to the method used for two terminal networks that a certain Hermitian form[4] (34) is positive definite. The steps are: first, to derive an expression for energy delivered to the network (25); second, to assume a special excitation (26) analogous to the one used before; and third to show that the energy in this case is a certain Hermitian form involving only elements of the impedance matrix of the network (28), (34). Because the energy delivered is always positive, the Hermitian form must be positive definite.

Call one terminal a reference terminal and let the currents into and the voltages at the other terminals be called, respectively,

$$i_1(t),\ i_2(t),\ \cdots i_n(t),\ v_1(t),\ v_2(t),\ \cdots v_n(t). \qquad (22)$$

Then let $\|i_j(t)\|$ be a vector (i.e., a $1\times n$ matrix) having components $i_1(t)$, $i_2(t)$, $\cdots$, and let $\|v_j(t)\|$ be a vector having components $v_1(t)$, $v_2(t)$, $\cdots$. Then $\|i_j(t)\|$ and $\|v_j(t)\|$ are connected by a matrix relation

$$\|v_j(t)\|=\int_{-\infty}^{\infty}\|k_{jk}(t-\tau)\|\ \|i_k(\tau)\|d\tau, \qquad (23)$$

which is analogous to (1). Carrying the analysis through exactly as before, we arrive at

$$e(u)=\frac{1}{2\pi}\int_{-\infty}^{\infty}\|Z_{jk}(j\omega)\|\ \|I_j(\omega)\|\ \|I_k^*(\omega)\|e^{j\omega u}d\omega \qquad (24)$$

and

$$E=e(0)=\frac{1}{2\pi}\int_{-\infty}^{\infty}\|Z_{jk}(j\omega)\|\ \|I_j(\omega)\|\ \|I_k^*(\omega)\|d\omega, \qquad (25)$$

where $\|Z_{jk}\|$ is the well-known impedance matrix. If now we let

$$i_n(t)=c_n i_0(t-\tau_n), \qquad (26)$$

i.e., i_0 multiplied by c_n and delayed by τ_n [i_0 being defined in (19)], then

$$E=Re\Big[\sum_{i,j} c_i c_j Z_{ij}(\sigma+j\omega)e^{j\omega(\tau_i-\tau_j)}\Big], \qquad (27)$$

from which it follows that

$$Re\Big[\sum_{i,j} c_i c_j Z_{ij}(\sigma+j\omega)e^{j\omega(\tau_i-\tau_j)}\Big]\geq 0 \qquad (28)$$

for all real c's and positive τ's.

If we let

$$\omega\tau_j=\varphi_j \qquad (29)$$

$$c_j e^{j\omega\tau_j}=u_j=a_j+jb_j \qquad (30)$$

$$Z_{ij}'=Z_{ji}'^*=\tfrac{1}{2}[Z_{ij}+Z_{ji}^*]=R_{ij}^s+jX_{ij}^a, \qquad (31)$$

then alternative equivalent forms are:

$$\sum_{i,j} c_i c_j[\cos(\varphi_i-\varphi_j)R_{ij}-\sin(\varphi_i-\varphi_j)X_{ij}]\geq 0 \qquad (32)$$

$$\sum_{i,j}[(a_i a_j+b_i b_j)R_{ij}^s+(a_i b_j-a_j b_i)X_{ij}^a]\geq 0 \qquad (33)$$

$$\sum_{i,j} u_i^* u_j Z_{ij}'\geq 0 \qquad (34)$$

for all a_i, b_i, c_i, u_i, and φ_i, where the a's, b's, and c's and φ's are arbitrary real numbers and the u's are arbitrary complex numbers. (The restriction of φ to positive numbers is no restriction at all.) In (31) and (34), * denotes a complex conjugate.

If the matrix of impedances is assumed to be symmetric, this condition reduces to

$$\sum_{i,j} c_i c_j R_{ij}\geq 0$$

for all real c_i, c_j. This is a necessary condition for realizability, and has been stated several times before.[5]

From (31) it follows that (34) is a Hermitian form, and the condition is simply that the Hermitian form (34) is positive definite. From (31) it also follows that the elements of the form Z_{ij}' depend only on the symmetric components of resistance, R_{ij}^s, and the skew-symmetric components of reactance, X_{ij}^a. The skew-symmetric components of resistance, R_{ij}^a, and the symmetric components of reactance, X_{ij}^s, are not involved.

Any one of these inequalities is sufficient to guarantee that $E\geq 0$ for all signals. Take, for example, an arbitrary

[4] A Hermitian form is an expression $F=\sum x_i h_{ij} x_j^*$ (where * denotes the complex conjugate) in which the matrix of coefficients h_{ij} has the property

$$h_{ij}=h_{ji}^*.$$

See G. Birkhoff and S. McLane, *A Survey of Modern Algebra* (Macmillan, New York, 1946), pp. 256–257.

[5] McMillan, reference 2. McMillan refers to earlier announcements of this result by Y. Ono (1946), himself (1948), and M. Bayard (1949).

set of signals $\|i_n(t)\|$. Let $I_j(\omega)=a_j(\omega)+jb_j(\omega)$. Direct computation shows that

$$Re(\|Z_{ij}(j\omega)\|\ \|I_j(\omega)\|\ \|I_j^*(\omega)\|)$$
$$=\sum_{i,j}[[a_i(\omega)a_j(\omega)+b_i(\omega)b_j(\omega)]R_{ij}(j\omega)$$
$$+[a_i(\omega)b_j(\omega)-a_j(\omega)b_i(\omega)]X_{ij}(j\omega)]\geq 0, \quad (35)$$

by (33) [which is equivalent to (34)]. Hence (25), which is the integral of (35) over frequency [the imaginary part of (25) is zero by symmetry], is also non-negative.

Hence we can deduce a converse. Imagine a network having $(n+1)$ terminals. Suppose it satisfies the following hypotheses.

(a″) It is linear.

(c″) No voltages appear between any pair of terminals before a current is fed to the network.

(d″) The Hermitian quadratic form (34) is positive definite.

Then

(b″) If currents of any wave form are fed to the terminals of the network, the total energy delivered to the network is not negative.

Inequalities (28) and (34) are the analog of (21). The reactive components do not appear for any network obeying reciprocity, because in that case the anti-symmetric parts are identically zero. Hence they do not appear in (21), for any two-terminal device obeys reciprocity.

As a practical test, any particular Hermitian form can be reduced to diagonal form.[6]

$$\sum_i y_i^* y_i Z_{ii}'' \quad (36)$$

where all the mixed terms vanish. The form is positive definite if and only if all the numbers Z_{ii}'' are non-negative. For the case $n=2$ (a three-terminal network) this can be done by "completing the square" as follows:

$$\sum_{i,j} u_i^* u_j Z_{ij}'$$
$$=u_1^*u_1Z_{11}'+u_1^*u_2Z_{12}'+u_2^*u_1Z_{21}'+u_2u_2^*Z_{22}' \quad (37)$$
$$=\left[u_1+u_2\frac{Z_{12}'}{Z_{11}'}\right]\left[u_1^*+u_2^*\frac{Z_{21}'}{Z_{11}'}\right]Z_{11}'$$
$$+u_2^*u_2\frac{Z_{11}'Z_{22}'-Z_{12}'Z_{21}'}{(Z_{11}')} \quad (38)$$
$$=y_1^*y_1Z_{11}+y_2^*y_2\frac{Z_{11}'Z_{22}'-Z_{12}'Z_{21}'}{(Z_{11}')}, \quad (39)$$

where

$$y_1=u_1+u_2\frac{Z_{12}'}{Z_{11}'}$$
$$y_2=u_2. \quad (40)$$

The condition reduces then to

$$Z_{11}'\geq 0$$
$$Z_{11}'Z_{22}'-Z_{12}'Z_{21}'\geq 0, \quad (41)$$

(which imply, incidentally, that $Z_{22}'\geq 0$) or simply, by (31),

$$R_{11}\geq 0$$
$$R_{22}\geq 0 \quad (42)$$
$$R_{11}R_{22}-(R_{12}^s)^2-(X_{12}^a)^2\geq 0.$$

F. B. Llewellyn[7] gives a condition which must be satisfied in order that a two terminal-pair network always have positive input and output impedances for passive terminations. This condition must be satisfied whenever (42) is satisfied. The following argument shows that it is. Llewellyn's condition is

$$4(R_{11}R_{22}+X_{12}X_{21})(R_{11}R_{22}-R_{12}R_{21})$$
$$-(R_{12}X_{21}-R_{21}X_{12})^2\geq 0. \quad (43)$$

In our notation, an equivalent form is

$$[R_{12}R_{22}-(R_{12}^s)^2-(X_{12}^a)^2][R_{11}R_{22}+(X_{12}^s)^2$$
$$+(R_{12}^a)^2]+(R_{12}^sR_{12}^a+X_{12}^sX_{12}^a)^2\geq 0. \quad (44)$$

The second factor of the first term and the second term are both inherently non-negative. By (42) the first factor of the first term is positive. Hence Llewellyn's condition is satisfied. However, Llewellyn's condition does not imply ours: his is a condition for stability, ours for passivity. All passive networks (according to our definition) are stable, but not all stable networks are passive. Llewellyn himself points out that his condition does not guarantee all the attributes of a passive system.

Gewertz's condition, quoted by Llewellyn, is made under the hypotheses of reciprocity. In our notation, it is

$$R_{11}R_{22}-(R_{12}^s)^2\geq 0. \quad (45)$$

Our formula (42) reduces to this when the nonreciprocal term X_{12}^a is removed. This implies that a reciprocal network which is stable under all passive terminations (Gewertz's hypothesis) is passive, but a nonreciprocal network stable under all passive terminations (Llewellyn's hypothesis) need not be passive. The conclusion is made more plausible by the following heuristic consideration. In a given network, a signal traveling

[6] G. Birkhoff and S. McLane, reference 4, Chap. IX.

[7] F. B. Llewellyn, Proc. Inst. Radio Engrs. **40**, No. 3, 281 (March, 1952).

through the network, reflected at the terminals, and returning is stable, provided the signal which reappears is smaller than the one that went in. If the network is reciprocal, this implies that it suffers a loss in both directions. If the network is nonreciprocal, it may increase in one direction of transmission and decrease in the other. Now if we determine the direction of preferred transmission, and provide an external low-loss feedback path for the return transmission, the network plus a passive external feedback path may be unstable. This is just the situation which prevails, for example, in a normal vacuum-tube amplifier.

The relation (34) can be used to derive other unobvious consequences. For example, suppose that for all i, j

$$R_{ij}=0. \tag{46}$$

Then by choosing all a's and b's equal to zero except

$$a_j=1, \qquad b_j=\pm 1, \tag{47}$$

it follows from (33) that

$$X_{ij}^a \geq 0, \qquad X_{ij}^a \leq 0. \tag{48}$$

Hence

$$X_{ij}^a=0. \tag{49}$$

That is, if the resistance terms in the matrix of impedances vanish, then so do the nonsymmetric reactance terms, or, in other words, if the matrix of impedances of a passive network contains no resistances, it obeys reciprocity.

Also, for example, if the matrix of impedances of a network is skew-symmetric, as in a gyrator, then the reactive term must vanish if it is to be passive.

In summary, if we accept hypotheses (a″), (b″), and (c″) as the definition of a passive linear network, then a necessary and sufficient condition that a network be linear passive is that the Hermitian form

$$\sum_{i,j} u_i^* u_j Z_{ij}'$$

be positive definite. For networks having three terminals this condition reduces to a form which can be compared with similar formulae of Gewertz and Llewellyn. Other nontrivial consequences can be derived from it.

The helpful criticism of J. G. Linvill is gratefully acknowledged.

Erratum Provided by Author

Ernest Kuh has criticized the transition from equation (23) to equation (24), saying that the indicated analysis fails. However, he has proved the result by another means.

23

Reprinted from *IRE Trans. Circuit Theory*, **CT-4**(3), 117–119 (1957)

The *A* Matrix, New Network Description*

THEODORE R. BASHKOW†

Introduction

A SYSTEM of ordinary linear differential equations is commonly treated in the form $F + dy/dt = Ay$, where y and F are column matrices and A is a constant matrix. The solution of the set of equations involves the eigenvalues of the A matrix, *i.e.*, the zeros of the polynomial in λ given by the determinant $|A - \lambda I|$, where I is the unit matrix and λ a scalar. Any set of high-order differential equations can be converted to such a set of first-order equations and consequently, it is possible to convert the set of second-order equations of the usual mesh or nodal analysis to this form. However, such manipulations do not always utilize a satisfactory set of variables. It will be shown that a new method of network analysis leads naturally to a set of first-order differential equations $F + dy/dt = Ay$ in which the y's are certain of the original voltage and current variables and the F's the voltage and current sources. The A matrix consists of scalar elements which are combinations of the inductances, capacitances or resistances of the networks. It will be seen that the natural frequencies of the network are the eigenvalues of the A matrix.

In order to arrive at this set of equations, it is necessary to use voltages across capacitances and currents through inductances as the set of independent variables. Therefore, all networks will be considered to consist of branches containing a single element; inductance, capacitance, or resistance.

The Proper Tree

A network tree is a subgraph connecting all of the network nodes, but having no closed paths. The remaining branches are called links. It has been shown[1] that the branch currents of a tree of the network can be expressed in terms of the link currents. Conversely, the link voltages can be expressed in terms of the tree branch voltages.

A proper tree of a network is now defined as one whose branches contain *every* capacitive element of the network or *every* capacitive element plus resistive elements. A given network, however, may have some of its capacitive elements in a closed loop, or perhaps all trees containing every capacitance also contain inductances. Any capacitances and inductances which prevent the formation of a proper tree will be called excess. In what follows it will be assumed that a network with excess elements has had capacitances added *across* each excess inductance and inductances placed *in series* with each excess capacitance.

After such treatment any network must contain a proper tree and it is this tree to which future reference is made. It is also assumed that all current sources have been open-circuited and voltage sources short-circuited.

Current Matrix

Given a network, (modified, if necessary, as indicated above), consider the proper tree and assign a current direction to each tree branch and to each link. Any link placed on the tree becomes part of a unique closed path and thus each link current is equivalent to a mesh current flowing in this closed path. It is now possible to form a table expressing branch currents in terms of link currents.

If the rows are labeled i_a for the tree branch currents and the columns I_α for the link (mesh) currents, then the table entries are as follows.

If a tree branch is in a given mesh, enter a $+1$ if tree branch and mesh currents are in the same direction. Enter a -1 if the currents are in opposite direction. If the tree branch is not in the mesh enter a 0. As an example consider the network and tree of Fig. 1.

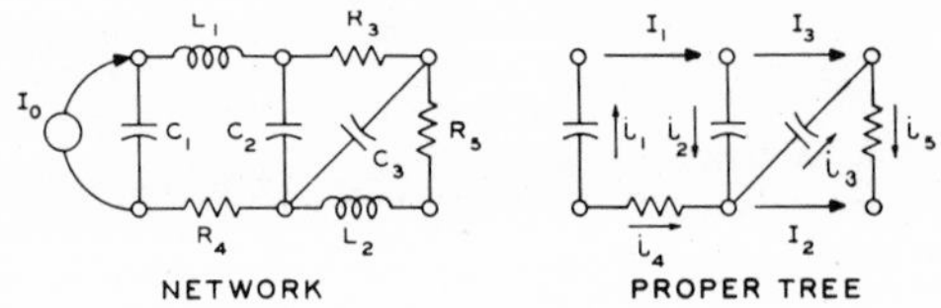

$$\begin{bmatrix} I_0 + C_1 dV_1/dt \\ C_2 dV_2/dt \\ C_3 dV_3/dt \\ G_4 V_4 \\ G_5 V_5 \\ L_1 dI_1/dt \\ L_2 dI_2/dt \\ R_3 I_3 \end{bmatrix} = \begin{bmatrix} 0 & 0 & 0 & 0 & 0 & +1 & 0 & 0 \\ 0 & 0 & 0 & 0 & 0 & +1 & 0 & -1 \\ 0 & 0 & 0 & 0 & 0 & 0 & -1 & -1 \\ 0 & 0 & 0 & 0 & 0 & -1 & 0 & 0 \\ 0 & 0 & 0 & 0 & 0 & 0 & -1 & 0 \\ -1 & -1 & 0 & +1 & 0 & 0 & 0 & 0 \\ 0 & 0 & +1 & 0 & +1 & 0 & 0 & 0 \\ 0 & +1 & +1 & 0 & 0 & 0 & 0 & 0 \end{bmatrix} \begin{bmatrix} V_1 \\ V_2 \\ V_3 \\ V_4 \\ V_5 \\ I_1 \\ I_2 \\ I_3 \end{bmatrix}$$

COMBINED MATRIX EQUATION

$$\begin{bmatrix} I_0 + C_1 dV_1/dt \\ C_2 dV_2/dt \\ C_3 dV_3/dt \\ L_1 dI_1/dt \\ L_2 dI_2/dt \end{bmatrix} = \begin{bmatrix} 0 & 0 & 0 & +1 & 0 \\ 0 & -G_3 & -G_3 & +1 & 0 \\ 0 & -G_3 & -G_3 & 0 & -1 \\ -1 & -1 & 0 & -R_4 & 0 \\ 0 & 0 & +1 & 0 & -R_5 \end{bmatrix} \begin{bmatrix} V_1 \\ V_2 \\ V_3 \\ I_1 \\ I_2 \end{bmatrix}$$

A MATRIX EQUATION

Fig. 1—Example of circuit analysis by new method.

* Manuscript received by the PGCT, October 18, 1956.
† Bell Telephone Labs., Inc., Murray Hill, N. J.
[1] J. L. Synge, "The fundamental theorem of electrical networks," *Quart. Appl. Math.*, vol. IX, p. 113; July, 1951.

Table I which corresponds[2] follows.

TABLE I

Tree Branch Currents	Link (Mesh) Currents I_1	I_2	I_3
i_1	+1	0	0
i_2	+1	0	−1
i_3	0	−1	−1
i_4	−1	0	0
i_5	0	−1	0

The same information can be given by the matrix equation

$$\begin{bmatrix} i_1 \\ i_2 \\ i_3 \\ i_4 \\ i_5 \end{bmatrix} = \begin{bmatrix} +1 & 0 & 0 \\ +1 & 0 & -1 \\ 0 & -1 & -1 \\ -1 & 0 & 0 \\ 0 & -1 & 0 \end{bmatrix} \begin{bmatrix} I_1 \\ I_2 \\ I_3 \end{bmatrix} \quad \text{or} \quad i = CI \tag{1}$$

where C is the current matrix.

Voltage Matrix

Label the voltage across each link v_α where v_α is a positive voltage drop in the same direction as the link current. Similarly, the voltage across each tree branch is labeled V_a. Now note that the matrix relating link voltages to tree branch voltages is the negative transpose of the previously found current matrix[3] or (for the example)

$$\begin{bmatrix} v_1 \\ v_2 \\ v_3 \end{bmatrix} = \begin{bmatrix} -1 & -1 & 0 & +1 & 0 \\ 0 & 0 & +1 & 0 & +1 \\ 0 & +1 & +1 & 0 & 0 \end{bmatrix} \begin{bmatrix} V_1 \\ V_2 \\ V_3 \\ V_4 \\ V_5 \end{bmatrix} \quad \text{or } v = -C^tV. \tag{2}$$

$-C^t$ is the voltage matrix. Its transpose is that part of a cut set schedule pertinent to the link voltages.[4]

Combined Matrix

Each branch of the proper tree, by construction, is either a capacitance or a resistance. Each tree branch current, i_a, is therefore $C_a dV_a/dt$ or $G_a V_a$. Similarly, each link is an inductance or a resistance. Therefore, each link voltage, v_α, is either $L_\alpha dI_\alpha/dt$ or $R_\alpha I_\alpha$. With these substitutions the previously found current and voltage equations can be formed into a single matrix equation as shown.

$$\begin{bmatrix} C_1\, dV_1/dt \\ \vdots \\ C_r\, dV_r/dt \\ G_{r+1}V_{r+1} \\ \vdots \\ G_n V_n \\ L_1\, dI_1/dt \\ \vdots \\ L_s\, dI_s/dt \\ R_{s+1}I_{s+1} \\ \vdots \\ R_m I_m \end{bmatrix} = \begin{bmatrix} 0_{\,(n \times n)} & C_{\,(n \times m)} \\ -C^t_{\,(m \times n)} & 0_{\,(m \times m)} \end{bmatrix} \begin{bmatrix} V_1 \\ \vdots \\ V_r \\ V_{r+1} \\ \vdots \\ V_n \\ I_1 \\ \vdots \\ I_s \\ I_{s+1} \\ \vdots \\ I_m \end{bmatrix} \tag{3}$$

The combined matrix of this equation is of necessity square and skew symmetric. If there are no resistances or excess reactive elements in a network then the combined matrix with all rows divided by the proper L or C is the A matrix defined in the Introduction.

At this point it is necessary to digress slightly to incorporate the network sources. A current source, I, connects two nodes of the network and can be treated as an additional link current. This current must flow either in the same or opposite direction as the currents in the tree branches which form its path through the network. If it is in the same direction as a tree branch current add $-I$ to the left-hand side of that particular row of (3). If it flows in the opposite direction add $+I$ to the left-hand side. Obviously, all contributions due to current sources will be in the 1st n rows of (3) since only these are current equations. A voltage source, E, can only appear in a branch of the network. Therefore, it will aid or oppose the previously determined link currents. If it aids a particular link current add $-E$ to the left-hand side of that row of (3). If it opposes a link current add a $+E$ to the left-hand side. All contributions due to voltage sources will be in the last m rows of (3) since these are the only voltage equations.

Elimination of "Unwanted" Variables

To reduce (3) in the general case to the desired form eliminate all variables V_{r+1} to V_n and I_{s+1} to I_m by expressing them in terms of the remaining $r + s$ variables. The resulting combined matrix is still square but is not skew symmetric.

Across each excess inductance L_α there is an added capacitance C_α. It is necessarily true that

$$L_\alpha\, dI_\alpha/dt = \pm V_\alpha$$

and

$$C_\alpha\, dV_\alpha/dt = \sum_1^r G_i V_i + \sum_1^s k_j I_j \mp I_\alpha. \tag{4}$$

The first of these equations states that the voltage across L_α and C_α must be of the same magnitude. The second

[2] The transpose of this table is that part of a tie set schedule pertinent to the tree branch current. E. A. Guillemin, "Introductory Circuit Theory," John Wiley and Sons, Inc., New York, N. Y. p. 14, 1953.

[3] Synge, *loc. cit.* This will be found in slightly different form.

[4] Guillemin, *loc. cit.*

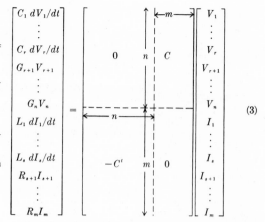

notes that the link current in L_α can only flow through the tree branch containing C_α. At least one of the I_j must appear in the expression. If all were missing it would imply that C_α is not connected to the rest of the network. The G_i are conductances and the k_j pure numerics (not necessarily unity) that may appear as a result of the previous elimination step. Since C_α is not part of the original network let it go to zero. Then from (4)

$$\pm I_\alpha = \sum_1^r G_i V_i + \sum_1^s k_j I_j$$
$$\frac{dI_\alpha}{dt} = \pm \frac{V_\alpha}{L_\alpha}. \tag{5}$$

Differentiating the first of (5) and substituting the second,

$$\pm V_\alpha = L_\alpha \left[\sum_1^r G_i \frac{dV_i}{dt} + \sum_1^s k_j \frac{dI_j}{dt} \right].$$

However, dV_i/dt and dI_j/dt are already known in terms of the undifferentiated variables. Thus both I_α and V_α can be eliminated.

Similarly in series with each excess capacitance C_a there is an added inductance L_a. Necessarily then

$$C_a\, dV_a/dt = \pm I_a$$
$$L_a\, dI_a/dt = \sum_1^s R_j I_j + \sum_1^r k_i V_i \mp V_a. \tag{6}$$

The first of (6) states that the current through C_a and L_a must be of the same magnitude. The second notes that the voltage across C_a only appears in the mesh containing L_a. Some, but not all, of the V_i can be missing since L_a is in series with C_a and at least one other capacitance must exist to close the mesh. If L_a goes to zero, both I_a and V_a can be eliminated as in the previous case. In general, if the original network has R reactive elements and l of these are excess, then the final combined matrix will be of order $R - l$ (except in the case of ideal transformers, see below). This matrix with all rows divided by the proper L or C is the A matrix.[5]

In case there are two-winding ideal transformers in the network, the proper tree must include one winding of each transformer, with a current i_a. The other winding must be a link, with current I_b. For k-ideal transformers there are k pairs of linear constraints of the form

$$i_a = I_b/n$$
$$V_a/n = v_b.$$

If these are substituted in the combined matrix equation both I_b and V_a can be expressed in terms of the other variables. Ideal transformers may alter the previous argument to the following extent. The voltage, V_α, across an added capacitance may appear in the second expression of (4). Similarly the current I_a in an added inductance may appear in the second expression of (6). These variables may then be directly substituted in the first of (4) and (6) respectively without a step of differentiation. The final combined matrix will then be of order R instead of $R - l$.

[5] No account was taken of the sources here, but these do not alter the argument except to slightly complicate (4) to (6).

It is apparent that active elements which act as sources proportional to other voltages or currents in the network can also be included. These sources may be proportional to the allowed variables and can be included immediately. If they are functions of the unwanted variables then the necessary transformation will be found in one of the elimination steps.

Network Functions

The method of network analysis just described gives a set of first-order differential equations

$$F_i + \frac{dy_i}{dt} = \sum_{j=1}^n a_{ij} y_j \qquad (i = 1, 2, \cdots, n) \tag{7}$$

which can be written in matrix form as

$$F + \frac{dy}{dt} = Ay. \tag{8}$$

Let $d/dt = \lambda$ and (8) becomes the algebraic matrix equation

$$F = (A - \lambda I)y \tag{9}$$

where I is the unit matrix. By Cramer's rule the solution to (9) is

$$y_j = \sum_{i=1}^n (-1)^{(i+j)} \frac{F_i \mid (A - \lambda I)_{ij} \mid}{\mid A - \lambda I \mid}. \qquad (j = 1, 2, \cdots, n). \tag{10}$$

$|(A - \lambda I)_{ij}|$ is the determinant of the submatrix of $(A - \lambda I)$ obtained by deleting row i and column j. The y's of (10) are voltages across capacitances and currents through inductances and the F's are voltage and current sources or are proportional to them. The polynomial $| A - \lambda I |$ is therefore the characteristic equation of the network and its zeros, the eigenvalues of A, are the natural frequencies of the network.

If a single source is assumed, say a voltage E, then the scalar E can be factored out of the right-hand side of (10). Each ratio y_i/E is therefore a network function. In some cases only one component of the column matrix F is nonzero and the numerator of each network function reduces to a single term $| (A - \lambda I)_{ij} |$.

Conclusion

The A-matrix description relates electric networks in a natural way to a canonical mathematical form. The establishment of this connection is a necessary first step for the exploitation of mathematical results not heretofore used in circuit theory.

Acknowledgment

The author is greatly indebted to C. A. Desoer for pointing out that the derivation of the A matrix could be simplified by the use of network trees.

VII
Surveys and Historical Summaries

Editor's Comments on Papers 24 and 25

24 Belevitch: *Summary of the History of Circuit Theory*

25 Darlington: *A Survey of Network Realization Techniques*

The final two papers in this volume are historical surveys. There are too few such surveys, especially surveys approaching the quality of these. A few words will explain why these two were chosen.

In 1962 the Institute of Radio Engineers celebrated its 50th year with a special anniversary issue of the *Proceedings of the IRE*. Of the many papers contained in this issue—which totaled nearly 1000 pages—the one that was devoted to the history of circuit theory was prepared by Vitold Belevitch, and it remains one of the most detailed and lucid available, covering the circuit theory field through 1960 (Paper 24).

The second volume of the *IRE Transactions on Circuit Theory* was devoted to a single-topic special issue on the subject of "Realization Techniques," for which Sidney Darlington served as guest editor. The introductory paper for this issue was prepared by Darlington (Paper 25). Not only does it survey the realization techniques available at that time, including those contributed in that issue, but it also lists a number of unsolved problems, some of which have remained unsolved.

The lead paper in another special issue of the *IRE Transactions on Circuit Theory* contains another historical survey by Belevitch (1958), which is more detailed than that given in Paper 24 but only for the field of filter design. On the subject of filter design by computer, a survey has been given by Szentirmai (1971). Finally, the very interesting historical background for the development of the idea of feedback is outlined by Bode (1960) and is highly recommended.

References

Belevitch, Vitold. "Recent Developments in Filter Theory," *IRE Trans. Circuit Theory*, **CT-5**, 236–252 (1958).

Bode, H. W. "Feedback—The History of an Idea," *Proc. Symp. Active Networks and Feedback Systems* (Polytechnic Institute of Brooklyn), **10**, 1–17 (1960).

Szentirmai, George. "Computer Aids in Filter Design: A Review," *IEEE Trans. Circuit Theory*, **CT-18**, 35–40 (1971). This paper also appears in George Szentirmai, ed., *Computer-Aided Filter Design*, IEEE Press, New York, 437 pp., 1973.

24

Reprinted from *Proc. IRE*, **50**(5), 848–855 (1962)

Summary of the History of Circuit Theory*

V. BELEVITCH†, FELLOW, IRE

Summary—**After a brief survey of the state of circuit theory before World War I, the various directions of its development in the last 50 years are discussed, mainly in relation with applications to communication engineering. The early period of network design (1920–1925) was followed by the beginning of synthesis (1926–1935). The next steps were the development of feedback amplifier theory and insertion loss filter theory. The numerous new directions of research started during and after the second war are briefly mentioned. Finally recent progress is reported in formal realizability theory and in topological synthesis. A last section deals with nonlinear and linear variable circuits.**

Introduction

Although circuit theory is more than 100 years old (Ohm's law, 1827; Kirchoff's laws, 1845), it seems that no systematic account of its historical development has ever been written. The present essay attempts to cover the last 50 years, the fiftieth anniversary of the IRE being taken as an excuse to exclude the more distant past. This limitation is also justified by the development of circuit theory itself, which shifted from a steady to an accelerated progress a few years before World War I, simultaneously with the expansion of communication technology following the invention of the vacuum tube (de Forest's audion, 1906). This growth of circuit theory is directly testified by the number of articles published per year, which remained below unity till 1910 and increased from 5 to 25 in the period 1920–1940; after a drop during World War II, the increase continued, and a figure of 100 was exceeded in 1954.

We start with a brief survey of the state of the theory just before World War I and discuss in separate sections the various directions of its development up to the present days, mainly in relation with applications to communication engineering. Due to space restrictions, bibliographic references are omitted altogether: at least 200 important contributions out of a total of some 2000 would deserve mention. Important authors and dates are simply quoted in the main text; the dates generally refer to publications in regular scientific journals, for it was materially impossible to search through patents, theses and reports.

Circuit Analysis Before 1914

Long before 1914 circuit theory had emerged, from general electromagnetic theory, as an independent discipline with original concepts and methods. The conception of a circuit as a system of idealized lumped elements is already firmly established—drawings of Leyden jars and rheostats have gradually disappeared in favor of the now familiar graphical symbols. This assumes, at least implicitly, that a resistor is considered as a 2-terminal *black box* defined by the relation $v = Ri$, rather than as a physical device made of metal or carbon. This abstract point of view becomes

* Received by the IRE, May 22, 1961.
† Comité d'Etude des Calculateurs Electroniques, Brussels, Belgium.

more prominent in the modern developments, where the laws defining the elements, and the interconnection constraints (Kirchoff's laws) are explicitly taken as postulates. With this approach, *network theory*[1] only studies the properties of systems of elements and becomes a purely mathematical discipline dealing with abstract structures generated by various sets of postulates. Another consequence of this point of view is that the physical devices themselves are no longer studied by network theory but by what is sometimes called *device theory;* complicated physical devices are then naturally described by their *equivalent circuits.*

Conventional network elements are linear and time-independent; moreover, the stored electric and magnetic energies, and the dissipated power, are positive definite quadratic forms, and the constraints are instantaneously workless. This reduces the problem of network analysis to the classical theory of small vibrations in dissipative mechanical systems. The use of complex variables to combine the amplitudes and phases of harmonic steady states, and the separation of transients into normal modes, all familiar in analytical dynamics, were naturally taken over by circuit analysts. In particular, Steinmetz (1894) vulgarized the use of complex quantities in electrical engineering. Some specifically electric properties were, however, also established; we mention Kirchoff's topological rules (1847) and their extension by Feussner (1902–1904), the so-called Thévenin's theorem (Helmholtz, 1853), the star-delta transformation (Kennelly, 1899) and the concept of duality (Russel, 1904).

The impedance concept was fluently used by Heaviside, with p treated both as a differential operator and as an algebraic variable. The use of complex values for p is physically justified by Campbell in 1911. Although the relation between Heaviside's operational calculus and integration in the complex plane was only clarified after 1916, the mathematical tools allowing to derive transient behavior from harmonic response were available, since all is needed for lumped circuits is Heaviside's formula based on a partial fraction expansion. It took, however, 20 more years to introduce the terminology of poles and zeros into common engineering practice.

Although network analysis is, in principle, already separated from line theory, both branches remain in close contact in their later evolution, so that a few words about the latter are not out of place. Classical line theory has been worked out in detail by Heaviside, and such concepts as matching, reflection, iterative impedance, etc. began to be transferred from continuous to discrete structures, probably through the intermediary of artificial lines, a subject to which Kennelly devoted a book in 1917. Although the general concept of quadripole, or 2 port, as a black box characterized by its voltage equations, does not explicitly appear before 1921 (Breisig), attenuators and artificial lines calibrated in "equivalent length of standard line" were used earlier for telephone transmission measurements (the decibel was only introduced in 1924). Finally, Heaviside's invention of inductive loading successfully applied by Pupin led Campbell, through the theory of iterative structures (1903), to the invention of the electric filter to be discussed in the next section.

Network Design, 1920–1925

The progress in circuit theory during this period is connected with the development of long distance telephony in more than one respect. First, the theory of loaded lines is closely related with the invention of the electric filter, which found an immediate application in carrier telephony (first realiztion in 1918, Pittsburgh-Baltimore). Secondly, the design of bidirectional amplifiers (two-wire repeaters) immediately raised several problems of network synthesis (hybrid coils, balancing networks, filters, equalizers), not to mention the question of over-all stability. Although satisfactory engineering solutions were rapidly found to all these problems, their theoretical foundation was insufficiently systematic—the art of design, with its cut and try procedures had not yet matured into modern synthesis.

The electric filter was invented during World War I, independently by Campbell and Wagner. The first filters were iterative ladder stuctures, although lattice sections (in particular all-pass sections) are discussed by Campbell (1920–1922). After Zobel's invention of m-derivation (1923–1924), the catalogue of elementary sections was sufficiently extended to cover most practical requirements. In the spirit of line theory, a filter was designed as a cascade connection of sections with matched image-impedances, but the difference between image attenuation and insertion loss was thoroughly estimated by Zobel.

To Zobel are also due the understanding of the ideal transmission conditions (frequency independent attenuation and linear phase), the design of constant impedance equalizers (1928) including a method of approximation to prescribed frequency characteristics, the discovery of bandwidth conservation in the low-pass /band-pass transformation, and the first correction for mutual bridging effects in a parallel connection of two filters (x-derivation).

Simultaneously with the development of filter theory, the general concept of a quadripole, with its impedance, admittance and chain (ABCD) matrices was introduced in Germany and France, and the rules for computing the matrices of series, parallel and tandem combinations of quadripoles were discovered. One distinguished

[1] Our restriction of the term *network* (as opposed to *circuit*) to situations where reference is made to an idealized model conforms with the IRE Standards ("IRE Standards on Circuits: Definitions of Terms for Linear Passive Reciprocal Time Invariant Networks, 1960" (60 IRE 4.S2), Proc. IRE, vol. 48, p. 1609; September, 1960). Both terms are in fact synonymous in common usage. Another distinction has sometimes been made which reserves *network* for systems with free terminals or terminal-pairs (Campbell and Foster, 1920); in this terminology, an n-port becomes a circuit when terminations are specified at all ports.

between 2-terminal pair networks and 3-terminal networks (grounded quadripoles), and the combination rules were correctly restricted to cases where no "longitudinal voltage" appears. Here again, the separation of a general disturbance into a transversal and a longitudinal wave (using both conductors as "go circuit" and the earth as "return") was familiar in line theory, and effectively used in long distance telephony (phantom circuits) since 1903. A similar separation into three modes (symmetrical components) for three-phase power circuits was introduced by Fortescue in 1918.

The concept of n-terminal-pair network, or n-port, does not seem to have appeared during this period. It occurs, however, implicitly in a 1920 paper by Campbell and Foster, which is probably the first publication on network synthesis in the true sense—the energy relations in matched nondissipative 4 ports are discussed, the biconjugacy of the networks is proved, and a complete enumeration of all realizations is given; moreover, the circuits (which include the familiar hybrid coil) are explicitly treated as composed of ideal transformers, a new network element whose theory is established.

Acoustical telephone repeaters (telephone+microphone) were used before the invention of the vacuum tube, and difficulties due to singing were experienced. It is not easy to trace the evolution during the first war, and the post war publications already show a well-established technique. This is testified by the Campbell-Foster paper of 1920, by the design formulas for balancing networks (Hoyt, 1923–1924), and by the recognition of the fact that singing is essentially limited by line irregularities (Crisson, 1925).

The Beginning of Synthesis, 1926–1935

Although Foster's proof of his reactance theorem (1924) is already a transition from the methods of analytical dynamics to those of modern network synthesis, the first paper dealing explicitly with the realization of a one-port whose impedance is a prescribed function of frequency is Cauer's 1926 contribution, based on continuous fraction expansions (also studied by Fry, 1926). With Cauer's and Foster's theorems, the synthesis problem for one ports containing two kinds of elements only was solved. The analogous problem for general one ports was solved by Brune (1931) and led to the concept of positive real function. Pomey (1928) proved that the real part of the impedance matrix of a general passive 2-port was positive definite at real frequencies. The remaining developments of this period deal mainly with nondissipative n-ports; although Gewertz (1933) found a synthesis method for general 2-ports (containing all three kinds of elements), the general problem for n-ports was only solved after World War II and will be discussed in another section. The concentration on LC networks is closely connected with filter theory and its engineering interest. Another aspect of network synthesis, the approximation problem, made also its appearance during this period; the maximally flat approximation was used by Butterworth (1930) in the design of multistage amplifiers; simultaneously and independently, Cauer realized the optimal character of the Chebyshev approximation and solved the approximation problem for an important class of image-parameter filters. Finally, it should be remarked that the canonical structures obtained as solutions of the various synthesis problems made a free use of ideal transformers; the much more difficult problem of synthesis without transformers was not of paramount interest for communication applications and has only been treated recently.

The simplest network after the one port is the symmetric 2-port, which involves two frequency functions only. Geometrically symmetric 2-ports were treated by Bartlett (1927) and Brune (1932), whereas Cauer (1927) and Jaumann (1932) found a number of canonical circuits for all symmetric 2-ports. Dissymetric, and, in particular, antimetric 2-ports were studied by Cauer, who also extended Foster's theorem to LC n-ports (1931) and showed (1932–1934) that all equivalent LC networks could be derived from each other by the linear transformations considered by Howitt (1931). Certain classes of symmetric n-ports were studied by Baerwald (1931–1932).

Cauer's first book on filter design (1931) contains tables and curves for the Chebyshev approximation to a constant attenuation in the stop-band of an image parameter low-pass filter, as well as frequency transformations for other filter classes. The solution of the approximation problem involved rational functions whose extremal properties were established by Zolotareff in 1877 and which reduce to ordinary Chebyshev polynomials when elliptic functions are replaced by trigonometric functions. Cauer's presentation of his design data was based on canonic structures, practically less convenient than ladder structures, and was not accepted in engineering circles before it was realized that the statement and the solution of the approximation problem were of interest in themselves, for most of Cauer's results could easily be transferred to the ladder structure. The systematic theory of image parameter filters was further developed by Bode (1934) and Piloty (1937–1938), thus placing Zobel's earlier results in a clearer perspective. The particular problems raised by crystal filters were studied by W. P. Mason (1934–1937).

Feedback Amplifiers, 1932–1945

The construction of oscillators was one of the first applications of vacuum tubes, and the amplification increase due to a positive feedback below the threshold of oscillation is mentioned in several early publications on radio-engineering. The elementary, but erroneous, physical reasoning based on the round and round circulation of the signal in the feedback loop yielded the geometric series $\mu(1+\mu\beta+\mu^2\beta^2+\cdots)$ for the effective amplification. This gave the impression that instability must occur for all $|\mu\beta|>1$, for the series then diverges.

Stable behavior with a negative feedback exceeding 6 db was thus apparently forbidden, and such low values have little practical interest. It was soon recognized that the theory based on the geometric series was contradicted by experience, and the expression $\mu/(1-\mu\beta)$ was used even for $|\mu\beta| > 1$ without theoretical justification until Nyquist (1932) proved his famous stability criterion and showed the error in the older theory—the physical reasoning based on the loop circulation is only correct when transients are taken into account, the terms in the geometric series then becoming convolution products; the steady-state formula is obtained as the limit after infinite time; the transient series is convergent, but not always uniformly, so that the limit operation and the summation cannot be interchanged. The case of conditional stability, predicted by Nyquist's theory, was experimentally confirmed (Peterson, *et al.* 1934) and the advantages of negative feedback were systematically discussed by Black (1934).

The domestication of negative feedback made it possible to design the wide-band highly linear amplifiers required for multichannel carrier telephony. Although various designs for interstage networks were proposed, for instance by Wheeler and Percival, the limitations due to parasitic capacities, and the way of overcoming them by an optimum over-all design, were only clarified by Bode in 1940, with the help of the integral relations between attenuation and phase.

The fact that the real and imaginary parts of physical network functions could not be specified independently from each other was first stated in connection with the ideal filter paradox. Küpfmüller's treatment (1926) of the transient response to a unit-step led to the well-known sine-integral embodying a response preceding the stimulus. Restrictions imposed by causality to physical response functions are mentioned by Y. W. Lee (1932) in connection with a method of synthesis for arbitrary transfer functions based on Fourier transforms, a method patented by Lee and Wiener (1938). Physical approximations to the ideal filter response, both in amplitude and phase, are discussed by Bode and Dietzold (1935). Explicit integral relations between real and imaginary parts of various network functions were studied independently by Cauer (1932–1940), Bayard (1935) and Leroy (1935–1937), but similar relations were known earlier in the theory of optical dispersion (Kramers-Kronig, 1926).

Bode extended the relations to the case where either component is specified in a partial frequency range, and worked out their consequences for input, output and interstage networks with prescribed parasitic capacities. He also computed the maximum obtainable over-all feedback in terms of band width and asymptotic loop transmission. The stability criterion for multi-loop amplifiers was obtained by Llewellyn, and the effect of feedback on impedances was discussed by Blackman (1943). Bode introduced the concepts of return reference and sensitivity in his classical book (1945).

Insertion Loss Filters and Related Problems

The limitations of image-parameter theory first appeared in connection with the design of filter groups, a problem frequently encountered in carrier telephony. Zobel's procedure of x-derivation, already mentioned, was first replaced by a more systematic method of impedance correction (Bode, 1930). An image-parameter theory of constant impedance filter pairs was developed by Brandt (1934–1936), Cauer (1934–1937) and Piloty (1937–1939), and it was recognized that this also yielded a solution to the equivalent problem of open-circuit filter design.

A completely new approach to the whole problem is contained in Norton's paper (1937) on constant impedance filter pairs, where the method of design starting from a prescribed insertion loss is established. The general synthesis problem for a reactance 2-port with prescribed insertion loss was solved independently by Cocci (1938–1940), Darlington (1939), Cauer (1939–1941) and Piloty (1939–1941). These contributions establish the canonic realization of a reactance 2-port as a ladder structure with mutual inductances restricted to adjacent or nearly adjacent arms and, as a consequence, the possibility of realizing an arbitrary passive one port by a network containing one resistance only. The approximation problem for insertion loss filters was reduced to the similar problem for image-parameter filters. Finally, Darlington also devised a method for precompensating the dissipative distortion.

As already mentioned, the prewar evolution of network theory was closely related with the development of wire communication, and the perfection reached by filter and feedback amplifier theory around 1940 made possible the design to strict tolerances which is required in long distance telephone equipment. On the contrary, the easier narrow-band problems of radio-engineering were treated by elementary circuit analysis, and it is only with the advent of video and pulse techniques that the theory of wide-band multi-stage amplifiers (without over-all feedback) underwent a systematic development, mainly during World War II. Another direction of war-time evolution was the extension of filter techniques to higher frequencies, leading to transmission line filters and microwave networks. Both directions have influenced classical filter theory and, as a consequence, common mathematical methods are now used in an extended field.

In microwave applications, the classical description of network performance in terms of voltages, currents, impedance and admittance matrices, was naturally replaced by a description based on transmitted and reflected wave amplitudes, leading to the concept of scattering matrix (Montgomery, Dicke and Purcell, 1948) taken over from general physics. This concept is also of interest in the field of lumped networks, where it was introduced independently and simultaneously by Belevitch. The scattering formalism allowed an easier presentation of insertion loss filter theory, and has been of great help in other applications to be discussed later.

The war-time progress in amplifier design is described in the book of Valley and Wallmann (1943), but similar work has been done independently in Germany, namely by Cauer (posthumous publications). Most input and inter-stage circuits are actually ladder filters, terminated or open-circuited, so that filter and amplifier problems are closely related. Explicit formulas for the element values of various important classes of ladder filters have been investigated by many workers, but it was recently realized that most results had been anticipated by Takahasi (paper in Japanese, 1951; English adaptation, 1960). The design of input and output circuits is also related with the broad-band matching problem, which consists in constructing a 2-port which transforms a given frequency-dependent output load (for instance a resistance shunted by a parasitic capacitance) into a pure input resistance. A rigorously constant resistance can be obtained by a lossy 2-port, but it is practically important to know the approximation to matching in a given frequency range obtainable from a lossless 2-port, and to see how the transmission loss varies with the degree of impedance equalization for lossy 2-ports. The broad-band matching problem with lossless 2-ports was solved by Fano (1950) for important classes of load impedances. The relations for lossy 2-ports were obtained by Carlin and La Rosa (1952–1955) with the help of the scattering formalism.

Distributed amplifiers, which overcame the limitations imposed by parasitic capacities to stagger-tuned circuits, had been invented by Percival in 1937, but were only practically exploited after 1948; the problem of their optimum design is still unsolved, although some progress has been recently achieved.

The scattering formalism proved useful in the design of various classes of n-ports of interest in telephone applications. Belevitch (1950–1955) discussed matched n-ports with equal losses between any couple of ports, and biconjugate n-ports for $n>4$. Dosoer (1958) and Oswald (1958) established design methods for a class of filter 4-ports (invented by Darlington in 1938) used in submerged repeaters.

The methods of best approximation used in filter and amplifier design were applied to other problems, namely to the design of various classes of delay networks. The maximally flat approximation for the group delay was obtained by Thomson (1949), and the Chebyshev approximation by Ulbricht and Piloty (1960). The Chebyshev approximation to a constant phase difference between two all-pass 2-ports (of interest for polyphase modulation) was obtained independently by Darlington, Orchard and Saraga (1950), and a mathematically equivalent problem on the phase angle of an impedance (of interest in feedback amplifier design) was solved by Baumann (1950). For problems having no exact analytic solution, Darlington (1952) described a procedure based on a series expansion in Chebyshev polynomials, whereas the potential analogy and the related electrolytic tank technique were practically applied at least since 1945 (Hansen and Lundstrom).

New Trends in Post-War Evolution

The field of application of circuit theory extended in so many new directions during and after World War II that only the major tendencies can be outlined. Several new developments lie on the borderline between circuit theory and other disciplines (information theory and noisy systems, electronic computers, automatic control), and will not be discussed. The present section attempts a brief classification of the new ideas, concepts and methods, in relation with their engineering applications; the next two sections review in detail the postwar developments of network theory *stricto sensu*. Nonlinear and linear variable circuits are treated separately in the last section.

Pulse techniques, already mentioned, raised various synthesis problems in the time domain. Although every problem stated in the time domain is, in principle, equivalent to a problem in the complex frequency domain, the approximation requirements are often difficult to translate from one domain into the other, and the convergence properties in the two domains may be quite different. Various mathematical methods, such as Laguerre functions (Lee, 1932), time series expansions (Lewis, 1952), complex integration (Guillemin and Cerillo, 1952), have been used, but there seems to have been little fundamental progress; for instance, the problem of the steepest monotonic response in presence of a given parasitic capacity is not completely solved, although important contributions have been published, namely by Zemanian (1954) and Papoulis (1958).

Microwave circuits have also been previously mentioned. The realization of a microwave gyrator (Hogan, 1952), based on the Faraday effect in ferrites, justified the interest of the ideal gyrator concept (Tellegen, 1948) and promoted theoretical research on nonreciprocal networks initiated by Tellegen (1948–1949). Theoretical work on the synthesis of passive n-ports (both reciprocal or not) is discussed in the next section. Its practical importance is due to the fact that, without the limitations of reciprocity, better performances can be obtained (interstage networks; Tellegen, 1951) or otherwise impossible behavior can be achieved (design of circulators: Carlin, 1955).

Progress has been slower in the theory of active networks because it was more difficult to represent the real physical devices by simple ideal elements; needless to say, even greater difficulties await the circuit theorists of the future, with the advent of integrated microelectronic devices. Classical amplifier theory dealt in fact with passive networks separated by ideal unilateral buffer stages, and amplifiers with large negative feedback were used to realize the active transfer function $-1/\beta$ involving only the transfer ratio β of the passive feedback network. With the advent of the transistor

(Bardeen and Brattain, 1948), practical design problems became much more difficult due to the internal feedback of the device. The fundamental limitations of active non-unilateral devices, at a fixed frequency, were progressively understood; for instance, the conditions for intrinsic stability were established by Llewellyn (1952), power invariants were found by S. J. Mason (1954), and noise invariants by Haus and Adler (1956–1959). The analysis of complex feedback structures was clarified by the use of the flow-graph notation (S. J. Mason, 1953). After the introduction of gyrators, it appeared more convenient to represent active devices by an equivalent circuit containing gyrators and (positive and negative) resistors, so that negative resistance remained the only new element in the theory of active networks. Although various negative resistance effects (arc discharge, dynatron, etc.) had been known for many years, practical and economic devices simulating linear negative resistors became available only with solid-state components (negative impedance converter: Linvill, 1953; tunnel diode: Esaki, 1958), and this favored the adoption of negative resistance as the basic element in theoretical work. Recent results in the theoretical field are summarized at the end of the next section.

Returning now to the synthesis of passive reciprocal networks, we briefly comment on the progressive separation between formal realizability theory (where ideal transformers are freely accepted) and topological synthesis (where even mutual inductances are excluded). The recent interest in topological methods originated from several distinct fields. First, mutual inductances, and even self-inductances, are difficult to realize at very low frequencies; this stimulated research on RC circuits, mainly for servomechanism applications, although similar problems arose earlier in the design of RC-oscillators (invented by van der Pol and van der Mark in 1934). Secondly, even at higher frequencies, it may be economical to replace inductances by capacitances combined with positive and negative resistances; this possibility was known theoretically before 1930, but became only practical with the availability of solid-state devices. Topological problems also arose in the design of contact networks, and some recent developments in the theory of switching are related with various fundamental problems of conventional network theory. Finally, the treatment of network problems on electronic computers asked for a more complete and detailed algebraization of all topological notions and raised various enumeration problems. Topological analysis was recently discussed by many workers, such as Bryant, Okada, Percival, Seshu and Watanabe; for combinatorial and enumeration problems, we refer to the book of Riordan (1958).

Formal Realizability

The synthesis of passive reciprocal n-ports has been achieved by three methods. The first one extended Gewertz's procedure for 2-ports and consisted in the successive extraction of reduced impedance matrices; the process is heavy and laborious, and only of historical interest; it enabled, however, Oono (1946) and Bayard (1949) independently to prove that any positive real impedance or admittance matrix is realizable, thus showing that RLC elements and the ideal transformer constitute a *complete* system of passive reciprocal elements. After a discussion of the 2-port case by Leroy and Belevitch (1949), it appeared that the first method was actually a disguised extension of Darlington's process for one ports, *i.e.*, the realization of an n-port as a reactive $2n$-port closed on n resistors. The second method consists in a direct application of this idea; it was used by Bayard (1950) and with the scattering formalism by Belevitch (1951). The third method extends Brune's process for one ports and arrives to a structure containing the minimum number of reactive elements; synthesis by this method was achieved independently by Oono (1948), Mac Millan (1948–1952) and Tellegen (1953) and contributed to a satisfactory definition of the degree of a rational matrix. In an important paper (1954), Oono and Yasuura rediscuss the synthesis by the second method, using the scattering matrix both for reciprocal and nonreciprocal n-ports and solve completely the equivalence problem.

In connection with the design of narrow-band unsymmetrical band-pass filters, Baum (1957–58) introduced a new fictitious passive element, the imaginary resistance (or frequency-independent reactance), and showed its interest in various synthesis problems. Using this concept, Belevitch obtained a simple derivation of Brune's for one-ports (1959) and extended Tellegen's process to nonreciprocal n-ports (1960).

Extensions to active networks were delayed due to difficulties arising with certain pathological n-ports admitting none of the conventional matrix descriptions (Tellegen, 1954; Carlin, 1955). These difficulties were circumscribed by Oono (1960), and by Carlin and Youla (1960). The latter showed that any active n-port is realizable as a passive nondissipative $3n$-port closed on n positive and n negative resistors. In a companion paper, Youla and Smilen (1960) establish the gain-bandwidth limitations due to parasitic capacitances in negative-resistance amplifiers and derive design formulas for optimum amplifiers.

Topological Synthesis

Logically, but not historically, the first problem arising in this field is the one of discriminating between constraints which are realizable without ideal transformers and general workless reciprocal constraints. In synthesis problems, constraints generally appear under the form of prescribed incidence relations between loops and branches, and the problem is to determine the necessary and sufficient conditions under which a prescribed incidence matrix corresponds to a graph. A necessary condition is well known (the matrix must be totally unimodular, *i.e.*, have all its minors of all orders equal to 0 or ± 1), but is certainly not sufficient, a classical counter-example being the dual of the constraints in a nonplanar graph (Foster, 1952). Sufficient conditions were found by Tutte (1958–1959). Various algorithms have recently been devised which either prove a given matrix to be unrealizable, or lead to a unique realization, within trivial isomorphisms. The first such algorithm is due to Gould (1958) and was applied to the synthesis of contact networks.

The next problem is the synthesis of resistance n-ports, and of resistance networks having $n+1$ terminals (grounded n-ports). It was known for some time that dominancy (any diagonal element not smaller than the sum of the moduli of all elements in the same row) was sufficient, but not necessary, for the admittance matrix of an n-port. Necessary and sufficient conditions for a 3-port with prescribed impedance or admittance matrix were obtained by Tellegen (1952) by expressing that the network cannot yield current nor voltage gain under any set of open- or short-circuit conditions. A nontrivial extension of Tellegen's approach led Cederbaum (1958) to the condition of paramountcy (any principal minor not smaller than the modulus of every minor based on the same rows), which is weaker than dominancy. Paramountcy is necessary for admittance and impedance matrices, but its sufficiency is not established for impedance matrices with $n>3$. For $(n+1)$-terminal networks, the synthesis is equivalent to to a congruence transformation into a positive diagonal matrix by means of a realizable constraint matrix. An algorithm yielding a unique (except for trivial variants) transformation, if possible at all, has been found by Cederbaum, but this algorithm accepts totally unimodular transformation matrices, which are not necessarily realizable. Related synthesis procedures were described by Guillemin (1960) and Biorci (1961).

For two-element kind one ports, the canonical structures of Foster and Cauer contain no transformers. It has long been thought that the transformers appearing in Brune's process are unavoidable in the synthesis of general three-element kind one ports; a canonical realization without mutual inductance was, however, published by Bott and Duffin (1949). In spite of various small improvements (Pantell, 1954; Reza, 1954), the method is quite wasteful in elements; no further general improvement has been obtained, but procedures due to Miyata (1952) and Guillemin (1955) sometimes yield more economical realizations.

The synthesis of RC 2-ports was first treated by Guillemin (1949), who showed that the class of grounded RC 2-ports was practically not narrower than the class of general reciprocal 2-ports, in the sense that the modulus of the input-output voltage ratio of any 2-port of the second class can be arbitrarily approximated within the first class, except for a constant multiplier. Guillemin's synthesis was in terms of parallel ladders; Orchard (1951) established a synthesis by RC lattices and discussed the extraction of terminal resistances in order to realize prescribed insertion loss functions. The approximation problem was treated by Ozaki and Fujisawa (1953). Miscellaneous synthesis procedures were discussed by a number of authors, but the most general and complete results are due to Fialkow and Gerst in a half-dozen papers published between 1951 and 1955. These authors found the necessary and sufficient conditions for voltage ratios of grounded and nongrounded RC and RLC 2-ports, as well as for various restricted structures (ladder, lattice); they indicated canonic realizations and discussed the value of the constant multiplier which fixes the maximum available voltage gain. Articles by Cederbaum (1956) and Kuh and Paige (1959) bring certain additional precisions.

The synthesis of RC grounded 2-ports with prescribed admittance or impedance matrices is still an unsolved problem. A series-parallel synthesis procedure was described by Ozaki (1953), and sufficient conditions under which it succeeds are known, but have not been proved necessary. The problem progressed through contributions of Lucal (1955), Slepian and Weinberg (1958) and Adams (1958), but Darlington's conjecture (1955), stating that any RC grounded 2-port admits an equivalent series-parallel realization, is still unproved.

The problem of ladder synthesis of two-element kind 2-ports arose earlier in filter theory, and the important method of zero-shifting in cascade synthesis was introduced by Bader (1942). Necessary and sufficient conditions for ladder realizability, in the case of a prescribed impedance matrix, or prescribed insertion loss, are unknown in general, but have been established for important classes of networks (Fujisawa, 1955; Watanabe, 1958) and applied to filter design (Meinguet, 1958).

Active grounded RC networks have recently been discussed by Kinariwala (1959–1960) and Sandberg (1960).

Nonlinear and Linear Variable Circuits

In contrast with the maturity of linear network theory, it is often considered that the theory of nonlinear and of linear variable networks is still in its infancy: it is not yet completely separated from nonlinear mechanics and has not reached the stage of synthesis. The present brief review limited to purely electrical problems is intended to show, however, that there has been some systematic progress and that the theory in its present state is no longer a collection of odd results.

First, it is important to separate the problems raised by the unavoidable nonlinearities appearing in nominally linear systems, and the intentional nonlinear effects allowing performances unobtainable with linear systems. The calculation of nonlinear distortion is relatively elementary in the case of slightly nonlinear characteristics (vacuum tube amplifiers, carbon microphone, etc). A more difficult problem arose in connection with the cross-modulation due to hysteresis in loading coils (Kalb and Bennett, 1935). The computation of the modulation products generated by two harmonic signals turned out to be mathematically equivalent to the similar problem in an ideal rectifier, and the amplitudes were obtained as hypergeometric functions. Further progress is reported in later publications by Bennett and others, and the results proved useful in a number of problems involving sharp nonlinearities, such as overload effects in rectifier modulators and feedback amplifiers.

The typically useful properties of nonlinear, or linear variable systems are related with frequency conversion in a wide sense, including harmonic or subharmonic generation, and even simple oscillation, for an oscillator transforms dc power into power at its own frequency. The theory of the triode oscillator was developed by Appleton and van der Pol (1920–1926), and explained such effects as synchronization, amplitude hysteresis, etc. Subharmonic generations in the triode oscillator were analyzed by Mandelstam and Papalexi (1931). Although van der Pol had shown the continuity between quasi-harmonic and relaxation oscillations, these extreme types of behavior actually occur in different technical applications and continued to be treated in separated contexts. Starting from relaxation oscillators and multivibrators (Abraham and Block, 1919) a particular branch of electronic circuit technology developed in the direction of such applications as time bases, counters, logical circuits, wave form generators, etc. On the other hand, the main concern in the design of harmonic oscillators was frequency stability; the way it is affected by nonlinearities was deduced from the principle of harmonic balance (Groszkowski, 1933) and the analysis of linear effects led to the development of the numerous oscillator circuits bearing the names of their inventors. All the above problems are, however, treated by elementary circuit analysis and are more related with device technology than with circuit theory.

Frequency conversion by amplitude modulation raised only elementary problems as long as tubes were used. With the introduction of rectifier modulators, such as the Cowan and ring modulators, in the early thirties, more difficult problems appeared, due to the interaction of all products whose amplitudes thus finally depend on the load impedances at all frequences. In carrier systems, modulators normally operate between highly selective filters, and a corresponding small signal theory for rectifier modulators between selective terminations was developed after 1939 by such workers as Caruthers, Kruse, Stieltjes, Tucker and Belevitch. The conception of a linear variable network as a linear network with an infinite number of ports (treating the impedance presented to each modulation product as a separate termination) did not yield tractable solutions in the case of nonselective terminations, and the case of RC loads (Belevitch, 1950) was treated by a direct method. The general problem became recently of major importance for the optimum design of filters for pulse-time modulated systems; after contributions by Cattermole (1958) and Desoer (1958), an analysis leading to a finite set of linear equations was obtained by Fettweis (1959). Another application of rectifier modulators was treated by Miller (1959) who showed that better frequency dividers could be designed by separating the various nonlinear effects which occur simultaneously in the tube of ordinary oscillators—in the circuits of Miller, the tube is a linear amplifier and the necessary nonlinear effects are produced by rectifier circuits separated by selective filters.

Harmonic generation by rectifier circuits does not seem to have been treated until quite recently—Page (1956) published a theorem on the maximum harmonic efficiency, and Belevitch and Neirynck (1957) described optimal circuits. On the contrary, harmonic production by non-linear reactance is not subject to the same restrictions, and magnetic harmonic generators (Peterson, Manley and Wrathall, 1937) are widely used. The theory of frequency conversion in magnetic modulators is more recent—power relations in nonlinear reactances were discussed by Manley and Rowe (1956), Page (1957), Duinker (1957) and many others. The possibility of amplification and the related negative resistance effects (Hartley) were known however, around 1920 and applied quite early to frequency dividers (transformation of a 60 cps supply into a 20 cps telephone ringing current) and to magnetic amplifiers. The recent revival of interest in parametric amplification is due to the availability of solid-state nonlinear capacitances. On the other hand, resonant circuits with nonlinear inductances, and the phenomenon of ferroresonance were also discussed before World War I, and led to Duffing's equation (1918). New interest in related phenomena was raised by the invention of the parametron (Goto, 1955).

The analysis of complicated nonlinear oscillator circuits by the classical methods of nonlinear mechanics is practically impossible and various approximate engineering methods have been developed, the first one being the so-called equivalent linearization of Krylov and Bogoliubov (1937). During the Second World War, a method based on the describing function has been introduced separately in the U.S.A., the U.S.S.R. and in France. In this method, one derives, from the instantaneous response of the nonlinear element to a harmonic excitation or to a combination of such excitations, an amplitude response or a set of amplitude responses, which are functions of the incident amplitudes; the linear part of the circuit is characterized by its transfer function or matrix. With the additional hypothesis that the linear circuits are highly selective, a finite set of algebraic equations is obtained for the steady-state amplitudes, and it is no longer necessary to consider explicitly any differential equations. This also holds true for the stability analysis of the steady-states, which is performed by linear perturbation methods, thus using only the standard criteria of linear network theory (Hurwitz, Nyquist). This "filter method" (as it is called in Russia) allowed a much simpler derivation of the classical results of van der Pol and others, and is now being successfully applied to new problems.

25

Reprinted from *IRE Trans. Circuit Theory,* **CT-2**, 291–297 (Dec. 1955)

A Survey of Network Realization Techniques*

SIDNEY DARLINGTON†

I. Introduction

WITHIN the framework of network theory, as a whole, there are many known realization techniques. This is as it should be, since realization techniques are primarily design tools. An extensive kit of tools is needed, to perform efficiently the various jobs which are encountered in practice.

The following is a brief survey of realization techniques in general. In Sections II and III, important ingredients of realization techniques are discussed. In Section IV, some past trends in the development of realization techniques are noted. In Section V, some unsolved problems are described.

This paper is not intended to be an exhaustive survey of all aspects of all realization techniques. Only certain high points are considered. In particular, the list of references covers only a small fraction of all the significant and useful contributions which have, in fact, been made.

II. The Classical Model of Realization Techniques

Many realization techniques fit a well-defined pattern, or model, which we shall call the "classical" model. The prototype of the classical model is R. M. Foster's "Reactance Theorem" [1]. Let us examine the essential ingredients of Foster's technique.

Foster's technique relates to a particular class of networks—namely, the class of all two-terminal networks made up of ideal (linear, lumped) reactance elements. It also relates to a particular external characteristic of these networks—namely, the driving-point impedance, thought of as a function of frequency.

The techniques begins with a set of necessary and sufficient conditions, applied to the impedance function. These define the class of all functions which can be realized as the driving-point impedances of two-terminal networks of reactances. It then turns out that all functions, within this class, can be realized with the familiar parallel connection of resonant circuits (or, alternatively, with the series connection of antiresonant circuits). To say this another way, the parallel connection may be used to realize *any* impedance which is realizable with *any* two-terminal configuration of reactance elements. Finally, the element values of the parallel connection can be found directly from a partial fraction expansion of the admittance function (reciprocal of the impedance).

* Received by the PGCT, August 5, 1955.
† Bell Telephone Laboratories, Inc., Murray Hill, N. J.

The essential ingredients of the technique may be summarized as follows:

1. A class of networks (two-terminal reactance networks).
2. A class of functions (impedances of networks in the network class).
3. Necessary and sufficient conditions, which define the function class in mathematical terms.
4. A "canonical configuration," or subclass of the network class, which is sufficient for the realization of the entire function class.
5. A straightforward technique for finding the element values of the canonical configuration, given any specific function within the function class.

Note that the concept of "equivalent networks" is inherent in item 4. There are many different networks producing identical impedance functions. That is why a subclass of the network class can correspond to the entire function class.

The function class may be thought of as an abstract function of the network class, in the sense of the usual theory of real variables.

a. Examples of Techniques Which Fit the Classical Model

Since the publication of Foster's Reactance Theorem, many other realization techniques have been fitted to the same general model. They differ in regard to the network class, the function class, or the canonical configuration (network subclass). Some (but by no means all) of these are listed in Table I (next page). Note that each member of a function class may have to be a set of functions. For example, each member may be set of driving-point and transfer impedances which determine the external characteristic of an n-terminal network.

III. Techniques Which Do Not Fit the Classical Model

Many important and useful techniques do not fit the classical model, or at least do not fit it exactly. Some of them are examined below.

a. Everyday Building-Block Techniques

A very important example is the everyday design of constant-resistance equalizers [2]. The network is usually a tandem connected sequence of constant-resistance sections, with matched image impedances. A "building-block" technique is used, which combines the interpolation and realization parts of the over-all design problem. In everyday applications, each section is quite simple. Furthermore, it is usually designed in

TABLE I

EXAMPLES OF THE CLASSICAL MODEL

Items	Author(s)	Ref.	Network Class		Function Class	Canonical Networks
			Terminals	Elements		
1.	Foster	[1]	2	2 Kinds	Y	S, P
2.	Cauer	[5]	2	2 Kinds	Y	Ladder
3.	Brune	[14]	2	RLCM	Y	S, P, Tand.
4.	Darlington	[4]	2	RLCM	Y	1R
5.	Cauer	[17]	2 Pair	LCM	(Y)	S, P
6.	Bode	[18]	2 Pair	LC	Image Par.	Lat., Tand.
7.	Darlington	[4]	2 Pair	LCM	Ins. Loss	Tand.
8.	Gewertz	[19]	2 Pair	RLCM	(Y)	S, P, Tand.
9.	Tellegen	[20]	2 Pair	LCG	(Y)	
10.	Dietzold	[21]	2 Pair	RCA	Trans. Func.	Feedback
11.	Cauer	[22]	n Pair	LCM	(Y)	S, P
12.	Oono	[23]	n Pair	RLCM	(Y)	Min. R's
13.	McMillan	[24]	n Pair	RLCM	(Y)	
14.	Tellegen	[25]	n Pair	RLCM	(Y)	Min. Els.
15.	Bott-Duffin	[3]	2	RLC	Y	Tree
16.	Reza	[13]	2	RLC	Y	Bridges
17.	Westcott	[15]	2	RLCD	Y	Tr. or Brdg.
18.	Ozaki	[26]	3	Min. Ph. RC	Y_{12} Only	S, P
19.	Weinberg	[27]	4 Bal.	RLC	Y_{12} Only	Lattice
20.	Fialkow-Gerst	[16]	4 Bal. or 3	2 Kinds	Y_{12} Only	
21.	Fialkow-Gerst	[28]	4 Bal. or 3	RLC	Y_{12} Only	

4 Bal.: Two terminal pairs, balanced.
R, C, L: Resistances, Inductances, Capacities.
M: Mutual Inductances (and sometimes transformers).
G, A, D: Gyrators, Active elements, Dissipation.
Y: Impedance or Admittance Functions.
Y_{12}: Transfer Impedances or Admittances.
(Y): Complete sets of impedances or admittances.
Image Par.: Image impedances and image transfer constants.
Ins. Loss: Insertion losses.
Trans. Func.: Transfer functions (voltage ratios).
Min. Ph.: Minimum phase networks.
S, P, Tand., Lat.: Series, parallel, tandem sections, lattices.
Min. R's, Min. Els.: Minimum Numbers of resistances, elements.

such a way as to avoid mutual inductances, without recourse to the redundant elements of Bott and Duffin [3]. This last feature makes it very difficult to define, in precise terms, either a network class or a function class. In spite of the lack of precisely stated areas of applicability, the method is, of course, a very useful one.

In the case of image parameter *filters*, made up of cascade connected sections, it is not so difficult to define definite network and function classes. In everyday filter design, however, the designer may not, in fact, be conscious of this point of view. He has at hand a knowledge of sections sufficient for his needs, and may not seek a complete catalog of all possible sections.

b. "Incomplete" Techniques

From the standpoint of the classical model, the everyday design of equalizers may be said to be "incomplete." To complete it, one must define more definitely the network and function classes. Other useful realization techniques are incomplete, in the same sense. (Some of these, of course, may be completed in the future.)

An example is Lucal's method of designing three-terminal RC networks, described in this issue. We do not know, at present, the full extent of the appropriate function class. Lewis's paper, which is concerned with the boundaries of function classes, is another example.

Still another example is the design of ladder type filters on an insertion loss basis [4]. We designed filters of this sort for years, without knowing the exact function class which can be realized without mutual inductances. Fujisawa's paper in this issue defines the function class for the special network class consisting of mid-series LC ladders.

Incomplete techniques are considered further in Section V, which discusses various unsolved problems.

c. Network Transformations

A practical fault of the classical model is as follows: As a design technique, it applies only to a subclass of the complete network class. Any function in the function class can be realized, but the "canonical" network arrived at may not be attractive from the practical standpoint. One remedy is to cover the function class with various canonical subclasses of the same complete network class, any of which can be chosen at will. (Recall the canonical reactance two-terminal networks of Foster [1] and of Cauer [5].) Another way is to solve the classical problem for more restricted network classes

(See Section V.) Still another is to develop theories of equivalent networks, so that initial network designs can be transformed into more attractive "equivalent" designs.

We have no really comprehensive theory of equivalent networks. On the other hand, many interesting and useful interrelationships between networks are known.

A very simple and very useful relation is Norton's [6] equivalence, involving L networks and ideal transformers. It is illustrated in Fig. 1.

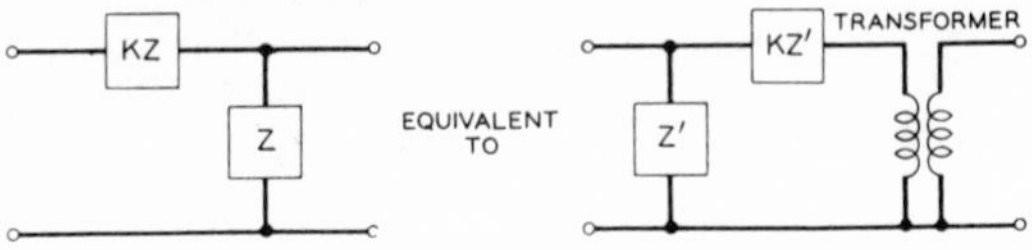

Fig. 1

Norton's equivalence is a very special case of the theory of affine transformations, as applied to networks by Howitt [7]. These transformations take the following form:

$$\overline{Y} = A'YA.$$

Here, Y is a matrix representation of the original network, and $\overline{Y}$ is a corresponding representation of the transformed network. A is a matrix which determines the transformation, and A' is the transpose of A. Matrix A is subject to simple restrictions which insure that the external characteristics of the network will not be changed.

The affine transformations do not describe all network transformations. Furthermore, they sometimes lead to nonphysical networks. On the other hand, they represent a useful tool; and they probably deserve more use than they have so far enjoyed.

1. *An Equivalence Theorem.* The following theorem was recently proved by the author, by means of a series of relatively simple affine transformations. The theorem will be used in Sections III*d* and V*b*.

Assume a network with m external nodes (including the reference node) and n internal nodes. Let the elements be resistances and condensers only, and let the nodes be those of the usual nodal analysis. Then the general branch has admittance $g_{ij}+c_{ij}p$, where g_{ij} and c_{ij} must be non-negative.

The theorem asserts that the above network may be replaced by an equivalent network with the following properties: The two networks are indistinguishable so long as only the m external terminals are accessible. The new network has the same number of internal nodes, n. Each branch admittance between two internal nodes, or between one external and one internal node, is either g_{ij} *or* $c_{ij}p$, but never $g_{ij}+c_{ij}p$ (and we do not permit more than one branch between any one pair of nodes). The admittance branches between pairs of external nodes, however, may still be $g_{ij}+c_{ij}p$. All the g_{ij} and c_{ij} are still non-negative.

The equivalence is illustrated in Fig. 2. In Fig. 2(a), any of the branches indicated may have an admittance of the form $g_{ij}+c_{ij}p$. In Fig. 2(b), each of the branches labeled "1 *EL*" is either g_{ij} or $c_{ij}p$, but not both. Of these, of course, part will usually be g_{ij} and part will be $c_{ij}p$.

Note that branches which do not occur at all in the original network may appear in the transformed network, but that the number of nodes remains the same. Mathematically, we may say that the transformed network may correspond to a complete graph, even though the original network does not.

There is an analogous theorem for networks of three kinds of elements. Here, each internal branch in the transformed network contains only two of the three kinds of elements.

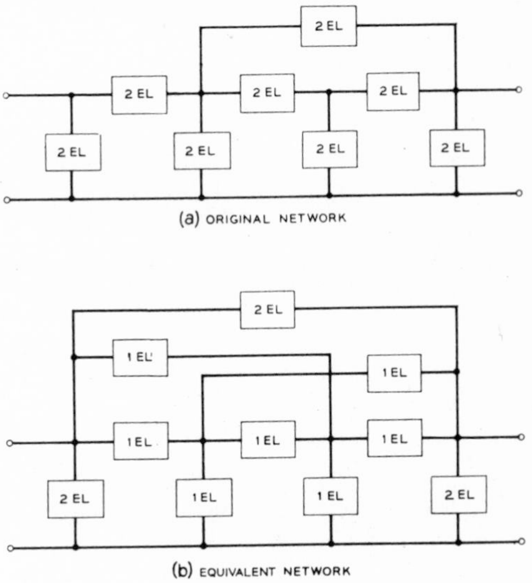

Fig. 2

d. Topology

There is a close correspondence between various properties of networks and the abstract mathematical theory of linear graphs (a part of topology). Graph theory was first applied to network theory by Kirchoff himself [8]. It is a useful tool, which has an important place in realization techniques. It is likely to become increasingly important as we get further away from the simple patterns of the "canonical" configurations associated with our present techniques. Fundamentals of graph theory, bearing on the realization problem, are discussed by Seshu, in his paper in this issue.

The mathematical theory of linear graphs considers graphs of various general sorts. Both oriented and unoriented graphs have been found of interest, in network theory. The theorem stated in the previous section indicates that signed graphs may also be of interest. Signed graphs are linear graphs in which each line is assigned to one or the other of two (abstract) categories, usually designated + and −. These may be compared with the purely conductive and purely capacitive branches of the theorem.

e. Loss-Phase Relations and Other Integrals

There is a substantial body of theory which does not assume a finite network, and yet is important in the realization of finite networks. This theory relates to any functions which are analytic in the right half of the complex frequency plane. As applied by Bode [9], and others, it sets limits on what we can hope to do with any number of elements, in any configuration.

Examples are the loss-phase relations, the integral defining resistance efficiency, etc.

f. Computation Techniques

In Section II, item 5 mentions a "straightforward" technique for finding the element values of the canonical configuration. Sometimes, notably in filter theory, the original straightforward technique has turned out to be practically difficult. The amount of labor involved has been great, and small differences between large numbers have degraded the accuracy of computations. As a result, additional effort has been needed to devise improved techniques for finding element values, with no changes in either the network classes or the function classes.

In special cases, simple formulas have been found, which give the element values directly. Examples are Norton's [10] and Bennett's [11] formulas for maximally flat filters, Dishal's [12] for Tchebycheff filters, and Green's [29] for both these types of filters when they are to be terminated in general unequal resistances.

More generally, explicit formulas for the element values have not been found; but various improved techniques for arriving at the element values have, in fact, been devised. An example is Guillemin's technique for tandem decomposition, described in this issue.

IV. Some Past Trends

The building-block methods were the first realization techniques by which complicated equalizers and filters could be designed. For a while, every bright young network engineer hoped to achieve fame by inventing new and better building blocks.

Foster's "Reactance Theorem" introduced the classical model of a realization technique in 1924. Eventually, it was followed by other techniques, fitted to the same model. At first, the trend was to generalize either the network class or the function class. This trend is shown by entries 1 through 14 in Table I. With few exceptions these generalizations shared one serious weakness. Their network classes, and also their canonical network subclasses, included many networks with mutual inductances, which are generally unpalatable to the engineers who must build the networks. At first, this was a necessary evil. It is generally much more difficult to develop a realization technique for network classes which exclude mutual inductances. (Two-terminal networks of reactances are an exception.) It was necessary to start by solving some of the easier problems, which were themselves sufficiently formidable.

Eventually a new trend developed in the direction of techniques for more restricted network classes, which assure more practical networks. The notable example, of course, is the Bott and Duffin theorem [3]. For a long time it was believed that mutual inductances could not be excluded from dissipative two-terminal networks without restricting the corresponding function class. This was disproved by the Bott and Duffin theorem, which shows that all positive real functions can be realized as impedances of two-terminal networks, without mutual inductances. It depends, of course, on the concept of redundant elements, and very many redundant elements are sometimes needed.

Other techniques for restricted network classes have also been developed. Examples are Miyata's and Kuh's methods of designing special classes of two-terminal networks, described in this issue.

As more and more complicated problems have been solved, another trend has been in the direction of more advanced mathematical disciplines. In the design of Campbell-Zobel filters, hyperbolic functions were the important mathematics. In the classical realization techniques of the 1930's, function theory was most important. Now forms of abstract mathematics, such as matrix algebra, topology, and linear vector spaces, are being used increasingly. For example, Belevitch uses a "scattering matrix," in his paper in this issue, to simplify the analysis of reactance four-poles terminated in resistances.

With the advent of large scale computing machines, extensive, but routine calculations have become less objectionable. The calculation of the zeros of a polynomial furnishes an example. Fifteen years ago it was possible to calculate the zeros of a polynomial, of moderate degree, but the calculation was still long and laborious. Today, the same calculation can be made quickly and easily, on a general purpose computing machine of only moderate size.

V. Some Unsolved Problems

In spite of the great progress which has been made, realization techniques are still hampered by many unsolved problems. A few of these are described briefly below.

a. Two-Terminal Impedances—"Price List"

We need to know more about what positive real functions can be realized as two-terminal impedances, without mutual inductances and without excessive numbers of redundant elements. Reza's [13] transformation of the Bott and Duffin configuration sets an upper bound on the number of redundant elements, which we know we will not have to exceed. This upper bound is very high, however, except for positive real functions of quite low degree.

One question is as follows: Is the above upper bound the *least* upper bound, when all possible configurations are considered? When the degree of the positive real function is high, is there perhaps always one configuration or another which is simpler than Reza's canonical configuration?

A related question stems from the degree of arbitrariness inherent in the Bott and Duffin method of decomposition, and related methods. At each stage in a decomposition, the Bott and Duffin method can be applied either in the series or in the parallel manner. Furthermore, instead of removing resistances in the manner of Brune [14], one can remove dissipation in the manner described by Wescott [15]. Is there an optimum *strategy*, in the choice of a particular decomposition, at each stage, which will optimize the chance of simplifications at later stages? (Theoretically, one can carry out all possible decompositions and compare the results. Practically, of course, this becomes entirely too laborious.)

Another approach to the problem is to find subclasses of the class of all positive real functions, such that all functions in each subclass can be realized with networks of reasonable complexity. Miyata's and Kuh's papers, in this issue, furnish examples. More such subclasses are needed. In this connection, a very difficult question which has not been answered is the following: What is the class of all positive real functions which can be realized without mutual inductances and with no more than, say, k redundant elements?

The information needed may be compared with a price list. We should be able to look up the rock bottom price on any positive real function, in terms of network complexity. We should be able to determine what positive real functions can be bought for a given price.

b. Multi-Terminal Networks Without Mutual Inductances

In recent years, there has been much interest in three- and four-terminal networks without mutual inductances. For the most part, however, the results obtained apply only to the realization of transfer impedances, transfer admittances, or transfer voltage ratios, in the absence of any specifications restricting the driving-point functions. In other words, the members of the function class, of Section II, have been transfer functions, rather than complete sets of driving-point and transfer functions. Even so, the analytical problems have been formidable. The most notable results are probably those obtained by Fialkow and Gerst [28].

In many applications, only a transfer function is specified in advance, and the choice of the driving-point functions is, in fact, left up to the network designer. In many others, however, all three driving-point and transfer functions are specified in advance. Important examples are constant-resistance phase-shifting networks and filter networks terminated at both ends in resistances. For these, we do also need realization techniques for function classes in which the members are complete sets of driving-point and transfer functions, to be realized with network classes which are classes of three-terminal networks. Existing techniques are frequently useful, but are unnecessarily restricted.

1. *Three-Terminal Networks of Two Kinds of Elements.* Lucal, in this issue, considers the design of three-terminal networks, made up of two kinds of elements, with no mutual inductances. He assumes that all three impedances or admittances are specified in advance. On the other hand, he restricts himself to a special subclass of the complete network class—namely, networks which correspond to a series and parallel type of decomposition. Furthermore, he does not establish, completely, the corresponding function class. The extension of Lucal's results to a complete realization technique, fitting the classical model, is a very interesting but very difficult unsolved problem.

Two crucial questions are the following: First, exactly what is the function class for three-terminal networks of two kinds of elements, when the members of the function class are to be complete sets of driving-point and transfer admittances? The conditions established by Fialkow and Gerst [16], for transfer functions only, are still necessary. Apparently, however, they are no longer sufficient. Second, what is a canonical subclass of the complete network class sufficient for the realization of all functions in the function class? Lucal does not claim that his series-parallel connections define a *canonical* subclass. It is this author's personal belief, however, that, in fact, they do.

The specific equivalence theorem which needs to be proved (or disproved) is as follows: Given a three-terminal network of two kinds of elements, without mutual inductances, there exists a similar network, equivalent to the first, which has the series-parallel configuration. This theorem has *not* been rigorously proved. On the other hand, no counter examples have been found.

The theorem stated in Section III*c* offers some encouragement. The $g+cp$ branches between the external terminals form a Π, in parallel with the rest of the network. They can be removed without prejudicing the series-parallel equivalence theorem. This leaves a network in which each branch is either a resistance *or* a condenser. A measure of the complexity of general networks of this sort, is the number of internal nodes. (Note that the series-parallel type equivalent network may have to have more nodes than the original network.) When there is only one internal node, in the original network, the proof of the equivalence theorem is trivial. When there are two internal nodes, the proof is quite difficult. The author has examined the two-node case in some detail. While no formal proof has yet been written out, it is almost certain that one can be. The methods which have been used, however, are extremely laborious. They cannot reasonably be applied to networks with as many as three internal nodes. Clearly what is needed is some entirely new method of analysis. (It is just possible that the series-parallel equivalence theorem holds for networks with one and two internal nodes, but not for more complicated networks.)

c. Techniques for New Network Classes

Various unsolved problems can be grouped under this heading.

One example is the realization of networks in which certain of the element values are specified in advance. Here the network class must be such that all its members incorporate the specified elements.

Another example is the realization of various combinations of elements of the newer types. By newer types is meant such elements as transistors, gyrators, transmission lines, waveguides, and waveguide barriers. Much has already been done along these lines. The paper by Ozaki and Ishii, in this issue, is one such contribution.

Eventually, we should have systematic realization techniques for very general networks of time-varying and nonlinear elements.

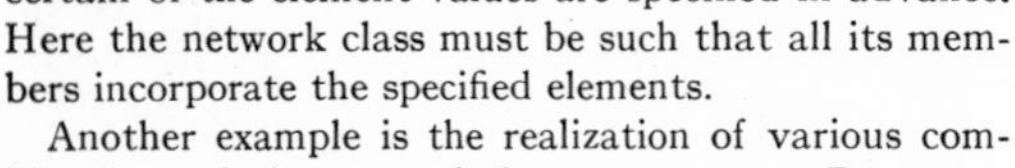

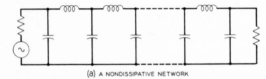

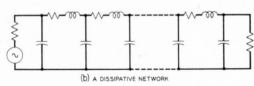

Fig. 3

d. Computational Problems

Many of the computations encountered in applying realization techniques, are still excessively laborious. While much has been done to reduce the labor, much more needs to be done.

One interesting question in this regard is the following: Norton [10], Bennett [11], Dishal [12], and Green [29] have obtained simple, explicit, formulas for the element values of certain special types of filters, even though they are designed on the insertion loss basis. Can similar formulas be derived for other special types of insertion loss filters?

Another question relates to the two circuits shown in Fig. 3. In Fig. 3(a), the reactance network is nondissipative. In Fig. 3(b), the inductances are dissipative, with the dissipation equivalent to series resistances. Within certain restrictions on the amount of dissipation, and on the relative sizes of the terminations, the two circuits can produce the same insertion loss. When the inductances are to be equally dissipative (equal Q's), we can compute the element values of the dissipative network from those of the nondissipative network. The calculation is long and laborious, but it is feasible. When the inductances are to be unequally dissipative, or when the values of the series resistances are specified in advance, we merely know that the dissipative network exists. We do not know how to compute its element values, in any reasonable way.

A network transformation theory which preserves transfer impedances, but not driving-point impedances might help solve problems of this sort.

Bibliography

[1] Foster, R. M., "A Reactance Theorem." *Bell System Technical Journal*, Vol. 3 (1924), pp. 259–267.
[2] Zobel, O. J., "Distortion Correction in Electrical Circuits with Constant Resistance Recurrent Networks." *Bell System Technical Journal*, Vol. 7 (1928), pp. 438–534.
[3] Bott, R., and Duffin, R. J., "Impedance Synthesis without Use of Transformers." *Journal of Applied Physics*, Vol. 20 (1949), p. 816.
[4] Darlington, S., "Synthesis of Reactance 4-Poles." *Journal of Mathematics and Physics*, Vol. 18 (1939), pp. 257–353.
[5] Cauer, W., "Die Verwirklichung von Wechselstromwiderstanden vorgeschriebener Frequenzabhängigkeit." *Archiv für Elektrotechnic*, Vol. 17 (1926), pp. 355–388.
[6] Norton, E. L., United States Patent No. 1,708,950. Issued April 16, 1928 ("Electric Wave Filter").
[7] Howitt, N., "Equivalent Electrical Networks," (PROCEEDINGS OF THE IRE, Vol. 20 (June, 1932), pp. 1042–1051.
[8] Kirchoff, G., "Ueber die Auflösung der Gleichungen, auf welche man bei der Untersuchungen der linearen Vertheilung galvanischer Strome geführt wirt." *Annalen der Physik und Chem.*, Vol. 72 (1847), pp. 497–508.
[9] Bode, H. W., "Network Analysis and Feedback Amplifier Design." New York. D. Van Nostrand Co., 1945.
[10] Norton, E. L., "Constant Resistance Networks with Applications to Filter Groups." *Bell System Technical Journal*, Vol. 16 (1937), pp. 178–193.
[11] Bennett, W. R., United States Patent No. 1,849,656. Issued March 15, 1932 ("Transmission Network").
[12] Dishal, M., "Two New Equations for the Design of Filters." CONVENTION RECORD OF THE IRE (1955), Part 5.
[13] Reza, F. M., "Supplement to Brune Synthesis," AIEE, *Communications and Electronics*, No. 17, pp. 85–90; March, 1955.
[14] Brune, O., "Synthesis of a Finite Two-Terminal Network whose Driving-Point Impedance is a Prescribed Function of Frequency." *Journal of Mathematics and Physics*, Vol. 10 (1931), pp. 191–236.
[15] Westcott, J. H., "Driving-Point Impedance Synthesis with Maximally Lossy Elements." *Proceedings of the Symposium on Modern Network Synthesis*, Polytechnic Institute of Brooklyn, 1955.
[16] Fialkow, A. D. and Gerst, I., "The Transfer Function of General Two-Terminal-Pair RC Networks." *Quarterly of Applied Mathematics*, Vol. 10 (1952), pp. 113–127.
[17] Cauer, W., "Untersuchungen über ein Problem, das drei positiv definite quadratische Formen mit Streckenkomplexen in Beziehung Setzt." *Math. Ann.*, Vol. 105 (1931), pp. 86–132.
[18] Bode, H. W., "A General Theory of Electric Wave Filters." *Journal of Mathematics and Physics*, Vol. 13 (1934), pp. 275–362. Summary in *Bell System Technical Journal*, Vol. 14 (1935), pp. 211–214.
[19] Gewertz, C. M., "Synthesis of a Finite, Four-Terminal Network from Its Prescribed Driving-Point Functions and Transfer Function." *Journal of Mathematics and Physics*, Vol. 12 (1933), pp. 1–257. Also in book form, "Network Synthesis." Baltimore, Williams & Wilkins, 1933.
[20] Tellegen, B. D. H., "Synthesis of Passive, Resistanceless Four-Poles That May Violate the Reciprocity Relation." *Philips Research Report.*, Vol. 3 (1948), pp. 321–337.

[21] Dietzold, R. L., U. S. Patent No. 2,549,065, Issued April 17, 1951 ["Frequency Discriminative Electric Transducer"].
[22] Cauer, W., "Äquivalenz von $2n$-Polen ohne Ohmsche Widerstande." *Nachrichten von der Gesellschaft der Wissenschaften* (*N.F.*), Vol. 1 (1934), pp. 3–33.
[23] Oono, Y., "Synthesis of a Finite $2n$-Terminal Network by a Group of Networks, Each of Which Contains Only One Ohmic Resistance." *Journal of Mathematics and Physics*, Vol. 29 (1950), pp. 13–26.
[24] McMillan, B., "Introduction to Formal Realizability Theory." *Bell System Technical Jour.*, Vol. 31 (1952), pp. 217–279, 541–600.
[25] Tellegen, B. D. H., "Synthesis of $2n$-Poles by Networks Containing the Minimum Number of Elements." *Journal of Mathematics and Physics*, Vol. 32 (1953), pp. 1–18.
[26] Ozaki, H., "Synthesis of Three-Terminal Network Without Ideal Transformer." *Technology Reports of the Osaka University*, Vol. 3, Osaka, Japan; 1953.
[27] Weinberg, L., "New Synthesis Procedures for Realizing Transfer Functions of RLC and RC Networks." Massachusetts Institute of Technology, Research Laboratory Electronics, Technical Report 201; 1951.
[28] Fialkow, A., and Gerst, I., "Transfer Functions of Networks Without Mutual Reactance." *Quarterly of Applied Mathematics*, Vol. 12 (1954), pp. 117–131.
[29] Green, E., "Synthesis of Ladder Networks to Give Butterworth or Chebychev Response in the Pass Band." *Institute of Electrical Engineers*, London, Monograph no. 88, January 15, 1954.

Author Citation Index

Subject Index